Edited by
Satishchandra B. Ogale,
Thirumalai V. Venkatesan, and
Mark G. Blamire

Functional Metal Oxides

Related Titles

Collins, T.T., Reynolds, D.D., Litton, C.C. (eds.)

Zinc Oxide Materials for Electronic and Optoelectronic Device Applications

2011
ISBN: 978-0-470-51971-4

Geckeler, K.E., Nishide, H. (eds.)

Advanced Nanomaterials

2010
ISBN: 978-3-527-31794-3

Duò, L., Finazzi, M., Ciccacci, F. (eds.)

Magnetic Properties of Antiferromagnetic Oxide Materials

Surfaces, Interfaces, and Thin Films

2010
ISBN: 978-3-527-40881-8

Bruce, D.D., Bruce, D.W., O'Hare, D., Walton, R.I. (eds.)

Functional Oxides

2010
Print ISBN: 978-0-470-99750-5

Kumar, C.S.S.R. (ed.)

Nanostructured Oxides

2009
ISBN: 978-3-527-32152-0

Edited by Satishchandra B. Ogale, Thirumalai V. Venkatesan, and Mark G. Blamire

Functional Metal Oxides

New Science and Novel Applications

Verlag GmbH & Co. KGaA

The Editors

Prof. Satishchandra B. Ogale
National Chemical Laboratory (CSIR-NCL)
Physical & Materials Chemistry
Dr. Homi Bhabha Road, Pashan
411 008 Pune
India

Prof. Thirumalai V. Venkatesan
National University of Singapore
Faculty of Engineering
4 Engineering Drive 3, Block A
117576 Singapore
Singapore

Prof. Mark G. Blamire
University of Cambridge
Department of Materials Science
Pembroke Street
CB2 3QZ Cambridge
United Kingdom

Library of Congress Card No.: applied for

British Library Cataloguing-in-Publication Data
A catalogue record for this book is available from the British Library.

Bibliographic information published by the Deutsche Nationalbibliothek
The Deutsche Nationalbibliothek lists this publication in the Deutsche Nationalbibliografie; detailed bibliographic data are available on the Internet at <http://dnb.d-nb.de>.

© 2013 Wiley-VCH Verlag GmbH & Co. KGaA, Boschstr. 12, 69469 Weinheim, Germany

Print ISBN: 978-3-527-33179-6
ePDF ISBN: 978-3-527-65489-5
ePub ISBN: 978-3-527-65488-8
Mobi ISBN: 978-3-527-65487-1
oBook ISBN: 978-3-527-65486-4

Cover Design Grafik-Design Schulz, Fußgönheim
Typesetting Laserwords Private Ltd., Chennai, India
Printing and Binding Markono Print Media Pte Ltd, Singapore

Printed in Singapore
Printed on acid-free paper

Contents

Preface

Complex functional oxides exhibit a fascinating range of behaviors: from colossal magnetoresistance and high-temperature superconductivity to ferroelectricity and multiferroicity. Oxides therefore can potentially provide solutions to many of the technological challenges facing the world, including lower energy consumption devices and the shift to renewable sources of energy. Unlike their semiconducting counterparts such as silicon, GaAs, oxides have more degrees of freedom in terms of the properties at the molecular level, which when exploited could lead to unusual functionalities. For example, the oxygen octahedra in the perovskite oxides can affect the material properties dramatically when the octahedral tilt angles are changed by a variety of means, strain, for example. Indeed, the molecular orbitals of the atoms play a very strong role in determining the properties at oxide interfaces and these could be affected by strain and, electric and magnetic fields. Hence, their functional behavior is intimately coupled to structure and stoichiometry. On one hand, this implies that the oxide properties can be readily tuned by several means, but on the other, it also makes oxides more sensitive to precise processing conditions and strain within device structures than the metals or semiconductors that they might aim to replace. With advances in growth techniques in recent years and in situ monitoring of growth, the field of oxide research is moving steadily to the domain of realistic oxide electronics. In the course of this research, several novel phenomena are being constantly discovered, which are challenging our understanding about the structure–property relationships in these systems. Integration of oxide systems with nonoxide materials such as other classes of inorganic materials (sulfides, nitrides), functional carbon (graphene, CNT), and organics (molecules, polymers) is promising to emerge as a frontier area of research, with significant implications for applications in the key areas of energy, environment, and health. In the field of energy, the ability to tune the position of the conduction and valence bands of the material has a crucial bearing on our ability to use the material in applications such as photocatalysis, water splitting, or CO_2 sequestration. Very often in such energy applications, the chemical stability of the material becomes the limiting factor as to whether a material may be deployed or not. Once again, oxides emerge as winners here, as many of the oxides are quite stable in corrosive environments.

The application and understanding of oxides require a multidisciplinary approach and the problems would be best solved by teams of chemists, physicists, materials, and device engineers working together. This book provides an overview of the current state of developments and suggestions as to how the field might move forward. Several eminent experts worldwide and their collaborators/students have contributed to this effort through their chapters. The editors greatly appreciate their time and effort in putting together such excellent chapters without which the book would not have been possible. We believe that this book will prove to be a good reading for young students joining research as well as for young research scientists working in the field. The editors have tried their best to see that several areas of current interest are adequately covered. The intent is to transmit the current excitements in the field to the readership and we hope that the book will serve this purpose.

One of us, SBO, would like to acknowledge the strong support of his family in all his endeavors including this book effort, the parent institute CSIR (Director General Prof. Sameer Brahmachari) and the laboratory CSIR-NCL (Director Dr. Pal and past Director Dr. Sivaram), and students/collaborators. MGB would like to thank the past and present members of the Device Materials Group, University of Cambridge, for their enthusiasm and commitment to novel electronic materials and his wife for support away from the laboratory. TVV would like to thank his team at NUSNNI-NanoCore for their extraordinary dedication to this field of research, his wife for overlooking his not being able to be with her all the time, and his daughter for coming over to Singapore to finish her high schooling here.

Finally, and most importantly, we would like to thank the Wiley publishing team including Dr. Esther Levy with whose invitation this book got initiated and was later followed up by Dr. Gudrun Walter, Ms. Lesley Belfit, and their colleagues. Their patience and pursuance were important for the successful conclusion of this process.

CSIR-NCL, India   *Satishchandra B. Ogale*
NUS, Singapore   *Thirumalai V. Venkatesan*
Cambridge, UK   *Mark G. Blamire*

List of Contributors

Ariando
National University of Singapore
NUSNNNI-NanoCore
Singapore 117576
Singapore

and

National University of Singapore
Department of Physics
Singapore 117542
Singapore

Mark G. Blamire
University of Cambridge
Department of Materials Science
Pembroke Street
Cambridge CB2 3QZ
UK

M. B. H. Breese
National University of Singapore
NUSNNNI-NanoCore
Singapore 117576
Singapore

and

National University of Singapore
Department of Physics
Singapore 117542
Singapore

and

National University of Singapore
Singapore Synchrotron Light
Source
Singapore 117603
Singapore

Ching-Jung Cheng
University of New South Wales
School of Materials Science and
Engineering
Sydney
New South Wales 2052
Australia

J. M. D. Coey
Trinity College
School of Physics
Dublin 2
Ireland

S. Dhar
National University of Singapore
NUSNNNI-NanoCore
Singapore 117576
Singapore

and

National University of Singapore
Department of Electrical and
Computer Engineering
Singapore 117576
Singapore

Mehmet Egilmez
University of Cambridge
Department of Materials Science
Pembroke Street
Cambridge CB2 3QZ
UK

Y. P. Feng
National University of Singapore
NUSNNNI-NanoCore
Singapore 117576
Singapore

and

National University of Singapore
Singapore Synchrotron Light
Source
Singapore 117603
Singapore

Atsushi Fujimori
University of Tokyo
Department of Physics
7-3-1 Hongo
Bunkyo-ku
Tokyo 113-0033
Japan

Tomoteru Fukumura
University of Tokyo
Department of Chemistry
7-3-1 Hongo
Bunkyo-ku
Tokyo 113-0033
Japan

K. Gopinadhan
National University of Singapore
NUSNNNI-NanoCore
Singapore 117576
Singapore

and

National University of Singapore
Department of Electrical and
Computer Engineering
Singapore 117576
Singapore

Michael Grätzel
Laboratoire de Photonique et
Interfaces
Ecole Polytechnique Fédérale de
Lausanne (EPFL)
Institut des Sciences et Ingénierie
Chimiques
Station 6
1015 Lausanne
Switzerland

Arunava Gupta
University of Alabama
Center for Materials for
Information Technology
2007 Bevill Building
Tuscaloosa, AL 35487
USA

Hans Hilgenkamp
National University of Singapore
NUSNNNI-NanoCore
Singapore 117576
Singapore

and

University of Twente
Faculty of Science and
Technology and MESA+ Institute
for Nanotechnology
7500 AE
Enschede
The Netherlands

Isao H. Inoue
National Institute of Advanced
Industrial Science and
Technology
Tsukuba 305-8562
Japan

Daisuke Kan
University of Maryland
Department of Materials Science
and Engineering
College Park
MD 20742
USA

and

Kyoto University
Institute for Chemical Research
Gokasyo
Uji
Kyoto 611-0011
Japan

Masashi Kawasaki
University of Tokyo
Quantum Phase Electronics
Center and Department of
Applied Physics
7-3-1 Hongo
Bunkyo-ku
Tokyo 113-8656
Japan

and

RIKEN Center for Emergent
Matter Science
2-1 Hirosawa
Wako
Saitama, 351-0198
Japan

Weinan Lin
Nanyang Technological
University
Division of Physics and
Applied Physics
School of Physical and
Mathematical Sciences
Singapore 637371
Singapore

James Lourembam
Nanyang Technological
University
Division of Physics and
Applied Physics
School of Physical and
Mathematical Sciences
Singapore 637371
Singapore

Lily Mandal
Centre of Excellence in Solar
Energy
Physical and Materials
Chemistry Division
National Chemical Laboratory
(CSIR-NCL)
Pune 411008
India

Valanoor Nagarajan
University of New South Wales
School of Materials Science and
Engineering
Sydney
New South Wales 2052
Australia

Mohammad K. Nazeeruddin
Laboratoire de Photonique et
Interfaces
Ecole Polytechnique Fédérale de
Lausanne (EPFL)
Institut des Sciences et Ingénierie
Chimiques
Station 6
1015 Lausanne
Switzerland

Satishchandra B. Ogale
Centre of Excellence in Solar
Energy
Physical and Materials Chemistry
Division
National Chemical Laboratory
(CSIR-NCL)
Pune 411008
India

Prahallad Padhan
Indian Institute of Technology
Madras
Department of Physics
Chennai 600036
Tamilnadu
India

Karin M. Rabe
Rutgers University
Department of Physics and
Astronomy
Piscataway
NJ 08854
USA

Pavle V. Radovanovic
University of Waterloo
Department of Chemistry
200 University Avenue West
Waterloo
Ontario N2L 3G1
Canada

Ramamoorthy Ramesh
University of California at
Berkeley
Department of Physics
Berkeley, CA
USA

and

University of New South Wales
School of Materials Science and
Engineering
Sydney, NSW
Australia

and

University of California at
Berkeley
Department of Materials Science
and Engineering
Berkeley, CA
USA

Jason W. A. Robinson
University of Cambridge
Department of Materials Science
Pembroke Street
Cambridge CB2 3QZ
UK

A. Roy Barman
National University of Singapore
NUSNNNI-NanoCore
Singapore 117576
Singapore

and

National University of Singapore
Department of Physics
Singapore 117542
Singapore

A. Rusydi
National University of Singapore
NUSNNNI-NanoCore
Singapore 117576
Singapore

and

National University of Singapore
Department of Physics
Singapore 117542
Singapore

and

National University of Singapore
Singapore Synchrotron Light
Source
Singapore 117603
Singapore

Frédéric Sauvage
Université de Picardie Jules Verne
Laboratoire de Réactivité et
Chimie des Solides
CNRS UMR7314
33 rue Saint-Leu
80039 Amiens Cedex
France

and

Réseau sur le Stockage
Electrochimique de l'Energie
(RS2E)
FR CNRS3459
France

Akihito Sawa
National Institute of Advanced
Industrial Science and
Technology
Tsukuba 305-8562
Japan

Lukas Schmidt-Mende
University of Konstanz
Department of Physics
Universitätsstr. 10
78464 Konstanz
Germany

Jan Seidel
Lawrence Berkeley National
Laboratory
Materials Sciences Division
Berkeley, CA
USA

and

University of California at
Berkeley
Department of Physics
Berkeley, CA
USA

and

University of New South Wales
School of Materials Science and
Engineering
Sydney, NSW
Australia

Ichiro Takeuchi
University of Maryland
Department of Materials Science
and Engineering
College Park
MD 20742
USA

M. Venkatesan
Trinity College
School of Physics
Dublin 2
Ireland

T. Venkatesan
National University of Singapore
NUSNNNI-NanoCore
Singapore 117576
Singapore

and

National University of Singapore
Department of Electrical and
Computer Engineering
Singapore 117576
Singapore

and

National University of Singapore
Department of Physics
Singapore 117542
Singapore

Hiroki Wadati
University of Tokyo
Department of Applied Physics
and Quantum-Phase Electronics
Center (QPEC)
2-11-16 Yayoi
Bunkyo-ku
Tokyo 113-0032
Japan

Umesh V. Waghmare
Jawaharlal Nehru Centre for
Advanced Scientific Research
Theoretical Sciences Unit
Bangalore 560 064
Karnataka
India

Jonas Weickert
University of Konstanz
Faculty of Physics
78457 Konstanz
Germany

Tom Wu
Nanyang Technological
University
Division of Physics and
Applied Physics
School of Physical and
Mathematical Sciences
Singapore 637371
Singapore

Hongjun Xu
Trinity College
School of Physics
Dublin 2
Ireland

Part I
Magnetic Oxides

Functional Metal Oxides: New Science and Novel Applications, First Edition.
Edited by Satishchandra B. Ogale, Thirumalai V. Venkatesan, and Mark G. Blamire.

1
Introduction to Magnetic Oxides

J. M. D. Coey, M. Venkatesan, and Hongjun Xu

Oxides are ubiquitous. The Earth's crust and mantle are largely made up of compounds of metal cations and oxygen anions. Looking at the composition of the crust in Figure 1.1, we see that oxygen is the most abundant element and the most common metals are aluminum and silicon. Most rocks are therefore aluminosilicates. The next most abundant element, and the only transition metal other than titanium to feature among the top ten, which account for over 99% of the crust, is iron (Table 1.1). Remarkably, the same electronic configuration, $2p^6$, is shared by five of the top ten ions, which account for 92% of the atoms in the crust. Usually, only iron, with its two common charge configurations, Fe^{2+} ($3d^6$) and Fe^{3+} ($3d^5$), forms ions with a partially filled shell containing electrons of unpaired spin that exhibit a net *magnetic* moment. At 2.1 at. % (5.7 wt.%), iron is 40 times as abundant as all the other magnetic elements put together; the runners up – manganese, nickel, and cobalt – trail far behind.

For over 20 centuries, up until about 1740 [1], the only useful permanent magnets known to man were lodestones. These prized natural magnetic rocks were largely composed of impure *magnetite*, the black spinel-structure oxide Fe_3O_4 with a ferrimagnetic structure that had been magnetized by a fortuitous lightning strike [2]. The other common rock-forming iron oxide is *hematite*, the reddish corundum-structure sesquioxide αFe_2O_3. Hematite is also magnetically ordered, in a canted antiferromagnetic structure, but its magnetization is about 200 times weaker than that of magnetite.

The weak remanent magnetism imparted to rocks as they cooled in the Earth's magnetic field has allowed us to read the record of fluctuations of the magnitude and direction of the field at the Earth's surface. The remanence is largely due to segregated nano crystallites of titanomagnetite in the rock [3]. The Earth's field is a precious shield that has protected us from the solar wind and allowed life to develop on our planet over the past 3.5 billion years. We learn from the magnetic record that it has reversed numerous times on a geological timescale. The tectonic movements of the plates were thereby pieced together, leading to the first unified theory of Earth sciences. It was actually a quest to understand the magnetism of rocks and baked clay that motivated Louis Néel to formulate the molecular field theory of ferrimagnetism [4], completing the theory of antiferromagnetism that

Functional Metal Oxides: New Science and Novel Applications, First Edition.
Edited by Satishchandra B. Ogale, Thirumalai V. Venkatesan, and Mark G. Blamire.
© 2013 Wiley-VCH Verlag GmbH & Co. KGaA. Published 2013 by Wiley-VCH Verlag GmbH & Co. KGaA.

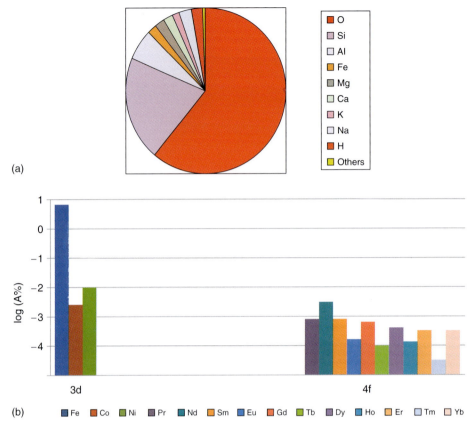

Figure 1.1 (a) Abundance of elements in the Earth's crust in atomic % and (b) abundance of magnetic elements.

Table 1.1 Elements in the Earth's crust (atomic %).

Ion	Abundance	Configuration
O^{2-}	60.7	$2p^6$
Si^{4+}	20.6	$2p^6$
Al^{3+}	6.1	$2p^6$
Na^+	2.6	$2p^6$
$Fe^{2+/3+}$	2.1	$3d^{6/5}$
H^+	2.1	$1s^0$ no electrons
Ca^{2+}	1.9	$3p^6$
Mg^{2+}	1.8	$2p^6$
K^+	1.5	$3p^6$
Ti^{4+}	0.3	$3p^6$

he had developed in his thesis, which was based on the original molecular field theory of ferromagnetism that his mentor and PhD supervisor Pierre Weiss had formulated in 1906, 30 years previously.

Ferrimagnetic oxides are quite strongly magnetic and insulating when the only iron ions they contain are ferric ions. This was a winning combination at the time because there was a need for magnetic materials to be used at microwave frequencies. Metallic alloys based on iron, or even magnetite itself (which contains both Fe^{2+} and Fe^{3+}), were unsuitable because of prohibitive eddy-current losses associated with their conductivity. Oxides from the cubic structural families of spinels and garnets, especially nickel–zinc ferrite, manganese–zinc ferrite, and yttrium–iron garnet (YIG), proved to be the champions here.

Another winning combination was ferrimagnetism and strong uniaxial anisotropy. This was the basis of the success of the hexagonal ferrites with the magnetoplumbite structure, which now account for 90% of the tonnage of permanent magnets produced worldwide. Discovered at the Philips laboratory in the Netherlands around 1950, these were the first true permanent magnets, in the sense that their coercivity could exceed their magnetization, which meant that they could remain magnetized whatever be their shape. A nice, square hysteresis loop is the icon of permanent magnetism. Individual magnet ownership has progressed from one or two per person 60 years ago to a 100 times as many today.

Magnetic recording, which demanded less permanently magnetized materials that could be remagnetized in the reverse direction, as required, spurred the next big development. This included the production of acicular (needlelike) particles of the other ferric sesquioxide, spinel-structure γFe_2O_3, which is known as *maghemite*. Subsequently, Fe_2O_3 was surface-doped with cobalt, and then acicular CrO_2, the only simple oxide that is a ferromagnetic metal was developed. These micrometer-sized particles made up the magnetic media on hard and floppy disks and tapes used for audio, video, and data recording for several decades. However, the relentless march of progress characterized by the magnetic version of Moore's law, which states that the areal density of recorded information doubles every 18–24 months, has now rendered particulate oxide media obsolete. Thin film metallic media are now used to store everything that is downloaded from the Internet..

Intellectually too, scientific interest in magnetic oxides has waxed and waned. The first Golden age was the 1950s and the 1960s, when the ferrites were explored and their properties optimized [5]. The phenomenology of the exchange interactions among localized electrons in 3d shells was systematically investigated [6]. Discovery of the first family of ferromagnetic oxides, the mixed-valence manganites, also dates from this period [7]. The interest in them revived when the copper oxide high-temperature superconductors were found to have a similar, perovskite-related structure, and the magnetoresistance of thin film mixed-valence manganites was shown to be "colossal". Ferromagnets have a higher magnetization than ferrimagnets, but the double-exchange mechanism that allows them to order magnetically above room temperature entails electrons hopping among localized cores ($3d^3$ for manganites and $3d^5$ for magnetite) so that the oxides are conducting and not insulating. The manganites illustrate nicely the progression from studies on bulk ceramics to single crystals to thin films and other nanostructures which marks the historical evolution of research in magnetic oxides.

We may now be entering a new Silver age for oxide research, where multifunctionality, control of defects, interfaces, and thin-film device structures are the new challenges [8]. These ideas are discussed in the later chapters of the book. Here we do the groundwork, by introducing some of the basic ideas and magnitudes relating to structure and properties of magnetic oxides [9, 10]. SI units are used consistently throughout this chapter, but tables of cgs conversions can be found in textbooks on magnetism [11]. The merits of using SI units are compelling. Not only is it possible to check that equations are dimensionally correct, but ideas about the shape of $M(H)$ hysteresis loops, for example, clear when M and H are measured in the same units (ampere per meter).

1.1
Oxide Structures and Crystal Chemistry

Oxides are ionic compounds where small, highly charged metal cations are embedded in a lattice of oxygen anions. The O^{2-} has its stable $2p^6$ closed-shell configuration with an ionic radius $r_O = 140$ pm. Frequently the oxygen forms a dense-packed or distorted dense-packed lattice, either face-centered cubic (fcc) ABCABC or hexagonal close packed (hcp) ABABAB. There are two types of interstices in these structures, illustrated in Figure 1.2, which may be occupied by cations. The smaller one is the tetrahedral site with four oxygen neighbors, which can accommodate, without distortion, a spherical cation of radius $r_{tet} = ((3/2)^{1/2} - 1)r_O = 32$ pm. The

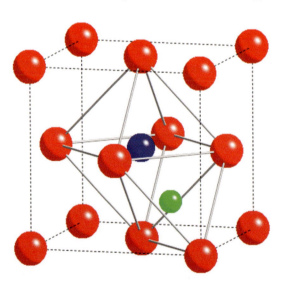

Figure 1.2 A close-packed oxygen lattice showing octahedral (blue) and tetrahedral (green) cation sites.

larger one is the octahedral site with six oxygen neighbors that can accommodate a spherical cation of radius $r_{oct} = ((2)^{1/2} - 1)r_O = 58$ pm. Here r_O is taken as 140 pm.

Table 1.2 lists the ionic radii of common 3d cations, and others that may be incorporated in oxide structures. All the tetrahedral and most of the octahedral cations create some distortion of the site. There are two tetrahedral interstices, and one octahedral interstice for each oxygen in dense-packed lattices; so that only a fraction of them can ever be filled. The larger cations such as Sr^{2+} or La^{3+} have ionic radii comparable to that of oxygen, and they often substitute for oxygen ions in the dense-packed lattice with a 12-fold oxygen coordination. The ionic radii of the trivalent rare earths in a 12-fold coordination decrease monotonically from 136 pm for La^{3+} to 114 pm for Lu^{3+}. Divalent Eu^{2+} has a stable half-filled shell $4f^7$ with $r = 140$ pm. Ionic radii increase with coordination number; they are about 25% smaller for the rare earths in 6-fold coordination than in a 12-fold coordination.

A variety of defects can be found in oxide structures [12]. Point defects include vacant oxygen or metal sites, and metal cations in interstitial sites, which are unoccupied in the perfectly-ordered structure. The former may trap one or two electrons, in which case the defect is known as an *F center*. Planar defects include grain boundaries in polycrystalline material, and missing planes of oxygens such as those found in the Magnéli phases Ti_nO_{2n-1}, which are based on rutile-structure TiO_2. Defects modify the electronic structure and may influence the optical and magnetic properties.

The ionic picture is, of course, an oversimplification. The chemical bond between metal and oxygen has part-ionic and part-covalent character, governed by the electronegativity of the atoms involved. The covalent character is more pronounced in tetrahedral sites and for cations (such as V^{4+}) with a high formal charge state.

Next we briefly present ten representative structures encountered in magnetic oxides. Of course there are many more than ten oxide structures, but these examples serve to illustrate the structural principles and cover the most common materials. We refer to them by the name of a mineral type, which is not necessarily an oxide. In each case a picture of the structure is included in Figure 1.3, and structural information on specific nonmagnetic oxides is provided in Table 1.3.

Table 1.2 Ionic radii of cations in oxides.

Fourfold tetrahedral	pm	Sixfold octahedral	pm	Sixfold octahedral	pm	12-fold substitutional	pm
Mg^{2+}	53	$Cr^{4+}\ 3d^2$	55	$Ti^{3+}\ 3d^1$	67	Ca^{2+}	134
Zn^{2+}	60	$Mn^{4+}\ 3d^3$	53	$V^{3+}\ 3d^2$	64	Sr^{2+}	144
Al^{3+}	42	$Al^{3+}\ 54$	—	$Cr^{3+}\ 3d^3$	62	Ba^{2+}	161
$Fe^{3+}\ 3d^5$	52	$Mn^{2+}\ 3d^5$	83	$Mn^{3+}\ 3d^4$	65	Pb^{2+}	149
Si^{4+}	40	$Fe^{2+}\ 3d^6$	78 (61)	$Fe^{3+}\ 3d^5$	64	Y^{3+}	119
—	—	$Co^{2+}\ 3d^7$	75 (65)	$Co^{3+}\ 3d^6$	61 (56)	La^{3+}	136
—	—	$Ni^{2+}\ 3d^8$	69	$Ni^{3+}\ 3d^7$	60	$Gd^{3+}\ 4f^7$	122

Values in brackets indicate a low-spin state.

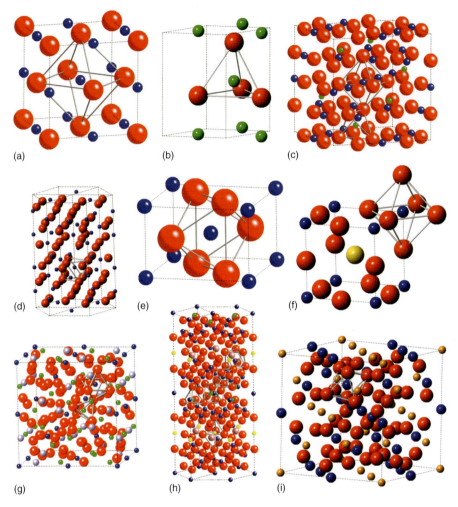

Figure 1.3 Some common oxide structure types (a) halite, (b) wurzite, (c) spinel, (d) corundum, (e) rutile, (f) perovskite, (g) garnet (h) magnetoplumbite, and (i) pyrochlore.

- *Halite.* The NaCl structure is composed of two interpenetrating fcc sublattices, with cations and anions alternating along [100] directions. Cations are octahedrally coordinated by anions and vice versa.
- *Wurzite.* The ZnS structure is composed of two interpenetrating hcp sublattices, so that cations and anions alternate along the 001 direction. Here the cations are tetrahedrally coordinated by anions and vice versa. The structure lacks a center of symmetry, and compounds with this structure may be piezoelectric or pyroelectric.
- *Spinel.* The structure of $MgAl_2O_4$ is based on an fcc array with 32 oxygen ions where $1/8$ of the tetrahedral interstices (A-sites) are occupied by Al and $1/2$ the octahedral interstices (B-sites) are occupied by a mixture of Mg and Al. This

Table 1.3 Properties of nonmagnetic oxides.

Compound	Structure	Lattice parameters (pm)	E_g (eV)	ε_0
MgO	Halite	$a_0 = 421$	7.8	9.7
ZnO	Wurzite	$a = 325; c = 521$	3.4	8.2
$MgAl_2O_4$	Spinel	$a_0 = 808$	5.6	8.6
Al_2O_3	Corundum	$a = 476; c = 1299$	8.8	10
TiO_2	Rutile	$a = 459; c = 294$	3.0	110

E_g is the energy gap and ε_0 is the dielectric constant.

is known as the *inverse cation distribution*. The lattice parameter is $a_0 = 808$ pm. The B-sites form a lattice of corner-sharing tetrahedra, which leads to frustration of antiferromagnetic superexchange in spinel ferrites with the normal cation distribution (Fe^{3+} on B-sites).

- *Corundum (sapphire).* This structure of Al_2O_3 is based on an hcp oxygen array, where Al occupies 2/3 of the octahedral interstices. The structure is rhombohedral, but it is often indexed on a larger hexagonal cell with 32 oxygen ions and $a = 476$ nm, $c = 1299$ nm.
- *Rutile.* This structure of TiO_2 can be regarded as a distorted hcp oxygen array, where Ti occupies 1/2 of the octahedral interstices. Each hcp sheet is deformed into a centered cubic array; so the structure is tetragonal with just four oxygen ions in the unit cell and $a = 459$ pm, $c = 256$ pm. TiO_2 can also crystallize in the anatase structure.
- *Perovskite.* The $CaTiO_3$ structure is pseudocubic with $a_0 = 379$ pm. Here the Ca and O ions together form an fcc lattice, and the Ti ions occupy 1/4 of the octahedral interstices, coordinated only by oxygen. The cell tends to be rhombohedrally or orthorhombically distorted, depending on the cation:oxygen radius ratios.
- *Double perovskite.* A variant of the perovskite structure has a unit cell doubled in all three directions, with 24 oxygen atoms. There are now two different octahedrally coordinated cations in the structure, which form a NaCl-type superlattice.
- *Garnet.* Pyrope, $Ca_3Fe_2Si_3O_{12}$, has a big cubic unit cell with $a_0 = 1145$ pm containing 96 oxygen atoms. The calcium is coordinated by eight oxygens; iron is in octahedral sites and silicon is in tetrahedral sites; $Y_3Fe_5O_{12}$ is known as YIG.
- *Magnetoplumbite.* $PbFe_{12}O_{19}$ has a tall hexagonal cell with $a = 589$ pm, $c = 2309$ pm, containing 76 oxygen ions. The structure can be considered as an hexagonal ABABAABABA stack of oxygen and lead ions, with the iron sites coordinated only by oxygen. There are three octahedral sites, a tetrahedral site and a five-coordinated trigonal bipyramidal site.
- *Pyrochlore.* This is a mineral with the ideal formula $Ca_2Nb_2O_7$. The structure is derived from that of fluorite (CaF_2), where the fluorine anions form a simple cubic array, with calcium in alternate body-centers, forming an fcc cation array. In pyrochlore, the lattice parameter is doubled, 7/8 of the fluorine sites are occupied

by oxygen but $1/8$ are vacant, and the structure includes an array of corner-sharing Ca tetrahedra. There are 56 oxygen ions in a cubic cell with $a_0 = 1008$ pm.

We conclude this section with a few remarks on the electronic structure of oxides. Normally, the oxides are insulators, with a valence band derived from the filled $2p^6$ oxygen levels, and a conduction band derived from unoccupied metal orbitals. Examples where no unpaired transition- element electrons are present are MgO, ZnO, $MgAl_2O_4$, Al_2O_3, and TiO_2. This is also a list of commonly-used substrates for thin-film growth of the magnetic counterparts. Properties of these baseline nonmagnetic oxides are given in Table 1.3. All have a wide bandgap and are optically transparent as single crystals and the powders are white.

Schematically, the electronic structure of the magnetic 3d metal oxides TO_x comprises a filled $2p^6(O)$ valence band separated from a 4s(T) conduction band by a primary energy gap E_g of several electron volts. The occupied $3d^n$ level of the transition-metal ion may then lie in the gap, or else it may lie below the top of the 2p band, provided that the $3d^{n+1}$ level lies above the top of the band. The value of n defines the valence state of the cation. The cations at the beginning, middle, and end of the 3d series tend to be quadrivalent, trivalent, and divalent, forming dioxides, sesquioxides, and monoxides, respectively. Examples are TiO_2 ($n = 0$), Fe_2O_3 ($n = 5$), and NiO ($n = 8$). On progressing across the 3d series, the occupied 3d level moves from a position high in the gap to a position below the top of the 2p band, owing to the stabilizing effect of the increasing charge on the transition-metal nucleus. Moreover, d bands tend to become narrower on moving across the series for the same reason.

The structure more frequently encountered when a transition metal is present in the structure, is for the $3d^n$ level to fall in the gap. Narrow d bands are formed by 3d(M) – 2p(O) hybridization, which frequently leads to optical absorption in the visible – leading to the observed colors (red for hematite, green for YIG, and black for magnetite) [14].

When the 3d level lies in the gap and the energy of the excitation $2(d^n) \rightarrow d^{n-1} + d^{n+1}$ is less than the 3d or 4d bandwidth, the oxide is a d-*band metal*. Examples include CrO_2 and $SrRuO_3$. However, if the energy of this excitation exceeds the d bandwidth, the material may be a *Mott insulator*, but when $3d^n$ level lies below the top of the 2p band, the relevant low-energy electronic excitation is $p^6d^n \rightarrow p^5 + d^{n+1}$. According to whether this is less than or greater than the d bandwidth, the material is a p/d *metal* or a *charge-transfer insulator* [13] (Figure 1.4).

There are a few ferromagnetic metal oxides where the 3d bands are spin split to the extent that a spin gap appears in either the $\uparrow$ or $\downarrow$ subband, and the electrons at the Fermi level are completely spin-polarized. Such materials are known as *half metals*. Stoichiometric half-metals exhibit a spin moment per unit cell which is an integral number of Bohr magnetons.

Oxides may not be precisely stoichiometric, but small deviations often have little influence on the electronic properties because the electrons or holes create a distortion of the lattice, forming immobile polarons that have a large effective mass.

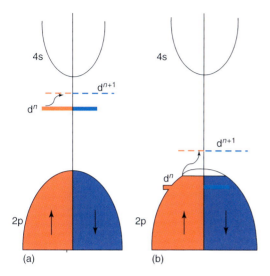

Figure 1.4 Electronic structure of a 3d oxide for an early transition metal (a) and a late transition metal (b). The low-energy d–d and p–d electronic excitations are shown.

1.2
Oxide Growth

1.2.1
Polycrystalline Materials

The easiest methods of oxide growth involve physical or chemical reactions that yield bulk polycrystalline material.

1.2.1.1 Precipitation

Precipitation is a widely used wet-chemical technique for synthesizing ultrafine ceramic powders of simple or complex oxides with a narrow particle size distribution [15]. The method avoids complex steps such as refluxing of alkoxides, and it is faster than other techniques. Precipitation or coprecipitation from aqueous salt solutions (nitrate, sulfate, chloride, perchlorate, etc.) by fine control of the pH by adding NaOH or NH_4OH solutions yields a poorly-crystallized hydroxide or an intimate mixture of the hydrated oxides. The particle size of the precipitate is strongly dependent on pH and the molarity of the precursor solution; it is usually converted to oxide by heating in air. The coprecipitation method offers simple and rapid preparation of complex oxides with control of particle size and good overall homogeneity.

1.2.1.2 Sol–Gel

The sol–gel method is based on the hydrolysis and condensation of metal–organic precursors in organic solvents [16]. The starting materials used in the preparation of the sol are usually inorganic metal salts or metal organic compounds such as metal alkoxides $M(OR)_z$ where R is an alkyl radical ($R = CH_3$, C_2H_5, etc.). In a typical sol–gel process, the precursor is subjected to a series of hydrolysis and polymerization reactions to progressively form a colloidal suspension, known as the *sol*. Further processing leads to the *gel*, from which it possible to make materials in different forms such as ultrafine or spherical-shaped powders. Thin films are produced by dip coating. All are normally subject to post-deposition heat treatment.

1.2.1.3 Solid-State Reaction

Solid-state diffusion is often used to make complex oxides from an intimate mixture of microcrystalline precursor powders [17]. Neither a solvent medium nor controlled vapor-phase interactions are involved. Sometimes known as *shake 'n' bake*, it is the most widely used method for the preparation of complex oxides from a mixture of solid starting materials. Interdiffusion takes place at an appreciable rate at high pressure and/or high temperature (1000–1500 °C for several days). The rate of the solid-state reaction also depends on the ambient atmosphere, structural properties of the reactants, their surface area, reactivity, and the thermodynamic free energy change associated with the reaction. Several cycles of grinding the powder and refiring at progressively higher temperatures are usually necessary to achieve a pure phase. Solid-state synthesis is used to create ceramic powders or dense polycrystalline sintered masses. It may be advantageous to use precursors such as carbonates or oxalates, which decompose at low temperature to give fine-grained oxides.

1.2.1.4 Combustion Synthesis

The term covers a group of methods related to solid-state synthesis and coprecipitation where a fuel is incorporated in the mixture to be fired [18]. An exothermic reaction promotes uniform heating of the solid mass, and the production of oxide powders with a narrow crystallite size distribution in the range 10–100 nm. The solution method typically involves dissolving the precursors such as metal nitrates (the oxidizer) and a compound such as urea or glycine (the fuel) in water, followed by a self-sustaining reaction between the dried constituents, initiated at a relatively low temperature of 500 °C. Microwave irradiation may be used to modify the reaction conditions using a simple microwave oven.

1.2.2
Single Crystals

Single crystals, which may range in size from micrometers to centimeters, are needed for detailed characterization of the physical properties including the elementary excitations (magnons, phonons, and excitons). The key to growing them is control of the nucleation process.

1.2.2.1 Bridgeman Method

The Bridgeman method is a slow, controlled freezing process taking place under liquid–solid equilibrium conditions by allowing the solid–liquid interface to move slowly until the entire molten charge is solidified [19, 20]. The growth takes place in a temperature gradient, and the idea is to create a single nucleus from which a single crystal will grow. The method involves melting polycrystalline material in a crucible with a pointed end and slowly cooling it from the bottom where a seed crystal nucleates at the tip. The crystal grows progressively up the length of the crucible. Either the crucible itself or a tube furnace with a temperature gradient can be moved. Compared to other growth methods, the Bridgeman method is rather simple but it cannot be applied if the system decomposes before it melts, or to oxides of elements with a high vapor pressure.

1.2.2.2 Czochralski Method

The Czochralski process is widely used to grow large single-crystal boules of semiconductors such as silicon, germanium, and gallium arsenide, but it can also be applied for many oxide crystals including Al_2O_3 (sapphire), $LaAlO_3$ (LAO), $Y_3Fe_5O_{12}$ (YIG), $Y_3Al_5O_{12}$ (YAG), and $Gd_3Ga_5O_{12}$ (GGG) [21]. Single-crystal material is pulled out of a slightly undercooled melt by dipping in a single-crystal seed and then slowly withdrawing it. The seed crystal rod is rotated at the same time as it is drawn out, and by precisely controlling the atmosphere, temperature gradient, rate of pulling, and speed of rotation, it is possible to extract a large, single-crystal oxide boule from the melt, which may then be sliced and polished to make substrates for thin film growth of other oxide materials.

1.2.2.3 Zone Melting (Image Furnace)

In the *float-zone* technique, the sample is in the form of a vertical polycrystalline rod, clamped only at its ends, a short segment of which is melted by a local heating [22]. The molten zone is suspended as a drop between the two solid parts of the rod, and it is moved along the rod by slow motion of either the heater or the rod itself. The optical floating zone technique, which makes use of an infrared image furnace, has been extensively utilized to grow single crystals of oxides. Early designs had one or two mirrors, but now, four mirrors are generally used to obtain more uniform sample heating. Ellipsoidal mirrors are used to focus the light from a halogen or xenon lamp onto the sample to produce the molten zone, making the technique suitable for both conducting and nonconducting materials. Optical heating is particularly convenient and efficient for oxides that absorb easily in the infrared. The sample is protected from its environment by a large diameter, clear quartz tube, which prevents evaporated material from settling on the mirror and allows control of the growth atmosphere and gas pressure around the growing crystal [23]. Optimizing crystallization rate, atmosphere, gas pressure, and temperature is the key to achieving stable growth and, good crystal quality. This technique is commonly used to produce pyrochlore, perovskite, and double-perovskite single crystals. Zone melting also helps to purify the oxide.

1.2.2.4 Flux Method

The flux method involves crystal growth by slow cooling of a high-temperature solution. It is suitable for growing crystals of incongruently-melting compounds, but virtually any stable oxide may be grown from a suitable solvent. The flux and the oxide or its constituents are melted in a platinum crucible, which is then cooled extremely slowly. Times of order a month may be required to complete the crystal growth cycle. Oxides (B_2O_3, Bi_2O_3, BaO, and PbO), hydroxides (KOH and NaOH), or halides such as PbF_2 can be used as solvents [24]. However, eutectics, found in binary ($PbO-PbF_2$, Li_2O-MoO_3, $Li_2O-B_2O_3$, etc.) or ternary diagrams, are generally preferred owing to their low temperature of melting and low viscosity. Flux growth is useful whenever the melting temperature is high and when the vapour pressure at the melting point is elevated. The main disadvantage of the technique is the low-growth rate, more than 100 times slower than for crystals pulled from the melt. Crystals may take weeks to grow.

1.2.2.5 Chemical Vapor Reactions

Transport of material in the gas phase is used to grow crystals in a sealed quartz tube that is placed in a temperature gradient in a tube furnace [25]. A powder of the oxide to be grown is included with a transport agent such as I_2 or $TiCl_2$ with which it can react. When the reaction is exothermic, the oxide is transported from the cool zone of the furnace to the hot zone where the compound decomposes and the oxide crystals grow. The temperature gradient must be carefully controlled, and the process may take several days, but the quality of the small oxide crystals is often excellent.

1.2.3
Thin Films

Magnetic oxide thin films can be prepared by physical or chemical methods. In physical deposition methods, a source of material is separated by a distance d from the substrate, which is often heated in the range $400-1000\,°C$ to facilitate growth [26]. Some variants are indicated in Figure 1.5. At low pressure, the atomic species from the source arrive at the substrate without collision, but at higher pressure, they are thermalized by collision with the gas atoms in the chamber. At room temperature, $\lambda = 6/P$, where λ is the mean free path of the atom in millimeters and P is the pressure in pascals. Oxide thin films are used as tunnel barriers, and as functional elements in thin film stacks (ferromagnetic, antiferromagnetic, ferroelectric oxides). The methods are now described in more detail. *In situ* measurement of the thicknesses of the thin films may be achieved using optical reflectometry or a quartz crystal monitor.

1.2.3.1 Physical Methods

Thermal Evaporation Resistive evaporation is a commonly-used vacuum deposition process in which electrical energy is used to heat a boat containing the charge or a filament that heats the material to be deposited up to the point of evaporation [27, 28]. The vapour condenses in the form of a thin film on the cold substrate surface. The method is restricted to materials with moderately low melting points to

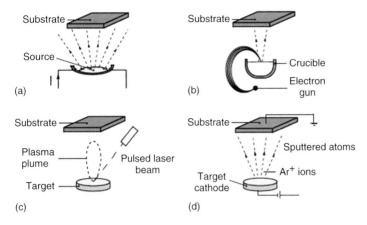

Figure 1.5 Methods for preparing thin films by a vapor or plasma condensing on a substrate: (a) thermal evaporation, (b) e-beam evaporation, (c) pulsed-laser deposition, and (d) sputtering.

avoid contamination by the boat, which is usually made of graphite, molybdenum, or tungsten.

E-Beam Evaporation   Electron beam (e-beam) evaporation is a variant that is used, both for research and on an industrial scale, for making thin film coatings [28]. The technique involves bombarding a target of the material to be evaporated with a beam of high-energy electrons that may be swept over the surface of the target in a specific pattern with the help of beam focussing coils. The target is placed either directly in a water-cooled crucible or in a crucible liner that is made of a different material such as graphite or tungsten, which has a higher melting point than the target and does not form an alloy with it. As a result of the localized melting of the target, the material evaporates and is transported to a substrate. Another geometry utilizes a conducting target in the shape of a thin rod and the e-beam is electrostatically accelerated toward the end of the rod. As the material is evaporated, the rod is fed manually or automatically to keep a constant deposition rate. This geometry is used for materials with a very high melting point such as molybdenum or tungsten. E-beam evaporation can be used in a reactive environment to make thin films of oxides by evaporating a metallic target in the presence of reactive gas. In our case, the oxygen is either directly released into the evaporation chamber or introduced from a low-energy divergent-beam plasma ion source that is directed at the substrate. The latter technique, also known as ion-beam-assisted deposition (IBAD), produces more stoichiometric and better quality oxide films. Very high deposition rates can be achieved by e-beam evaporation compared to sputtering, which makes it advantageous for coating thick films. This technique is widely used to produce SiO_2, Al_2O_3, and some transition-metal oxides (TiO_2, HfO_2, and ZrO_2).

Molecular Beam Epitaxy (MBE)   Molecular beam epitaxy (MBE) is a sophisticated version of vacuum evaporation [29]; it is a method of laying down layers of materials a few atoms thick in ultrahigh vacuum (UHV). Molecular beams of the constituent elements are generated from heated sources and travel without scattering to a substrate where they combine to form an epitaxial film. The most common type of MBE source is the effusion cell (K-cell). The growth rate depends on the flux of material in the molecular beams, which can be controlled by the evaporation rate and, most importantly, switched on and off with shutters in a fraction of the time required to grow one monolayer. Typical growth rates are a monolayer per second or a micrometer per hour. MBE can produce high-quality layers with very abrupt interfaces and good control of thickness, doping, and composition. Because of the high degree of control, MBE is a valuable tool for development of sophisticated electronic and optoelectronic devices, but it is not used suited for industrial production.

Sputtering Sputtering is the preferred industrial thin film vacuum deposition technique, but it is also widely used in research laboratories. Sputtered films exhibit excellent, reproducibility, uniformity, density, purity and adhesion. It is possible to make oxides, nitrites, and other compounds of precise composition by reactive sputtering from metal targets [28, 30]. In *dc sputtering*, substrates are placed in the vacuum chamber, and it is evacuated to high vacuum before a low pressure (0.05–1 Pa) of the process gas, usually argon, is introduced. Sputtering starts when a negative potential of a few hundred volts is applied to the target material to be deposited, causing a plasma or glow discharge. Positively charged Ar^+ ions generated in the plasma collide with the negatively biased target. The momentum transfer ejects atomic-scale particles from the target, which traverse the chamber and are deposited as a thin film on the surface of the substrate. A magnetic field is usually created near the target surface by means of an arrangement of permanent magnets, known as a *magnetron*, in order to improve the ionization efficiency. Oxygen is mixed with the argon sputtering gas to produce oxides from metal targets. Alternatively, to make oxide or other insulating films directly, the radio-frequency method of *rf sputtering* is employed. Here the power supply commonly operates at 13.56 MHz. For part of the cycle, Ar ions bombard the target; for the rest of the cycle, electrons neutralize the build up of positive charge. Electrons also ionize the argon to create the plasma. Sputtering systems often have multiple targets, which permit the fabrication of complex thin film stacks used for spin electronic applications. An argon pressure of 0.02 Pa is usually sufficient to maintain a radio-frequency discharge.

Pulsed Laser Deposition (PLD) Pulsed laser deposition (PLD) is the most versatile physical method to deposit small high-quality metallic and insulating films in the laboratory; it has been intensively developed since 1987 when it was applied to grow high-temperature superconductors [31]. The technique employs high-power UV laser pulses (typically $\sim 10^8 \times W\,cm^{-2}$) to ablate ionized material from the surface of a ceramic target in vacuum. This ablation event produces a transient, highly- luminous directional plasma plume that expands rapidly away from the target surface. The plume deposits material onto the substrate and a thin film is formed. The plume is stoichiometrically similar to the target, and so thin films of roughly the same composition as of the target can be easily produced. Uniform thickness and composition can be achieved by rastering the laser spot across the target surface and/or moving the substrate during deposition. Targets often usually dense ceramic disks of 10–25 mm in diameter. The laser commonly used for oxide growth is a KrF excimer laser ($\lambda = 248$ nm), but frequency-doubled or tripled Nd-doped yttrium aluminum garnet (YAG) lasers are also used. The energy of the beam can range from 100 to 500 mJ per pulse, and typical fluences of $1-5$ J cm^{-2} are incident on the target. The repetition rate varies between 1 and 20 Hz. The vacuum chamber can be partially filled with oxygen or nitrogen to allow deposition at higher pressures.

A problem with the method is that liquid droplets or particulates may contaminate the plume and settle on the surface of the film. This can be avoided by off-axis deposition, where the substrate is parallel to the plume. Alternatively, the droplets may also be trapped by rapidly rotating mechanical filters, which feature in the PLD tools that are being developed for the industrial production of ferroelectric films.

1.2.3.2 Chemical Methods

Chemical Vapor Deposition (CVD)   The growth of thin films by chemical vapor deposition (CVD) is an industrially significant process with good stability and reproducibility, which is used in a wide array of applications. CVD involves depositing a solid film from a gaseous molecular precursor [27]. Different energy sources, precursor gases, and substrates are used, depending on the desired product, but the precursors must be volatile, yet stable enough to be able to be delivered to the reactor where the volatilized precursor (such as silane, an organometallic, or a metal coordination complex) is passed over a heated substrate. Thermal decomposition of the precursor produces a thin film deposit, and ideally, the ligands associated with the precursor are cleanly lost to the gas phase as reaction products. Pressure and temperature are the important variables.

Chemical Vapor Transport   The method, akin to that used to grow single crystals, entails the reversible conversion of nonvolatile elements and chemical compounds into volatile derivatives [25]. The volatile derivative migrates throughout a sealed reactor, typically a sealed, evacuated glass tube heated in a tube furnace. Because the tube is in a temperature gradient, the volatile derivative reverts to the parent solid, which is deposited as a thin film on the substrate and the transport agent is released at the opposite end of the tube to where it originated.

Atomic Layer Deposition (ALD)   Atomic layer deposition (ALD) is based on sequential, self-limiting surface chemical reaction [32]. This unique growth technique can provide atomic layer control and allow ultrathin conformal films to be deposited on very high aspect ratio structures. ALD deposits films using pulses of gas that produce one atomic layer at a time. Within fairly wide process windows, the deposited film thickness is only dependent on the number of deposition cycles providing extremely high uniformity and thickness control. ALD reactions are typically carried out in the range 200–400 °C. In particular, ALD is currently used in the semiconductor industry for high-k gate dielectrics (HfO_2, ZrO_2).

Dip Coating   This is the main wet-chemical method to produce wide variety of oxide thin films such as ZnO, SnO_2, TiO_2, and $(Sn_{1-x}In_x)O_2$ (ITO). A substrate is dipped into a liquid coating solution of the chemical precursor and then gently withdrawn at a controlled rate. The thickness is determined by the balance of forces at the stagnation point on the liquid surface. A faster withdrawal rate pulls more fluid up onto the surface of the substrate before it has time to flow back into the solution. The thickness is also affected by fluid viscosity, density, and surface tension. Finally, the coating is cured by a conventional thermal treatment, or else UV, or IR irradiation.

Spray Pyrolysis   A water-based precursor solution is sprayed through a nozzle onto a substrate where the atomized solution is dried and the metal oxide film is formed. It provides an easy way to dope any element in a required ratio through the solution medium. This method is convenient for preparing pinhole-free, homogenous, smooth thin films of oxides such as TiO_2 with controllable thickness. Spray pyrolysis can also be used to produce oxide powders.

1.3
Magnetic Properties of 3d and 4f Ions

One of the stranger truths of Nature is that magnetism is intimately associated with angular momentum. It is the angular momentum of charged particles, specifically that of the electrons that is responsible for the magnetism of solids. In the Bohr model, the z-component of orbital angular momentum l is quantized in units of $\hbar$. Regarding the Bohr atom as a loop of quantized current $I = -el/2\pi r^2 m_e$, the magnetic moment $m = I\mathcal{A} = -(el/2\pi r^2 m_e)\pi r^2 = -(e/2m_e)l$. Here, $-e$ and m_e are the charge and mass of the electron; the magnetic moment and angular momentum are oppositely aligned because of the electron's negative charge. The proportionality factor between magnetic moment and angular momentum, here $-(e/2m_e)$, is known as the *gyromagnetic ratio*. However, angular momentum is quantized, with eigenvalues for the z-component being equal to $m_l\hbar$ where $m_l = 0$, ± 1, ± 2, $\ldots$ $\pm l$, so the orbital magnetic moment is also quantized, in units of the Bohr magneton

$$\mu_B = \frac{e\hbar}{2m_e}. \tag{1.1}$$

This quantity, equal to 9.27×10^{-24} A m^2, is the fundamental unit of atomic-scale magnetism.

Besides angular momentum of orbital origin, the electron also possesses an intrinsic angular momentum s known as *spin*, a property common to all fermions. The corresponding gyromagnetic ratio for the electron is $-(e/m_e)$, so that $m = -(e/m_e)s$.

The two possible values for the spin moment of the electron are $m_z = -2\, m_s \mu_B = \mp 1$ μ_B where $m_s = \pm^1\!/_2$. The two spin states are often represented by arrows $\uparrow$ and $\downarrow$ and the *g-factor*, defined as the ratio of the magnitude of the magnetic moment in units of Bohr magneton to the magnitude of the angular momentum, in units of $\hbar$, is precisely 1 for orbital moments and practically equal to 2 for spin moments. The Zeeman Hamiltonian for the electron in a magnetic field **B** is therefore represented by the Hamiltonian

$$\mathcal{H}_Z = -\boldsymbol{m}\!\cdot\!\mathbf{B} = \left(\frac{\mu_B}{\hbar}\right)(l_z + 2s_z)B \tag{1.2}$$

From the electron's point of view, it sees the nucleus orbiting around it, creating a magnetic field proportional to m_l, which interacts with the electronic spin moment. The celebrated spin–orbit interaction that is responsible for much that is important and useful in magnetism is represented by the Hamiltonian

$$\mathcal{H}_{so} = -\lambda \boldsymbol{l}\!\cdot\!\boldsymbol{s} \tag{1.3}$$

On passing from a single electron to a multielectron ion, the angular momentum of any filled (1s, 2s, 2p, etc.) shell is zero because both the spin moments and the orbital moments cancel each other. The moments residing in a partially filled 3d, 4d, 4f shell can be inferred from Hund's rules, namely: (i) maximize $S = \Sigma_i m_{si}$ for the *i* electrons in the shell, remembering that the orbitals can only accommodate one $\downarrow$ and one $\uparrow$ electron each, (ii) then maximize $L = \Sigma_i m_{li}$ consistent with S, and (iii) finally couple L and S to form the total angular momentum J so that $J = L - S$ if the shell is less than half full and $J = L + S$ if the cell is more than half full. When it is exactly half full, $\Sigma_i m_{li} = 0$; so $J = S$.

The magnetic moment of the free ion can be written as $g\mu_B J$, where the g-factor

$$g = \frac{3}{2} + \frac{[S(S+1) - L(L+1)]}{2J(J+1)} \tag{1.4}$$

Two examples:

1) Fe^{2+} ($3d^6$) has a spin occupancy of the 3d orbitals ⬆⬇ ↑ ↑ ↑ ↑ giving $S = 2$. The orbital sum is $L = 2$, hence $J = L + S = 4$ and $g = 3/2$.
2) Nd^{3+} ($4f^3$) has orbital occupancy of the 4f orbitals ↑ ↑ ↑ giving $S = 3/2$, $L = 6$ and $J = L - S = 9/2$ and $g = 4/3$.

The ground state of an ion is denoted by the term $^{2S+1}X_J$, where $2S+1$ is the spin degeneracy, and $X = S, P, D, F, \ldots$ denotes the value of $L = 0, 1, 2, 3, \ldots$. The terms for free Fe^{2+} and Nd^{3+} ions are $^5D_4$ and $^4I_{9/2}$, respectively. The moment is now represented as $\boldsymbol{m} = -(\mu_B/\hbar)(\mathbf{L} + 2\mathbf{S})$. The Zeeman and spin-orbit interactions are represented respectively by the Hamiltonians

$$\mathcal{H}_Z = -\boldsymbol{m}\!\cdot\!\mathbf{B} = \left(\frac{\mu_B}{\hbar}\right)(\mathbf{L}_z + 2\mathbf{S}_z)B \tag{1.5}$$

and

$$\mathcal{H}_{so} = -\Lambda \mathbf{L}\!\cdot\!\mathbf{S} \tag{1.6}$$

The above discussion concerned only the magnetism of free atoms or ions. About two-thirds of the atoms in the periodic table, and their isoelectronic free ions, have unpaired electrons and a net magnetic moment. However, when the ions are placed in the crystalline environment of a solid oxide, there are some drastic changes due to the influence of the oxygen neighbors. This is the *crystal-field* or *ligand-field interaction*. For 3d and 4d ions, it is stronger than the spin–orbit interaction, whereas for the 4f ions, which are well shielded from the outside world by the large filled 5s and 5p shells, the crystal field is screened so that it is only a perturbation on $\mathcal{H}_0 + \mathcal{H}_{so}$. (Table 1.4). For the rare earths, J is therefore a good quantum number, and the main effect of $\mathcal{H}_{cf}$ is to introduce single-ion anisotropy, whereby the ion may have one or more easy axes of magnetization in the crystal lattice.

For the 3d ions, the outer shell is unshielded from the environment, and the crystal-field interaction is stronger than the spin–orbit coupling. As a result, J is not a good quantum number. Orbital motion is impeded and the orbital moment is quenched. S is the good quantum number. From a magnetic viewpoint, the 3d ions in solids may be regarded as spin-only ions, with $m = g\mu_B S$ and $g \approx 2$. This is a welcome simplification, and it is supported by measurements of the paramagnetic susceptibility of crystals with dilute paramagnetic ions. The molar susceptibility of the ions is given by the Curie law $\chi_{mol} = C_{mol}/T$

$$\chi_{mol} = \mu_0 N_0 p_{eff}^2 \mu_B^2 / 3kT \tag{1.7}$$

where N_0 is Avogadro's number. In numerical terms,

$$C_{mol} = 1.571 \times 10^{-6} p_{eff}^2 \tag{1.8}$$

The effective Bohr magneton number p_{eff} agrees with $g\sqrt{[J(J+1)]}$ for 4f ions, but for 3d ions, it is closer to $2\sqrt{[S(S+1)]}$. Some discrepancies appear, especially for Co^{2+}, where spin–orbit coupling restores a significant orbital contribution to the moment, and Eu^{3+}, which has a $J = 0$ ground state, but a low-lying $J = 1$ excited multiplet.

We focus first on the 3d ions, in order to explain how the crystal field influences both the electronic structure and magnetic properties of oxides. We begin with the *one-electron model* [33, 34], which ignores the onsite 3d–3d Coulomb interactions. This is a good approximation for d^1, d^4, d^6, and d^9 ions when Hund's first rule applies, as these ions have a single electron or hole outside an empty, a half-filled, or a filled shell.

When the ion is found in an undistorted tetrahedral or octahedral site in an oxide, the local environment has *cubic* symmetry unlike the *spherical* symmetry of

Table 1.4 Relative magnitudes of energy terms for 3d and 4f ions.

	$\mathcal{H}_0$	$\mathcal{H}_{so}$	$\mathcal{H}_{cf}$	$\mathcal{H}_Z$ in 1 T
3d	$1-5 \times 10^4$	10^2-10^3	10^4	1
4f	$1-6 \times 10^5$	$1-5 \times 10^3$	$\approx 3 \times 10^2$	1

a free ion. The free ion eigenfunctions ψ_0, $\psi_{\pm 1}$, and $\psi_{\pm 2}$, where the subscripts denote m_l, must be replaced by suitable linear combinations, which reflect the cubic symmetry. They are

$$\psi_{xy} = (\frac{-i}{\sqrt{2}})(\psi_2 - \psi_{-2})$$

$$\psi_{yz} = (\frac{i}{\sqrt{2}})(\psi_1 + \psi_{-1})$$

$$\psi_{zx} = (\frac{-1}{\sqrt{2}})(\psi_1 - \psi_{-1})$$

$$\psi_{x^2-y^2} = (\frac{1}{\sqrt{2}})(\psi_2 + \psi_{-2})$$

$$\psi_{z^2} = \psi_0 \tag{1.9}$$

This basis set of d orbitals is illustrated in Figure 1.6.

We consider the influence of a crystal field due to an octahedron or tetrahedron of oxygen neighbors on the 3d wave functions. Considering the disposition of the oxygen orbitals in the octahedron, it is obvious that the xy, yz, and zx orbitals are degenerate – they are labeled as t_{2g} orbitals. It is less obvious that the $x^2 - y^2$ and z^2 orbitals, labeled as e_g, are degenerate, but they are clearly higher in energy because the electron density is maximum near the negatively charged anions. The crystal-field splitting is illustrated in Figure 1.7.

In the tetrahedral site, the splitting is reversed. The e orbitals are lower and the t orbitals are higher (the "g" subscript is dropped when there is no center of symmetry). The splitting in cubic coordination is similar (Figure 1.7).

The splitting cannot be entirely explained in terms of the electrostatic potential due to oxygen anions; about half is attributable to the different overlap of the t and e orbitals with the lower-lying 2p oxygen orbitals which introduces, in an octahedral site for example, greater bonding–antibonding splitting for the 2p-e_g σ-bonds than the 2p-t_{2g} π-bonds. This is known as the *ligand-field effect*. The magnitude of Δ_{oct} is of order 1 eV (Figure 1.8).

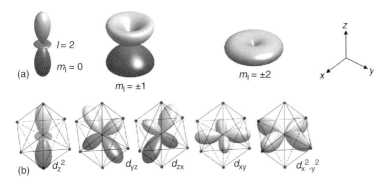

Figure 1.6 The 3d orbitals for a free ion (a) and the orbitals for the ion in a cubic crystal field (b).

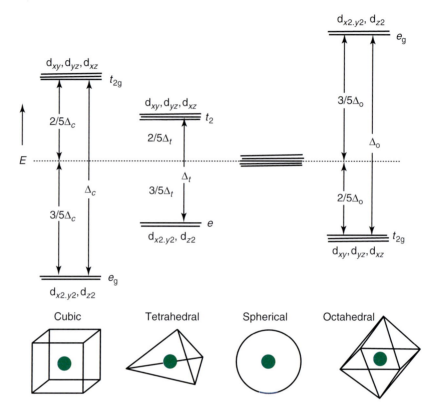

Figure 1.7 Crystal-field splitting for an ion in sites with octahedral, tetrahedral, and cubic coordination.

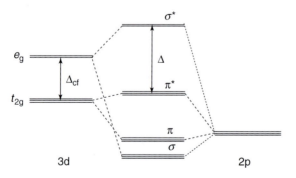

Figure 1.8 Illustration of the splitting of the t_{2g} and e_g orbitals in an octahedral environment. Δ_{cf} is the crystal-field splitting calculated in a point charge model. The extra ligand-field splitting due to the hybridization of 3d and sp orbitals gives the total splitting Δ.

The influence of a distortion, trigonal, tetragonal, or monoclinic, is to raise the degeneracy of the one-electron energy levels, as shown in Figure 1.9 for tetragonal distortion. The splitting preserves the center of gravity of the sets of orbitals.

The crystal-field stabilization energy, which contributes to the site preference of ions in crystal structures such as spinel where they have a choice, is calculated relative to the unsplit level. For example, for Cr^{3+} ($3d^3$) in octahedral sites, it is $3 \times (2/5)\Delta_{oct} = 1.2\,\Delta_{oct}$ whereas for Fe^{3+} ($3d^5$), it is zero. The cation radius and the $2p$–$3d$ mixing also influence site preference.

The *Jahn–Teller effect* is the tendency of some ions to spontaneously deform their local environment in order to improve their crystal-field stabilization energy. A $3d^3$ ion on an octahedral site gains nothing from a distortion of the oxygen octahedron, because of the center of gravity rule, but a $3d^4$ ion will tend to induce a tetragonal deformation that splits the e_g levels, lowering the energy of the occupied orbital, and hence of the ion (Figure 1.9). The Jahn–Teller effect is strong for d^4 and d^9 ions in octahedral coordination (Mn^{2+}, Cu^{2+}) and for d^1 and d^6 ions in tetrahedral coordination (V^{4+}, Co^{3+}). If the local strain is ε, the energy change $\Delta E = -A\varepsilon + B\varepsilon^2$, where the first term is the crystal-field stabilization energy D_{cfse} and the second term is the increased elactic energy. The J-T distorsion may be static or dynamic.

In the one-electron picture, the energy of an ion is obtained by populating the lowest orbitals with the available electrons. Hund's first rule implies that the five ↓ orbitals lie above the five ↑ orbitals by an amount $\mathcal{J}_H$, the on-site exchange energy

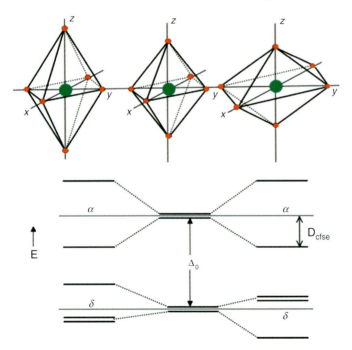

Figure 1.9 Influence of a tetragonal elongation or flattening of an octahedral site on the one-electron energy levels.

which is $\approx 1.5\,\mathrm{eV}$. In some cases, however, $\Delta_{oct} > \mathcal{J}_H$, which means that the t_{2g} orbitals are filled before the e_g orbitals. This produces an ion in a *low-spin* state where Hund's rules no longer apply. In oxides, low-spin states are most frequently found for Co^{2+}, which has $S = 2$ in the high-spin state and $S = 1/2$ in the low-spin state.

Although the one-electron picture is relatively easy to grasp and can be readily related to energy-band calculations, it does not do justice to the many-electron interactions in 3d ions, especially for high-spin ions with two, three, seven, or eight 3d electrons, which are in an F-state. Ions with five electrons are in an S state, while those with one, four, six, or nine electrons are in a D state. Here S, D, and F refer to the values of L (0, 2, or 3) given by Hund's second rule. The excited states of the $3d^n$ shell that are probed in optical transitions have been calculated from crystal-field theory, and they are represented on Tanabe–Sugano diagrams [14].

Two interesting electronic effects that may be observed for 3d electrons in oxides are *charge order* and *orbital order* [35]. The first arises when a particular lattice site is occupied by an ion in a mixture of two different valence states. Examples are Mn^{4+} and Mn^{3+} or Fe^{3+} and Fe^{2+}. The average 3d occupancy is nonintegral, for example, 3.5 or 5.5, and it is possible for the extra 3d electron to hop rapidly among the d^3 or d^5 ion cores at a high enough temperature, or else it may settle on alternate sites with some lattice distortion in a charge-ordered state.

Orbital order arises when the 3d occupancy is integral, but in the undistorted, high-temperature phase, the electron occupies a degenerate orbital. An example is Mn^{3+} ($t_{2g}{}^3 e_g{}^1$). A lattice distortion such as tetragonal compression and expansion on alternate sites can lead to alternate occupancy of d_{z2} and d_{x2-y2} orbitals. The ordered states generally revert to disordered states at a high-temperature phase transition, where an entropy of approximately $R\ln 2$ is released, comparable to the entropy $R\ln(2S + 1)$ released in the vicinity of a magnetic disordering transition. These ordering effects are discussed further in Section 1.4.

Crystal-field effects on the rare-earth ions in oxides are related to the energy levels of the highly correlated $4f^n$ shell, rather than to those of the individual 4f orbitals. The crystal-field Hamiltonian is often written as $\mathcal{H}_{cf} = \Sigma_{n,m} B_n{}^m \hat{O}_n{}^m$, where $n = 0$, 2, 4, or 6 and $m \le n$; $\hat{O}_n{}^m$ are Stevens operators that are combinations of the angular momentum operators $\hat{J}_i$ and $\hat{J}^2$. The number and type of terms in the expansion depend on the site symmetry. Crystal-field parameters $B_n{}^m$ reflect both features of the 4f ion and features of the lattice site. These are separated in the more explicit formulation

$$\mathcal{H}_{cf} = \sum_{n,m} A_n{}^m \theta_n < r^n > \hat{O}_n{}^m \tag{1.10}$$

where the crystal-field coefficients $A_n{}^m$ parameterize the derivatives of the electrostatic potential due to the lattice at the rare-earth site, $<r^n>$ is the average over the 4f electron distribution and θ_n is the 2^n-pole moment of the ion – quadrupole for $n = 2$, hexadecapole for $n = 4$ and 64-pole for $n = 6$. The expressions for uniaxial and cubic symmetry are, respectively,

$$\mathcal{H}_{cf} = B_2{}^0 \hat{O}_2{}^0 \tag{1.11}$$

and

$$\mathcal{H}_{cf} = B_4{}^0 \hat{O}_4{}^0 + 5B_4{}^4 \hat{O}_4{}^4 \tag{1.12}$$

where the Stevens operators are

$$\hat{O}_2{}^0 = 3\hat{J}_z{}^2 - J(J+1)$$

$$\hat{O}_4{}^0 = [35\hat{J}_z{}^4 - 30J(J+1)\hat{J}_z{}^2 + 25\hat{J}_z{}^2 - 6J(J+1) + 3J(J+1)^3]$$

and

$$\hat{O}_2{}^4 = \tfrac{1}{2}[\hat{J}_+{}^4 + \hat{J}_-{}^4]$$

[36, 37]. The charge densities for the $4f^n$ ions are shown (with exaggerated asphericities) in Figure 1.10. The quadrupole moments θ_2 reflect the oblate or prolate shape of the charge distribution and follow the quarter-shell rule; they are negative for Ce, Pr, Nd, Tb, Dy, and Ho but positive for Sm, Er, Tm, and Yb. (Trivalent Eu has $J = 0$, and Pm is unstable; so it is not shown.)

The same formalism may be applied for the 3d ions, but the averages are over the 3d shell and the Stevens operators now involve **L** rather than **J** and only run up to $n,m = 4$ [37]. The uniaxial anisotropy energy can be represented by a term

$$\mathcal{H}_a = D\hat{J}_z{}^2 \tag{1.13}$$

for the rare earths, and a spin Hamiltonian

$$\mathcal{H}_a = D\hat{S}_z{}^2 \tag{1.14}$$

for the 3d ions. For non-S-state 3d ions in noncubic sites, D/k may be of the order of one Kelvin (0.1 meV). Single-ion magnetic anisotropy, expressed as an energy per ion, is a weak interaction. It determines the directon of sublattice magnetization in oxides.

1.4
Magnetic Interactions in Oxides

Exchange interactions are symmetry-constrained coulomb interactions that have the effect of coupling electronic spins. Intraionic exchange, also known as $\mathcal{J}_H$ or

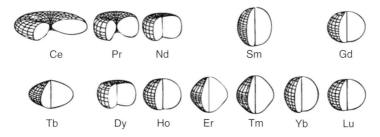

Figure 1.10 The charge densities for the trivalent rare-earth ions.

Hund's rule exchange, couples electrons in a partially filled shell on an ion so as to maximize the total spin (Hund's first rule). In oxides, it is normally greater than the crystal/ligand-field splitting Δ so that the $\uparrow$ orbitals are filled first, before beginning to fill the $\downarrow$ orbitals. If $\Delta > \mathcal{J}_H$, the ion enters a low-spin state, in conflict with Hund's first rule. The rules were formulated for free atoms and ions. The magnitude of $\mathcal{J}_H$ in the 3d series is about 1.0-2.0eV.

Interionic exchange, coupling the spins on adjacent ions, is much weaker. It depends on the overlap of exponentially decaying wave functions, and does not usually exceed 0.01 eV. However, it is responsible for the magnetic order of the ionic spins already created by the intraionic coupling. Interionic exchange between a pair of ions i, j is described by the Heisenberg Hamiltonian

$$\mathcal{H}_{ex} = -2\mathcal{J}_{ij}\mathbf{S}_i\cdot\mathbf{S}_j \tag{1.15}$$

Heisenberg exchange $\mathcal{J}$ may couple the pair of spins parallel or antiparallel, depending on its sign. The interaction is summed over all pairs of ions, but in oxides, the sum is largely restricted to a few nearest-neighbor shells. The exchange parameters $\mathcal{J}_{ij}$ determine the magnetic ordering temperature and the spin-wave dispersion relations. An important difference between ferromagnetic ($\mathcal{J} > 0$) and antiferromagnetic ($\mathcal{J} < 0$) interactions is that the latter are subject to geometric frustration owing to lattice topology [11].

1.4.1
Superexchange

Antiferromagnetic exchange interactions are responsible for the antiferromagnetic order in oxides such as NiO, $CaMnO_3$, or Fe_2O_3. There is little direct overlap of the wave functions of the nearest-neighbor cations, but they overlap strongly with the 2p orbitals of the neighboring O^{2-} anions. The oxygen bridges transmit a *superexchange* interaction via the hybridization between 2p orbitals and 3d orbitals. The interaction involves two virtual electron transfers. First an electron is transferred from the oxygen $2p^6$ shell to an adjacent M_1 ion, leading to a virtual d^{n+1} state, at an energy cost U. Then the 2p hole is then filled by electron transfer from another M_2 ion, which overlaps with the same 2p shell. This has the effect of coupling the spins of M_1 and M_2. A simple case is shown in Figure 1.11, where M_1 and M_2 are ions with a half-filled d shell (Fe^{3+}, Mn^{2+}), which overlap with the same 2p orbital. Then transfer of a p $\downarrow$ electron into an empty 3d $\downarrow$ state of M_1 (the 3d $\uparrow$ states are all full) leaves a 2p $\downarrow$ hole, which can only be filled by a 3d $\downarrow$ electron from M_2. Hence the configuration in Figure 1.11(b) is lower in energy than the configuration in Figure 1.11(a). Both electrons in the oxygen 2p orbital can then spread out into unoccupied 3d orbitals when the ion spins are antiparallel; the superexchange interaction $\mathcal{J}$ is therefore negative [38]. Since superexchange interactions involve simultaneous virtual transfer of two electrons with the instantaneous formation of a $3d^{n+1}2p^5$ excited state, the interaction is of order $-2t^2/U$, where t is the p–d transfer integral and U is the on-site 3d Coulomb interaction. The transfer integral is of the order of 0.1 eV, and the on-site Coulomb interaction is 2 eV or more so the interaction

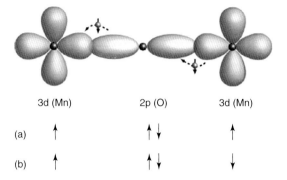

3d (Mn)	2p (O)	3d (Mn)
(a) ↑	↑ ↓	↑
(b) ↑	↑ ↓	↓

Figure 1.11 A typical superexchange bond. Antiferromagnetic configuration (b) is lower in energy than ferromagnetic configuration (a).

is of order 0.01 eV ($\sim$100K). $\mathcal{J}$ depends sensitively on the interatomic separation, and also on the M$-$O$-$M bond angle, θ_{12}, varying as $\cos^2\theta_{12}$.

The occupancy and orbital degeneracy of the 3d states is the critical factor in determining the strength and sign of superexchange. There are many possible cases to consider and they were discussed by Goodenough in his book [6]. The results were summarized in the Goodenough–Kanamori rules, which were reformulated by Anderson, in a simpler way that makes it unnecessary to consider the oxygen.

1) When two cations have lobes of singly occupied 3d orbitals that point toward each other, giving large overlap and hopping integrals, the exchange is strong and antiferromagnetic ($\mathcal{J} < 0$). This is the usual case, for $120°$–$180°$ M$-$O$-$M bonds.
2) When two cations have an overlap integral between singly occupied 3d orbitals which is zero by symmetry, the exchange is ferromagnetic and relatively weak. This is the case for $\sim$90° M$-$O$-$M bonds.
3) When two cations have an overlap between singly occupied 3d orbitals and empty or doubly occupied orbitals of the same type, the exchange is also ferromagnetic, and relatively weak.

Superexchange is more commonly antiferromagnetic than ferromagnetic, because the overlap integrals are likely to be larger than zero.

1.4.2
Double Exchange

This interaction arises between 3d ions that have both localized and delocalized electrons [39]. Unlike ferromagnetic superexchange, a mixed-valence configuration is required for double exchange. In a mixed-valence manganite such as $(La_{1-x}A_x)MnO_3$, A$=$Ba, Ca, Sr, the two Mn valence states are imposed by the charge states of the other ions in the compound, La^{3+}, A^{2+}, and O^{2-}. The d^3 core electrons for the octahedrally coordinated Mn ions are localized in a narrow $t_{2g}\uparrow$ band, but the fourth d electron inhabits a broader $e_g\uparrow$ band, hybridized with

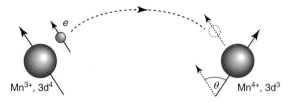

Figure 1.12 The double exchange interaction. The fourth delocalized electron in Mn^{3+} ion hops to its neighboring Mn^{4+} without changing its spin.

oxygen, where it can hop from one d^3 core to another, as shown in Figure 1.12. The configurations $d^3{}_i d^4{}_j$ and $d^4{}_i d^3{}_j$ on adjacent sites i and j are practically degenerate. On each site, there is a strong on-site Hund's rule exchange coupling $\mathcal{J}_H \approx 2\,eV$ between t_{2g} and e_g electrons, which is very much greater than the hopping energy. Electrons can hop freely if the core spins are parallel, as the delocalized electron does not have to change its spin direction, but when they are antiparallel, there is a large energy barrier due to the Hund's rule interaction. The ability to hop increases the extent of the d wavefunction and reduces the kinetic energy. Hence, the overall energy saving leads to ferromagnetic alignment of the spins of neighboring ions.

If the quantization axes of adjacent sites are misaligned by an angle θ, the eigenvector of an electron in the rotated frame is $(\cos\theta/2, \sin\theta/2)$. The transfer integral t therefore varies as $\cos(\theta/2)$. Double exchange is ferromagnetic because the transfer is zero when the ions on adjacent sites are antiparallel, $\theta = \pi$.

Another common double-exchange pair is Fe^{3+} and Fe^{2+}, which are d^5 and d^6 ions, respectively. The d^5 configuration is a half-filled, $\uparrow$ d shell, and the sixth, $\downarrow$ d electron occupies the bottom of a $t_{2g}\uparrow$ band when the ion is octahedrally coordinated by oxygen, where it can hop directly from one $d^5 \uparrow$ core to another.

The effective Hamiltonian of double exchange also has the characteristic scalar product $(\mathbf{S}_i \cdot \mathbf{S}_j)$; however, it contains not only this term but also higher powers of it. The highest power is determined by the magnitude of the localized spins:

$$\mathcal{H}_{DE} \sim \sum_{n=0}^{2s} \mathcal{J}_n(S)(\mathbf{S}_i \cdot \mathbf{S}_j)^n \tag{1.16}$$

When $\mathcal{J}_n(S)$ for $n \geq 2$ is not negligible, then $\mathcal{H}_{DE}$ is not of Heisenberg type [40]. However, the bilinear term, which always dominates, is always ferromagnetic.

Double exchange differs clearly from superexchange in the following respect: in superexchange, the electrons do not actually move between the two metal cations – the occupancy of the d shell of the two metal ions remains the same. In double exchange, the electrons are itinerant; they hop between the positive ions via the intermediate ligand (oxygen); this results in the material displaying from electrical conductivity as well as magnetic exchange coupling.

1.4.3
Antisymmetric Exchange

In crystals with uniaxial or lower symmetry, a weak interaction of the form

$$\mathcal{H}_{Dm} = -\boldsymbol{D} \times (\mathbf{S}_i \times \mathbf{S}_j) \tag{1.17}$$

was proposed by Dzyaloshinskii and Moriya. The vector $\boldsymbol{D}$ lies along the axis of symmetry, but its magnitude is only about 1% of $\mathcal{J}$. The effect of the interaction is to cant the spins away from the antiferromagnetic axis, producing a weak intrinsic ferromagnetic moment in a direction perpendicular to the antiferromagnetic axis. The most famous example is hematite, αFe_2O_3.

1.4.4
Direct Exchange

Direct exchange is dominant in metals but relatively unimportant in oxides. Two examples where it is operative are the d-band metals CrO_2 and $SrRuO_3$. Direct electron hopping in the partly filled t_{2g} band of Cr in the rutile structure is a ferromagnetic interaction; two of the three t_{2g} orbitals that form the π^* band (Figure 1.8) are filled, and the other is empty [41].

1.4.5
Orbital Order

We mentioned in Section 1.3 that interionic electrostatic interactions can also give rise to orbital order [42]. When there is a single occupied d orbital at each 3d ion site, antiparallel spin alignment on adjacent sites favors electron hopping and thus reduces the energy of the system (antiferromagnetic superexchange). However, if two degenerate orbital states are available, it is possible for the electron spins to hop to their neighbors with parallel spins, filling the other orbital, which may also lower the energy. In this case, spin order (ferromagnetism) is possible on the condition that different degenerate neighboring 3d orbitals are occupied in an alternating array (Figure 1.13a). By analogy with spin order, two possible choices of orbital can be described by a pseudospin T. Consider, for example, a $3d^4$ ion with one electron in the degenerate $d_{x^2-y^2}$ and d_{z^2} states in an octahedral site; when $d_{x^2-y^2}$ is occupied, the orbital order parameter is $T_z = 1/2$ and when d_{z^2} is occupied, it is $T_z = -1/2$ [43]. The pseudospin interaction $-2\mathcal{K} \, T_i \cdot T_j$ is antiferromagnetic and there is an interaction between the spin and pseudospin, of S and T, between different ions. This exchange interaction is represented by the following generalized Heisenberg Hamiltonian:

$$\mathcal{H} = -2\sum_{ij}[J_{ij}(T_i, T_j)S_iS_j + K_{ij}(T_i, T_j)] \tag{1.18}$$

The interactions $\mathcal{J}_{ij}$ and $\mathcal{K}_{ij}$ originate from the quantum mechanical process with intermediate virtual states (Figure 1.13).

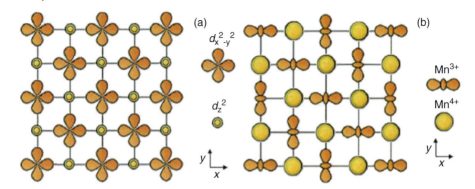

Figure 1.13 Examples showing orbital order of the x^2-y^2/z^2 type (a) and both orbital order of the $x^2/y^2/z^2$ type and charge order projected onto an MnO_2 plane (b).

In more general cases, the transfer integral t_{ij} depends on the direction of the bond ij relative to the pair of orbitals $d_{x^2-y^2}$ or d_{z^2}. This gives rise to anisotropy of the Hamiltonian in pseudospin space as well as in real space. In many respects, analogies can be drawn between S and T, in spite of the anisotropy in T space. However, one aspect that is special to T is Jahn–Teller coupling. When the two apical O atoms move toward the ion, the energy of d_{z^2} becomes higher than $d_{x^2-y^2}$, and the degeneracy is lifted. The Jahn–Teller effect is represented by the following Hamiltonian for a single octahedron:

$$\mathcal{H}_{JT} = -g(T_x Q_2 + T_z Q_3) \qquad (1.19)$$

where (Q_2, Q_3) are the coordinates for the displacements of O atoms surrounding the transition-metal atom and g is the elastic coupling constant. Whenever long-range orbital order exists, $<T_x> 0$ and $<T_z> 0$, the Jahn–Teller distortion is always present.

1.4.6
Charge Order

In a mixed-valence oxide, the repulsive coulomb interactions among the extra, mobile electrons will force them apart. If the site occupancy is a rational fraction such as $1/2$, there is a possibility that they will adopt a nonconducting, regular charge-ordered arrangement—a phenomenon known as *Wigner crystallization*. Charge ordering occurs in $La_{1-x}Ca_xMnO_3$ with $x=0.5$, below the critical temperature $T_c = 160$ K; where a charge-ordered antiferromagnetic state has been found.

The original example of a charge-ordering transition was magnetite, Fe_3O_4. Verwey proposed in 1939 that the electrons belonging to Fe^{2+} in B-sites of the spinel lattice undergo a charge-ordering transition below 120 K, forming a superlattice, which is now known as the *Verwey transition*. In fact, charge ordering is never this simple. The ionic charge states are not integral; they are close to

$n \pm \varepsilon$, where ε is of the order of 0.1. Recently, the charge-ordered structure has been solved using diffraction of synchrotron radiation on a small, untwined crystal of magnetite [44]. The complex, charge-ordered structure is represented by the superposition of many different charge density waves.

1.5
Concentrated Magnetic Oxides

Oxides exhibit a wide range of electric, magnetic, optical, and structural properties [9, 10, 11]. They can be insulating, semiconducting, metallic, ferroelectric, piezoelectric, ferromagnetic, ferrimagnetic, antiferromagnetic, or superconducting. A short account of each of the most common magnetic oxides is given in this section, and some related compounds are mentioned. Properties are summarized in Table 1.5. The dominant magnetic coupling in transition-metal oxides is normally antiferromagnetic superexchange, leading to antiferromagnetic or ferrimagnetic order. Ferromagnetism is usually associated with mixed valence and double exchange.

1.5.1
αFe_2O_3

The most abundant magnetic oxide is αFe_2O_3 (hematite). It is an insulating, deep-red rock-forming mineral with the rhombohedral corundum (sapphire) structure. Strong antiferromagnetic superexchange between the $3d^5$ ferric cations mediated by the $2p^6$ O^{2-} anions leads to an exceptionally high Néel temperature, $T_N = 960$ K. The easy axis below $T_M = 265$ K is the rhombohedral 111 axis, which corresponds to the c-axis. of the hexagonal close-packed oxygen lattice. Above T_M, known as the *Morin transition* temperature, the balance of lattice dipole and magnetocrystalline anisotropy energies changes sign, and the moments lie in the basal plane [45, 46]. The in-plane moments are no longer strictly collinear, because of the weak Dzyaloshinskii–Moriya interaction, Eq. 1.17, where the interaction vector $\mathcal{D}$

Table 1.5 Structural and magnetic properties of magnetic oxides [11].

Material	Structure		Order	m_0 (μ_B fu^{-1})	$\mu_0 M_s$ (T)	T_C or T_N (K)	Transport
NiO	Cubic	$Fm3m$	Af	—	—	525	Insulator
Fe_2O_3	Rhombohedral	$R3c$	cAf	0.005	0.003	960	Insulator
Fe_3O_4	Cubic	$Fd3m$	Ferri	4.0	0.60	860	Half metal
$Y_3Fe_5O_{12}$	Cubic	$Ia3d$	Ferri	5.0	0.18	560	Insulator
$BaFe_{12}O_{19}$	Hexagonal	$P6_3/mmc$	Ferri	20.0	0.48	740	Insulator
CrO_2	Tetragonal	$P4_2/mnm$	Ferro	2.0	0.49	396	Half metal
$La_{0.7}Sr_{0.3}MnO_3$	Rhombohedral	$R3m$	Ferro	3.6	0.55	360	Half metal
Sr_2FeMoO_6	Orthorhombic	$P4/mmm$	Ferri	3.6	0.25	425	Half metal
$Tl_2Mn_2O_7$	Cubic	$Fd3m$	Ferro	4.2	0.40	118	Semimetal

($\approx$0.1 K) must lie along the c-axis for symmetry reasons. The average value of $\mathcal{J}$ is $\approx$20 K and the slight canting of the antiferromagnetic sublattices at an angle of order $\mathcal{D}/\mathcal{J}$ produces a weak resultant magnetization of 2.5 kA m^{-1}. Although Fe^{3+} is an S-state ion with no orbital moment and no magnetocrystalline anisotropy to first order, the off-diagonal crystal-field terms mix some of the orbital character of the excited states into the ground site, leading to magnetocrystalline anisotropy. The magnetic dipole field at the Fe sites is about 1 T (only in cubic lattices does it sum to zero), giving an energy expression Eq. (1.14) with $\mathcal{D} \approx 0.4$ K at room temperature. The two contributions have slightly different temperature dependences, varying as $<Sz^2>$ and $<Sz^2>$, which leads to the spin reorientation at the Morin transition when their sum changes sign [46].

Ti, V, Cr, and Mn also form sesquioxides with the corundum structure. Of these, the most interesting is Cr$_2$O$_3$, which has a different antiferromagnetic stacking below its Néel point $T_N = 306$ K. It exhibits no spin reorientation, but it is *magnetoelectric*. This means that a small magnetic moment can be induced by an electric field.

1.5.2
Fe$_3$O$_4$

Magnetite is the other common rock-forming iron oxide, and the most famous magnetic mineral. It is a black conductor, with the inverse spinel structure. The octahedral sites {} are occupied by an equiatomic mixture of ferric and ferrous iron with rapid Fe^{2+}–Fe^{3+} electron hopping, while the tetrahedral sites [] are exclusively occupied by ferric cations. The formula can be written as [Fe^{3+}]{Fe^{2}Fe^{3+}}O$_4$. The octahedral and tetrahedral sublattices are coupled antiparallell by superexchange, giving a ferrimagnetic structure with a net moment of about 4 μ_B and a room-temperature magnetization of 480 kA m^{-1}. The Curie temperature is 860 K. The mobile electron associated with Fe^{2+} $\downarrow$ on the octahedral sites occupies a $\frac{1}{6}$-filled minority-spin t_{2g} band; the average B-site configuration is 3d^n $(t_{2g}{}^3 e_g{}^2)^\uparrow (t_{2g}{}^{0.5})^\downarrow$ with $<n> = 5.5$; so magnetite was expected to be a half metal with a spin gap in the majority density of states [47]. However, there is little evidence for an exceptionally high-spin polarization at room temperature. The 3d conduction electrons interact electrostatically with the oxygen anions to form small polarons with large effective mass, which move among the B-sites by thermally activated hopping [48].

Magnetite undergoes its celebrated Verwey transition at 120 K [44, 49]. This is an insulator–metal transition where the conductivity decreases abruptly by a factor of $\sim$100 and the symmetry of the crystal is lowered from cubic to monoclinic. Verwey's original model of regular charge ordering of Fe^{3+} and Fe^{2+} ions on octahedral sites [49] was oversimplified, and a complex ordering of fractional charges on these sites with iron trimer motifs has now been established, which is described by the superposition of 128 atomic displacement waves [44].

There is an important family of insulating *spinel ferrites*, of which the main members are ZnFe$_2$O$_4$, MgFe$_2$O$_4$, MnFe$_2$O$_4$, CoFe$_2$O, NiFe$_2$O$_4$, and Li$_{0.5}$Fe$_{2.5}$O$_4$. Only the first of these has the normal cation distribution; the others are inverse with Fe^{3+} on tetrahedral sites. Lithium ferrite has the highest Curie temperature of 943 K. Cobalt ferrite has by far the greatest cubic magnetocrystalline anisotropy and magnetostriction, with [100] easy axes, whereas all the others have [111] easy axes. Nickel–zinc ferrite and manganese–zinc ferrite are widely used as cores in high-frequency inductors.

γFe_2O_3 also has the spinel structure, with vacant octahedral sites; the formula may be written as $[Fe]\{Fe_{5/3}\square_{1/3}\}O_4$. The compound is a known insulator, with a moment of $3.3\,\mu_B$ per formula. The structure converts to αFe_2O_3 on heating above 800 K. Acicular $Co{:}\gamma Fe_2O_3$ powder is still produced for particulate magnetic recording media.

1.5.3
NiO

Nickel oxide has the cubic NaCl structure with a tiny rhombohedral distortion. It is an antiferromagnetic charge-transfer insulator [50] with a Néel temperature of 525 K [51]. The Ni^{2+} ions have a $3d^8$ configuration with $S=1$. Moments are arranged in ferromagnetic (111) planes perpendicular to the 111 direction, with the spins in adjacent planes oriented antiferromagnetically [51]. Nonstoichiometric nickel oxide $Ni_{1-\delta}O$ exhibits p-type semiconducting properties owing to the presence of holes in the oxygen 2p band. Stoichiometric NiO was used for exchange bias of early spin-valve structure.

Other antiferromagnetic monoxides with the NaCl structure are MnO, FeO, and CoO. Their Néel temperatures are below room temperature. EuO is an interesting insulating ferromagnet, with $4f^7$, $S = 7/2$ ions and a Curie temperature of 69 K.

1.5.4
$Y_3Fe_5O_{12}$

Garnets are cubic oxides with a large unit cell and general formula $<A_3>[B_2]\{X_3\}O_{12}$, where the brackets, $<>$, [], and {} denote sites with eightfold (cubic), tetrahedral, and octahedral coordination [52]. Pyrope, $<Mg_3>[Al_2]Si_3\}O_{12}$, was a natural example. The magnetic garnets are ferrites with formula $R_3Fe_5O_{12}$, where the antiparallel octahedral and tetrahedral sublattices are filled with Fe^{3+} ions, yielding a net ferromagnetic moment of $5\,\mu_B$ per formula and a net magnetization of $140\,kA\,m^{-1}$ for the iron. R is a trivalent rare-earth ion – Y^{3+} in the case of YIG. This material is a green insulator with a Curie temperature of 560 K and little magnetocrystalline anisotropy. It is widely used for microwave components such as filters and circulators on account of its very narrow ferromagnetic resonance linewidth and excellent insulating properties. YIG is vital for modern radio-frequency communications.

Iron garnets form for the whole series of rare earths from Pr to Lu. The rare-earth sublattice is weakly coupled to iron in the sense that the moment of Gd and the other magnetic heavy rare earths couple antiparallel to the resultant iron moment of $5\,\mu_B$ per formula at low temperature. Because of the weak coupling, the rare-earth magnetization falls off much more rapidly than that of the net iron magnetization with increasing temperature. As a consequence, there is a temperature, known as the *compensation temperature* T_{comp}, where the two exactly cancel. Its value for $Gd_3Fe_5O_{12}$ (GIG) for example is 285 K. A feature of the magnetism near the compensation temperature is that the coercivity grows as the magnetization disappears. Some data for GIG are shown in Figure 1.14. The explanation is that the coercivity, which depends on the defects and microstructure of the oxide, is usually a small fraction (<25%) of the anisotropy field H_a,

$$H_a = 2K/\mu_0 M_s \tag{1.20}$$

where K is the anisotropy constant and M_s is the net magnetization. As the magnetization falls to zero, the anisotropy field and the coercivity tend to diverge,

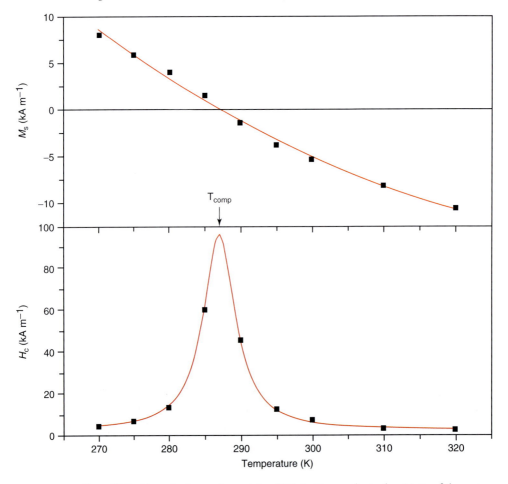

Figure 1.14 Magnetization and coercivity of $Gd_3Fe_5O_{12}$ powder in the vicinity of the compensation temperature at 285 K.

as seen in Figure 1.14b. Hematite, which has a very weak magnetization above T_M, exhibits substantial coercivity for a similar reason.

Garnets doped with Bi^{3+} ions exhibit a substantial magneto-optic Faraday effect on account of the large spin–orbit coupling associated with heavy atoms. Thin films were used for magnetooptic recording based on compensation-point writing where a spot of the film was heated well above T_{comp} and cooled in a field greater than the coercivity at the elevated temperature. At ambient temperature, which is close to T_{comp}, the increase in coercivity ensured the stability of the recorded bit [52]. The method is now obsolete, but heat-assisted magnetic recording is being developed to write on modern hard disc media which have very high coercivity at room temperature.

1.5.5
$SrFe_{12}O_{19}$

Strontium and barium ferrite crystallize in the hexagonal magnetoplumbite structure. The large hexagonal cell includes six cation sites, one for the large divalent alkaline earth and five for the ferric ions, three of them octahedral, one tetrahedral, and one trigonal bipyramidal site with fivefold oxygen coordination. The compound is ferrimagnetic, with a Curie temperature of 740 K [53]. In the unit cell, there are eight ferric ions on the majority sublattice and four on the minority sublattice, leading to a net moment of $20\,\mu_B$ per formula at $T = 0$ and a room-temperature magnetization of $380\,kA\,m^{-1}$. Even though all the magnetic ions are all ferric, there is sufficient anisotropy arising from the bipyramidal and distorted octahedral sites in the structure to produce uniaxial anisotropy of $K_1 = 300\,kJ\,m^{-3}$, which ensures that the magnetic hardness parameter, defined as

$$\kappa = (K_1/\mu_0 M_s^2)^{1/2} \tag{1.21}$$

is 1.35, sufficient to make a true permanent magnet. The hexagonal ferrites were the first low-cost materials that could be manufactured into magnets of any desired shape. Nowadays, around a million tonnes of these sintered and bonded ceramic magnets are produced annually.

1.5.6
$(La,Sr)MnO_3$

Many magnetic and superconducting oxides crystallize in the perovskite structure, or in its derivatives. The basic cubic cell with a lattice parameter $a_0 \approx 390\,pm$ contains just two cations: one is usually an alkaline earth or rare-earth cation, and the other a 3d cation. The mineral perovskite $CaTiO_3$ is uncommon in the Earth's crust, but silicate perovskites such as $MgSiO_3$ make up much of the mantle. From the point of view of magnetism, the mixed-valence manganites [7] have for long been of great interest. Much studied are the solid solutions between $La^{3+}Mn^{3+}O_3$ and $Sr^{2+}Mn^{4+}O_3$ or $Ca^{2+}Mn^{4+}O_3$, leading to mixed-valence compounds such as $La_{1-x}Sr_xMnO_3$. Each of the end members is antiferromagnetic and insulating ($LaMnO_3$ is a canted antiferromagnet [7]), but the solid solutions may be ferromagnetic and conducting. Optimally doped $La_{1-x}Sr_xMnO_3$ with $x = 0.3$ is a half metal with the greatest Curie temperature of 380 K, and a rhombohedrally distorted perovskite structure. The substitution of Sr for La creates a mixture of $Mn^{4+}(t_{2g}^3)^\uparrow$ and $Mn^{3+}(t_{2g}^3 e_g)^\uparrow$ on the B-sites of the structure [7] and the hopping $e_g^\uparrow$ electron mediates ferromagnetic coupling by double exchange. The ferromagnetic moment is consistently found to be slightly less than the 3.7_B, which would be expected if $(La_{0.7}Sr_{0.3})MnO_3$ were ideally half metallic, and it seems that both mobile $e_g^\uparrow$ electrons and immobile $t_{2g}^\uparrow$ electrons may be present at E_F. The compound is therefore classified as a transport half metal [54]. Residual resistivity is as low as $4 \times 10^{-7}\,\Omega\,m$, but a feature of the mixed-valence manganites is their colossal magnetoresistance. The resistivity may decrease by as much as 98% when a large field is applied close to T_C.

Elsewhere in the phase diagram, one can find different antiferromagnetic phases, and compounds that exhibit charge order and orbital order. There are other families of magnetic oxides with structures related to perovskite. These include the rare-earth orthoferrites $RFeO_3$ with an orthorhombic cell, which are antiferromagnets with Néel temperatures of 620–740 K. They form with all rare earths from La to Lu, and exhibit weak ferromagnetism with a moment of about $8\,kA\,m^{-1}$ owing to the Dzyaloshinkii–Moriya interaction. Other transition metals that form magnetically ordered perovskite-type materials are Ti, Cr, Co, and Ni. The possibilities for creating interesting solid solutions by substitution are almost endless but possibilities for practical applications are very limited because of the relatively low Curie temperatures. Cobalt in these materials is often found in a low-spin state, indicating that $\Delta > \mathcal{J}_H$.

Another big family are the double perovskites, which have a $(2, 2, 2)a_0$ cubic cell. Here a model compound is Sr_2FeMoO_6 (SFMO), which has a Curie temperature of 436 K [55]. The iron and molybdenum show NaCl-type order on the transition-metal sites of the perovskite lattice, and the single delocalized electron of the molybdenum mixes with the empty 3d states to provide strong ferromagnetic coupling among the iron spins. The moment is less than the $4\,\mu_B$ per formula expected for a stoichiometric half metal owing to Fe/Mo antisite defects.

1.5.7
CrO_2

Chromium dioxide has the tetragonal rutile structure, where the Cr^{4+} ions are octahedrally coordinated by oxygen [56]. It is the only binary oxide that is a ferromagnetic metal, with a Curie temperature of 396 K. Band structure calculations show that CrO_2 is a d-band half metal with a $\downarrow$ spin gap $\Delta^\downarrow$ of 0.5 eV in the minority-spin band. The conductivity at room temperature is $3 \times 10^5\,S\,m^{-1}$. The magnetization corresponds to an integral moment of 2.0_B per formula as expected for a half metal. Andreev reflection measurements on CrO_2 – superconductor point contacts give a spin polarization, $P \approx 90\%$ at temperatures of about 2 K [57]. Acicular CrO_2 powders were once used for magnetic video tapes.

Other magnetic rutile-structure oxides include VO_2, which is an antiferromagnet with a metal–insulator transition at 343 K [56].

1.5.8
$Tl_2Mn_2O_7$

The pyrochlore manganite $Tl_2Mn_2O_7$ is a cubic compound with an interesting electronic structure and unexpected magnetic properties [58]. There are eight formula units per unit cell. The Mn^{4+} ions are octahedrally coordinated by oxygen, and they form a corner-sharing tetrahedral array, similar to that of the B-sites in the spinel structure. In an ionic picture, the compound would be an insulator with Mn^{4+} ($3d^3$) and Tl^{3+} ($5d^{10}$) cations. Only the former bear a magnetic moment, $3\,\mu_B$, and they couple via Mn–O–Mn superexchange bonds. Antiferromagnetic Mn^{4+}–O–Mn^{4+} superexchange is highly frustrated by the three-membered rings of the tetrahedral array. The compound, however, is a ferromagnetic semimetal with $T_c = 118$ K and a small number of heavy holes at the top of a narrow band of mainly $t_{2g}{}^\uparrow$ character and an equal number of mobile electrons in a broad band of mixed Tl (6s), O (2p), and Mn (3d) character [59, 60]. The number of carriers has been estimated at 0.005 per manganese. The mobile electrons are expected to dominate the conduction, while the heavy holes will be easily localized by any impurities or disorder that may be present in the compound. $Tl_2Mn_2O_7$ can also be regarded as a half metal in so far as the heavy holes do not contribute significantly to the conduction.

More recently, the tetrahedral frustration in pyrochlore-structure compounds with rare earths has helped to develop the concept of *spin ice*. $Dy_2Ti_2O_7$ and $Ho_2Sn_2O_7$ are examples. The dysprosium has a [111] easy axis, and in each tetrahedron, the moments of two Dy^{3+} ions point inwards and two point outwards, analogous to hydrogen bonds in ice. Introducing defects in the frustrated lattice has the effect of creating widely separated positive and negative magnetic "charges," which are free to move in the lattice. These have been described as magnetic "monopoles," although they are always formed in pairs [61], and Maxwell's equation. $\nabla \cdot \boldsymbol{B} = 0$ is not violated!

1.6
Dilute Magnetic Oxides

A dilute magnetic oxide consists of a nonmagnetic oxide, such as those listed in Table 1.3, in which a small fraction of magnetic 3d cations has been introduced. The general formula of the oxide is MO_η, where η is an integer or rational fraction; so the formula of dilute magnetic oxide is

$$(M_{1-x}T_x)O_\eta$$

where $x < 10\%$. In view of our discussion of magnetic interactions in oxides in Section 1.4, we would not anticipate any long-range magnetic order in such a material when x is below the percolation threshold x_p. The value of x_p is $\sim 2/Z_c$, where Z_c is the cation–cation coordination number; for example, Z_c is 12 for ZnO and 10 for TiO_2 (rutile), as may be seen from Figure 1.3.

A dilute two-dimensional lattice with $Z_c = 4$ is illustrated in Figure 1.15, where it can be seen that random occupancy leads to a distribution of isolated T ions, some nearest-neighbor pairs and a few larger clusters. The dimers and clusters are expected to couple antiferromagnetically, with a net moment for odd-membered clusters. When $x < 0.1$, the magnetic entities respond independently to an applied magnetic field, leading to a paramagnetic response that can be modeled approximately as a sum of Curie and Curie–Weiss terms

$$\chi = C_1/T + C_2/(T-\theta) \tag{1.22}$$

where C_1 and C_2 are Curie constants (Eq. (1.8)), and θ is negative with magnitude <100 K, representing the strength of a single antiferromagnetic exchange bond. The predicted paramagnetic behavior has been found in well-crystallized samples, for example, in Co-doped ZnO crystals where the Co^{2+} substitutes for Zn on tetrahedral sites, and $\theta = 65$ K [62].

Over the past 10 years, beginning with the 2001 Science report by Matsumoto *et al* [63] on thin films of Co-doped anatase TiO_2 produced by PLD, there has been a spate of reports of ferromagnetic-like behavior in thin films and nanocrystalline samples of dilute magnetic oxides. Most remarkably, the ferromagnetism is found at room temperature, and the corresponding Curie temperature must be >400 K. Table 1.6 reports some of the early results on various oxide systems [64]. It can be seen that they may be semiconducting (ZnO), insulating (TiO_2), or metallic ($(La,Sr)TiO_3$), and the effect is seen for a variety of 3d cations.

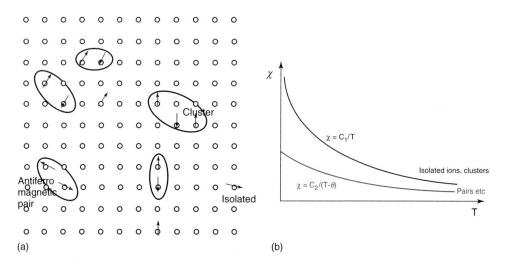

Figure 1.15 Dilute magnetic oxide below the percolation threshold. (a) Pairs and small clusters of magnetic ions are coupled by antiferromagnetic superexchange. (b) This leads to a paramagnetic susceptibility of the form given by Eq. 1.22.

Table 1.6 Reports of ferromagnetic oxide thin films with T_C above room temperature (Ref. [63]).

Material	E_g (eV)	Substitutions	Moment (μ_B) per 3d ion	T_C (K)
TiO$_2$	3.2	V – 5%	4.2	>400
		Co – 7%	0.3	>300
		Co – 1%–2%	1.4	>650
		Fe – 2%	2.4	300
SnO$_2$	3.5	Fe – 5%	1.8	610
		Co – 5%	7.5	650
ZnO	3.3	V – 15 %	0.5	>350
		Mn – 2.2%	0.16	>300
		Fe 5%,Cu 1%	0.75	550
		Co – 10%	2.0	280–300
		Ni – 0.9%	0.06	>300
CeO$_2$	2.0	Co – 3%	5.8	>750
In$_2$O$_3$	3.7	Fe – 5%	1.4	>600
		Cr – 2%	1.5	900
ITO	3.5	Mn – 5%	0.8	>400
(La,Sr)TiO$_3$	—	Co – 1.5%	2.5	550

The excitement engendered by these results was partly due to their quite un-expected nature, and partly because it was hoped that these films and related nanoparticles would turn out to be the long-anticipated room-temperature ferro-magnetic semiconductors. The dilute magnetic oxides were thought to be *dilute magnetic semiconductors* (DMS). A DMS is a semiconducting material where 3d or 4f dopants order ferromagnetically under the influence of long-range exchange interactions to produce a uniform ferromagnetic state. The textbook example is $(Ga_{1-x}Mn_x)As$, where ferromagnetic ordering of Mn^{2+} ions occurs which occupy a level below the top of the As 4p valence band, with a value of T_C as high as 170 K for doping at the solubility limit $x = 0.07$ [65]. The material is magnetically homogeneous on the scale of a few nanometers [66].

The possibility of high-temperature ferromagnetism in ZnO doped with 5% of manganese was proposed by Dietl *et al.* [67]. The ferromagnetism was thought to be like that in (Ga,Mn)As, mediated by spin-polarized holes in the ligand p band. How-ever, with the exception of ZnO, where the cations are in tetrahedral coordination and the bonding has significant covalent character with 2p holes that can exhibit reasonable mobility, oxides do not usually make good semiconductors. The carriers tend to be trapped and form small polarons owing to their Coulomb interaction with the oxygen anions. This is why many oxides may be off-stoichiometric, yet remain insulating.

There is a voluminous literature on the magnetic properties of dilute magnetic oxides, which is reviewed in subsequent chapters in this book. For doped ZnO alone, there are over 1000 references in a recent review [68]. Here we restrict our discussion to just three points:

- The dilute magnetic oxides are not dilute magnetic semiconductors, as defined above.
- They are magnetically inhomogeneous, and only a small fraction of the volume is ferromagnetically ordered.
- The magnetism (when it is not due to impurity phases or experimental artifacts) is closely associated with defects in the material.

Some magnetization curves of thin films of dilute magnetic oxides are shown in Figure 1.16. What is remarkable about the data is that the curves are essentially anhysteretic and that they show very little temperature dependence between room temperature and 4 K. However, it should be emphasized that the superposition of curves obtained at different temperatures is achieved after correcting the data for a diamagnetic signal arising from the substrate, or from the oxide matrix in the case of bulk samples of nanoparticles. Since this cannot be precisely determined in the thin films it is normal practice to correct the data simply to give zero high-field slope which means that any temperature-dependent Curie law paramagnetism may be overlooked. However, the lack of temperature dependence of the nonlinear magnetization curves, like those in Figure. 1.16 entirely excludes the possibility of

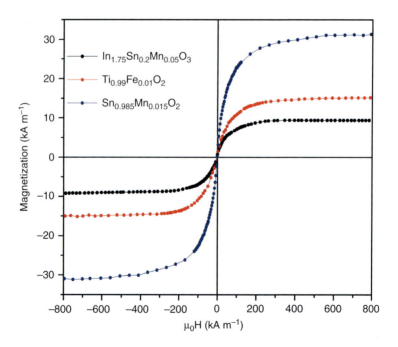

Figure 1.16 Some representative magnetization curves of thin film of dilute magnetic oxides.

superparamagnetism, for which the magnetization curves should superpose when M/M_s is plotted versus (H/T).

The first point is established by the observation that the dopant ions are not necessarily magnetically ordered when the sample is in its ferromagnetic state. The evidence includes (i) X-ray magnetic circular dichroism (XMCD) data on the cobalt in Co-doped ZnO [69], (ii) ^{57}Fe Mossbauer spectra of Fe-doped TiO_2 [70], and (iii) the Curie-law susceptibility observed for Mn in ITO and other dilute oxides when the data are not corrected to give zero high-field slope at each temperature [71]. All these measurements indicate that most, if not all, of the dopant ions remain *paramagnetic* down to liquid-helium temperatures, as expected from Eq. 1.22. The practice that we followed in Table 1.6 of relating the moment to the dopant, quoting it, for example, as $2.0\,\mu_B$ per Co ion is therefore misleading because it is not necessarily the 3d dopant that is ferromagnetically ordered. In fact, evidence that this approach was mistaken emerged quite early on, when moments per dopant exceeding the greatest possible spin-only value began to be reported, especially in lightly-doped material [72–75].

The second point can be inferred from the magnetization curves. If the samples were uniformly magnetized on a microscopic scale, as we expect for a DMS, the very small value of the magnetization ($\approx 10\,kA\,m^{-1}$ in thin films or $\approx 10\,A\,m^{-1}$ in nanoparticles) should give rise to a large value of the anisotropy field, even in cubic materials, and therefore produce substantial coercivity. Such coercivity, in excess of $100\,kA\,m^{-1}$, is observed in powders of rare-earth garnets near their compensation temperature (Figure 1.14). These materials exhibit a magnetization of 1–$10\,kA\,m^{-1}$, similar to that of the dilute magnetic oxides, yet no coercivity is found in the latter and thus we infer that they are unlikely to be uniformly magnetized.

Another serious problem with the idea that dilute magnetic oxides are homogeneously magnetized dilute ferromagnetic semiconductors is the high value of the Curie temperature. Few insulating oxides are ferromagnetic; they usually order antiferromagnetically or ferrimagnetically on account of the antiferromagnetic superexchange. Furthermore, very few oxides order magnetically above 1000 K, and nothing is known to have a higher Curie temperature than cobalt (1360 K). It stretches the imagination to believe that a material with 5% of magnetic ions could have a Curie temperature much above 400 K. The magnetic energy density varies as x or $\sqrt{x}$; so these high Curie temperatures cannot be achieved with uniformly magnetized material [64].

We can, however, do better, and estimate the volume fraction that is magnetically ordered from magnetization curves like those in Figure 1.16 [76]. The analysis is based on a comparison of the average magnetization of the sample M_s deduced directly as the saturation magnetization, with the magnetization of the ferromagnetic regions M_f deduced from the field H_0 obtained by extrapolating the slope of the magnetization curve at the origin to saturation. The argument is that the approach to saturation must be governed by dipolar interactions rather than by magnetocrystalline anisotropy – if magnetocrystalline anisotropy were involved, there would be a clear temperature dependence of the magnetization curves and eventual appearance of coercivity at low temperature. In fact the magnetization curves are practically anhysteretic, and show almost no temperature dependence between 4 K and room temperature. H_0 therefore represents the average magnitude of the demagnetizing field $H_0 = -\mathcal{N}M_f$, where the demagnetizing factor $\mathcal{N}$ is a number between 0 and 1, and M_f is the magnetization of the ferromagnetic regions. Often, there is little dependence of the magnetization on the field direction with respect to the plane of the film; so it is reasonable to take $\mathcal{N} \approx 1/3$, but the exact value is unimportant for our argument. We therefore deduce from the values of H_0 for the curves in Figure 1.16 that M_f is ~400 kA m^{-1} and M$_s$ ~20 kAm^{-1}. The ferromagnetic volume fraction in the films $f = M_s/M_f$ is therefore *only a few percent*. Applying the same argument to data on nanoparticles gives values that are much lower. Figure 1.17 is a scatter plot of very many samples of dilute and undoped magnetic oxides and nanoparticles.

The view that the magnetism is confined to a small portion of the sample volume is supported by the absence of any ferromagnetism in well-crystallized samples of dilute magnetic oxides [62, 77], and its appearance after treatments such as heating the sample under vacuum, which has the effect of creating defects and modifying the oxygen stoichiometry. Contamination of the samples by small quantities of ferromagnetic impurities such as iron or magnetite is sometimes a problem, and there are well-documented accounts of the perils that can arise when scrupulous care is not observed in the experiments and appropriate blank samples are not examined [78, 79]. The presence of such impurities may be flagged by the values of M_f. In systems such as Co-doped ZnO, the cobalt has a tendency to form dispersed cobalt nanoparticles, which are ferromagnetic [80]. It has been a bane that the lack of reproducibility of many of the results from different laboratories, and even

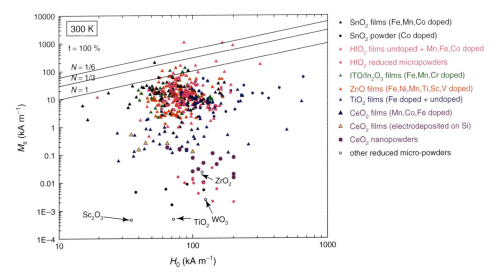

Figure 1.17 Compilation of data for thin film and nanoparticulate systems showing the correlation between M$_s$ and H_0 for different classes of material.

within the same laboratory, compounded by the fact that the magnetization of certain samples is found to decay with time [81]. The problems may be remedied by better control of the deposition conditions, for example, by electrochemical growth of oxides where the magnetic moment appears in a narrow range of conditions [82].

A defect-related origin of the ferromagnetism is supported by its observation in some samples of *undoped* material. These include films of HfO_2 [83, 84], ZnO[85, 86], TiO_2 [87], and In_2O_3 [74]. This phenomenon is known as d^0 *magnetism* [88], and it reinforces the idea that defects are necessary, but 3d dopants are not essential.

In order to arrive at a satisfactory explanation of the magnetic behavior of dilute magnetic oxides and d^0 systems, it is necessary to tie down the relevant defect structure. This is a challenge, as it is difficult to characterize and quantify unknown defects in solids, especially in thin films. There is an area where electronic structure calculations can help. The obvious places to look in a thin film are the surface, the interface with the substrate and the grain boundaries. For nanocrystalline material, a significant fraction of the atoms are at the surface or in grain boundaries. Evidence for a correlation between magnetism and grain size gas been reported for ZnO nanocrystallites [89]. Extended defects seem to be a more likely source of percolating magnetism than point defects.

The influence of defects on magnetism has been modeled in terms of a narrow, spin-split defect-related impurity band [90]. This has been shown to lead to a high Curie temperature, as spin-wave excitations are suppressed in the half-metallic band. This Stoner model of the magnetism, involving a small fraction of the sample volume, contrasts with the Heisenberg model used to treat the magnetism of ions with local moments in Section 1.4. Transition-metal dopants may nonetheless play a role, but as a reservoir of charge for the impurity band. When the ions exist in different charge states, they can help to ensure filling to the point where the Stoner criterion is satisfied. This phenomenon is known as *charge transfer ferromagnetism* [91]. Models for the magnetism of dilute magnetic oxides are discussed in Ref. [92].

A quite different and more speculative explanation is based on the suggestion that the moment is essentially of orbital origin, and that the electrons moving on the surface of a particle, or around a continuous grain boundary can develop a giant orbital moment, with giant paramagnetic susceptibility [93].

1.6.1
Magnetism of Oxide Surfaces, Interfaces, and Defects

There are inevitable changes in electronic structure at defects in crystals. The symmetry is lowered at the surface, or at the interface with a substrate. The surface may be charged or uncharged depending on the termination of plane involved. For example, the 001 surface of TiO_2 may present a plane of negatively charged oxygen or positively charged titanium ions. The reduced number of neighbors at the surface can produce band narrowing and localization. There are suggestions that the surfaces of some nonmagnetic oxides might even support magnetic order [94].

Some direct evidence that defects can lead to the weak high-temperature magnetism that we have been discussing is provided by experiments irradiating single crystals of TiO_2 [95], which show that the appearance of magnetism is associated with the tracks of radiation damage.

There are numerous calculations of the electronic structure of point defects in various kinds of oxides. A key parameter is the relative energy of singly and doubly-occupied defects. A single electron necessarily has an unpaired spin, but if the energy U needed to add a second electron is negative, the defects will tend to be unoccupied or occupied by two electrons with paired spins in an s-like orbital. The system is then nonmagnetic. However, if U is positive and the average electron occupancy of the defects is somewhat greater than one, the extra electrons may be delocalized and produce ferromagnetic coupling of the core spins. For example, a missing pair of Zn and O ions in ZnO might be a defect with this property [96].

Turning to extended defects, a very interesting model system is the planar interface between (001) $LaAlO_3$ and $SrTiO_3$, illustrated in Figure 1.18. Each oxide has a perovskite-type structure; the structure of $LaAlO_3$ may be considered as a sequence of alternating polar planes of $(LaO)^+$ and $(AlO_2)^-$, whereas that of $SrTiO_3$ consists of alternating charge-neutral planes of (SrO) and (TiO_2). The planar interface reconstructs to transfer 0.5 electron per interfacial unit cell into the empty Ti 3d band close to the interface, giving an electron density of 3×10^{18} m^{-2} [94]. The properties of the two-dimensional electron gas at the interface are quite unusual. There is a tendency toward phase segregation into ferromagnetic and superconducting regions [97]. The first evidence for magnetic order was indirect, from magnetoresistance measurements [98], but recent experiments have shown evidence of magnetization corresponding to magnetic moments of order $1\,\mu_B$ per unit cell persisting up to room temperature, and superconductivity below 60 K [99]. Analysis of the magnetization curve, assuming $\mathcal{N} = 1/3$ for the ferromagnetic regions, yields a thickness of the ferromagnetic interface of order 1 nm, or three unit cells.

The recent work on oxide–oxide interfaces demonstrates that they can show electronic properties completely different from that of either constituent, including room-temperature ferromagnetism. By gating, it is possible to control the properties

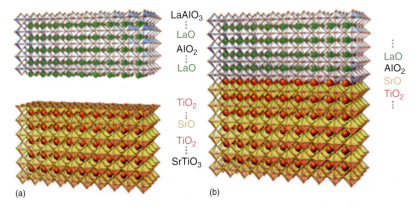

Figure 1.18 (a) Schematic of a TiO_2-terminated (001) $SrTiO_3$ single crystal before contact with an epitaxial $LaAlO_3$ overlayer. (b) Epitaxial $LaAlO_3$ on TiO_2-terminated $SrTiO_3$.

with an external electric field [100]. The results encourage a more systematic investigation of well-defined oxide interfaces in a quest to understand better the properties of dilute magnetic oxides and d^0 magnetism.

1.7
Conclusions

This introduction to magnetic oxides provides a background to subsequent chapters in the first half of the book. While the magnetic properties of concentrated magnetic oxides, and well-crystallized, paramagnetic dilute magnetic oxides are now well-understood, the magnetism of defective dilute magnetic oxides, d^0 magnets, and oxide–oxide interfaces is not. There are areas where new physics and materials science may emerge, as well as possible applications enabled by the high Curie temperature.

Acknowledgments

This work was supported by Science Foundation Ireland as part of the NISE project, contract number 10/INI/I3002.

References

1. Kloss, A. (1994) *Geschichte des Magnetismus*, VDE-Verlag, Berlin.
2. Blackman, M. (1983) The lodestone: a survey of the history and the physics. *Contemp. Phys.*, **24**, 319.
3. Dunlop, D.J. and Özdemir, Ö. (1997) *Rock Magnetism: Fundamentals and Frontiers*, Cambridge University Press, New York.
4. Kurti, N. and Néel, L. (1988) *Selected Works of Louis Néel*, Gordon and Breach, New York.
5. Wijn, H.P.J. and Smit, J. (1959) *Ferrites: Physical Properties of Ferrimagnetic Oxides in Relation to their Technical Applications*, John Wiley & Sons, Inc., New York.
6. Goodenough, J.B. (1963) *Magnetism and the Chemical Bond*, Wiley Interscience, New York.
7. Coey, J.M.D., Viret, M., and Von Molnar, S. (1999) Mixed-valence manganites. *Adv. Phys.*, **48**, 167.
8. Ogale, S.B. (2005) *Thin Films and Heterostructures for Oxide Electronics*, Springer, Berlin.
9. Dionne, G.F. (2009) *Magnetic Oxides*, Springer.
10. Craik, D.J. (1975) *Magnetic Oxides*, John Wiley & Sons, Inc., New York.
11. Coey, J.M.D. (2010) *Magnetism and Magnetic Materials*, Cambridge University Press, Cambridge.
12. Catlow, C.R.A. and Stoneham, A.M. (1981) Defect equilibria in transition metal oxides. *J. Am. Ceram. Soc.*, **64**, 234.
13. Zaanen, J., Sawatzky, G.A., and Allen, J.W. (1985) Band gaps and electronic structure of transition-metal compounds. *Phys. Rev. Lett.*, **55**, 418.
14. Burns, R.G. (1993) *Mineralogical Applications of Crystal Field Theory*, Cambridge University Press, Cambridge.
15. Patnaik, P. (2004) *Dean's Analytical Chemistry Handbook*, McGraw-Hill, New York.

16. Brinker, C.J. and Scherer, G.W. (1990) *Sol–Gel Science: the Physics and Chemistry of Sol–Gel Processing*, Academic Press.

17. West, A.R. (2007) *Solid State Chemistry and its Applications*, Wiley India.

18. Patil, K.C., Aruna, S.T., and Ekambaram, S. (1997) Combustion synthesis. *Curr. Opin. Solid State Mater. Sci.*, **2**, 158.

19. Scheel, H.J. (2000) Historical aspects of crystal growth technology. *J. Cryst. Growth*, **211**, 1.

20. Karas, G.V. (2005) *New Developments in Crystal Growth Research*, Nova Science Publication Inc., New York.

21. Brandle, C.D. (2004) Czochralski growth of oxides. *J. Cryst. Growth*, **264**, 593.

22. Balbashov, A.M. and Egorov, S.K. (1981) Apparatus for growth of single crystals of oxide compounds by floating zone melting with radiation heating. *J. Cryst. Growth*, **52**, 498.

23. Berthon, J., Revcolevschi, A., Morikawa, H. *et al.* (1979) Growth of wustite ($Fe_{1-x}O$) crystals of various stoichiometries. *J. Cryst. Growth*, **47**, 736.

24. Robertson, J.M. (1974) Growth of oxide single crystals–use and care of platinum apparatus. *Platinum Met. Rev.*, **18**, 15.

25. Scheafer, H. (1964) *Chemical Transport Reactions*, Academic Press, New York.

26. Ziese, M. and Thornton, M.J. (2001) *Spin Electronics*, Springer-Verlag, Berlin.

27. Sze, S.M. (2009) *Semiconductor Devices: Physics and Technology*, Wiley, New Delhi, India.

28. Hill, R.J. (1986) *Physical Vapour Deposition*, 2nd edn, Berkeley, CA.

29. Cho, A.Y. and Arthur, J.R. (1975) Molecular beam epitaxy. *Prog. Solid State Chem.*, **10**, 157.

30. Howson, R.P. and Ridge, M.I. (1981) Deposition of transparent heat-reflecting coatings of metal oxides using reactive planar magnetron sputtering of a metal and/or alloy. *Thin Solid Films*, **77**, 119.

31. Dijkkamp, D., Venkatesan, T., Wu, X.D. *et al.* (1987) Preparation of Y-Ba-Cu oxide superconductor thin films using pulsed laser evaporation from high Tc bulk material. *Appl. Phys. Lett.*, **51**, 619.

32. George, S.M. (2010) Atomic layer deposition: an overview. *Polymer*, **1550**, 125.

33. Burns, R.G. (1981) Intervalence transitions in mixed-valence minerals of iron and titanium. *Annu. Rev. Earth Planet. Sci.*, **9**, 345.

34. Chamritski, I. and Burns, G. (2005) Infrared- and Raman-active phonons of magnetite, maghemite, and hematite: a computer simulation and spectroscopic study. *J. Phys. Chem. B*, **109**, 4965.

35. Khomskii, D.I. and Sawatzky, G.A. (1997) Interplay between spin, charge and orbital degrees of freedom in magnetic oxides. *Solid State Commun.*, **102**, 87.

36. Ballhausen, C.J. and Gray, H.B. (1962) The electronic structure of the vanadyl ion. *Inorg. Chem.*, **1**, 111.

37. Hutchings, M.T. and Ray, D.K. (1963) Investigation into the origin of crystalline electric field effects on rare earth ions: I. Contribution from neighbouring induced moments. *Proc. Phys. Soc.*, **81**, 663.

38. Anderson, P.W. (1950) Antiferromagnetism. Theory of superexchange interaction. *Phys. Rev.*, **79**, 350.

39. Anderson, P.W. and Hasegawa, H. (1955) Considerations on double exchange. *Phys. Rev.*, **100**, 675.

40. Nolting, W. and Ramakanth, A. (2009) *Quantum Theory of Magnetism*, Springer-Verlag, Berlin.

41. Coey, J.M.D. and Venkatesan, M. (2002) Half-metallic ferromagnetism: example of CrO_2. *J. Appl. Phys.*, **91**, 8345.

42. Tokura, Y. and Nagaosa, N. (2000) Orbital physics in transition-metal oxides. *Science*, **288**, 462.

43. Feng, D. and Jin, G. (2005) *Introduction to Condensed Matter Physics*, World Scientific.

44. Senn, M.S., Wright, J.P., and Attfield, J.P. (2012) Charge order and three-site distortions in the verwey structure of magnetite. *Nature*, **481**, 873.

45. Morrish, A.H. (1994) *Canted Antiferromagnetism: Hematite*, World Scientific, Singapore.

46. Artman, J.O., Murphy, J.C., and Foner, S. (1965) Magnetic anisotropy in antiferromagnetic corundum-type sesquioxides. *Phys. Rev.*, **138**, A912.

47. Zhang, Z. and Satpathy, S. (1991) Electron states, magnetism, and the Verwey transition in magnetite. *Phys. Rev. B*, **44**, 13319.

48. Wohlfarth, E.P. and Buschow, K.H.J. (1980) *Ferromagnetic Materials: A Handbook on the Properties of Magnetically Ordered Substances*, North Holland, Amsterdam.

49. (a) Verwey, E.J.W. (1939) Electronic conduction of magnetite (Fe_3O_4) and its transition point at low temperatures. *Nature*, **144**, 327. (b) Verwey, E.J.W. and Haayman, P.W. (1941) Electronic conductivity and transition point of magnetite. *Physica*, **8**, 979.

50. Kuneš, J., Anisimov, V.I., Skornyakov, S.L. *et al.* (2007) NiO: correlated band structure of a charge-transfer insulator. *Phys. Rev. Lett.*, **99**, 156404.

51. Morin, F.J. (1959) Oxides which show a metal-to-insulator transition at the Neel temperature. *Phys. Rev. Lett.*, **3**, 34.

52. Gilleo, M.A. and Wohlfarth, E.P. (1980) Ferromagnetic insulators: garnets, in *Ferromagnetic Materials: A Handbook on the Properties of Magnetically Ordered Substances*, Vol. 2 Chapter 1, North Holland, Amsterdam.

53. Stäblein, H. (1982) *Handbook of Ferromagnetic Materials*, Vol. 3, North Holland, Amsterdam, p. 441.

54. Nadgorny, B., Mazin, I.I., Osofsky, M. *et al.* (2001) Origin of high transport spin polarization in $La_{0.7}Sr_{0.3}MnO_3$:direct evidence for minority spin states. *Phys. Rev. B*, **63**, 184433.

55. Kobayashi, K.I., Kimura, T., Sawada, H. *et al.* (1998) Room-temperature magnetoresistance in an oxide material with an ordered double-perovskite structure. *Nature*, **395**, 677.

56. Chamberland, B.L. (1977) The chemical and physical properties of CrO_2 and tetravalent chromium oxide derivatives. *Crit. Rev. Solid State Mater. Sci.*, **7**, 1.

57. Soulen, R.J., Byers, J.M., Osofsky, M.S. *et al.* (1998) Measuring the spin polarization of a metal with a superconducting point contact. *Science*, **282**, 85.

58. Shimakawa, Y., Kubo, Y., and Manako, T. (1996) Giant magnetoresistance in $Ti_2Mn_2O_7$ with the pyrochlore structure. *Nature*, **379**, 53–55.

59. Singh, D.J. (1997) Magnetoelectronic effects in pyrochlore $Tl_2Mn_2O_7$: role of Tl-O covalency. *Phys. Rev. B*, **55**, 313.

60. Imai, H., Shimakawa, Y., Sushko, Y.V. *et al.* (2000) Carrier density change in the colossal-magnetoresistance pyrochlore $Tl_2Mn_2O_7$. *Phys. Rev. B*, **62**, 12190.

61. Castelnovo, C., Moessner, R., and Sondhi, S.L. (2008) Magnetic monopoles in spin ice. *Nature*, **451**, 42.

62. Rao, C.N.R. and Deepak, F.L. (2004) Absence of ferromagnetism in Mn-and Co-doped ZnO. *J. Mater. Chem.*, **15**, 573.

63. Matsumoto, Y., Murakami, M., Shono, T. *et al.* (2001) Room-temperature ferromagnetism in transparent transition metal-doped titanium dioxide. *Science*, **291**, 854.

64. Coey, J.M.D. and Chambers, S.A. (2008) Oxide dilute magnetic semiconductors–fact or fiction? *MRS Bull.*, **33**, 1053.

65. Jungwirth, T., Wang, K.Y., Mašek, J. *et al.* (2005) Prospects for high temperature ferromagnetism in (Ga, Mn) as semiconductors. *Phys. Rev. B*, **72**, 165204.

66. Dunsiger, S.R., Carlo, J.P., Goko, T. *et al.* (2010) Spatially homogeneous ferromagnetism of (Ga, Mn) as. *Nat. Mater.*, **9**, 299.

67. Dietl, T., Ohno, H., Matsukura, F. *et al.* (2000) Zener model description of ferromagnetism in zinc-blende magnetic semiconductors. *Science*, **287**, 1019.

68. Pan, F., Song, C., Liu, X.J. *et al.* (2008) Ferromagnetism and possible application in spintronics of transition-metal-doped ZnO films. *Mater. Sci. Eng., R*, **62**, 1.

69. Tietze, T., Gacic, M., Schütz, G. *et al.* (2008) XMCD studies on Co and Li doped ZnO magnetic semiconductors. *New J. Phys.*, **10**, 055009.

70. Coey, J.M.D., Stamenov, P., Gunning, R.D. *et al.* (2010) Ferromagnetism in defect-ridden oxides and related materials. *New J. Phys.*, **12**, 053025.

71. Venkatesan, M., Gunning, R.D., Stamenov, P. *et al.* (2008) Room temperature ferromagnetism in Mn- and Fe-doped indium tin oxide thin films. *J. Appl. Phys.*, **103**, 07D135.

72. Ogale, S.B., Choudhary, R.J., Buban, J.P. *et al.* (2003) High temperature ferromagnetism with a giant magnetic moment in transparent Co-doped $SnO_{2-\delta}$. *Phys. Rev. Lett.*, **91**, 77205.

73. Venkatesan, M., Fitzgerald, C.B., Lunney, J.G. *et al.* (2004) Anisotropic ferromagnetism in substituted zinc oxide. *Phys. Rev. Lett.*, **93**, 177206.

74. Hong, N.H., Sakai, J., Poirot, N. *et al.* (2006) Room-temperature ferromagnetism observed in undoped semiconducting and insulating oxide thin films. *Phys. Rev. B*, **73**, 132404.

75. Fitzgerald, C.B., Venkatesan, M., Dorneles, L.S. *et al.* (2006) Magnetism in dilute magnetic oxide thin films based on SnO_2. *Phys. Rev. B*, **74**, 115307.

76. Coey, J.M.D., Mlack, J.T., Venkatesan, M. *et al.* (2010) Magnetization process in dilute magnetic oxides. *IEEE Trans. Magn.*, **46**, 2501.

77. Ney, A., Ollefs, K., Ye, S. *et al.* (2008) Absence of intrinsic ferromagnetic interactions of isolated and paired Co dopant atoms in $Zn_{1-x}Co_xO$ with high structural perfection. *Phys. Rev. Lett.*, **100**, 157201.

78. Grace, P.J., Venkatesan, M., Alaria, J. *et al.* (2009) The origin of the magnetism of etched silicon. *Adv. Mater.*, **21**, 71.

79. García, M.A., Fernández Pinel, E., De la Venta, J. *et al.* (2009) Sources of experimental errors in the observation of nanoscale magnetism. *J. Appl. Phys.*, **105**, 013925.

80. Dorneles, L.S., Venkatesan, M., Gunning, R. *et al.* (2007) Magnetic and structural properties of Co-doped ZnO thin films. *J. Magn. Magn. Mater.*, **310**, 2087.

81. Zukova, A., Teiserskis, A., van Dijken, S. *et al.* (2006) Giant moment and magnetic anisotropy in Co-doped ZnO films grown by pulse-injection metal organic chemical vapor deposition. *Appl. Phys. Lett.*, **89**, 232503.

82. Ackland, K., Monzon, L.M.A., Venkatesan, M. *et al.* (2011) Magnetism of nanostructured CeO_2. *IEEE Trans. Magn.*, **47**, 3509.

83. Venkatesan, M., Fitzgerald, C.B., and Coey, J.M.D. (2004) Thin films: unexpected magnetism in a dielectric oxide. *Nature*, **430**, 630.

84. Ran, J. and Yan, Z. (2009) Observation of ferromagnetism in highly oxygen-deficient HfO_2 films. *J. Semicond.*, **30**, 102002.

85. Zhang, Y. and Xie, E. (2010) Nature of room-temperature ferromagnetism from undoped ZnO nanoparticles. *Appl. Phys. A*, **99**, 955.

86. Mal, S., Nori, S., Jin, C. *et al.* (2010) Reversible room temperature ferromagnetism in undoped zinc oxide: correlation between defects and physical properties. *J. Appl. Phys.*, **108**, 073510.

87. Yoon, S.D., Chen, Y., Yang, A. *et al.* (2006) Oxygen-defect-induced magnetism to 880 K in semiconducting anatase $TiO_{2-\delta}$ films. *J. Phys. Condens. Matter*, **18**, L355.

88. Coey, J.M.D. (2005) d^0 ferromagnetism. *Solid State Sci.*, **7**, 660.

89. Straumal, B.B., Mazilkin, A.A., Protasova, S.G. *et al.* (2009) Magnetization study of nanograined pure and Mn-doped ZnO films: formation of a ferromagnetic grain-boundary foam. *Phys. Rev. B*, **79**, 205206.

90. Edwards, D.M. and Katsnelson, M.I. (2006) High-temperature ferromagnetism of sp electrons in narrow impurity bands: application to CaB_6. *J. Phys. Condens. Matter*, **18**, 7209.

91. Coey, J.M.D., Wongsaprom, K., Alaria, J. *et al.* (2008) Charge-transfer ferromagnetism in oxide nanoparticles. *J. Phys. D: Appl. Phys.*, **41**, 134012.

92. Evegeny, Y.T., Tsymbal, E.Y., and Zutic, I. (2010) *Spin Transport and Magnetism in Electronic Systems*, Taylor and Francis.

93. Hernando, A., Crespo, P., and Garcia, M.A. (2006) Origin of orbital ferromagnetism and giant magnetic anisotropy at the nanoscale. *Phys. Rev. Lett.*, **96**, 57206.

94. Millis, A.J. (2011) Moment of magnetism. *Nat. Phys.*, **7**, 749.

95. Zhou, S., Čižmár, E., Potzger, K. *et al.* (2009) Origin of magnetic moments in defective TiO_2 single crystals. *Phys. Rev. B*, **79**, 113201.

96. Chakrabarty, A. and Patterson, C.H. (2011) Defect-trapped electrons and ferromagnetic exchange in ZnO. *Phys. Rev. B*, **84**, 054441.

97. Bert, J.A., Kalisky, B., Bell, C. *et al.* (2011) Direct imaging of the coexistence of ferromagnetism and superconductivity at the $LaAlO_3$/$SrTiO_3$ interface. *Nat. Phys.*, **7**, 767.

98. Brinkman, A., Huijben, M., Van Zalk, M. *et al.* (2007) Magnetic effects at the interface between non-magnetic oxides. *Nat. Mater.*, **6**, 493.

99. Ariando , Wang, X., Baskaran, G., and Liu, Z.Q. (2011) Electronic phase separation at the $LaAlO_3$/$SrTiO_3$ interface. *Nat. Commun.*, **2**, 188.

100. Caviglia, A.D., Gariglio, S., Reyren, N. *et al.* (2008) Electric field control of the $LaAlO_3$/$SrTiO_3$ interface ground state. *Nature*, **456**, 624.

2
Magnetic/Multifunctional Double Perovskite Oxide Thin Films

Prahallad Padhan and Arunava Gupta

2.1
Introduction

In some materials, particularly in a number of functional oxides, two or more intrinsic functional properties occur simultaneously. They exhibit a variety of interesting and, in some cases, unique physical properties. Among the multifunctional oxides, the electronic and magnetic properties of the double perovskite oxides have recently attracted much scientific and technological interest. The structure of the B-site-ordered double perovskite oxides, $A_2B'B''O_6$, is derived from that of the simple perovskites, ABO_3, by having the B-site cation alternately occupied by two different cationic species, B' and B''. Depending on the relative size and oxidation state, the B' and B'' ions can be crystallographically completely ordered, making up a rock salt-type lattice in which each B' cation has only B'' cations as nearest cation neighbors and vice versa. Accordingly, the unit cell of B-site-ordered double perovskite is eight times larger volume than the primitive cubic cell of single perovskites. Such perfect cation ordering is not always achieved, and in many instances, there is some degree of mixing between the B' and B'' cations [1–3]. In general, the smaller the difference in the size and charge between the two B-site cations, the lower is the equilibrium degree of order at the B site. Thermodynamically, the highest degree of order is obtained using low synthesis temperatures. However, cation diffusion is slow at low temperatures and the degree of order is kinetically controlled. Thus, to obtain highly ordered double perovskite samples, in practice, relatively long synthesis periods at moderate temperatures are required [1–4]. Double perovskite oxide systems with various combination of A, B', and B'' cations have been synthesized and characterized. Their crystalline distortions and physical properties depend sensitively on the size and valences of the A, B', and B'' cations. Furthermore, band structure calculations indicate that some of the systems are half-metallic ferromagnets with relatively high Curie temperature T_c [5]. Local-spin-density calculations also predict half-metallic characteristics in some of the antiferromagnetic double perovskites [6].

The ferromagnetic half-metals exhibit large low-field room-temperature magnetoresistance (MR) [7]. In addition, some of the ferromagnetic double perovskites

Functional Metal Oxides: New Science and Novel Applications, First Edition.
Edited by Satishchandra B. Ogale, Thirumalai V. Venkatesan, and Mark G. Blamire.
© 2013 Wiley-VCH Verlag GmbH & Co. KGaA. Published 2013 by Wiley-VCH Verlag GmbH & Co. KGaA.

with magnetic ordering near room temperature exhibit insulator-like charge transport [8, 9]. For example, La_2NiMnO_6 (LNMO) is an ordered double perovskite that is a ferromagnetic semiconductor with a Curie temperature of 280 K. Recent studies of the material in the bulk have revealed large magnetic-field-induced changes in the resistivity and dielectric properties at temperatures as high as 280 K [10]. This is a much higher temperature than previously observed for such a coupling between the magnetic, electric, and dielectric properties in other ferromagnetic semiconductors. Substitution at the A site can also lead to ferroelectric behavior in the double perovskites. Azuma *et al.* [11] have succeeded in synthesizing the "designed" compound Bi_2NiMnO_6 (BNMO) in the bulk under high pressure with ferroelectric and ferromagnetic transition temperatures of 485 K and 140 K, respectively. Materials that are simultaneously magnetic and ferroelectric, referred to as *multiferroic*, are rare in nature and most of them are antiferromagnets with a small response to an external magnetic field. More recently, it has been reported that the B-site-ordered double perovskite Sr_2MgMoO_6 is a promising mixed oxide-ion/electron conductor with an excellent tolerance to sulfur that provides direct electrochemical oxidation in dry methane at $800\,°C$ [12, 13]. Moreover, La-doped Sr_2MgMoO_6 performs somewhat better on natural gas [14]. Although the study of double perovskites is still in its infancy, it is clear that a number of them display two or more functional properties with rich and fascinating fundamental physics, and therefore present opportunities for a wide range of potential applications.

A particularly promising research direction is the synthesis of artificial crystalline structures of the double perovskite oxides in the form of thin films, which allows us to go beyond equilibrium materials in exploring new properties, developing new functionalities, and analyzing fundamental physical processes. Therefore, a great deal of effort is underway to synthesize thin films of the double perovskite oxides, including stabilization of metastable structures. Thin films provide an excellent opportunity to control the strain, chemical heterogeneity, and artificial structuring in order to realize new or enhanced properties, which are absent or impossible to obtain in the form of bulk [15–19]. It is also possible to overcome the natural preference for disorder or low-dimensional ordering in some of the materials by controlling the growth parameters [20]. Such materials may pave the way for device structures such as ferroelectric field-effect transistors, controllable dielectric microwave devices, high-frequency RF modules, magnetoresistive sensors, or electrically controllable and readable magnetic random access memories.

With this brief introduction on magnetic/multifunctional double perovskite oxides, we proceed to explore the important requirements for synthesizing high-quality oxide thin film using the pulsed laser deposition (PLD) technique, followed by characterization of their morphology, surface, electronic structure, transport, and magnetic properties of thin films of various magnetic/multifunctional double perovskite oxides. We end the chapter with a discussion of potential applications and future scope of magnetic/multifunctional double perovskite oxide thin films.

2.2
Thin-Film Deposition

A wide range of physical properties can emerge in functional oxides because of the extreme sensitivity of electron interactions to apparently small structural changes, such as bond angles and lengths. Subtle structural changes arising from lattice-mismatch-induced strain during heteroepitaxial film growth and/or from defects can thus be exploited for their realization. Even for structurally perfect functional oxides materials, the surfaces and interfaces have properties that can differ significantly from the bulk. However, for many practical applications, it is important to grow thin films of functional oxides that are as close as possible to that of its bulk form. Furthermore, to help understand and interpret the measurements on thin films, most (but not all) theoretical studies assume that there are no defects either at the interfaces or in the bulk. It is, however, quite challenging to grow thin films that approach this ideal condition. Nevertheless, understanding and controlling the subtle structural changes in complex oxides is essential to unraveling the basic science underlying their properties. Thin-film deposition plays a crucial role in modern device fabrication processes. The deposition of thin films of functional oxides, because of their greater structural complexity, lower surface mobility, and propensity for defect formation, represent a significant challenge. However, the development of new deposition and characterization techniques has enabled the growth of high-quality oxide films, at least for research-scale requirements.

2.2.1
The Requirement for Epitaxy

When a thin film is deposited on a single-crystalline substrate, the interfacial energy is minimized across the interface in an attempt to merge the symmetries of the substrate and the growth material. Thus, it is energetically favorable for the film material to crystallographically align itself so as to match the substrate-bonding symmetry and periodicity. For very thin films, the growth is elastically strained to have the same interatomic spacings as those of the substrate. The interface is, therefore, coherent with atoms on either side lining up. This alignment of crystallographic atomic positions in the single-crystal substrate with those in the film is referred to as *epitaxy*. Epitaxial growth leads to a single out-of-plane growth orientation, in addition to an in-plane orientation or texture. Consequently, epitaxy occurs for any combination of film and substrate having some degree of match, provided that the substrate symmetry is not masked during film growth by interfacial disorder. In practice, although the choice of suitable single-crystal substrates is limited, the only fundamental criterion for good epitaxy appears to be small fractional mismatch in the periodicities of the two materials along the interfaces. It is well known that real substrate surfaces consist of many defects, such as adatoms, atomic clusters and islands, terraces and edges and vacancies, and so on. The surface of the substrate plays a critical role in achieving control of film growth on an atomic scale as it acts as an initial template for crystal growth. The morphology of the substrate surface can affect the growth mode and even the crystal structure of the film. Surface physical properties, for example, are highly dependent on surface treatment procedures. Therefore, appropriate substrate surface preparation before film deposition is very important and has become essential for modern surface science and thin-film technology. Rigorous substrate treatment not only enables the removal of surface contaminants but also allows the control of the surface defects density and microstructure. The as-polished commercially available substrates are typically rinsed with organic solvents and then subjected to ultrasonic agitation in distilled water and organic solvents in order to remove contaminants. Oxide substrates, such as $SrTiO_3$ (STO), are then often treated in a NH_4F:HF buffer solution followed by a thermal treatment at 950 °C in flowing oxygen to achieve surfaces with atomically smooth steps and terraces [20, 21].

2.2.2
Interfaces, Strain, and Defects

The epitaxial growth of a thin-film material, which is compositionally and/or structurally different from the substrate material, is referred to as *heteroepitaxy*. The heteroepitaxy and lattice mismatch in thin-film systems often lead to structural distortion. If the lattice mismatch is very small, then the heterojunction interfacial structure is essentially similar to that for homoepitaxy (i.e., thin-film material being the same as the substrate). However, differences in film and substrate chemistry, surface energy, and coefficient of thermal expansion can strongly influence the physical properties and perfection of the interface. A lattice mismatch as small as possible is generally preferred and in fact can be achieved through careful selection of the substrate or buffer layer material. However, the films can grow strained if the difference between the film and substrate lattice parameters is not very small. In general, the in-plane film lattice expands or contracts to accommodate the crystallographic difference with the substrate. In an epitaxial ultrathin film, the material is elastically strained to have the same interatomic spacing as that of the substrate. As the film thickness increases, the total elastic strain energy increases correspondingly, eventually exceeding the energy associated with a relaxed structure consisting of an array of the so-called misfit dislocations. In the case of epitaxial growth, elastically strained structure usually exists only up to a certain critical film thickness, beyond which misfit dislocations are introduced. The misfit dislocations lie in planes parallel to the interface. For oxide films, the strain is often observed to relax much more slowly than in simple metals and semiconductors, considerably beyond the critical thickness. Stacking faults that are perpendicular to the surface of the substrate are frequently observed in the relaxed films [22, 23]. This phenomenon can occur between film and substrate pairs composed of different materials that have the same crystal structure. Also, the lattice parameter differences are an order of magnitude larger than in the case of lattice-matched heteroepitaxy. In oxides, surfaces and interfaces have an even greater tendency than the bulk to develop nonstoichiometry, strain, or structural defects, with a correspondingly larger modification of properties. It has been demonstrated in a number of instances that the imposition of strain in the multifunctional thin film through lattice mismatch with the substrate modifies the structural [24–26] and physical properties [27, 28]. The tremendous sensitivity of the properties of oxides to defects and strain is greatly amplified at interfaces. The sensitivity of many oxides to strain indicates that epitaxy provides a powerful tool to control their electronic and magnetic properties.

2.2.3
Pulsed Laser Deposition

The development of techniques for the deposition of high-quality thin films of multicomponent oxides emerged in parallel with the research on oxide high-T_C superconductors. Among them, the widely adopted PLD technique provides very good stoichiometry control, optimized oxygen thermodynamics and a high degree of structural perfection for thin-film growth of a wide range of multifunctional oxide materials. In PLD, ultrashort pulses of ultraviolet light are used to ablate material from a target, often in a background atmosphere of oxygen for the growth of oxides. The atomic and molecular species, which are transported and adsorbed on the substrate surface, undergo a series of oxidation, recombination, and rearrangement reactions that eventually lead to nucleation and growth of the film. The short pulse duration ensures that thermal heating of the target is minimized and thus thermal evaporation, which would tend to select the more volatile species, is minimized. In general, reasonably high-quality epitaxial growth can be achieved merely by maintaining an optimum background gas pressure and substrate temperature, and ensuring that a stoichiometric flux of cationic species arrive at the substrate by suitably adjusting the laser fluence and target–substrate distance.

2.2.4
Control of Deposition Parameters

Efficient and stoichiometric ablation of the target material requires nonequilibrium excitation of the ablated volume to temperatures well above that required for evaporation. This generally requires that the laser pulse be of short duration, reasonably high in energy density, and is strongly absorbed by the target material. Each ablation pulse will typically provide material sufficient for the deposition of only a submonolayer of the desired phase. The amount of film growth per laser pulse depends on the distance between the target and substrate, the laser fluence, oxygen pressure, and substrate temperature. Although the target–substrate separation is generally fixed in a deposition system, the laser fluence essentially controls the cation stoichiometry ratio [29, 30] of complex oxides such as the multifunctional double perovskites. On the other hand, oxygen background pressure and substrate temperature play a critical role in obtaining double perovskite thin films with physical properties closely approaching those of the bulk material [23, 31–37]. Thin-film deposition conditions of functional oxides vary widely with changes in the composition of the functional oxides. However, large variations in the deposition conditions are also reported for multifunctional oxides with the same basic components [38, 39]. Using optimized deposition conditions, one can control the deposition of the materials at an atomic layer level by monitoring and controlling the number of laser pulses using *in situ* techniques such as reflection high-energy electron diffraction (RHEED).

2.2.5
Stoichiometry Control with Volatile Constituents

Thin-film growth of functional oxide consisting of volatile components with accurate stoichiometry control, optimized oxygen thermodynamics, and high structural perfection is quite challenging. Although early work on sputtering demonstrated the feasibility of high-T_C oxide superconductor film growth, problems of stoichiometry control associated with selective resputtering of the more volatile components led to the widespread adoption of the PLD technique for a wide variety of functional oxides. The short pulse duration ensures that equilibrium thermal heating of the target is low, and thus thermal evaporation, which would tend to select the more volatile elements, is minimized. Since the plasma plume associated with the ablation process in the presence of a background gas contains mostly relatively low-energy species, there is negligible resputtering of the growing film. During the growth of thin films of functional oxides consisting of a volatile element, for example, Bi, the oxidation/reduction processes make them susceptible to point defects, in the form of metal/oxygen stoichiometry ratio changes. In general, these thin films are grown under more stringent control of oxygen pressure and substrate temperature. In addition to the control of these deposition parameters, the volatility of the element is frequently compensated using a target with volatile element(s) content being higher than the stoichiometric value [37, 40]. Control of background oxygen and substrate temperature at around the structural transition temperature is the most effective means of controlling the cation/anion ratio [18, 40]. Overall, the laser fluence, oxygen pressure, substrate temperature, and the distance between the target and substrate play important roles in determining volatile element stoichiometry in multicomponent oxides. Although the ablated target contains volatile species, for example, Bi, high-quality stoichiometric thin films of multicomponent oxides have been achieved using optimized deposition parameters [18, 31, 40, 41].

2.2.6
Control of Defect Concentration

In the case of epitaxial thin films, the lattice mismatch, morphology of the substrate surface, volatile element(s) of the target, laser fluence, substrate temperature, and oxygen pressure can result in structural defects. Primarily, minimization of lattice mismatch and achievement of atomically flat substrate surface – since it produces dislocations and associated recombination centers – can help reduce the defect concentration in epitaxial thin films. Moreover, the optimization of laser fluence, substrate temperature, and oxygen pressure play a crucial role in achieving control over defect concentration. Control of oxygen incorporation during film growth, rather than postannealing, is the most effective way of controlling the cation/anion ratio. Despite maintaining an adequate oxygen background during film growth, many perovskites still exhibit oxygen nonstoichiometry. In addition to controlling equilibrium defects by adjusting the growth temperature and postannealing treatments, both the growth kinetics and growth mode play important roles in determining the concentration of nonequilibrium defects [23, 42]. In contrast to high-rate, high-supersaturation growth of oxides by standard PLD, which results in island growth and also rather defective materials, low-rate growth has been demonstrated to often produce higher quality film with physical properties closely approaching those of the bulk material [21].

2.3
Structure and Morphology

In general, the physical properties of multifunctional oxides thin films are strongly influenced by their microstructural characteristics, such as crystalline and amorphous state, crystallographic orientation, crystallite size, strain, and so on. Among the analytical methods especially suitable for characterization of thin films, X-ray diffraction plays an important role because it is nondestructive, noncontact, and often provides quantitative results.

2.3.1
Phase and Epitaxy: X-Ray

The X-ray diffraction technique can be used to verify the presence of defects, parasitic phase, structural distortion, and ordering of B-site ions in thin films of the double perovskites, in addition to the study of their epitaxy, structural phase, film thickness, and surface roughness. The presence of interference fringes in the low angle X-ray diffraction scans of thin films enable accurate determination of their thickness and surface roughness [39, 42]. On the other hand, the high-angle $\theta - 2\theta$ X-ray diffraction profiles of thin films grown on oriented single-crystal substrates provide information regarding their crystallographic structure and growth orientation. The presence of (001)-oriented X-ray reflections of the film and the substrate are observed for thin films of a wide variety of materials, indicating epitaxial or textured growth of the double perovskite phase. On the basis of the angular positions of the reflection peaks, the out-of-plane lattice parameter of the film can be calculated. The out-of-plane lattice parameter of Sr_2FeMoO_6 (SFMO) films is observed to be significantly larger than in the bulk [21, 22, 34]. The larger c-axis value has been attributed to several factors, including changes in the ordering of the Fe and Mo atoms in the lattice [34, 43, 44], thermal strain induced by differential thermal expansion on cooling from the growth temperature [22], lattice mismatch (for SFMO films grown on STO) [21], structural film defects [21], and cation off-stoichiometry [30]. For example, Sanchez *et al.* [21] have reported the variation of *c*-axis length for SFMO on STO as a function of deposition temperature, deposition rate, and film thickness. Their results indicate that the *c*-axis length decreases and approaches the bulk value with increasing growth temperature and decreasing deposition rate. Moreover, as the thickness of SFMO film is increased, the film peak positions gradually shift toward the bulk value. However, the out-of-plane lattice parameter does not match the bulk value even for a 600 nm thick film [22, 32]. In the case of La_2CuSnO_6 (LCSO) thin films, the out-of-plane lattice parameter shows a tendency to increase with increasing deposition temperature [17], a trend opposite to that observed for SFMO on STO [32, 34, 44]. Similarly, an expanded out-of-plane lattice parameter compared to the pseudo-cubic bulk value (0.3876 nm [45]) has been observed for thin films of LNMO grown at low oxygen pressures. However, as the oxygen pressure increases, the out-of-plane lattice parameter steadily decreases before leveling off at higher pressures. The variation of lattice parameter with oxygen pressure is explained in terms of variations in the defect concentration due to oxygen vacancies and the varying energy and flux of the species impinging on the surface of the growing film [23].

In most cases, the epitaxial growth of the double perovskite phase is strongly dependent on the choice of substrate and deposition conditions. Under nonoptimal growth conditions, parasitic phases are observed based on the X-ray diffraction patterns of the films [28, 38, 46]. For example, an unidentified parasitic phase has been observed in the $\theta - 2\theta$ X-ray diffraction pattern of Sr_2CrWO_6 (SCWO) grown on (001)-oriented $LaAlO_3$ (LAO) substrate, while films grown on (001)-oriented STO substrate are single phase and epitaxial [38]. However, Geprags *et al.* [46] have observed parasitic phases in thin films of SCWO grown at oxygen pressure higher than 6.6×10^{-4} mbar on (001)-oriented STO. In contrast, X-ray diffraction profiles of thin films (with deposition time <20 min) of SFMO grown on STO, LAO, and MgO indicate epitaxial growth. But as the film thickness increases, those deposited on MgO indicate single-phase growth, but an additional $SrMoO_4$ phase is observed for films grown on STO and LAO substrates [26]. Postannealing the SFMO film grown on MgO in oxygen results in reduction of the SFMO peak intensities along with the appearance of two new peaks at 29.7° and 61.5°, attributed to the (004) and (008) planes of $SrMoO_4$ [42]. Indeed, Sanchez *et al.* [21] have observed the parasitic $SrMoO_4$ phase in thin films of SFMO grown on STO with oxygen pressure higher than 10^{-4} mbar. For Bi_2FeMnO_6 (BFMO) thin films grown on MgO and STO substrates, the presence of γ-Fe_2O_3 phase is observed with either high substrate temperatures or low oxygen pressures, while the films tend to be Bi-rich and favor growth of Bi_2O_3 for low substrate temperatures or high oxygen pressures [37]. In fact, growth of single-phase BFMO thin films occurs only over a fairly narrow process window. For example, background oxygen pressure changes of ±0.2 mTorr away from 1 mTorr leads to the formation of significant amounts of secondary phases. In the case of LNMO films, La deficiency is induced when operating at high deposition temperatures and high oxygen pressures [36]. It has been reported that excess oxygen is accommodated by formation of vacancies in $LaMnO_3$, preferentially at the La site [47]. The La deficiency can induce the formation of secondary phases, as reported by Dass *et al.* [9] However, if the growth temperature is low and oxygen pressure is relatively high, LNMO films show evidence of two distinct ferromagnetic phases [36].

2.3.2
Surface Morphology: Scanning Electronic Microscopy and Atomic Force Microscopy

The scope of atomic force microscopy (AFM) applications includes high-resolution examination of surface topography; compositional mapping of heterogeneous samples; and studies of local mechanical, electrical, magnetic, and thermal properties. These measurements can be performed on scales ranging from hundreds of micrometers down to a few nanometers.

The surface morphology of double perovskite oxide thin films can be controlled by substrate-induced stress and deposition conditions. AFM analysis shows that SCWO films grow epitaxially on STO substrates with similar crystalline perfection and very smooth surface morphology as the doped manganites. However, the surface morphology of epitaxial SCWO thin films on other substrates, such as $NdGaO_3$ (NGO), $MgAl_2O_4$, and LAO, exhibit much rougher surface and a more distorted crystal structure [39]. This is caused by the larger lattice mismatch of these substrates for SCWO when compared with STO. Similar changes in surface morphology of SFMO thin films grown on STO, LAO, and MgO substrates have also been observed [48]. Interestingly, Manako *et al.* [15] have observed an atomic-scale step-and-terrace structure with 0.4 nm step height, which corresponds to single perovskite unit cell height of SFMO grown on (001) STO. This is consistent with the simple expectation from the crystal structure of SFMO that all the perovskite sheets are equivalent, with the same number of FeO_6 and MoO_6 octahedra arranged in the form of a checker board. As shown in Figure 2.1, the steps are essentially identical over the entire scan region of the sample. On the other hand, films grown on (111)-oriented STO substrates show triangular-shaped grains. A spiral structure of atomic-scale steps is observed for individual grains, indicating that the growth mode of the film is two-dimensional spiral growth mediated by screw dislocations. It is noteworthy that every step seen in the image has a height of about 0.45 nm, which corresponds to the (111)-d spacing of the SFMO unit cell and not half of the value. This implies that the growth of (111)-oriented film occurs with the chemical formula as same as that of a growth unit and that there exists a preferred surface termination of B-site cations [15].

Borges *et al.* [49] have observed a granular structure with grain size in the range of 100–200 nm for SFMO thin films grown on STO substrates at various deposition temperatures. In addition, these films show the existence of L-shaped holes with a depth of ~12 nm. Auger electron mapping indicates a higher concentration of iron in the holes. Jalili *et al.* [26] have shown using SEM and AFM that

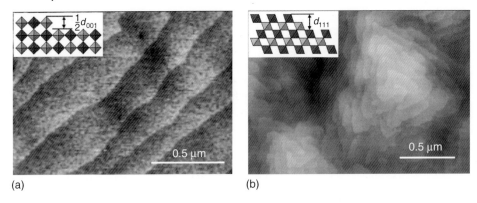

(a)   (b)

Figure 2.1 Atomic force microscopic image for Sr_2FeMoO_6 films epitaxially grown on (a) $SrTiO_3$ (001) and (b) $SrTiO_3$ (111) substrates. Insets are schematic representation of step structures [15].

the surface morphology of thinner films of SFMO grown on STO, LAO, and MgO substrates depends sensitively on the nature of the substrate. But as the deposition time is increased, the surface morphology of film becomes independent of the substrate, with formation of randomly distributed nanoparticles (<20 nm in size). The presence of nanoparticles is explained on the basis of particulate formation, possibly due to changes in the surface condition of the SFMO target after extended ablation and also due to the formation of secondary phases such as Fe_2O_3, MoO_2, and $SrMoO_4$, or a-axis aligned SFMO grains [26]. The surface morphology and surface roughness of Sr_2CrReO_6 (SCRO) films grown on STO have been found to depend on the substrate temperature used during deposition. The film grown at 700 °C shows an average grain (or domain) size of about 50 nm, while the film grown at 800 °C possess grain (or domain) sizes of about 300–500 nm. The surface roughness (rms) over an area of 1.5×1.5 mm^2 are 0.4 and 1.2 nm for films grown at 700 and 800 °C, respectively [50]. The effect of deposition rate on the surface morphology of SFMO thin film has been studied by varying the ablation rate from the target (25 and 15 Å min^{-1}) [21]. The film grown at a higher deposition rate shows a rather granular morphology with mean roughness around 100 Å, while films grown at a lower deposition rate exhibit a much smoother surface morphology (mean roughness around 20 Å). The higher deposition rate leads to SFMO films containing much higher density of structural defects. This translates into more efficient strain relaxation with increasing film thickness.

2.3.3
Interfaces and Defects: Transmission Electron Microscopy

Transmission electron microscopy (TEM) is widely used to investigate the growth morphology of thin films in cross section. It is also used extensively to study the quality of the film–substrate interface. In the case of strained thin films, the TEM images of substrate–film interface show a sharp and coherent epitaxial relationship with no evidence of misfit dislocations or other structural defect [35, 46, 51]. However, the presence of such features in the TEM images strongly depends on the degree of strain and deposition parameters. As explained in Section 2.2, ultrathin films are elastically strained to have the same interatomic spacing as that of the substrate. Such pseudomorphic growth of a 25 nm thick SFMO film on STO substrate has been confirmed using TEM [22]. The thickness up to which pseudomorphic growth persists is associated with the so-called critical film thickness, which strongly depends on the deposition parameter. For example, the TEM image of LNMO deposited at 100 mTorr partial pressure of oxygen on STO substrate shows pseudomorphic growth of about 5 nm near the interface that appears essentially defect-free [23]. When the film reaches a critical thickness, defects form in the LNMO film as a result of strain relaxation. Subsequently, defects are generated that extend all the way through the sample until they reach the surface, as shown by the arrows in the Figure 2.2. On the other hand, for the film grown at 800 mTorr oxygen, the structural quality is excellent, being essentially free of any major defects throughout the cross section of the film. Thus, a higher oxygen background concentration appears to have a significant influence in improving the structural quality of LNMO films [23]. The TEM image of a 600 nm thick SFMO film exhibits an array of stacking fault and misfit dislocations associated with the relaxation of the substrate-induced strain [22].

The LNMO thin films grown on STO under 100 and 800 mTorr of background oxygen have been examined by phase-contrast high-resolution TEM. At a thickness of 5–10 nm, the strain in the film is relaxed, nucleating numerous defects in the remaining thickness of the film. The LNMO layer grown at 100 mTorr oxygen is essentially single phase, and analysis of fast Fourier transform spectra reveals an *I*-type structure [51]. Symmetry considerations suggest that a long-range cationic ordering is not compatible with this structure and may thus indicate the presence of antisite defects. On the other hand, the LNMO thin film grown in 800 mTorr oxygen indicates that the sample is structurally biphasic at room temperature. In addition to the same *I*-type crystal structure, numerous regions of the layer crystallize in a secondary structure characterized by a new set of reflections with even *I*-index. This phase has also been observed by Singh *et al.* [51] and labeled as *P*-type owing to the breaking of the centered symmetry of the *I* phase.

Scanning transmission electron microscopy (STEM) provides additional information about the interfaces and the ordering of the B-site ions when high-resolution *Z*-contrast images are captured [17]. The STEM image of an LCSO thin film grown on $(LaAlO_3)_{0.3}-(Sr_2LaTaO_6)_{0.7}$ (LSAT), acquired using a high-angle annular dark-field detector (HAADF) is shown in Figure 2.3. In this image, the brightness of the atoms is approximately proportional to $Z^{1.7}$, where Z is an atomic number. Therefore, the elements in the images can be specified from the intensity of the atoms. The obvious difference in the intensity of the Cu and Sn is clear evidence of the Cu/Sn ordering in the layered configuration. Furthermore, the

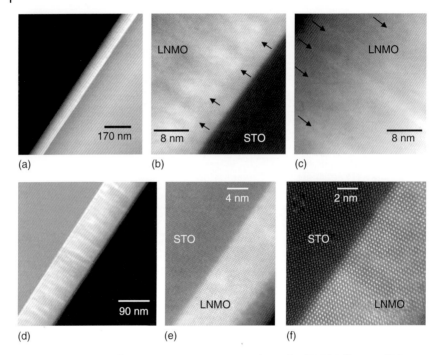

Figure 2.2 Low-magnification cross-section STEM images of La$_2$NiMnO$_6$ (LNMO) films on SrTiO$_3$ (STO) substrate grown under (a) 100 mTorr and (d) 800 mTorr O$_2$. High-resolution STEM images near the interface region showing the epitaxial growth of LNMO films on STO substrates grown under (b) 100 mTorr and (e) 800 mTorr O$_2$. High-resolution STEM images of LNMO films grown under (c) 100 mTorr and (f) 800 mTorr O$_2$ [23].

intensity profile reveals that the LaO–CuO$_2$–LaO distance is 3.8 Å, while that of LaO–SnO$_2$–LaO is 4.7 Å. These values are quite close to those observed in bulk LCSO. A similar study has been conducted for thin film of La$_2$CoMnO$_6$ (LCMO), and an excellent unit-cell to unit-cell epitaxial relationship between the film and the substrate is observed, consistent with epitaxial growth of LCMO film. The film–substrate interface is sharp and coherent with no evidence of formation of secondary phases or any significant interdiffusion over large distances. The only observed defects are the areas of brighter contrast with halos, which is attributed to twin boundaries. They originate from the interface region and run parallel to the growth direction all the way up to the film surface. The average spacing of the twin boundaries varies but is typically of the order of 15 nm [35].

Functional oxide films provide an exciting opportunity to create attractive functionalities and multifunctional devices. A key issue for better understanding and developing these functionalities is to investigate interfacial electronic structures and properties using nondestructive characterization and testing methods.

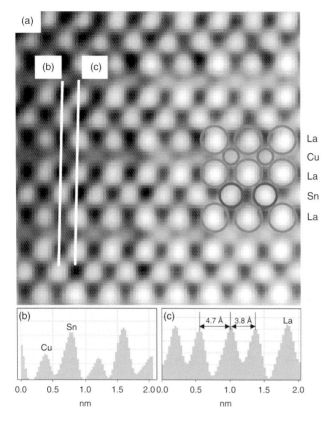

Figure 2.3 (a) STEM–HAADF image of a layered double perovskite La_2CuSnO_6 film grown on a $(LaAlO_3)_{0.3}-(Sr_2LaTaO_6)_{0.7}$ substrate. Circles represent La, Sn, and Cu atoms. Intensity profiles along solid lines labeled (b) and (c) are shown in (b) and (c), respectively [17].

2.4
Electronic Structure

2.4.1
Raman Spectroscopy

Raman spectroscopy has been widely used to study the crystal symmetry, internal stress, cation disorder, local lattice distortion, oxygen deficiency, and the presence of impurity phases in the materials. This technique has been used to study double perovskite thin films [23, 52–55] and primarily observed orthorhombic (*Pbnm*, with *c* being the long axis), rhombohedral ($R\bar{3}$), and monoclinic ($P12_1/n1$) structures. The experimentally observed ordered $R\bar{3}$ and $P12_1/n1$ structures can be obtained from the parent $Fm\bar{3}m$ by rotation of $B'O_6$ and $B''O_6$ around the cubic $[111]_c$ or $[001]_c$ and $[110]_c$ directions. In the ideal $Fm\bar{3}m$ structure four Γ-point phonons ($A_{1g} + E_g + 2F_{2g}$) are Raman allowed. But symmetry analysis shows that for the $R\bar{3}$ structure, eight ($4A_g + 4E_g$) modes are Raman allowed, whereas in the $(P12_1/n1)$ structure, the number of Raman modes increases to 24 $(12A_g + 12B_g)$. The polarized Raman spectra [24] of thin films of three different double perovskites (LCMO, LNMO, and BNMO) are shown in the Figure 2.4. All the spectra are dominated by two main features: (i) a high-frequency band that peaks in the 600–670 cm^{-1} range, pronounced much more strongly with parallel polarization of the incident and scattered light and (ii) a broader band at a lower frequency (460–540 cm^{-1}), which can exhibit a complex thin-film structure, as a rule of comparable intensity for the crossed and parallel scattering configurations. Iliev *et al.* [54] have performed lattice dynamical calculations for LCMO and confirmed that the 645 cm^{-1} corresponds to stretching (breathing) vibrations, whereas the modes near 490 cm^{-1} are of mixed type, involving both antistretching and bending motions. The important point here is that the symmetry of the stretching mode is different in the disordered and ordered phases, which

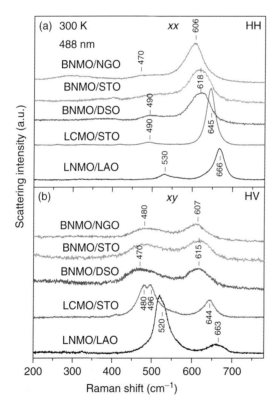

Figure 2.4 (a,b) (Color online) Comparison of the Raman spectra of $Bi_2NiMnO_6/NdGaO_3$, $Bi_2NiMnO_6/SrTiO_3$, $Bi_2NiMnO_6/DyScO_3$, $La_2CoMnO_6/SrTiO_3$, and $La_2NiMnO_6/LaAlO_3$ as obtained with 488 nm excitation at 300 K. *x* and *y* denote two orthogonal directions rotated by 45° with respect to the Mn–O–Ni/Co chains. The spectra are vertically shifted for clarity. The *xy* vertical scale is enlarged by a factor of 3 compared to that of the *xx* spectra [24].

makes it possible to determine the correct structure from the polarized Raman spectra. Polarized Raman measurements have also been carried out on thin films of several double perovskite systems to assess the relative contribution of incomplete B-site ordering versus other types of defects formed at low oxygen concentrations. The spectra are taken with exact xx, $x'x'$, yy, xy, and $x'y'$ scattering configurations, where $x||$ [100], $x'||$ [110], $y||$ [010], and $y'||[\bar{1}10]$, with the first and second letters referring to the polarization direction of the incident and the scattered light, respectively. The B-site-ordered and disordered structures belong to different symmetry groups, and the corresponding Raman spectra, if taken in exact scattering configurations, obey different Raman polarization selection rules. In the ordered structures ($P12_1/n1$ or $R\bar{3}$), the symmetric stretching Raman mode between 650 and 700 cm^{-1} is allowed in parallel xx or yy, but forbidden with crossed xy scattering configuration. On the other hand, the antisymmetric stretching mode between 500 and 550 cm^{-1} is allowed in xy but forbidden with the xx or yy configuration. The fully disordered structure belongs to the $Pbnm$ space group, where the polarization selection rules are just the opposite. The symmetric stretching mode should be present in the cross-polarized xy spectra and the antisymmetric stretching mode in the xx or yy spectra [23].

The high-frequency band corresponds to stretching (breathing) vibrations of the Ni(Co)/MnO$_6$ octahedra, whereas the lower lying complex band is in the frequency range of antistretching and bending motions. The Raman spectra profiles of thin films of LCMO, LNMO, and BNMO are similar, although at 300 K, the structures of LCMO/LNMO ($P12_1/n1$) and BNMO ($C2$) are different. This is confirmed by the fact that the profile of the Raman spectra of BNMO is practically insensitive to the $P12_1/n1 \leftrightarrow C2$ structural transition at $\approx$458 K [24]. The Raman spectra of LCMO/STO at room temperature are consistent with ($P12_1/n1$) symmetry, with significant contribution from ordered rhombohedral phase. The weak line of B_g symmetry seen in the xy spectrum at 640 cm^{-1} can be associated with only the ($P12_1/n1$) phase. On the other hand, the temperature-dependent $x'y'$ spectra do not show the peak structure of detectable intensity [54]. This suggests that unlike bulk LCMO (where the structure transition is found at $\approx$600 K) [45], the monoclinic structure of LCMO/STO films is sustained up to at least 800 K. It is plausible to expect that the strain and deposition conditions affect the delicate balance between monoclinic and rhombohedral phases, the latter being evidenced by a rather diffuse transition in bulk LCMO where the two phases coexist over a 100 K wide temperature range [45]. Truong et $al.$ [56] have also observed two modes (at 483 and 498 cm^{-1}) in the Raman spectra in the same xy configuration of LCMO on STO at room temperature. However, on cooling the film to 10 K, the Raman spectrum of LCMO/STO shows two additional modes (470, 484, 499, and 515 cm^{-1}). The additional modes are explained by the splitting of single excitation of disordered bulk into two excitation modes due to cation ordering. As mentioned earlier, the Raman spectra of LNMO/STO at room temperature are consistent with ($P12_1/n1$) symmetry with significant contribution from ordered rhombohedral phase. However, as the temperature increases above room temperature, the markers of the $P12_1/n1$ phase are no longer pronounced and the intensity of the spectra as a whole gradually decreases, partly due to line broadening. No clear phonon anomaly that may be indicative of sharp monoclinic-to-rhombohedral structural transition is observed [56].

In general, the change in phonon frequency (ω) with temperature ($\omega(T)$) is a sum of several contributions owing to (i) anharmonic scattering ($\Delta\omega_{anh}$), (ii) spin-phonon coupling ($\Delta\omega_{as-ph}$), (iii) lattice expansion and/or contraction ($\Delta\omega_{latt}$), and (iv) phonon renormalization ($\Delta\omega_{ren}$), resulting from electron–phonon coupling [57]. The temperature dependence of the main phonon modes of LCMO/STO deviates at $T < T_C$ from the standard dependence $\omega_{anh}(T) = \omega_0 - C\left(1 + 2/\left(\exp\left(-\hbar\omega_0/2KT\right) - 1\right)\right)$, where ω_0 and C are adjustable parameters. However, the frequency shift of the most pronounced Raman modes in LCMO or LNMO film scales with $\frac{M^2(T)}{M_{max}^2}$ as measured on the same film, thus confirming that anomalous phonon softening is due to spin-phonon interactions in the ferromagnetic state [54, 55]. The contribution of $\Delta\omega_{ren}$ can be neglected in the case of BNMO as it is a ferroelectric material and the carrier concentration is low. Nevertheless, the roles of $\Delta\omega_{as-ph}$ and $\Delta\omega_{latt}$ are pronounced mostly below T_{CM} and near T_{CE}, respectively. Above T_{CM} and far from T_{CE}, $\omega(T)$ is expected to be governed by the anharmonic phonon–phonon scattering.

The $\omega(T)$ of BNMO on NGO show the contribution of two $\omega_{anh}(T)$ characterized by different ω_0 and C, one in the temperature range of 140 and 420 K, and the other above 490 K, respectively, which points to the existence of a structural transition between 420 and 490 K. This is consistent with the observation of a structural transition at 485 K in bulk BNMO [11]. Another visible anomaly in the experimental $\omega(T)$ dependence is the deviation from $\omega_{anh}(T)$ toward lower frequencies for $T < 140$ K. Such phonon softening below the ferromagnetic ordering temperature has also been reported for LCMO [53, 54] and LNMO [55, 58] and has been ascribed to magnetic-order-induced phonon renormalization.

The interesting observation is that the antisymmetric stretching mode in LNMO thin films, in general, occurs at a higher frequency (blueshift) while symmetric stretching vibrations mode appears at a lower frequency (redshift) than the corresponding mode in the bulk. Moreover, the spectra show that the blue- and redshift of the Raman modes increase for films deposited at low oxygen pressures. The shift of the antisymmetric and symmetric stretching vibrations mode in opposite directions is likely to be related to the fact that the antisymmetric stretching vibrations mode in the ($P12_1/n1$) structure primarily involves in-plane antisymmetric stretching vibrations, whereas the symmetric stretching vibrations mode is fully symmetric and involves both in-plane and out-of-plane stretching vibrations of the NiO_6 and MnO_6 octahedra. The increase in the out-of-plane lattice parameter in response to the decreasing oxygen pressure, which is discussed in the previous section, will result in a decrease in only the symmetric stretching vibrations mode frequency. The observed broadening and intensity decrease in the Raman spectral features with a decrease in the oxygen pressure has a reasonable explanation in larger lattice distortions with increasing oxygen deficiency, incomplete ordering at B' and B'' sites, and coexistence of non-Jahn–Teller Ni^{2+}, Mn^{4+} and Jahn–Teller Ni^{3+}, Mn^{3+} ions. At relatively low concentrations, the local distortions act as scattering centers for phonons, thus shortening the phonon lifetimes and broadening the Raman lines corresponding to zone-center normal modes. The stronger distortions in the oxygen sublattice, as a whole, result in the breakdown of translation symmetry, and phonons from the whole Brillouin zone can contribute to the scattering. This gives rise to broad bands, reflecting the "smeared" one-phonon density of states, which has maxima at frequencies corresponding to those of zone boundary phonons [23]. The decrease in peak intensity with reduction in film thickness is in large part due to the reduced scattering volume. The frequencies of the antisymmetric and symmetric stretching modes of the thinner films are appreciably higher than the corresponding phonon frequencies in bulk LNMO. The blueshifts are related to compressive stress in the LNMO thin films, resulting both from the lattice mismatch between the film and the substrate and from the thermal expansion coefficient difference [23].

The systematic decrease in the stretching mode frequency in the LNMO–LCMO–BNMO sequence can be qualitatively explained by the increase in the averaged B–O bond length ($B =$ Ni, Co, Mn) from 1.963 Å in LNMO [45] to 1.978 Å in LCMO [45] and 1.994 Å in BNMO [11]. These bond length values, however, are somewhat different in the case of thin films, owing to the strain resulting from the film–substrate lattice mismatch. This explains the variation in the main Raman band position [24] in BNMO/NGO, BNMO/STO, and BNMO/DyScO$_3$, as shown in Figure 2.4.

2.4.2
X-Ray Absorption Spectroscopy

X-ray absorption spectroscopy (XAS) is a powerful structural technique to investigate the short-range environment around selected atomic constituents in condensed matter. The X-ray absorption spectra are analyzed using dipole selection rules of soft X-ray absorption to determine the final state and magnitude of intensity, for the initial electron-occupied state. The dipole selection rules are very effective in determining which of the $2p^5 3d^{n+1}$ final states are reached and with what intensity, starting from a particular $2p^6 3d^n$ initial state, unique to soft X-ray absorption. Such a technique has been used to study the valence state and spin character of the electrons in the initial states of different elements of the thin films of several double perovskites systems.

Guo *et al.* [59] have measured the X-ray absorption spectra for Mn and Ni in LNMO. The spectra of Mn correspond to on-site transitions of the form $2p^6 3d^n \rightarrow 2p^5 3d^{n+1}$ and present two groups of multiplets, namely, the $L_2(641 - 645$ eV) and $L_3(652 - 656$ eV) white line regions, split by the spin—orbit interaction of the Mn 2p core level. The line shape of the spectrum depends strongly on the multiplet structure given by the Mn 3d–3d and 2p–3d Coulomb and exchange interactions, as well as by the local crystal field effects and hybridization with the O 2p ligands. The comparison of Mn $L_{2,3}$ XAS spectra of LNMO with that of CaMnO$_3$, LaMn$_{0.5}$Ni$_{0.5}$O$_3$, and LaMnO$_3$ reveals that the Mn in LNMO is essentially in Mn^{4+} state. However, the line shape of L_2 edge of Ni XAS spectra of LNMO is very similar to that of NiO rather than to that of $RNiO_3$, where R = Pr and Nd. The presence of small satellites in the Ni L_2 edge is well understood in terms of a covalent ground state of mainly $Ni^{2+}(3d^8)$ character plus an anion-dependent fraction of the $3d^9L$ and $3d^9L^2$ configurations, where L stands for an anion (ligand) hole. In thin films of LNMO, the oxidation states of Ni ions is Ni^{2+}, consistent with the observation of the Mn^{4+} state in the Mn $L_{2,3}$ XAS spectra, that is, fulfilling the charge balance requirement. Kitamura *et al.* [60] have also studied the XAS of LNMO on (100) STO and found that the Ni^{2+} and Mn^{4+} states are dominant, which preferably leads to $Ni^{2+}–O–Mn^{4+}$ ferromagnetic superexchange interaction in LNMO. The XAS spectra of various thin films of LNMO reveal that Ni^{2+} and Mn^{4+} states are dominant regardless of the oxygen pressure and substrate temperature during thin-film growth [36].

Gray *et al.* [61] have compared the Cr and Fe $L_{3,2}$ XAS spectra of the artificial La_2FeCrO_6 (LFCO) double perovskite that is grown as a single unit cell $LaFeO_3/LaCrO_3$ superlattice on STO (111) to that of $Cr^{3+}(Cr_2O_3)$ and Fe^{3+} ($\alpha − Fe_2O_3$) standards. The Cr/Fe L edge begins around 570/700 eV and extends to 590/730 eV. On the basis of the analogous local electronic structure and strong resemblances in spectral features, the spectrum for the LFCO can be interpreted based on calculated spectra of Cr_2O_3. The ground state of Cr is high-spin t_{2g}^3 with 3+ valence. Cr_2O_3 has a crystal field splitting parameter 10 Dq = 2.0 eV and a small distortion from octahedral symmetry. In the same manner, the Fe L edges of the LFCO and $\alpha − Fe_2O_3$ can be compared. Fe ions are in the high-spin state $t_{2g}^3 e_g^2$ and have a valency of 3+. The crystal field splitting parameter for $\alpha − Fe_2O_3$ is 10 Dq = 1.45 eV and the ligand-field symmetry is O_h.

Di Trolio *et al.* [30] have studied the Fe 2p X-ray absorption spectrum of SFMO films grown on STO substrate. The X-ray absorption spectrum of SFMO shows the $L_3(\sim705 − 715)$ eV and $L_2(\sim715 − 730)$ eV edges, corresponding to the dipole transitions from the core Fe $2p_{3/2}$ and $2p_{1/2}$ levels, respectively, to the empty Fe 3d states. In general, for the iron oxides, the L_3 absorption edge for Fe^{2+} ions in the octahedral symmetry consists of a main peak at lower energy accompanied by a weaker peak at higher energy, whereas for Fe^{3+} ions, the intensity ratio between the two peaks is reversed. So the authors have considered the L_3 edge line shape as being representative of the Fe ion valence state and compare the Fe 2p as well as Mo 3p spectra with the corresponding spectra of polycrystalline SFMO. The Fe 2p as well as Mo 3p spectra of SFMO thin films closely reproduce the corresponding spectra of polycrystalline SFMO. The L_3 edge spectrum of the SFMO grown at 6 J cm^{-2} laser fluence exhibits an altered ratio between the low-energy (~708 eV) and high-energy (~710 eV) peaks, wherein the results are 20% lower than the SFMO grown at 1.6 J cm^{-2} laser fluence. The relative increase of the spectral intensity in correspondence with the high-energy peak originates from Fe ions in a chemical environment different from SFMO. Such ions can be related to metastable surface phases that produce iron oxides when exposed to the atmosphere and/or to the formation of spurious phases diluted in the perovskite matrix, as was reported earlier.

2.4.3
X-Ray Photoemission Spectroscopy

X-ray photoemission spectroscopy (XPS) can measure composition and valence state of elements near the surface for complex oxide films. Because of high-surface sensitivity of the XPS technique, one must remove any surface contamination, and extreme care should be taken to avoid lattice disruption that alters the film stoichiometry and charge state. So the cleavage of bulk films, and ion sputtering to remove multiple surface layers require *in situ* analysis following the growth and without any air exposure. Cleaved films are difficult to measure by all except for microscopic techniques because their cross-sectional thicknesses are typically well below a micrometer. Furthermore, they also require cleavage of the substrate, unless the films can delaminate, for example, by *in situ* bending. Ar^+-ion sputtering can remove surface contamination and probe the outermost layers of these oxides. However, the sputtering process can disrupt the lattice, changing the stoichiometry and the atomic charge states.

X-ray photoemission studies have been conducted on the surfaces of thin films of various multifunctional double perovskite oxides. As an example, the X-ray photoemission spectra of the valence band and 2p core level of Ni and Mn for LNMO films [23], grown on STO at 200 and 400 mTorr O_2 are shown in Figure 2.5. The entire valence band region exhibits the characteristics of extensively hybridized Mn 3d–O 2p and Ni 3d–O 2p orbitals. The 2p core-level spectra of the LNMO film grown at 200 mTorr oxygen pressure shows a shift in the binding energy of the Mn $2p_{3/2}$ and Ni ($2p_{3/2}$ and $2p_{1/2}$) peaks relative to the film grown at

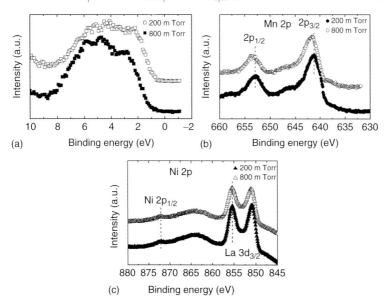

Figure 2.5 XPS measurement of La_2NiMnO_6 thin films grown under 200 and 800 mTorr O_2: (a) valence band spectra, (b) Mn 2p XPS spectra, and (c) Ni 2p XPS spectra [23].

800 mTorr. There is a general trend that the core-level binding energy increases with increasing oxidation state of a given ion, provided that the ions are located in similar coordination environments. Thus, the shift in core-level binding energy indicates that the oxidation state of Mn in the sample grown under 800 mTorr oxygen is higher than that of the one grown under 200 mTorr. This suggests that charge disproportionation of the $Mn^{4+} - Ni^{2+} \rightarrow Mn^{3+} - Ni^{3+}$ type likely occurs with decreasing oxygen pressure during film growth. The binding energy of Ni $2p_{3/2}$ and $2p_{1/2}$ in the sample grown at 800 mTorr is found to be somewhat larger than that for the one grown at 200 mTorr, indicating that the oxidation state of Ni in the sample grown at 800 mTorr is lower than that in the one grown at 200 mTorr. The spin states of Ni in the two samples are in accordance with the spin states of Mn, which adjusts to ensure charge neutrality and oxygen stoichiometry.

The X-ray photoemission in the entire valence band region of the thin film of LCMO/STO exhibits the characteristics of extensively hybridized Co 3d–O 2p and Mn 3d–O 2p orbitals [62]. The core-level spectrum of Co 2p exhibits the $2p_{3/2}$ and $2p_{1/2}$ spin-orbit coupling at 780.01 and 795.68 eV. The comparison of peaks positions of Co in LCMO and $LaCoO_3$ core-level spectrum suggests that the oxidation state of Co ions in the LCMO film is 2+. However, the core-level Mn 3s spectrum shows two peaks at 88.51 and 82.72 eV. It is well known that the 3s core level of the Mn ion exhibits an exchange splitting, which results from the exchange interaction between the Mn 3s and Mn 3d electrons, and the magnitude of the splitting depends on the valence state of the Mn ions. The Mn 2p core-level spectrum exhibits $2p_{3/2}$ and $2p_{1/2}$ spin–orbit coupling at 641.94 and 652.95 eV, respectively. The positions of the $2p_{3/2}$ peak in bulk Mn_2O_3 and MnO_2 are at 641.3 and 642.2 eV, respectively. This suggests that the oxidation state of Mn in the LCMO film is likely 4+, considering that charge neutrality is consistent with the deduced 2+ Co valence.

The XPS spectra of Fe and Mn in BFMO reveal the usual $2p_{3/2}$ and $2p_{1/2}$ doublet arising from spin-orbital interaction [37]. No shoulder peak is observed around the Fe $2p_{3/2}$ peak in either sample. The binding energy of the Fe $2p_{3/2}$ level is 710.61 eV in BFMO. Moreover, a satellite peak is found about 8.0 eV above the Fe $2p_{3/2}$ principal peak. This satellite peak is considered to be characteristic of the oxidation state of Fe. Owing to different d orbital electron configurations during relaxation of the metal ions, Fe^{2+} and Fe^{3+} will show a satellite peak at 6 or 8 eV above their $2p_{3/2}$ principal peaks, respectively. The structure and similarity of the Fe 2p core-level spectra in BFMO films indicates that Fe is mainly in the 3+ valence state. On the other hand, in BFMO films, the Mn $2p_{3/2}$ principal peak has a binding energy of 642.2 eV. A shoulder peak below this energy originates from Mn^{2+} and indicates the presence of multiple valence states of Mn. In stoichiometric BFMO, the average valence state of the B-site cation is 3+, and as Fe is mainly present as 3+, this suggests that Mn must exhibit Mn^{2+}, Mn^{3+}, and Mn^{4+} valence states [37].

The XPS spectra of the valence band and O 1s region of SFMO have been studied after sputtering the sample to different depth. Sputtering did not appear to change the overall intensity or the profile of the doublet of X-ray photoemission spectra of Sr 3d and Fe 2p regions of the SFMO, which indicates that the $SrMoO_4$ overlayer and the SFMO film have similar local chemical environments [42, 63]. The existence of this doublet Fe $2p_{3/2}$ and $2p_{1/2}$ only in the as-deposited SFMO film and that after 10 s of sputtering strongly suggests that the nanoparticles on the surface of the as-grown film, as seen in the SEM image, are primarily Fe_2O_3. After sputtering for longer than 160 s, the lowest binding energy feature can be attributed to metallic Fe (Fe^0), which is produced by reduction in the higher Fe oxidation states due to Ar ion sputtering [42, 63]. The lack of a metallic Fe peak for the as-grown film confirms the high quality of the studied PLD-grown SFMO film and the absence of any significant amount of Fe phase in the sample, in contrast to an earlier work that reported an additional metallic Fe peak for the as-deposited film [64]. Furthermore, the nearly identical intensity evolution with respect to sputtering time found for both the as-grown and postannealed films suggests that the cubic nanoparticles seen in the postannealed film also consist of Fe_2O_3, which has the Fe^{3+} oxidation state [42]. The XPS spectra of the Mo 3d region as a function of sputtering time for as-grown SFMO films before and after annealing have been studied, both of which depict evolution of rather complex sets of peaks [42, 63]. The similarity in O 1s, Sr 3d, Fe 2p, and Mo 3d depth profile of as-grown and postannealed films supports the formation of $SrMoO_4$ overlayers on the surfaces of both the as-grown and postannealed films. Furthermore, the intensity ratios of the 3d features of Mo^{4+}, Mo^{5+}, and Mo^{6+} for SFMO remain unchanged with sputtering depth after 160 s, supporting the model that the observed $SrMoO_4$ secondary phase is formed predominantly on the surface and not in the subgrain boundaries of the as-grown film [42].

Nechache *et al.* [31] have studied the XPS of thin films of Bi_2CrFeO_6 (BFCO) deposited at various oxygen pressure and substrate temperatures. The XPS analysis of the surface of the film deposited at 1.2×10^{-2} mbar and 600 °C reveals that the square-shaped outgrowths seen in SEM images are Bi-rich, Cr-deficient, and Fe-poor. Their overall chemical composition corresponds to that of $Bi_{7.8}Cr_{0.62}O_{12+x}$. The XPS analysis also shows that the oxidation state is 3+ for each cation. The Fe 2p core-level spectra of the films deposited at 680 °C under vacuum indicate the presence of Fe^{2+} and Fe^{3+} ions. The contribution of Fe^{2+} in the films is gradually reduced with increasing oxygen pressure. Comparing the data with results obtained on $BiFeO_3$ thin films, the authors have concluded that the film prepared at 1.2×10^{-2} mbar and 680 °C is nearly stoichiometric BFCO.

2.5
Physical Properties

2.5.1
Magnetization

The magnetization (*M*) of multifunctional double perovskite oxides depends strongly on the arrangement of the B-site ions. The most significant effect of B-site disorder is to reduce the net magnetization of the sample. One possibility is that the B-site disorder destroys the specific spin arrangement of *B'* and *B''* sublattices without any significant effect on the individual magnetic moments at these sites. This can be achieved by transforming the ferromagnetic coupling between $B'-O-B''$ to an antiferromagnetic coupling between $B'-O-B''$ driven by the superexchange. Alternatively, the magnetic moments at each individual site may decrease owing to the different chemical environments induced by disorder, without affecting the nature of the spin order within the B' and B'' sublattices. In reality, the situation may be a combination of both these effects, with a simultaneous reduction in the magnetic moment at different sites as well as a change in the nature of the magnetic coupling between different sites. The functionalities of the double perovskites related to magnetization have great potential for technical applications. Thus, it is very important to understand the fundamental magnetic properties of these oxides. Several techniques have been used to study the magnetization of multifunctional double perovskite oxide thin films.

2.5.1.1 Ferromagnetic Resonance

Ferromagnetic resonance (FMR) is a very useful technique for studying magnetic thin films because the FMR characteristics depend sensitively on the magnetic properties of the film that are influenced by structure and morphology. So far, this technique has not been used to study thin films of the double perovskites, except for LNMO. Kazan *et al.* [65] have studied the magnetization of LNMO thin films using FMR technique. Two distinct FMR modes are observed in both in-plane and out-of-plane geometries of the thin films grown on STO and NGO. The angular dependence of resonance field shows two different four-fold in-plane magnetic anisotropies. The easy axis of the first mode lies along the (110) axis of STO, while the easy axis of the second mode lies along the (100) axis of STO. These unusual angular dependences reveal that the two main FMR modes exhibit 'antiphase' behavior, that is the minimal resonance field of the first mode corresponds to the maximal field for the second one and vice versa. This may imply the coexistence of two magnetic phases or at least two magnetic orderings with different effective magnetization in LNMO thin films. The coexistence of two different magnetic ordering has indeed been observed in polycrystalline samples of $LaNi_{0.5}Mn_{0.5}O_3$ [66]. The crystal structure of the two phases is found to be different: orthorhombic for the low-T_C phase and rhombohedral for the high-T_C phase. The presence of two structural phases has also been observed in thin films of LNMO [23]. The high-symmetry crystal phase (*I*-type) corresponds to the crystal structure of the low oxygen pressure sample. The second phase is a lower symmetry state (*P*-type) that can have B-site ordering. The LNMO grown under optimal conditions is found to be rhombohedral ($R\bar{3}$) at high temperatures and it transforms to a monoclinic structure at low temperatures. These two structures typically coexist at room temperature and they have been confirmed using Raman spectroscopy [55]. Two different FMR resonance modes with two-fold in-plane magnetic anisotropy are observed in LNMO/NGO, similar to the LNMO/STO film. The easy axes of the first FMR mode lies along the (010) direction, while the second FMR mode lies along the (100) direction of the film. A possible reason for the differences in the

FMR spectra of the samples on different substrates may be because of the presence of varying epitaxial strain attributable to the lattice–substrate mismatch.

2.5.1.2 Kerr Spectroscopy

The magneto-optical Kerr effect (MOKE) in ferromagnets is an important phenomenon and has been extensively utilized to study distinctive properties related to low-dimensional magnetism. This technique has not yet been much exploited to study the magnetization of thin films of double perovskite oxides. Rudiger *et al.* [67] have measured the out-of-plane Kerr hysteresis loop of (001)-oriented 150 nm thick SFMO film grown on (001)-oriented LAO. During measurement, the SFMO film is magnetized to saturation with an out-of-plane field of 1.5 T in both field directions to eliminate birefringence effects. The Kerr hysteresis loop at room temperature shows significant reduction in the saturation moment (M_S). However, magnetic saturation is essentially achieved with 1.5 T magnetic field at 10 K for photon energies of 1.4 and 4.0 eV. The authors have interpreted the polar Kerr rotation spectrum in terms of possible "charge transfer" transitions that are derived from the spin-polarized density of states. The overall intrinsic polar Kerr rotation is <0.045° in the photon energy range investigated. This value is significantly smaller than that observed for other half-metallic ferromagnets.

2.5.1.3 Superconducting Quantum Interference Device

Superconducting quantum interference device (SQUID) has been widely used to measure the magnetization of various double perovskite oxide thin films. For SFMO thin films, a wide range of M_S value is reported [30, 32, 68]. Even for SFMO films grown on the same substrate, there are variations in the reported M_S values. For example, Westerburg *et al.* [34] have obtained an M_S value of 4 μ_B fu^{-1} for SFMO thin films grown on STO under Ar atmosphere. Di Trolio *et al.* [29, 30] have reported M_S of 3.4 μ_B fu^{-1} for films prepared by PLD on STO substrate at low laser fluence, whereas Shinde *et al.* [32] have reported M_S of 3.28 μ_B fu^{-1} The different values of M_S for SFMO grown on STO under different conditions suggest that the magnetization of SFMO is highly sensitive to the deposition parameters. In addition, postannealing treatment of SFMO thin films grown on MgO reduces the M_S from 3.4 to 1.4 μ_B fu^{-1} with negligible change in the observed coercive field (H_C) (~0.56 kOe) [42]. This indicates that postannealing in O_2 has no significant effect on the grain formation in the bulk and suggests that the development of $SrMoO_4$ overlayer and antiferromagnetic α-Fe_2O_3 nanoparticles during the postannealing in O_2 are responsible for the observed reduction in M_S. On the other hand, the magnetization in SFMO films does not saturate even at low temperatures at 5 T field, which is attributed to the ferromagnetic clusters forming a spin glass as a result of random intergrain coupling [49]. However, the amount of antisite defects calculated from the M_S of films on STO and MgO is 5 and 16%, respectively. Deepak *et al.* [48] have calculated the concentration of antisite defects from the M_S at 5 K for SFMO films grown on STO, LAO, and MgO substrates. The M_S observed for thin film of SFMO on STO is the highest, while it is the lowest for the film on MgO. The M_S trend is in agreement with the variations in the lattice–substrate mismatch. The lattice distortion and sample inhomogeneities decrease the degree of ferromagnetic long-range order, causing a decrease in magnetization.

Sánchez *et al.* [21] have studied the variation of M_S of SFMO films with changes in the lattice parameter c, which can be taken as a measure of the strain level in the film. The study reveals that higher M_S values are always linked to lower values of the out-of-plane lattice parameter, that is, with the SFMO phase exhibiting a substantial relaxation of the substrate-induced strain. The observed higher value of magnetization for the strain-relaxed SFMO phase indicates that the Fe/Mo-ordered phase is incompatible with a strong deformation of the lattice. Therefore, a full strain relaxation is not necessarily linked to full magnetization of SFMO phase. Since the more strained layers would contribute to the film thickness but only little to the ferromagnetic moment, thus reducing the average magnetization, this may be associated with an inhomogeneous distribution of antisite disorder, with the highly disordered region present at the substrate interface. A marked decrease in the ordering temperature is also observed with elongation of the c axis. The M_S of SCRO thin film grown on STO also depends strongly on the deposition parameters [27, 46]. The magnetic hysteresis loop of SCRO film grown on STO shows M_S of $\sim 0.8\,\mu_B$ fu^{-1} with strong magnetic anisotropy and with antisite defects of <20% [46]. Czeschka *et al.* [27] have studied the magnetization of thin films of SCRO grown on $BaTiO_3$ (BTO). For the tetragonal and the rhombohedral phases of BTO, the magnetization values are essentially identical, whereas in the orthorhombic phase, a reduction is observed. This indicates that the reduced M_S might be due to the highly twinned BTO crystal in the orthorhombic phase. In the tetragonal phase of BTO, the SCRO film shows negligible in-plane magnetic anisotropy, as also observed in SCRO on STO [46]. In the orthorhombic phase of BTO, the situation is completely different: the H_C of the (100) and (110) loops differ by more than 1 T. In particular, the hysteresis loops at 200 K reveal an H_C of 0.87 T for the (100) orientation of the external field and a much larger value of 2.3 T for the (110) orientation. Thus, a huge in-plane magnetic anisotropy is present. On cooling into the rhombohedral phase of BTO, the situation changes again: the H_C and the magnetizations at 7 T for both in-plane hysteresis along (100) and (110) as well as the (001) loops are essentially identical. In other words, the magnetic behavior appears isotropic, with no evidence of shape or crystalline anisotropy contributions. This is remarkable, as it suggests that giant strain-induced anisotropies of more than 1 T are effective in the SCRO film, compensating the demagnetization field.

At 5 K, the magnetization of BFMO grown on STO is larger than that measured at room temperature but does not show any signature of saturation at 1 T field. The magnetization at 0.9 T is estimated to be 0.03 μ_B/B-site, which is far from the value expected if the Fe and Mn ions are ordered in the B sites, indicating that most of the B-site cations are disordered in this sample [37]. Thin films of LCMO/STO grown at 1 and 10 mTorr oxygen pressure show a single magnetic transition at 80 K. However, two ferromagnetic transitions at around 80 and 230 K are observed for the LCMO/STO film grown at oxygen pressure higher than 100 mTorr. On the other hand, field-cooled and zero-field-cooled temperature-dependent magnetization of all samples show spin glass-like behavior [35]. Thin films of LCMO/STO have also been reported to show double switching in the hysteresis loop over the entire ferromagnetic regime [69]. Singh *et al.* [69] have ruled out the possibility of one of the loops resulting from B-site cationic Co^{3+}/Mn^{3+} disorder structure or antisite defects, on the basis of the fact that no clear signature of a second FM phase transition at low temperature has been observed in their samples. Instead, they suggest that the double switching in the hysteresis loop arises from the presence of two magnetic domains possessing different coercivities. Such type of double switching in the magnetic hysteresis loops is also observed for films of LCMO/STO grown at 1, 100, 200, 400, and 600 mTorr O_2 pressure [35]. The M_S at first increases with increasing O_2 pressure, with the highest M_S obtained in the film grown at 200 mTorr and then it decreases slightly with further increasing O_2 pressure. Meanwhile, the H_C also shows a similar dependence on oxygen pressure. The maximum value of M_S measured at 10 K for the sample grown under 200 mTorr O_2 is 5.7 μ_B fu^{-1}, which is close to the theoretical spin-only value of 6 μ_B fu^{-1} for ferromagnetic ordering of Co^{2+} and Mn^{4+} ions. Thus, the coexistence of two ferromagnetic phase transitions can be directly associated with the presence of O_2 vacancies. As is to be expected, a larger concentration of O_2 vacancies is created in the films grown at low O_2 partial pressures (1 and 10 mTorr).

The influence of the variation in O_2 pressure on the Curie temperature is essentially negligible for LNMO films grown on LAO substrates [23]. However, M_S at 10 K decreases from 4.63 μ_B fu^{-1} as the O_2 pressure during growth is decreased from 800 mTorr. The lower value of M_S when compared to the theoretical spin-only value of 5 μ_B fu^{-1} indicates that the LNMO film grown even at 800 mTorr is not fully ordered. So, the amount of disorder increases as the O_2 pressure during the growth of LNMO decreases. Oxygen vacancies induce transfer of an electron from the e_g band of Ni^{2+} ions to neighboring Mn^{4+} ions, creating Ni^{3+} and Mn^{3+} Jahn–Teller ions, which cause local Jahn–Teller distortions of the octahedral sites. The Jahn–Teller distortions inhibit long-range ordering of Ni and Mn on different lattice sites, stabilizing vibronic FM Mn^{3+}–O–Mn^{3+}, Ni^{3+}–O–Mn^{3+}, and Ni^{3+}–O–Ni^{3+} superexchange interactions in an atomically disordered volume. As pointed by Dass and Goodenough [70], such vibronic superexchange provides a smaller stabilization when compared to the static ferromagnetic Mn^{4+}–O–Ni^{2+} superexchange, thus the magnetization is decreased.

For certain deposition conditions of LNMO grown on (111)-oriented STO, the films are polycrystalline and indicate the presence of impurity phases [58]. Some of these films exhibit single T_C at ~138 or ~275 K, while the other films show two T_C at ~140 and ~295 K. The observed T_C and M_S of LNMO on (111) STO grown under different conditions are explained based on the distribution of the cations present in LNMO. A broad T_C is observed for the LNMO thin film grown on (001)-oriented STO [59]. The magnetization is found to be proportional to $(T_C - T)^{0.31}$, where the exponent is close to that for a Heisenberg ferromagnet with short-range coupling. The magnetization of thin films of LNMO grown at 770 °C and 300 mTorr O_2 on (001)-oriented STO, LAO, and LSAT substrates exhibit T_C ~280 K despite slight changes in the structure [71]. The M_S at 10 K of the films deposited on STO and

LSAT substrates is about 4.6 μ_B fu^{-1}, which is close to the expected spin-only value of 5 μ_B for ferromagnetic ordering of Ni^{2+} ($S = 1$) and Mn^{4+} ($S = 3/2$) ions. The M_S value at 10 K of the film on LAO is slightly lower ($\sim$4.0 μ_B) than that of the films on STO and LSAT. This may be due to the presence of nonmagnetic regions near the defects in the crystal structure caused by lattice mismatch.

2.5.1.4 X-Ray Magnetic Circular Dichroism

The advantage of X-ray magnetic circular dichroism (XMCD) is its element specificity and the possibility to extract spin and orbital magnetic moments using the magneto-optical sum rules. In order to understand the nature of magnetic exchange in double perovskites, XMCD has been used to determine the value of the magnetic moment at the B' and B'' sites of double perovskites. The Re 5d magnetization curve in SCRO at $T \approx 10$ K measured by XMCD at the Re L_2 absorption edge is shown in Figure 2.6 [72]. A remarkably large H_C of about 1.27 T is observed. The XMCD data is in good agreement with the macroscopic magnetization measured using SQUID at 5 K (see inset of Figure 2.6). The M_S of 0.89 μ_B fu^{-1} is quite close to the value of 1 μ_B expected from a simple ionic picture, where the magnetic moments of Cr^{3+} and Re^{5+} are coupled antiferromagnetically. Guo *et al.* [59] have studied the XMCD of LNMO thin films. The XMCD of LNMO is largely negative at both the Ni and the Mn L_3 edges, indicating that the Ni^{2+} and Mn^{4+} ions are aligned ferromagnetically with Mn^{4+} − O − Ni^{2+} superexchange interactions. The XMCD intensity decreases as the temperature increases without any appreciable change in the spectral line shape, corresponding to a decrease in the magnetization with increasing temperature. By using the XMCD orbital sum rule and spin sum rule, the orbital magnetic moment and the spin magnetic moment values of the Mn 3d states in the ground state are estimated to be 2.6598 and 0.1055 μ_B/atom, respectively. Thus, the ratio of the orbital magnetic moment to the spin magnetic

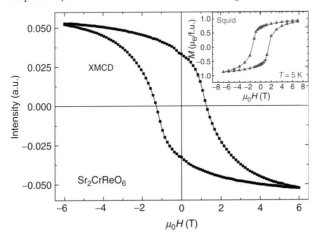

Figure 2.6 XMCD signal of Sr_2CrReO_6 (SCRO) film for various magnetic field measured at the ReL$_2$ edge at $\sim$10 K. The inset shows the field-dependent magnetization of the same sample measured by SQUID at 5 K [72].

moment of the Mn 3d state in the ground state is 0.039 at 100 K. This means that the orbital moment for the Mn^{4+} ions is mostly quenched. Indeed, for the $3d^3$ configuration in Mn^{4+} compounds, the majority t_{2g} shell is fully occupied without any orbital degree of freedom, and thus a practically quenched orbital moment is to be expected. The estimated value of orbital magnetic moment and the spin magnetic moment of the Ni 3d state in the ground state are 0.5215 and 0.2188 μ_B atom^{-1}. However, the ratio of orbital magnetic moment to spin magnetic moment of the Ni 3d states is 0.246, which is unexpectedly large for the half-filled $d^8 (t_{2g}^6 e_g^2)$ system. It should be noted that the magnetic moments of Ni obtained from the XMCD rules are rather ambiguous as the La M edge provides low XMCD signal. Moreover, the empty La 4f states can also be slightly polarized in principle. The value of the total magnetic moment of LNMO thin film at 100 K, estimated from the XMCD spectra, is 3.05 μ_B fu^{-1}, which is less than the value of M_S (4.0 μ_B fu^{-1}) measured at 100 K using a SQUID. One possible reason for this is the residual moment at O site, which is not included in the calculation. Moreover, by applying a 0.6 T magnetic field along 30° angle with the b axis of the samples can lead to an additional reduction owing to anisotropy. Nevertheless, the T dependence of total magnetic moment estimated from the XMCD spectra agrees nicely with that of the M_S observed using a SQUID.

Gray *et al.* [61] have studied the XMCD of an artificial LFCO double perovskite that is grown as a single unit cell $LaFeO_3/LaCrO_3$ superlattice on (111)-oriented STO substrate. Comparing both the Cr and Fe 2p XMCD, the agreement between the polarities of the two spectra indicates the parallel alignment of magnetic moments on Cr and Fe ions in the presence of an external magnetic field. Both Cr and Fe show a maximum asymmetry of approximately − 2.6% at the L_3 edges. This is unexpected because, for a typical ferromagnet with an average magnetic moment of 3–5 μ_B per transition-metal site, the anticipated XMCD is 30–40%. The magnitude of the magnetic moments are estimated to be 0.31 μ_B Cr^{-1} and 0.21 μ_B Fe^{-1} for the superlattice in a magnetic field of 5 T and, when the magnetic field is lowered down to 0.1 T, there is no measurable XMCD signal at either edge, indicating that the moment at remanence is quite small. In addition to the Goodenough–Kanamori rule being ambiguous for $d^3 - d^5$ superexchange, the authors attribute the observed lower magnetic moment of LFCO to other possibilities, including (i) the growth of disordered thin film rather than the rock salt perovskite structure and (ii) the system consisting of patches of G-type antiferromagnetic $LaFeO_3$ and $LaCrO_3$, which will have no moment at zero field.

2.5.2
Transport Properties

The transport properties in single crystals and bulk ceramics of the double perovskite compounds can be significantly different from those in thin films owing to the substrate-induced stress. The influence of substrate-induced stress and deposition parameters on the transport properties of several double perovskite systems have been studied in some detail.

2.5.2.1 **DC Transport**

The temperature-dependent resistivity ($R(T)$) of epitaxial thin films of half-metallic SFMO exhibit a transition from a semiconductor-like to metal-like behavior [15, 21, 28, 32–34, 43]. In contrast, films deposited on yttria-stabilized zirconia (YSZ) substrate are polycrystalline and semiconducting down to the lowest measurement temperature (~10 K) [28]. A semiconductor-like behavior is also observed over the entire temperature range for SFMO films grown at a higher laser fluence [29]. The semiconducting temperature dependence of SFMO films is due to strong electron scattering occurring at the grain boundaries. Such grain boundaries can originate either from spurious phases or from the high degree of cation disorder at the Fe and Mo sites. Moreover, SFMO films exhibit lower resistivity for higher degree of cationic ordering as scattering processes at the AFM antisite disordered patches are reduced. Samples for which the presence of spurious insulating $SrMoO_4$ is detected show a strong increase in the residual resistivity. Shinde *et al.* [32] have studied the aging effect on the transport properties of SFMO/STO thin films grown in vacuum, argon, and oxygen. The aging effect is not observed in the $R(T)$ of SFMO/STO film grown in argon, while SFMO/STO films grown in vacuum strongly reflect an aging effect. The $R(T)$ of the latter changes with time, eventually attaining a stable value, but with undesirable properties. This indicates that SFMO/STO films grown in vacuum are oxygen deficient and have some type of defect-related instability.

There are only a few reports on the study of $R(T)$ of thin films of other double perovskite systems. The $R(T)$ of SCWO thin films grown on (110) NGO, (001) $MgAl_2O_4$, and (001) LAO substrates show insulator-like behavior, while films grown on (001) or (111) STO substrates exhibit metal-like behavior [39]. The observed $R(T)$ of SCWO thin films is attributed to the large lattice mismatch between the film and substrate. However, possible contribution from the conducting oxygen-deficient STO substrate cannot be ruled out. An interesting effect of structural transition is observed in the $R(T)$ of SCRO thin films grown on BTO substrates [27]. On cooling SCRO/BTO film from 300 K, the resistivity initially increases slightly and then suddenly drops at 285 K (i.e., the temperature at which the BTO crystal becomes orthorhombic). Between 285 and 191 K, the resistivity again increases continuously until it jumps to higher values at 191 K, when the BTO transforms into the rhombohedral phase. Further cooling leads to a steadily increasing resistivity. The authors have attributed the observed resistance jumps to strain-induced changes in the orbital configuration of SCRO.

2.5.2.2 **DC Transport in Magnetic Field**

The change in resistance in the presence of a magnetic field (i.e., MR) has only been reported for SFMO thin films. The MR of SFMO thin films depend strongly on the substrate [48], deposition parameter [21, 33], and postannealing conditions [32]. The MR at 70 K for the SFMO films grown on STO, LAO, and MgO substrates is maximum for the film on MgO, while it is minimum for the film on STO. This variation of MR is explained based on possible differences in the antisite defect density [48]. The antisite defects introduce antiferromagnetic ordering between the B-site ions, leading to a decrease in the spin-dependent scattering occurring at the grain boundaries or at magnetic domain boundaries with the application of a magnetic field, and hence leads to higher MR. In another study, the MR at 10 K of SFMO/STO grown at a faster deposition rate (25 Å min^{-1}) has been found to be higher (-7.3% at 2 T) than that of a film grown at a lower deposition rate (15 Å min^{-1}) [21]. This is ascribed to the higher density of structural defects in the film, which act as physical boundaries. The contribution of structural defects is usually very small in epitaxial films grown at lower deposition rates, which are essentially free of structural defects. Muduli *et al.* [28] have observed a relatively large MR of ~13% under 0.3 T field for polycrystalline films when compared to a very small ~2.5% MR in epitaxial films, which is presumably due to tunneling through grain boundaries. Epitaxial thin films of SFMO exhibit negative and hysteretic MR at low fields [33, 43]. It is more reasonable to associate the low-field MR of epitaxial thin films with conduction across antisite boundaries. However, no grain boundary MR is observed in as-deposited polycrystalline films of SFMO grown on highly polished polycrystalline STO [32]. Nevertheless, after annealing these films at $475 - 500\ °C$ in a $Ar + O_2$ mixture, a signature of low-field MR is observed. These results indicate that the low-field MR is not directly related to the cation disorder in SFMO films, but likely related to the presence of insulating regions, and the low-field MR originates from spin-dependent tunneling across these regions.

2.5.2.3 AC Transport

Besides the half-metallic characteristics for some of the double perovskites, a number of others exhibit semiconducting or insulating behavior. Very few of the latter have been studied using the AC-impedance technique. In this technique, the impedance (Z) and phase angle (θ) are measured at low AC voltage as a function of temperature and frequency. To study the nature of the capacitive response, in most cases, the film is simply modeled as a parallel network of a resistor (R) and an effective capacitor (C). The extracted capacitance is proportional to the frequency-dependent complex dielectric function $\varepsilon(\omega)$. The dielectric function can be expressed as the sum of the frequency-dependent real $\varepsilon_1(\omega)$ and imaginary $\varepsilon_2(\omega)$ parts, $\varepsilon(\omega) = \varepsilon_1(\omega) + i\varepsilon_2(\omega)$. The dissipation factor tanδ is then given by $\varepsilon_2(\omega)/\varepsilon_1(\omega)$, where δ is the phase difference between the applied electric field and the induced current. The tanδ and $\varepsilon_1(\omega)$ values are calculated from the effective C extracted from the Z and θ measurements as a function of frequency and/or temperature for LCMO [73], LNMO [19], and BNMO [18, 40].

The $\varepsilon_1(\omega)$ at 10 kHz for a 345 Å thick film of BNMO grown on 1.0 wt% Nb-doped STO (STNO) is somewhat higher [40] than that of the reported bulk value (~200) [11]. A complete analysis for the determination of the actual film dielectric constant requires separate inclusion of the impedance contributions from the interfaces of BNMO and electrodes (top and bottom). An intriguing feature observed for BNMO films reported in Ref. [40] is the rapid change of the effective dielectric constant (measured at 100 kHz) at the ferromagnetic transition temperature near 130 K when compared to the very small anomaly at T_C observed in Ref. [18]. Because of the leakage characteristics of thin BNMO films, the ferroelectric properties are not measured in Ref. [40]. However, the ferroelectric behavior of a 1000 Å thick BNMO thin film has been confirmed from $P-E$ hysteresis loop measurement at 7 K in Ref. [18], and is shown in Figure 2.7. Although the slightly distorted hysteresis loop observed includes the effect of leakage current, the saturated polarization is about 5 µC cm^{-2} above 80 kV cm^{-1}. Owing to an increase in the leakage current at elevated temperatures, the ferroelectric transition temperature of BNMO films has not been directly confirmed from ferroelectric hysteresis and dielectric measurements.

A qualitatively similar temperature dependence of $\varepsilon_1(\omega)$ is observed for thin films of LCMO and LNMO. In the case of LNMO, the drop-off temperature for $\varepsilon_1(\omega)$ with increasing frequency shifts to higher temperatures. This behavior is also reflected in the shift of the dissipation factor peak. Overall, the observed variations in $\varepsilon_1(\omega)$ and the peak in $\tan\delta(\omega)$ are typical signatures of relaxation behavior [19]. As the sample is cooled below room temperature, the decrease in thermal energy results in a lowering of the frequency dispersion, which becomes slower at temperatures below the peak temperature of the loss tangent curve. Ultimately, all dipolar motion freezes at a sufficiently low temperature (below 80 K) and the dispersion vanishes. The negligibly small variation in $\varepsilon_1(\omega)$ below 80 K may be due to a space-charge polarization at the LNMO/STNO interface. The frequency-dependent $\varepsilon_1(\omega)$ for an LNMO film at various temperatures is shown in Figure 2.8. At higher temperatures, the dielectric constant at low frequencies remains essentially constant up to a certain cutoff frequency, at which the oscillation period matches the intrinsic timescale of the system – the so-called relaxation time (τ) – and then rapidly decreases at higher frequencies. The Debye model has been used to model $\varepsilon_1(\omega)$ in terms of τ [74]. The fit to the dielectric constant of LNMO using a Debye relaxation behavior is indicated by the solid line in Figure 2.8. Although there is some ambiguity in τ at temperatures below 80 K, the model fits well with the measured $\varepsilon_1(\omega)$ of LNMO. The relaxation rate obtained from these fits drops rapidly with increasing temperature. This temperature-dependent relaxation rate indicates an increasing dipole density and a faster polarization process [19].

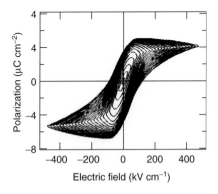

Figure 2.7 Ferroelectric hysteresis loop of a Bi_2NiMnO_6 (BNMO) thin film measured at 7 K and 5 kHz frequency [18].

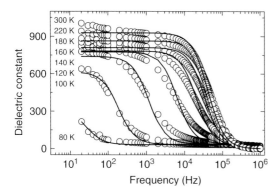

Figure 2.8 The frequency dependence of the dielectric constant ε_1 of a 850 Å thick La$_2$NiMnO$_6$ (LNMO) grown on 0.5 wt% Nb-doped SrTiO$_3$ at several temperatures. The solid lines are the fit to the Debye model [19].

2.5.2.4 AC Transport in Magnetic Field

Detailed studies of AC transport in double perovskite thin films in the presence of a magnetic field are limited. The temperature dependence of $\varepsilon_1(\omega)$ in the presence of a magnetic field for LCMO, LNMO, and BNMO thin films is qualitatively similar to that of their zero-field behavior [19, 40, 75]. Using the same parallel network of R and C with a complex dielectric function, the values of $\varepsilon_1(\omega)$ have been calculated for different magnetic fields from Z and θ measured at various frequencies and temperatures. The peak shift with magnetic field in the temperature-dependent $\tan\delta(\omega)$ of LNMO demonstrates the influence of the field on the relaxation of the dipole and charge particle transitions and indicates that electric dipole fluctuations are influenced by magnetic ordering and fluctuations [19].

The changes in $\varepsilon_1(\omega)$ with magnetic field [the magnetodielectric constant (MDC)] for various temperatures have been calculated using the relation MDC $= [\varepsilon_1(\omega,H) - \varepsilon_1(\omega,0)]/\varepsilon_1(\omega,0)$. The temperature-dependent MDC(MDC(T)) of LNMO grown on (001)-oriented 0.5 wt% Nb-doped STO and on SrRuO$_3$/SrTiO$_3$ is shown in Figure 2.9. The qualitative behavior of MDC(T) for both films is similar. The MDC at 0.5 T increases as the film is heated above 10 K, exhibiting a peak and then again deceases to a negligibly small value. The peak temperature shifts slightly to higher temperatures when the magnetic field is increased from 0.5 to 7 T, with a corresponding increase in the MDC. A similar shift of the peak temperature is observed in measurements with increasing frequency, as measured for both films. Although the qualitative behavior of MDC(T) for both the films is similar, the small difference in peak temperature and significant differences in MDC at 10 kHz are attributed to differences of the bottom and top electrodes in the two samples. While the MDC decreases with increasing excitation frequency, it increases with increasing magnetic field. Although the MDC ($\sim$11%) at 1 T for the 850 Å thick film of LNMO is similar to that observed in the bulk, the observed peak temperature is significantly lower than in the bulk. In addition, the LNMO film does not show an anomaly in $\varepsilon_1(\omega,H)$ in the vicinity of T_C as observed in the bulk($\sim$270 K for

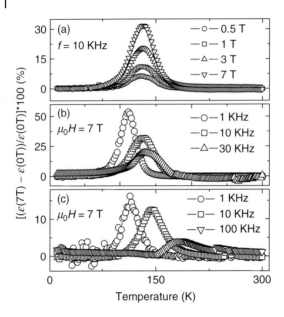

Figure 2.9 Temperature-dependent magnetodielectric constant of an 850 Å thick La$_2$NiMnO$_6$ (LNMO) grown on 0.5 wt% Nb-doped SrTiO$_3$ at several (a) magnetic fields and (b) frequencies. (c) Temperature-dependent magnetodielectric constant of a (3400 Å)La$_2$NiMnO$_6$/ (85 Å)SrRuO$_3$/SrTiO$_3$ sample at different frequencies [19].

the films when compared to the bulk value of ∼280 K) [10]. The differences in the MDC, peak temperature, and $\varepsilon_1(\omega,H)$ between the thin film and bulk LNMO are likely due to fundamentally different origin of the MDC in the two cases.

The log(τ) of LNMO varies linearly with inverse temperature, and its Arrhenius fit yields relaxation time (τ_0) and activation energy (U) values of 66 ns and 31.8 meV (367 K), respectively. The relaxation time and, hence, the activation energy at 7 T field is lower than its zero-field value ($\tau_0 = 72$ ns and $U = 30.9$ meV), indicating the presence of coupling between the spin and dipolar ordering, which is responsible for the observed MDC(T) in LNMO. However, as the MDC is not proportional to M^2 (M = magnetization) over the entire temperature range below T_C, phonon–spin coupling can be ruled out [75] as the origin of the magnetoelectric coupling, unlike the case of LCMO [73].

The dielectric constant and loss tangent at 100 kHz as a function of magnetic field have been determined, which are significantly faster than the characteristic resistor–capacitor (RC) time, at a temperature of 150 K, around which the RC time (∼600 Hz) changes negligibly. The "magnetoloss" is defined in terms of the loss tangent, analogous to the MDC. Both the MDC and magnetoloss increase with increasing magnetic field, suggesting that the variation is not due to the MR of LNMO or the LNMO–electrode interfaces [19]. Similar variation of magnetoloss and MDC with magnetic field is also observed for BNMO [40]. Furthermore, these values are independent of the orientation of magnetic field with respect to the out-of-plane excitation current, consistent with the fact that the observed magnetodielectric effect does not result from the accumulated space charge at the interface and does not originate from the interaction of the carriers with the magnetic field itself.

The exchange magneto-electronic interaction is of the form $\gamma P^2 M^2$, where P and M are the polarization and magnetization, respectively [76]. The coupling constant γ is typically a function of temperature. The exchange magneto-electronic interaction will result in MDC below the magnetic transition temperature. Since the ferroelectric transition temperature ($\sim$485 K) of BNMO is much larger than the magnetic transition temperature ($\sim$130 K), the dependence of the electric parameter on temperature near the magnetic transition temperature can be neglected. Thus, the MDC will simply be proportional to the square of the magnetization. At 150 K, the field-dependent MDC of BNMO overlaps well with the field-dependent M^2. The overlap of the field-dependent MDC and M^2 data indicates that the exchange interaction between electric dipole ordering and fluctuations and magnetic ordering and fluctuations is likely responsible for the observed MDC of BNMO [40].

2.6
Applications of Multifunctional Oxides

There is increasing interest in the development and characterization of multifunctional materials because they exhibit rich physical and chemical properties and offer exciting opportunities for applications. Some of the multifunctional double perovskites display mixed electronic and ionic conductivities. For example, the double perovskite Sr_2MgMoO_6 is a promising mixed oxide-ion/electron conductor with an excellent tolerance to sulfur, which can be used for direct electrochemical oxidation of dry methane at 800 $^\circ$C [12]. Moreover, La-doped Sr_2MgMoO_6 performs somewhat better on natural gas [14]. Thus these double perovskite oxides are potential candidates for solid oxide fuel cell (SOFC) devices, which can be used for either stationary or mobile generation of electrical energy from a gaseous fuel. The ordered double perovskites exhibiting half-metallic or insulating behavior and a high Curie temperature have attracted much attention because of their rich physics and potential applications. In this context, multifunctional materials that combine two or more ferroic functionalities are promising for applications in fields such as spintronics and nonvolatile data storage [77]. As recently demonstrated [78], ferromagnetic–ferroelectric functionalities can be advantageously used to encode information in terms of both electric polarization and magnetization, giving rise to a four-logic state memory. The coupling between magnetic and electrical properties leads to additional versatility for related devices, such as electric-field-controlled magnetic data storage.

The identification of suitable barrier materials for half-metallic oxide electrodes is of considerable interest in the field of spin-polarized transport applications owing to the high polarization of these oxides. However, in most cases, the high-spin polarization vanishes at room temperature owing to their low Curie temperature. Kobayashi *et al.* [8] have argued that double perovskites such as SFMO present a high enough Curie temperature that may allow 100% spin polarization at room temperature. Bibes *et al.* [79] have used this material and successfully fabricated micrometer-sized tunnel junctions using a process based on resistance-controlled indentation of a thin insulating resist layer covering this surface. However, the high-spin polarization of SFMO is only evidenced at low temperature, which may be due to the poor quality of the films. Moreover, it has been shown recently that Fe deficiency at the SFMO interface can account for a loss of MR in SFMO-based magnetic tunnel junctions [80]. Fix *et al.* [81] have used epitaxial SFMO as the bottom electrode to study the quality of STO tunnel barrier deposited by PLD on STO (001) substrate. It has been shown that the structural, surface, and electrical properties of the STO barrier grown on SFMO layer are excellent. These results are very promising for further study of magnetic tunnel junctions prepared by PLD using SFMO electrode. The possible use of the double perovskite oxide LNMO as a ferromagnetic barrier in $LaNiO_3/LNMO/LaNiO_3$ junctions for spin filter devices has been studied theoretically in the ballistic regime [82]. The results indicate that the spin polarization of the conductance increases with barrier thickness and the spin filter effect is weakened by hole doping into the $LaNiO_3$ electrodes.

2.7
Future Directions

A particularly promising research direction is the integration of the outstanding properties of double perovskite oxides in complex heterostructures to achieve materials with enhanced or extended functionalities. Such materials may pave the way for novel device structures. Notable examples are ferroelectric field-effect transistors, controllable dielectric microwave and radiofrequency (rf) devices, magnetoresistive sensors, and electrically controllable and readable magnetic random access memories. An obvious approach for the realization of oxide materials with improved functionality is the integration of two properties in one and the same material. Well-known examples are ferromagnetic semiconductors [83] combining semiconducting and ferromagnetic properties, or intrinsic multiferroics [84] displaying finite ferromagnetic and ferroelectric order parameter. However, the realization of such intrinsic multifunctional materials is challenging, in particular for room-temperature operation. Another promising approach for the realization of multifunctional materials is the integration of double perovskite oxides with different properties in artificial heterostructures. This approach has become feasible with the enormous progress in oxide thin-film technology over the past two decades. For example, presently laser molecular beam epitaxy (laser-MBE) employing *in situ* RHEED enables growth of oxide thin films with crystalline quality approaching those of traditional semiconductors [85]. Moreover, in close analogy to GaAs/AlAs heteroepitaxy, it is possible to grow complex heterostructures composed of different oxides on suitable substrates in a layer-by-layer or block-by-block mode [86].

Acknowledgments

The authors gratefully acknowledge support under NSF ECCS-1102263 and ONR N000140610226. The authors also acknowledge collaborations with H. Z. Guo, S. Street, P. LeClair (University of Alabama), S. J. Pennycook, M. Varela (Oak Ridge National Laboratory), M. N. Iliev (University of Houston), J. Ziang (Louisiana State University), M. A. Subramanian (Oregon State University), and other coworkers that resulted in some of the results reported here.

References

1. Woodward, P., Hoffmann, R.-D., and Sleight, A.W. (1994) *J. Mater. Res.*, **9**, 2118.
2. Shimada, T., Nakamura, J., Motohashi, T., Yamauchi, H., and Karppinen, M. (2003) *Chem. Mater.*, **15**, 4494.
3. Huang, Y.H., Lindén, J., Yamauchi, H., and Karppinen, M. (2004) *Chem. Mater.*, **16**, 4337.
4. Vasala, S., Lehtimaki, M., Huang, Y.H., Yamauchi, H., Goodenough, J.B., and Karppinen, M. (2010) *J. Solid State Chem.*, **183**, 1007.
5. Kobayashi, K.-I., Kimura, T., Tomioka, Y., Sawada, H., Terakura, K., and Tokura, Y. (1999) *Phys. Rev. B*, **59**, 11159.
6. Pickett, W.E. (1998) *Phys. Rev. B*, **57**, 10613.
7. Kobayashi, K.-I., Kimura, T., Sawada, H., Terakura, K., and Tokura, Y. (1998) *Nature (London)*, **395**, 677.
8. Wold, A., Arnott, R.J., and Goodenough, J.B. (1958) *J. Appl. Phys.*, **29**, 387.
9. Dass, R.I., Yan, J.-Q., and Goodenough, J.B. (2003) *Phys. Rev. B*, **68**, 064415.
10. Rogado, N.S., Li, J., Sleight, A.W., and Subramanian, M.A. (2005) *Adv. Mater.*, **17**, 2225.
11. Azuma, M., Takata, K., Saito, T., Ishiwata, S., Shimakawa, Y., and Takano, M. (2005) *J. Am. Chem. Soc.*, **127**, 8889.
12. Huang, Y.H., Dass, R.I., Xing, Z.L., and Goodenough, J.B. (2006) *Science*, **312**, 254.
13. Huang, Y.H., Dass, R.I., Denyszyn, J.C., and Goodenough, J.B. (2006) *J. Electrochem. Soc.*, **153**, A1266.
14. Ji, Y., Huang, Y.H., Ying, J.R., and Goodenough, J.B. (2007) *Electrochem. Commun.*, **9**, 1881.
15. Manako, T., Izumi, M., Konishi, Y., Kobayashi, K., Kawasaki, M., and Tokura, Y. (1999) *Appl. Phys. Lett.*, **74**, 2215.
16. Yin, H.Q., Zhou, J.-S., Zhou, J.P., Dass, R., McDevitt, J.T., and Goodenough, J.B. (1999) *Appl. Phys. Lett.*, **75**, 2812.
17. Masuno, A., Haruta, M., Azuma, M., Kurata, H., Isoda, S., Takano, M., and Shimakawa, Y. (2006) *Appl. Phys. Lett.*, **89**, 211913.
18. Sakai, M., Masuno, A., Kan, D., Hashisaka, M., Takata, K., Azuma, M., Takano, M., and Shimakawa, Y. (2007) *Appl. Phys. Lett.*, **90**, 072903.
19. Padhan, P., Guo, H.Z., LeClair, P., and Gupta, A. (2008) *Appl. Phys. Lett.*, **92**, 022909.
20. Chang, J., Lee, K., Jung, M.H., Kwon, J.-H., Kim, M., and Kim, S.-K. (2011) *Chem. Mater.*, **23**, 2693.
21. Sánchez, D., Garcia-Hernandez, M., Auth, N., and Jakob, G. (2004) *J. Appl. Phys.*, **96**, 2736.
22. Fix, T., Stoeffler, D., Colis, S., Ulhaq, C., Versini, G., Vola, J.P., Huber, F., and Dinia, A. (2005) *J. Appl. Phys.*, **98**, 023712.
23. Guo, H.Z., Burgess, J., Ada, E., Street, S., Gupta, A., Iliev, M.N., Kellock, A.J., Magen, C., Varela, M., and Pennycook, S.J. (2008) *Phys. Rev. B*, **77**, 174423.
24. Iliev, M.N., Padhan, P., and Gupta, A. (2008) *Phys. Rev. B*, **77**, 172303.
25. Boucher, R. (2005) *J. Phys. Chem. Solids*, **66**, 1020.
26. Jalili, H., Heinig, N.F., and Leung, K.T. (2010) *J. Chem. Phys.*, **132**, 204701.
27. Czeschka, F.D., Geprägs, S., Opel, M., Goennenwein, S.T.B., and Gross, R. (2009) *Appl. Phys. Lett.*, **95**, 062508.

28. Muduli, P.K., Budhani, R.C., Topwal, D., and Sarma, D.D. (2009) *J. Phys. Conf. Ser.*, **150**, 042132.

29. Di Trolio, A., Larciprete, R., Marotta, V., Testa, A.M., and Fiorani, D. (2006) *Phys. Status. Solidi. C*, **3**, 3229.

30. Di Trolio, A., Larciprete, R., Testa, A.M., Fiorani, D., Imperatori, P., Turchini, S., and Zema, N. (2006) *J. Appl. Phys.*, **100**, 013907.

31. Nechache, R., Harnagea, C., Carignan, L.-P., Gautreau, O., Pintilie, L., Singh, M.P., Menard, D., Fournier, P., Alexe, M., and Pignolet, A. (2009) *J. Appl. Phys.*, **105**, 061621.

32. Shinde, S.R., Ogale, S.B., Greene, R.L., Venkatesan, T., Tsoi, K., Cheong, S.-W., and Millis, A.J. (2003) *J. Appl. Phys.*, **93**, 1605.

33. Song, J.H., Park, J.-H., and Jeong, Y.H. (2005) *J. Appl. Phys.*, **97**, 046105.

34. Westerburg, W., Reisinger, D., and Jakob, G. (2000) *Phys. Rev. B*, **62**, R767.

35. Guo, H.Z., Gupta, A., Zhang, J., Varela, M., and Pennycook, S.J. (2007) *Appl. Phys. Lett.*, **91**, 202509.

36. Kitamura, M., Ohkubo, I., Kubota, M., Matsumoto, Y., Koinuma, H., and Oshima, M. (2009) *Appl. Phys. Lett.*, **94**, 132506.

37. Bi, L., Taussig, A.R., Kim, H.S., Wang, L., Dionne, G.F., Bono, D., Persson, K., Ceder, G., and Ross, C.A. (2008) *Phys. Rev. B*, **78**, 104106.

38. Venimadhav, A., Sher, F., Attfield, J.P., and Blamire, M.G. (2006) *Solid State Commun.*, **138**, 314.

39. Philipp, J.B., Reisinger, D., Schonecke, M., Opel, M., Marx, A., Erb, A., Alff, L., and Gross, R. (2003) *J. Appl. Phys.*, **93**, 6853.

40. Padhan, P., LeClair, P., Gupta, A., and Srinivasan, G. (2008) *J. Phys. Condens. Matter*, **20**, 355003.

41. Singh, M.P., Truong, K.D., Fournier, P., Rauwe, P., Rauwe, E., Carignan, L.P., and Ménard, D. (2008) *Appl. Phys. Lett.*, **92**, 112505.

42. Jalili, H., Heinig, N.F., and Leung, K.T. (2009) *Phys. Rev. B*, **79**, 174427.

43. Venimadhav, A., Sher, F., Attfield, J.P., and Blamire, M.G. (2004) *J. Magn. Magn. Mater.*, **269**, 101.

44. Asano, H., Ogale, S.B., Garrison, J., Orizco, A., Li, Y.H., Smolyaninova, V., Galley, C., Downes, M., Rajeswari, M., Ramesh, R., and Venkatesan, T. (1999) *Appl. Phys. Lett.*, **74**, 3696.

45. Bull, C.L., Gleeson, D., and Knight, K.S. (2003) *J. Phys. Condens. Matter*, **15**, 4927.

46. Geprags, S., Czeschka, F.D., Opel, M., Goennenwein, S.T.B., Yu, W., Mader, W., and Gross, R. (2009) *J. Magn. Magn. Mater.*, **321**, 2001.

47. Tofield, B.C. and Scott, W.R. (1974) *J. Solid State Chem.*, **10**, 183.

48. Kumar, D. and Kaur, D. (2010) *Physica B*, **405**, 3259.

49. Borges, R.P., Lhostis, S., Bari, M.A., Versluijs, J.J., Lunney, J.G., Coey, J.M.D., Besse, M., and Contour, J.-P. (2003) *Thin Solid Films*, **429**, 5.

50. Asano, H., Kozuka, N., Tsuzuki, A., and Matsui, M. (2004) *Appl. Phys. Lett.*, **85**, 263.

51. Singh, M.P., Grygiel, C., Sheets, W.C., Boullay, P., Hervieu, M., Prellier, W., Mercey, B., Simon, C., and Raveau, B. (2007) *Appl. Phys. Lett.*, **91**, 012503.

52. Guo, H., Burgess, J., Street, S., Gupta, A., Calvarese, T.G., and Subramanian, M.A. (2006) *Appl. Phys. Lett.*, **89**, 022509.

53. Singh, M.P., Truong, K.D., Laverdière, J., Charpentier, S., Jandl, S., and Fournier, P. (2008) *J. Appl. Phys.*, **103**, 07E315.

54. Iliev, M.N., Abrashev, M.V., Litvinchuck, A.P., Hadjiev, V.G., Guo, H., and Gupta, A. (2007) *Phys. Rev. B*, **75**, 104118.

55. Iliev, M.N., Guo, H., and Gupta, A. (2007) *Appl. Phys. Lett.*, **90**, 151914.

56. Truong, K.D., Laverdière, J., Singh, M.P., Jandl, S., and Fournier, P. (2007) *Phys. Rev. B*, **76**, 132413.

57. Granado, E., Garcia, A., Sanjurjo, J.A., Rettori, C., Torriani, I., Prado, E., Sanchez, R.D., Canerio, A., and Oseroff, S.B. (1999) *Phys. Rev. B*, **60**, 11879.

58. Truong, K.D., Singh, M.P., Jandl, S., and Fournier, P. (2009) *Phys. Rev. B*, **80**, 134424.

59. Guo, H., Gupta, A., Varela, M., Pennycook, S., and Zhang, J. (2009) *Phys. Rev. B*, **79**, 172402.

60. Kitamura, M., Ohkubo, I., Matsunami, M., Horiba, K., Kumigashira, H., Matsumoto, Y., Koinuma, H., and

Oshima, M. (2009) *Appl. Phys. Lett.*, **94**, 262503.

61. Gray, B., Nyung Lee, H., Liu, J., Chakhalian, J., and Freeland, J.W. (2010) *Appl. Phys. Lett.*, **97**, 013105.

62. Guo, H.Z., Gupta, A., Calvarese, T.G., and Subramanian, M.A. (2006) *Appl. Phys. Lett.*, **89**, 262503.

63. Rutkowski, M., Hauser, A.J., Yang, F.Y., Ricciardo, R., Meyer, T., Woodward, P.M., Holcombe, A., Morris, P.A., and Brillson, L.J. (2010) *J. Vac. Sci. Technol., A*, **28**, 1240.

64. Santiso, J., Figueras, A., and Fraxedas, J. (2002) *Surf. Interface Anal.*, **33**, 676.

65. Kazan, S., Mikailzade, F.A., Özdemir, M., Aktaş, B., Rameev, B., Intepe, A., and Gupta, A. (2010) *Appl. Phys. Lett.*, **97**, 072511.

66. Joseph Joly, V.L., Joy, P.A., Date, S.K., and Gopinath, C.S. (2002) *Phys. Rev. B*, **65**, 184416.

67. Rüdiger, U., Rabe, M., Güntherodt, G., Yin, H.Q., Dass, R.I., and Goodenough, J.B. (2000) *Appl. Phys. Lett.*, **77**, 2216.

68. Yin, H.Q., Zhou, J.-S., Dass, R., Zhou, J.-P., McDevitt, J.T., and Goodenough, J.B. (2000) *J. Appl. Phys.*, **87**, 6761.

69. Singh, M.P., Charpentier, S., Truong, K.D., and Fournier, P. (2007) *Appl. Phys. Lett.*, **90**, 211915.

70. Dass, R.I. and Goodenough, J.B. (2003) *Phys. Rev. B*, **67**, 014401.

71. Hashisaka, M., Kan, D., Masuno, A., Takano, M., Shimakawa, Y., Terashima, T., and Mibu, K. (2006) *Appl. Phys. Lett.*, **89**, 032504.

72. Majewski, P., Geprags, S., Sanganas, O., Opel, M., Gross, R., Wilhelm, F., Rogalev, A., and Alff, L. (2005) *Appl. Phys. Lett.*, **87**, 202503.

73. Singh, M.P., Truong, K.D., and Fournier, P. (2007) *Appl. Phys. Lett.*, **91**, 042504.

74. Daniel, V.V. (1967) *Dielectric Relaxation*, Academic Press, New York, pp. 4–19.

75. Weber, S., Lunkenheimer, P., Fichtl, R., Hemberger, J., Tsurkan, V., and Loidl, A. (2003) *Phys. Rev. Lett.*, **91**, 257208.

76. Kimura, T., Kawamoto, S., Yamada, I., Azuma, M., Takano, M., and Tokura, Y. (2003) *Phys. Rev. B*, **67**, 180401.

77. Smolenskii, G.A. and Chupis, I. (1982) *Sov. Phys. Usp.*, **25**, 475.

78. Gajek, M., Bibes, M., Fusil, S., Bouzehouane, K., Fontcuberta, J., Barthelemy, A., and Fert, A. (2007) *Nat. Mater.*, **6**, 296.

79. Bibes, M., Bouzehouane, K., Barthélémy, A., Besse, M., Fusil, S., Bowen, M., Seneor, P., Carrey, J., Cros, V., Vaurès, A., Contour, J.-P., and Fert, A. (2003) *Appl. Phys. Lett.*, **83**, 2629.

80. Fix, T., Bertoni, G., Ulhaq-Bouillet, C., Colis, S., Dinia, A., Verbeeck, J., and Van Tendeloo, G. (2007) *Appl. Phys. Lett.*, **91**, 023106.

81. Fix, T., Da Costa, V., Ulhaq-Bouillet, C., Colis, S., Dinia, A., Bouzehouane, K., and Barthélémy, A. (2007) *Appl. Phys. Lett.*, **91**, 083104.

82. Itoh, H., Ozeki, J., and Inoue, J. (2007) *J. Magn. Magn. Mater.*, **310**, 1994.

83. Ohno, H., Shen, A., Matsukura, F., Oiwa, A., Endo, A., Katsumoto, S., and Iye, Y. (1996) *Appl. Phys. Lett.*, **69**, 363.

84. Fiebig, M. (2005) *J. Phys. D*, **38**, R123.

85. Gupta, A., Gross, R., Olsson, E., Segmüller, A., Koren, G., and Tsuei, C.C. (1990) *Phys. Rev. Lett.*, **64**, 3191.

86. Opel, M., Geprags, S., Menzel, E.P., Nielsen, A., Reisinger, D., Nielsen, K.-W., Brandlmaier, A., Czeschka, F.D., Althammer, M., Weiler, M., Goennenwein, S.T.B., Simon, J., Svete, M., Yu, W., Huhne, S.-M., Mader, W., and Gross, R. (2011) *Phys. Status Solidi A*, **208**, 232.

Part II
Dopants, Defects and Ferromagnetism in Metal Oxides

Functional Metal Oxides: New Science and Novel Applications, First Edition.
Edited by Satishchandra B. Ogale, Thirumalai V. Venkatesan, and Mark G. Blamire.
© 2013 Wiley-VCH Verlag GmbH & Co. KGaA. Published 2013 by Wiley-VCH Verlag GmbH & Co. KGaA.

3
Magnetic Oxide Semiconductors: on the High-Temperature Ferromagnetism in TiO$_2$- and ZnO-Based Compounds

Tomoteru Fukumura and Masashi Kawasaki

3.1
Introduction

3.1.1
Diluted Magnetic Semiconductors

Semiconductors without magnetic elements are usually nonmagnetic (Figure 3.1a, center). On the other hand, magnetic compounds containing periodically arranged magnetic elements are ferromagnetic (Figure 3.1a, left) or antiferromagnetic. There is an intermediate type of these compounds termed *diluted magnetic semiconductors* ((Figure 3.1a, right) that are semiconductors doped with a small amount of transition metal. The doped transition metal ions are separately located in the substance, and the band carriers (electrons or holes) mediate exchange interaction between the spatially separated transition-metal ions through sp–d exchange interaction.

The first generation of the diluted magnetic semiconductors is Mn-doped II–VI compound semiconductors [2, 3]. Various II–VI semiconductors (ZnS, ZnSe, ZnTe, CdS, CdSe, CdTe, etc.) accept a large amount of Mn in the host, so that the bulk and thin film samples are available. Mn^{2+} ion substitutes divalent cation site with high-spin state, d^5, and hence the localized spins are introduced without carrier doping. The host semiconductor preserves its pristine crystal (hexagonal wurtzite or cubic zinc blende) structure and the energy band structure even with a significant amount of doped Mn ion and their localized spins. As a result, these compounds have both semiconducting and magnetic characters. The interplay between the band carriers and localized spins generates, for example, large magnetooptical and magnetoresistance effects. The giant magnetooptical effect in CdMnTe is used in an optical isolator device [4]. About the magnetism, the paramagnetism or the spin-glass state with short-ranged antiferromagnetic order was usually observed in those systems [2, 3].

Subsequently, modern thin film growth techniques such as molecular beam epitaxy opened the second generation of diluted magnetic semiconductors: Mn-doped III–V compound semiconductors (InAs, GaAs, etc.) [1]. Nonequilibrium process

Functional Metal Oxides: New Science and Novel Applications, First Edition.
Edited by Satishchandra B. Ogale, Thirumalai V. Venkatesan, and Mark G. Blamire.
© 2013 Wiley-VCH Verlag GmbH & Co. KGaA. Published 2013 by Wiley-VCH Verlag GmbH & Co. KGaA.

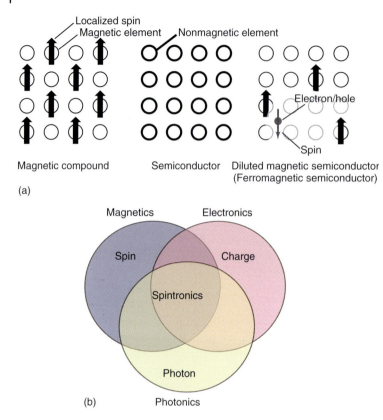

Figure 3.1 (a) Schematic structures of magnetic compound (left), nonmagneticsemiconductor (center), and ferromagnetic semiconductor (right). (Source: Modified from Ref. [1].) (b) Schematic concept of semiconductor spintronics.

of thin films at quite low-temperature growth enables the epitaxial stabilization of the host compounds with a much higher Mn content than the thermodynamic solubility limit. Here, Mn^{2+} ion substitutes trivalent cation site with high-spin state, hence the localized spins are introduced simultaneously with hole carriers. The significant amount of hole carrier ($>10^{19}$ cm^{-3}) mediates ferromagnetic exchange coupling between Mn spins (Figure 3.1a, right), and the Curie temperature (T_C) is around 190 K at present [5]. The emergence of ferromagnetism stimulated the research to demonstrate the concept of semiconductor spintronics, in which the interplay of spin, charge, and photon provides novel functionalities beyond modern electronics (Figure 3.1b) [6, 7]. However, in order to realize the device operation at room temperature, sufficiently high T_C above 500 K is necessary. The significant increase of T_C demands further material exploration of both host semiconductors and transition metals.[1]

1) Hereafter, we call ferromagnetic diluted magnetic semiconductor as ferromagnetic semiconductor.

3.1.2
Magnetic Oxide Semiconductors

Wide-gap oxide semiconductors show good electrical conductivity by doping electron carriers (up to $\sim 10^{21}$ cm^{-3}), and hence are used in various applications such as solar cells and transparent transistors. The chemical stability, ecological friendliness, and abundance in natural resources are beneficial for such applications. Various synthetic routes are available in the form of bulk, thin film, and nanoparticles. Even when limited to thin films, many growth methods are available such as sol–gel method, sputtering, pulsed laser deposition, and molecular beam epitaxy.

From the viewpoint of spintronics, the wide-gap semiconductors are expected to have heavy electron mass, which is advantageous for strong carrier-mediated exchange interaction such as Ruderman–Kittel–Kasuya–Yosida (RKKY) interaction, leading to higher T_C [8]. This was our motivation to develop magnetic oxide semiconductors as described in the following sections.

3.1.2.1 Development of Magnetic Oxide Semiconductor

First wide-gap magnetic oxide semiconductor discovered was wurtzite (Zn,Mn)O [8, 9], which was previously studied as a pigment [10]. The epitaxial thin film was grown on sapphire substrate by using pulsed laser deposition. The soluble Mn content was over 30% and the codoping of Mn and Al was possible, where Al served as an electron dopant.

Mn^{2+} ion substitutes Zn site with high-spin state, d^5 [11]. The lattice constants increased linearly with Mn content reflecting the Vegard's law (Figure 3.2a). Energy gap of ZnO is 3.2 eV, and broad in-gap absorption in (Zn,Mn)O increased with Mn content owing to spin forbidden d–d transition of Mn^{2+} ion (Figure 3.2b). (Zn,Mn)O represented a spin-glass behavior (Figure 3.2c). The exchange integral between the nearest neighbors derived from the Curie–Weiss plot of M versus T curve was -15 K, and the large antiferromagnetic exchange coupling was consistent with the small effective Mn content obtained from M versus H measurement. This value was larger than those in A_{1-x}Mn$_x$B (A = Cd, Hg, and Zn; B = Se, Te) reported in the literature [2, 12]. This result reflects that smaller anion size gives stronger antiferromagnetic exchange coupling for the II–VI group diluted magnetic semiconductors, later determined by photoemission spectroscopies [13, 14].

While ZnO:Al showed small negative magnetoresistance similar to that observed in an accumulation layer in ZnO [16], (Zn,Mn)O:Al showed a distinct magnetoresistance at low temperatures (Figure 3.2d) [8]. The magnetoresistance at 2 K was positive in a small magnetic field with a sharp peak and turned to be negative in a high magnetic field. The positive and the negative magnetoresistance were caused by the spin splitting enhanced by s–d exchange interaction [17] and the rise of the Fermi level in the redistributed majority-spin subband [18], respectively. This result represents that localized spins of Mn^{2+} ions interact with conducting carriers in ZnO.

By overviewing these results, the properties of (Zn,Mn)O were quite similar with those of other Mn-doped II–VI diluted magnetic semiconductors, hence was classified as a new member of Mn-doped II–VI diluted magnetic semiconductors,

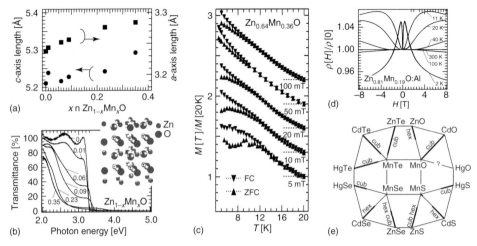

Figure 3.2 (a) Mn content (x) dependences of a- and c-axes lattice constants for $Zn_{1-x}Mn_xO$ [8]. (Reprinted with permission. Copyright 1999, American Institute of Physics.) (b) Transmittance spectra for $Zn_{1-x}Mn_xO$ with different x at room temperature. The inset shows wurtzite structure [8]. (Reprinted with permission. Copyright 1999, American Institute of Physics.) (c) Normalized magnetization (M) versus T curves for $Zn_{0.64}Mn_{0.36}O$ measured during zero-field cooling (triangle) and field cooling (inversed triangle) running in different magnetic fields. The curves are shifted vertically [9]. (Reprinted with permission. Copyright 2001, American Institute of Physics.) (d) Normalized magnetoresistance versus H curves for $Zn_{0.81}Mn_{0.19}O$ at different temperatures. The magnetic field was applied along the out of plane [8]. (Reprinted with permission. Copyright 1999, American Institute of Physics.) (e) Schematic digest of the $A_{1-x}Mn_xB$ (A, group II element; B, group VI element) magnetic semiconductors. (Source: Modified from W. Giriat and J. K. Furdyna in Ref. [3].) The bold lines represent the x range of their single phases, where "cub" and "hex" denote the cubic and hexagonal crystal structures, respectively [15]. (Reprinted with permission. Copyright 2005, IOP Publishing.)

in which the anion was oxygen (Figure 3.2e) [15]. Therefore, it was worthy to investigate the synthesis and properties of other transition-metal-doped ZnO as well as other transition-metal-doped oxide semiconductors.

3.1.2.2 Combinatorial Exploration of Magnetic Oxide Semiconductors

Combinatorial synthesis and high-throughput screening were applied to synthesize new magnetic oxide semiconductors and to examine their magnetic properties because of the high efficiency and systematic data acquisition [19]. For combinatorial growth of epitaxial thin film library of transition-metal-doped oxide semiconductors, a combinatorial pulsed laser deposition system was used. By controlling the patterned masks, the ceramic targets, and the laser pulses with a personal computer, three by three pixels were integrated on ~15 mm square substrate, in which the doping concentration of the transition metal was varied among each pixel [20]. In order to examine the magnetism of such combinatorial samples, standard magnetometer was not useful because it was impossible in principle to distinguish the magnetism of the pixels integrated on a substrate. Instead, scanning SQUID microscope (SQUID, superconducting quantum interference device) was used [21]. Such scanning probe microscopes were successfully used to observe a magnetic domain structure of (Ga,Mn)As [22, 23]. In these measurements, the stray magnetic field from magnetic domain structure was probed without applying external magnetic field. Therefore, background paramagnetic signals from the segregated secondary phase and substrate can be excluded, in contrast with the magnetometer measurement.

Figure 3.3a shows a photograph of combinatorial samples of ZnO doped with 3d transition metals, from Sc to Cu [24]. For each dopant, the solubility was determined with X-ray diffraction. Figure 3.3b–d shows magnetic circular dichroism (MCD), absorption, and cathodoluminescence spectra, respectively, of the 3d transition-metal-doped ZnO. It is obvious that each 3d dopant yielded in distinct features,

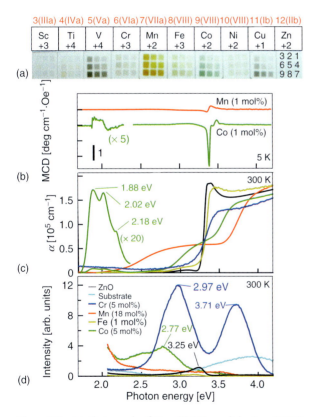

Figure 3.3 (a) Photograph of $Zn_{1-x}TM_xO$ (TM: 3d transition elements) combinatorial libraries with nine pixels integrated on 15×15 mm substrates. Estimated valence state of TM ions is shown in the upper panel [24]. (b) Magnetic circular dichroism (MCD) spectra for (Zn,Mn)O and (Zn,Co)O at 5 K [24]. (c) Absorption (α) and (d) cathodoluminescence spectra of Cr-, Mn-, Fe-, and Co-doped ZnO at 300 K [24]. (Reprinted with permission. Copyright 2001, American Institute of Physics.)

mainly because of different occupation number of 3d electrons for each transition-metal ion. It is noted that all 3d transition metals were successfully doped in ZnO with the help of combinatorial methodology, while Mn was often used as a dopant for diluted magnetic semiconductors, indicating the chemical tolerance of oxides to accept various dopants. Accordingly, it was useful to investigate the chemical trend of various properties such as magnetooptical effect [25], and to explore new combinations between the other oxide semiconductors and transition metals. During the materials exploration, the giant magnetooptical effect of (Zn,Co)O and the giant magnetoresistance of $(Sn,Mn)O_2$ were found, although those phenomena were observed only at low temperature [26, 27].

The extensive as well as fast materials exploration led to, eventually but instantly, the discovery of high-temperature ferromagnetism in $(Ti,Co)O_2$ for both anatase and rutile (Figure 3.4) [29, 30]. Co ions in TiO_2 were divalent (d^7) as described later. $(Ti,Co)O_2$ showed ferromagnetic magnetization even above 400 K (Figure 3.4b), and was semitransparent in the form of a thin film (the inset of Figure 3.4b). The magnetic image in zero magnetic field suggested the presence of maze-patterned magnetic domain structure with out-of-plane magnetization (Figure 3.4c), corresponding to a homogeneous ferromagnetic state without any secondary segregation, also verified by transmission electron microscope measurements.

The rutile and anatase structures are summarized by Tang [31] and Diebold [32] as follows. Crystal symmetry of rutile and anatase TiO_2 belong to D_{4h}^{14} ($P4_2/mnm$) ($a = 4.5937$ Å, $c = 2.9587$ Å) and D_{4h}^{19} ($I4_1/amd$) ($a = 3.7842$ Å, $c = 9.5146$ Å), respectively [33, 34]. The local site symmetry of the Ti cation and O anion for rutile is D_{2h} and C_{2v}, respectively [35], and that of Ti cation for anatase is D_{2d} [36]. Both rutile and anatase structures are composed of TiO_6 octahedra (Figure 3.4a), in which e_g levels of doped transition metal locate at higher energy levels than t_{2g} levels in contrast with the wurtzite and zinc blende types. The TiO_6 octahedron is connected with two edge-shared and eight corner-shared octahedra in rutile, and is connected with four edge-shared and four corner-shared octahedra in anatase. These differences lead to the difference of those electronic structures in rutile and anatase TiO_2 and may result in the difference of magnetic properties in rutile and anatase $(Ti,Co)O_2$ as described later.

3.1.2.3 Recent Status

The potential of magnetic oxide semiconductors as well as other doped wide-gap semiconductors for high T_C ferromagnetism was also proposed by mean field theory and first-principle calculation, although the promising compounds were predicted for p-type host, which is difficult to realize in oxide semiconductors [37, 38]. However, such possibility, in addition to experimental finding of the high-temperature ferromagnetism in $(Ti,Co)O_2$, boosted extensive search for high-temperature ferromagnetic semiconductors [39–45]. Recently, such a search was extended to compounds without magnetic elements [46–48].

So far, a lot of studies on high-temperature ferromagnetism in magnetic oxide semiconductors have been reported. The most interesting finding from such studies was the origin of the ferromagnetism, whether the ferromagnetism is attributed

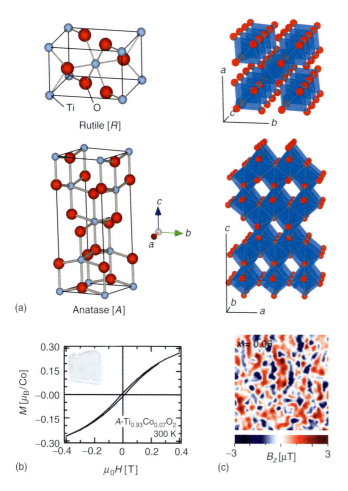

(a) Rutile [R] Anatase [A]

(b)

(c)

Figure 3.4 (a) Crystal structures of anatase and rutile TiO_2 [28]. TiO_6 octahedra in both structures are highlighted in right panels. (b) M versus H curve of anatase $Ti_{0.93}Co_{0.07}O_2$ at 300 K. Magnetic field was applied along the in-plane. The inset shows a photograph of the sample [29]. (c) Scanning SQUID microscope images (200 μm square) at 3 K for anatase $Ti_{0.94}Co_{0.06}O_2$ [29]. R and A stand for rutile and anatase, respectively, throughout all figures.

to a carrier-mediated mechanism, a defect-mediated mechanism, or an extrinsic effect. A recent study on the electric-field-induced ferromagnetism in $(Ti,Co)O_2$ at room temperature indicates the principal role of the carrier-mediated interaction [49], although it is too early to conclude that the high-temperature ferromagnetism in all magnetic oxide semiconductors is attributed to the same scenario. Drawing a historical outline of the magnetic oxide semiconductors is not the scope of this chapter as many review articles are available [39–45]. Instead, we tried to interpret the high-temperature ferromagnetism in TiO_2- and ZnO-based magnetic oxide

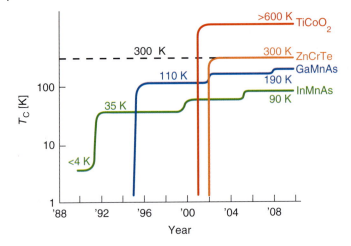

Figure 3.5 Evolution of the Curie temperatures for representative ferromagnetic semiconductors.

semiconductors from the viewpoint of the carrier-mediated mechanism. Figure 3.5 shows T_C of representative ferromagnetic semiconductors [50–52]. The T_C of $(Ti,Co)O_2$ is much higher than the others. Such high T_C could be utilized for room-temperature operation of semiconductor spintronic devices by elucidating the underlying mechanism.

3.2
Properties of $(Ti,Co)O_2$ [2)]

3.2.1
Thin Film Growth

The content of Co in $(Ti,Co)O_2$ films discussed in this chapter is far beyond the thermodynamic solubility limit [53], therefore, careful tuning of growth conditions is prerequisite to produce segregation-free epitaxial thin films by utilizing the nonequilibrium nature of film growth. Pulsed laser deposition with ultrahigh vacuum environment was used for this purpose because this method is well established for epitaxial thin film growth of oxides. Important factors to realize desired samples are as follows:

1) *Substrates*: Rutile and anatase thin films were grown on sapphire and perovskite (e.g., $LaAlO_3$ and $SrTiO_3$) substrates, respectively [54]. The lattice matching between thin film and substrate and the atomically flat step-and-terrace surface were important factors to achieve high solubility of Co, as

2) The samples in Section 3.2 were epitaxial thin films grown by pulsed laser deposition without notice.

two-dimensional growth promoted the formation of TiO$_2$ structure with high Co content, with a help of epitaxial stabilization. In other words, large lattice mismatch and rough substrate surface could result in three-dimensional growth, promoting the formation of a secondary phase such as Co metal.

2) *Laser ablation*: For the ablation of the oxide semiconductors, ultraviolet laser such as an excimer laser may be preferred to achieve high absorption at the target surface, resulting in a homogeneous supply of Ti–Co–O species due to the large energy gap of TiO$_2$ (3.2 eV). The various parameters such as laser power density and laser repetition rate must be optimized to realize high-quality films.

3) *Growth atmosphere*: The oxygen pressure during growth affected the amount of oxygen vacancy (electron donor) in TiO$_2$. Therefore, ultrahigh vacuum back pressure and finely controlled oxygen pressure were needed to control the electron density in a controlled manner, as demonstrated later.

4) *Growth temperature*: Low growth temperature avoided a roughening of film surface. An excessively reductive condition due to high growth temperature leads to the formation of a secondary phase such as Co metal.

5) *Buffer layer*: For heteroepitaxial growth, insertion of TiO$_2$ buffer layer was helpful to achieve the two-dimensional growth for both rutile and anatase thin films at low temperature.

Figure 3.6a–c represents the homoepitaxial growth of a high-quality rutile (Ti,Co)O$_2$ thin film/rutile TiO$_2$ buffer layer/rutile TiO$_2$ (110) substrate [55]. Owing to the use of the single-crystal substrate, the oscillation and streak pattern of reflection high-energy electron diffraction (RHEED) persisted till the end of growth, representing the two-dimensional growth of the segregation-free sample. The atomic force microscope image in the inset of Figure 3.6a shows the step-and-terrace structure of the grown film.

Figure 3.6d–i shows the results of microanalysis for a sample of anatase (Ti,Co)O$_2$ (001) thin film/anatase TiO$_2$ buffer layer/LaAlO$_3$ (100) substrate [49]. Bright-field scanning transmission electron microscopy (STEM) image (Figure 3.6d) and the energy-dispersive X-ray spectroscopy mapping (Figure 3.6e) show homogeneous distribution of Co atoms inside the thin film without any surface/interface segregation. The atomically resolved high-angle annular dark field-STEM (HAADF-STEM) images (Figure 3.6g,h) show the coherently arranged atoms both for Ti atoms in the buffer layer on the LaAlO$_3$ substrate and for (Ti,Co) atoms in the (Ti,Co)O$_2$ top layer. The uniform contrast of Ti and Co atoms in the top layer suggested the homogeneous distribution of Co atoms substituting for Ti sites. The atomic force microscope image (Figure 3.6i) represents no surface segregation. The epitaxial thin film growth of both rutile and anatase (Ti,Co)O$_2$ with optimizing the growth parameters of pulsed laser deposition resulted in the segregation-free thin films.

In addition, as a more industry-oriented method, sputtering was also used for thin film growth of rutile (Ti,Co)O$_2$. The epitaxial thin films were grown on the sapphire substrates [56]. In addition, the polycrystalline thin films and the magnetophotonic

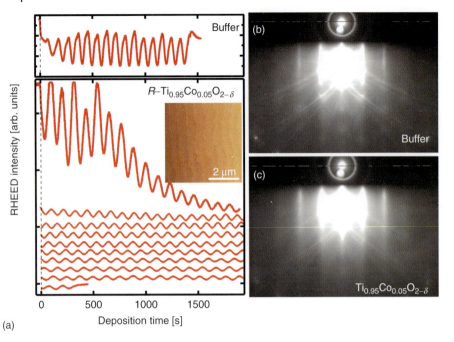

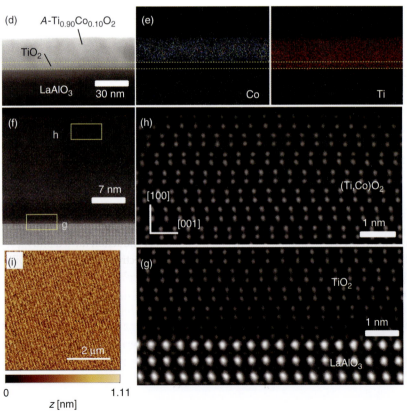

Figure 3.6 (a) Intensity oscillation of reflection high-energy electron diffraction (RHEED) during the growth of rutile TiO$_2$ buffer layer and rutile Ti$_{0.95}$Co$_{0.05}$O$_2$ thin film [55]. RHEED images taken for (b) TiO$_2$ buffer layer and (c) Ti$_{0.95}$Co$_{0.05}$O$_2$ thin film [55]. (d) A bright-field scanning transmission electron microscopic (STEM) image of an anatase Ti$_{0.90}$Co$_{0.10}$O$_2$ (001) thin film/TiO$_2$ (001) buffer layer/LaAlO$_3$ (100) substrate, and (e) the corresponding energy-dispersive X-ray spectroscopy mappings for Co and Ti [49]. (f) Cross section showing the location of images in (g) and (h) [49]. Atomically resolved high-resolution high-angle annular dark field (HAADF)-STEM images at the TiO$_2$/LaAlO$_3$ interface (g) and within the Ti$_{0.90}$Co$_{0.10}$O$_2$ thin film (h) [49]. (i) Atomic force microscope image of Ti$_{0.90}$Co$_{0.10}$O$_2$ thin film [49].

crystal structure that is (Ti,Co)O$_2$ thin film sandwiched by dielectric multilayers were grown on glass substrates [57]. In both cases, the carrier control was possible and the electric and magnetic properties were similar to those of epitaxial thin films grown by pulsed laser deposition.

3.2.2
Transport Properties

In both rutile and anatase TiO$_2$ films, the oxygen vacancy serves as the electron donor [31, 32]. The situation was the same for (Ti,Co)O$_2$. By changing the oxygen pressure during the growth, the amount of oxygen vacancy, that is, the electron density, was finely controlled. It is noted that the high-quality buffer layer was inevitable for reproducible and fine-tuning of the oxygen vacancy content in the film.

Figure 3.7a shows the oxygen pressure dependence of resistivity versus temperature curves for rutile (Ti,Co)O$_2$ [58]. By decreasing the oxygen pressure, the resistivity decreased monotonically. But the resistivity always increased with decreasing temperature representing hopping conduction. Figure 3.7b shows the mobility at 300 K for different samples [58]. The mobility ranged from 10^{-2} to 10^{-1} cm^2 V^{-1} s^{-1} irrespective of the Co content. As increasing electron density (more donors in films grown at low oxygen pressure) tends to degrade the mobility. Figure 3.7c shows the oxygen pressure dependence of resistivity versus temperature curves for anatase (Ti,Co)O$_2$ [59]. By decreasing the oxygen pressure, the resistivity decreased monotonically, and the hopping conduction turned into metallic conduction. Figure 3.7d shows the conductance, electron density, and mobility at 300 K as a function of oxygen pressure [59]. The mobility ranged from 10^0 to 10^1 cm^2 V^{-1} s^{-1} with a weak dependence on oxygen pressure and electron density. Relatively constant mobility in comparison with the electron density represents that the oxygen vacancy properly served as the electron donor in both systems. For the rutile and anatase, the resistivity at 300 K was comparable, but only anatase showed the metallic conduction at higher electron density due to the higher mobility by two orders of magnitude.

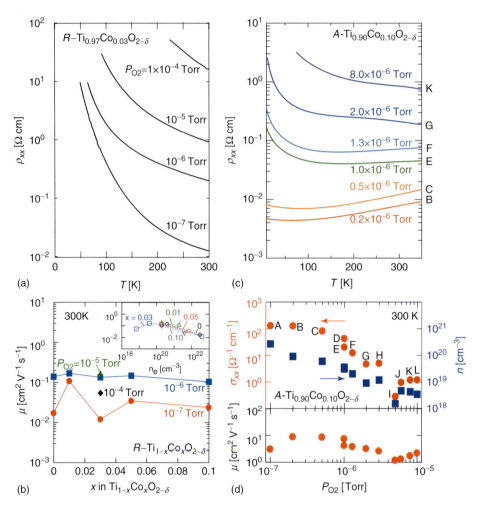

Figure 3.7 (a) Resistivity (ρ_{xx}) versus T curves for rutile $Ti_{0.97}Co_{0.03}O_2$, and (b) x dependence of mobility (μ) at 300 K for rutile $Ti_{1-x}Co_xO_2$ with different oxygen pressures (P_{O2}). $P_{O2} = 1 \times 10^{-7}$ (solid circle), 10^{-6} (solid square), 1×10^{-5} (solid triangle), and 10^{-4} Torr (solid diamond). The inset shows the electron density (n) dependence of μ at 300 K for rutile $Ti_{1-x}Co_xO_2$ with different x. $x = 0$ (open diamond), 0.01 (open triangle), 0.03 (open square), 0.05 (open circle), and 0.10 (open inverted triangle) [58]. (Reprinted with permission. Copyright 2008, IOP Publishing.) (c) ρ_{xx} versus T curves at 300 K and (d) P_{O2} dependences of conductivity (σ_{xx}) (solid circle) and electron density (n) (solid square) and μ at 300 K for anatase $Ti_{0.90}Co_{0.10}O_2$ with different P_{O2}. The samples in (c) and (d) are indexed with the alphabets (A–L), which are used also in Figure 3.11d and Figure 3.14 [59]. (Reprinted with permission. Copyright 2011, American Institute of Physics.)

3.2.3
Magnetic Properties

Magnetization measurement with a magnetometer is a fundamental probe of the magnetism but it is not always sufficient to investigate the magnetism of ferromagnetic semiconductors [15]. This is because the magnetometer does not distinguish the magnetic moments from intrinsic magnetization in the ferromagnetic semiconductor and those from extrinsic magnetization, for example, secondary phase segregations, magnetic surface contamination, and magnetic impurities in nonmagnetic substrates (Figure 3.8a) [60]. In the case of thin films, the magnitude of these signals is often comparable to the intrinsic signal. Accordingly, it is often difficult to prove the intrinsic ferromagnetism in new ferromagnetic semiconductors by the magnetometer measurement alone.

Alternatively, the intrinsic magnetism can be selectively evaluated by optical means. In ferromagnetic semiconductors, generally, s- or p-band carriers in the host semiconductors couple with d electrons of the doped transition-metal ion via the exchange interaction. Accordingly, the photoexcited band carriers show a magnetooptical effect owing to the spin-split feature of the energy band. MCD is the difference of the absorption coefficients for right-handed and left-handed circularly polarized lights, whose signal is proportional to the magnetization. MCD is also known to be proportional to the energy derivative of absorption coefficient reflecting the band structure of the host semiconductor [61]; hence, it is the largest at the absorption edge. In contrast with the magnetometer measurement, MCD signal from substrate is negligible because the circular dichroism is usually absent in the substrate. As shown schematically in Figure 3.8b, magnetic segregations could show MCD but such an MCD spectrum does not reflect the band structure of the host semiconductor. For example, ferromagnetic metal usually shows a featureless MCD [62]. Figure 3.8d shows the MCD and absorption spectra for (Zn,Co)O at 5 K and anatase (Ti,Co)O$_2$ at 300 K. ZnO is a direct transition semiconductor, thus the MCD spectrum of (Zn,Co)O showed a sharp peak at the absorption peak. The small MCD signal around 2 eV corresponded to the d–d transition of the Co^{2+} ion [24]. The magnetic field dependence of the MCD was linear at any photon energy (the inset of Figure 3.8d), representing that the sample was paramagnetic. On the other hand, (Ti,Co)O$_2$ showed broad and large MCD signal around and below the absorption edge. The magnetic field dependence of the MCD showed a ferromagnetic hysteresis loop (the inset of Figure 3.8d), where the shapes of hysteresis loop were identical at any photon energy. These behaviors are the hallmark of the intrinsic responses of a ferromagnetic semiconductor.

Moreover, the intrinsic magnetism can also be probed by electrical means. In conducting ferromagnets, generally, the Hall resistivity (ρ_H) is expressed as $\rho_H = R_0 B + R_S M$ (R_0, ordinary Hall coefficient; B, magnetic induction; R_S, anomalous Hall coefficient; and M, magnetization) [63]. The first term denotes the ordinary Hall effect due to the Lorenz force, the magnetic field dependence of which is proportional to the inverse of the carrier density. The second term denotes the anomalous Hall effect due to asymmetric carrier scattering in the presence

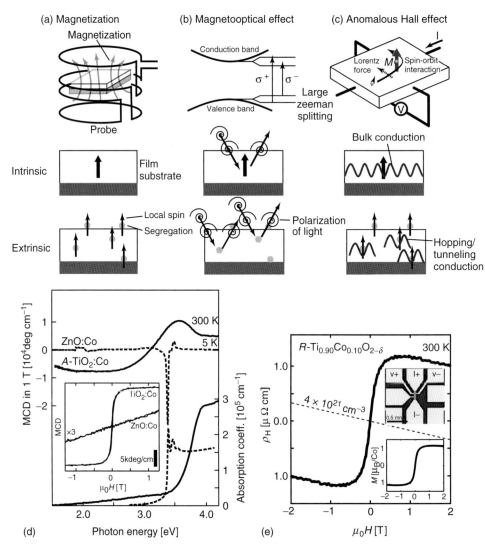

Figure 3.8 Schematic diagrams representing intrinsic and extrinsic origins of ferromagnetism for (a) magnetization, (b) magnetooptical effect, and (c) anomalous Hall-effect measurements. The upper panels represent the measurement schemes. The middle and bottom panels represent the intrinsic and extrinsic origins, respectively [15]. (d) Magnetic circular dichroism (MCD) and absorption spectra for paramagnetic (Zn,Co)O at 5 K (dashed line) and ferromagnetic anatase (Ti,Co)O_2 (solid line) at 300 K. The inset shows MCD versus H curves for (Zn,Co)O at 6 K and for (Ti,Co)O_2 at 300 K [15]. (e) The Hall resistivity ρ_H versus H curves for rutile (Ti,Co)O_2 at 300 K. The dashed line represents the ordinary part of the Hall resistivity, corresponding to the electron density of 4×10^{21} cm^{-3}. The upper inset is a photograph of the Hall bar for the measurements, and the bottom inset is M versus H curve for the same sample at 300 K [15]. (Reprinted with permission. Copyright 2005, IOP Publishing.)

of spin–orbit interaction, the magnetic field dependence of which is proportional to the magnetization (Figure 3.8c). While the anomalous Hall effect appears in the case of the bulk conduction in intrinsic ferromagnetism, the anomalous Hall effect may disappear in the case of tiny amount of magnetic segregation, owing to the lack of spin polarization in bulk. The exceptional case is the anomalous Hall effect in nonmagnetic compound heavily doped with magnetic element beyond a percolation threshold: the change in magnetoresistance with the magnetization until its saturation represented a hopping conduction of spin-polarized carriers between the adjacent segregation [64]. Figure 3.8e shows the magnetic field dependence of the Hall resistivity in rutile (Ti,Co)O$_2$. Magnetic-field-linear ordinary Hall resistivity (broken line) was superposed with the anomalous Hall resistivity, where the magnetic field dependence of the latter coincided with that of magnetization (the inset of Figure 3.8e). The magnetoresistance gradually changed with magnetic field irrespective of the saturation of the magnetization, and 3% of Co content was enough to show the anomalous Hall effect [65]. Accordingly, it is concluded that the anomalous Hall effect in (Ti,Co)O$_2$ is not originated from the hopping transport between magnetic segregations but is the intrinsic bulk character of ferromagnetic semiconductors.

For the magnetic oxide semiconductors, the role of carriers on the ferromagnetism is one of the central issues to be investigated. The magnetization, the magnetooptical effect, and the anomalous Hall effect are, respectively, the representative magnetic responses of static magnetic moments, interband photoexcited carriers, and itinerant carriers. Accordingly, the coincident magnetic field dependences in such independent measurements indicate that the observed magnetic responses are originated from single magnetic source. Thus, the observation of the anomalous Hall effect and the magnetooptical effect is useful criteria of the intrinsic ferromagnetism of ferromagnetic semiconductors.

3.2.3.1 Magnetization

Figure 3.9a,b shows *M* versus *H* curves at 300 K, respectively, for rutile and anatase (Ti,Co)O$_2$ with different electron density [58, 59]. The magnetization was almost zero for the lowest electron density and increased monotonically with increasing the electron density (see Figure 3.7d for the electron density for the samples shown in Figure 3.9b). The maximum saturation magnetization of the anatase was $\sim 2\,\mu_B$/Co, nearly twice as large as that of the rutile, and the hysteresis loop of the anatase was more clearly opened than that of the rutile. Figure 3.9c shows out-of-plane and in-plane *M* versus *H* curves at 300 K for anatase (Ti,Co)O$_2$ with different electron density [49]. For the lowest electron density, both the out-of-plane and in-plane magnetizations were negligible. By increasing the electron density, the out-of-plane magnetization was developed and was found to be larger than the in-plane one, representing the out-of-plane easy axis, which was also observed for the rutile [55]. These results represent that the increase in the electron density increased the magnitude of magnetization without a significant change in the magnetic anisotropy, which was useful to interpret the results of the electric-field-effect study described in Section 3.2.4.2.

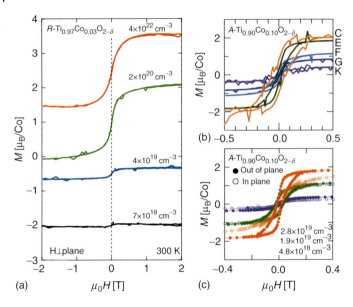

Figure 3.9 *M* versus *H* curves at 300 K for (a) rutile $Ti_{0.97}Co_{0.03}O_2$ and (b) anatase $Ti_{0.90}Co_{0.10}O_2$ with different electron densities. Magnetic field was applied along the out of plane. Each datum is shifted vertically in (a) [58, 59]. (Reprinted with permission. Copyright 2008, IOP Publishing; Reprinted with permission. Copyright 2011, American Institute of Physics.) (c) *M* versus *H* curves in out-of-plane and in-plane *H* at 300 K for anatase $Ti_{0.90}Co_{0.10}O_2$ with different electron densities [49].

3.2.3.2 Magnetic Circular Dichroism

Figure 3.10a shows the absorption and MCD spectra for rutile $(Ti,Co)O_2$ with different electron densities [58]. The absorption spectra showed a blueshift of absorption edge possibly owing to the Burstein–Moss shift and an increased absorption around 2 eV possibly as a result of the presence of Co d-level and/or oxygen vacancy for the sample with highest electron density. The two samples with the lower electron densities showed no MCD signal corresponding to the nonferromagnetic character. The other two samples with higher electron densities showed finite MCD signal with the ferromagnetic hysteresis (not shown), in which the negative MCD at visible range changed the sign around the absorption edge for higher photon energy. The blueshift of the MCD spectrum was coincident with that of the absorption spectrum, representing the close connection between the MCD spectra and the host energy band. Figure 3.10b shows the absorption and MCD spectra for anatase $(Ti,Co)O_2$ [62]. Here, $(La,Sr)_2AlO_4$ substrates were used instead of $LaAlO_3$ in order to avoid the circular dichroic signal from $LaAlO_3$ owing to the presence of twin crystal. A slight in-gap absorption and two absorption peaks beyond the absorption edge were observed, where the latter were also observed in pure TiO_2 corresponding to the critical points of the band structure [66]. It is noted that the anatase $(Ti,Co)O_2$ showed a large MCD signal with the ferromagnetic

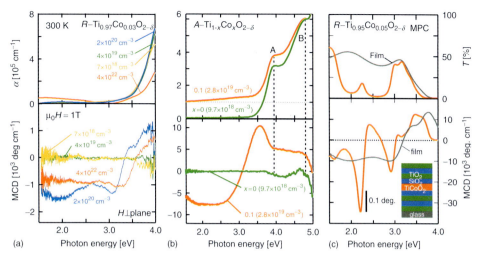

Figure 3.10 (a) Absorption (α) (upper panel) and magnetic circular dichroism (MCD) (bottom panel) spectra in 1 T at 300 K for rutile $Ti_{0.97}Co_{0.03}O_2$ with different electron density, and (b) those for anatase $Ti_{1-x}Co_xO_2$ ($x = 0$, 0.1). The magnetic field was applied along the out of plane. The electron density at 300 K is displayed. (b) The data for $x = 0.1$ are shifted vertically (upper panel), and the dashed lines correspond to the critical point energy in the band structure of anatase TiO_2 [58, 62]. (Reprinted with permission. Copyright 2008, IOP Publishing; Reprinted with permission. Copyright 2003, The Japan Society of Applied Physics.) (c) Transmittance (upper panel) and MCD (bottom panel) spectra for one-dimensional magnetophotonic crystal structure of $[TiO_2$ (51 nm)/SiO_2 (75 nm)]²/rutile $Ti_{0.95}Co_{0.05}O_2$ (98 nm)/$[SiO_2$ (75 nm)/TiO_2 (51 nm)]²/glass as illustrated in the inset. Magnetic field was applied along the out of plane. Data for $Ti_{0.95}Co_{0.05}O_2$ single-layer thin film on glass are also shown for comparison (gray curves) [57]. (Reprinted with permission. Copyright 2009, American Institute of Physics.)

hysteresis (not shown), $10^4 °cm^{-1}$ at maximum [62]. The MCD spectrum showed features similar to that of the rutile and had shoulder structures at the critical points, also representing the close connection with the host energy band. The magnetooptical effect as large as that in materials used in magnetooptical isolators with semitransparent nature at visible range is an interesting feature from the viewpoint of magnetooptical applications.

As described before, the rutile thin films were grown on glass substrates by sputtering. Hence, one-dimensional magnetophotonic crystal structure containing dielectric multilayers was grown easily (the inset of Figure 3.10c) [57]. In the structure, the electric field of light is magnified in the central magnetic layer leading to significant enhancement of the magnetooptical effect [67]. Figure 3.10c shows the transmittance and MCD spectra for $[TiO_2/SiO_2]^2$/rutile $(Ti,Co)O_2/[SiO_2/TiO_2]^2$ and single-layer $(Ti,Co)O_2$ thin film on glass substrates. The photonic bandgap was formed for $1.8-2.9$ eV with a resonant transmission owing to the interference at around 2.2 eV as seen in the transmittance spectrum. The maximum MCD was

$-34\,700^\circ\,\text{cm}^{-1}$ at 2.22 eV, exhibiting $\sim$380% enhancement in comparison with that of the single-layer film.

3.2.3.3 Anomalous Hall Effect

Since the anomalous Hall effect in $(\text{Ti},\text{Co})\text{O}_2$ was superposed with that of the ordinary Hall effect, information of both the magnetism and the electron density was obtained simultaneously. Figure 3.11a (thick solid curve) shows the magnetic field dependence of the Hall resistivity (ρ_H) at 300 K for rutile $(\text{Ti},\text{Co})\text{O}_2$ [58]. The

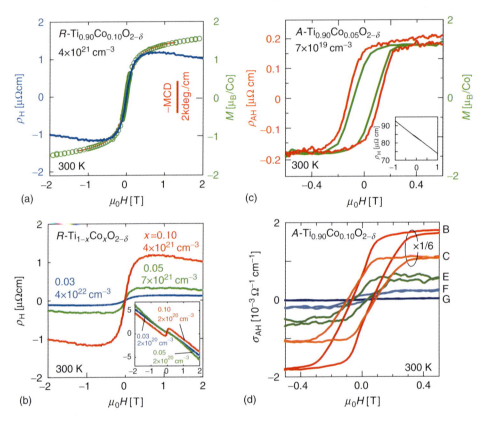

Figure 3.11 (a) Magnetic circular dichroism (MCD) (thin solid curve), Hall resistivity (ρ_H) (thick solid curve), and M (open symbol) versus H curves at 300 K for rutile $\text{Ti}_{0.90}\text{Co}_{0.10}\text{O}_2$. The sign of MCD is inverted for comparison [58]. (Reprinted with permission. Copyright 2008, IOP Publishing.) (b) ρ_H versus H curves at 300 K for rutile $\text{Ti}_{1-x}\text{Co}_x\text{O}_2$ with different x. Inset shows data for lower electron density [58]. (Reprinted with permission. Copyright 2008, IOP Publishing.) (c) Anomalous Hall resistivity (ρ_AH, dark gray) and M (light gray) versus H curves for anatase $\text{Ti}_{0.95}\text{Co}_{0.05}\text{O}_2$ at 300 K. Inset shows ρ_H versus H curve at 300 K [68]. (Reprinted with permission. Copyright 2007, American Institute of Physics.) (d) Anomalous Hall conductivity (σ_AH) versus H curves at 300 K for anatase $\text{Ti}_{0.90}\text{Co}_{0.10}\text{O}_2$ with different electron density. The magnetic field was applied along the out-of-plane for all the measurements [59]. (Reprinted with permission. Copyright 2011, American Institute of Physics.)

S-shape component corresponded to the anomalous Hall effect proportional to the magnetization. The magnetic field dependence was nearly the same as those of magnetization and MCD as seen in Figure 3.11a (open symbol and thin solid curve). At higher magnetic field, the amplitude of the Hall resistivity decreased linearly with the magnetic field owing to the contribution of ordinary Hall effect with the negative carrier polarity of electrons. The electron density and the Co-content dependences are shown in Figure 3.11b [58]. For the high electron density (Figure 3.11b), the contribution of ordinary Hall effect was weak, and the raw Hall resistivity data mostly represent the contribution of anomalous Hall effect that was enhanced by the increase in the Co content. For the low electron density (the inset of Figure 3.11b), the ordinary Hall effect was dominant, and the anomalous Hall effect was visible only around zero magnetic field, where the anomalous Hall effect also increased with the Co content.

In the case of anatase (Ti,Co)O$_2$, the ordinary Hall term was more dominant because ferromagnetism appeared for the electron density lower than that of rutile. The inset of Figure 3.11c shows the raw data of Hall resistivity of the anatase (Ti,Co)O$_2$. By subtracting the H-linear ordinary Hall effect, the anomalous Hall resistivity (ρ_{AH}) became visible with clear magnetic hysteresis which is in good coincidence with that of the magnetization Figure 3.11c [68].

In general, the anomalous Hall resistivity is monotonically increasing functions of magnetization and resistivity. For the ferromagnetic semiconductors, the magnetization (the resistivity) increases (decreases) with increase in the electron density n, where $\rho_{xx} = (en\mu)^{-1}$ (e: bare electron charge). Accordingly, the anomalous Hall resistivity could be a nonmonotonic function of the electron density on the verge of the transition from paramagnetic to ferromagnetic states. Instead, the anomalous Hall conductivity, $\sigma_{AH} = \rho_{AH}/(\rho_{xx}^2 + \rho_{AH}^2) \approx \rho_{AH}/\rho_{xx}^2$, is monotonically increasing functions of both magnetization and electron density, and hence is an appropriate quantity to describe the carrier-mediated ferromagnetism in the ferromagnetic semiconductors. Figure 3.11d shows the magnetic field dependence of the anomalous Hall conductivity at 300 K for anatase (Ti,Co)O$_2$ with different electron density [59]. For the lowest electron density, the anomalous Hall conductivity was not observed. With an increase in the electron density, the amplitude of anomalous Hall conductivity increased with enhanced hysteresis, indicating the enhanced magnetization on the verge of the transition from paramagnetic to ferromagnetic states. This result also indicates the carrier-mediated ferromagnetism in this compound, in combination with the result of electric-field-effect study as described in Section 3.2.4.2.

As seen in Figure 3.11d, the anomalous Hall conductivity monotonically increased with electron density and hence with longitudinal conductivity. Figure 3.12 shows the relation between the anomalous Hall conductivity and the longitudinal conductivity for the rutile and anatase (Ti,Co)O$_2$ [68]. The data include those at different temperatures and Co contents in addition to those from different research groups [65, 69–72]. This relation approximately follows the scaling law irrespective of the sample parameters, $\sigma_{AH} \propto \sigma_{xx}^{1.6}$, as was originally found for the

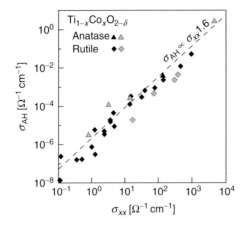

Figure 3.12 The relation between anomalous Hall conductivity (σ_{AH}) and longitudinal conductivity (σ_{xx}) for rutile and anatase (Ti,Co)O$_2$ thin films. The solid gray symbols correspond to data reported from other groups [68]. (Reprinted with permission. Copyright 2007, American Institute of Physics.)

rutile (Ti,Co)O$_2$ [65]. This relation implied that the magnetization increased monotonically (until the saturation) with increasing longitudinal conductivity, which promoted us to perform the electric-field-effect study on the ferromagnetism as described in Section 3.2.4.2.

3.2.4
Electric Field Effect

3.2.4.1 Electric Field Effect on Ferromagnetic Semiconductors
Electric field effect on ferromagnetic semiconductor enables to switch on and off the ferromagnetism by accumulating and depleting the carriers. This is in stark contrast with the ferromagnetic metals, in which the electric field effect is usually limited to change the magnetic anisotropy while preserving the magnetization amplitude [73, 74], except recent study on Co-ultrathin films [75]. The switching of the ferromagnetism in the ferromagnetic semiconductors was firstly demonstrated for (In,Mn)As [76]: the channel became paramagnetic (ferromagnetic) by applying positive (negative) gate voltages at low temperatures, where the hole was a charge carrier. Subsequently, the electric field effect was utilized for the magnetization reversal/rotation and the control of T_C [77–79]. The electric-field-effect studies on different compounds have been reported [80–87]. The magnetization measurement under electric field effect is a direct way but lacks information of the carrier conduction. Instead, the Hall effect is the most appropriate measurement for the electric-field-effect studies, as the electric field effect simultaneously modifies both carrier conduction and ferromagnetism, enabling the elucidation of the relation between carrier conduction and ferromagnetism.

Several technical issues of the electric-field-effect experiment have to be noted for careful inspection of the data. Use of piezoelectric material as a gate insulator may cause magnetoelastic effect [88]. In this case, the carriers may not play a significant role in the manipulation of ferromagnetism. For field-effect heterostructure devices, incidental bilayer transport may produce artificial anomalous Hall effect, which shows nonlinear magnetic field dependence despite the nonmagnetic character [89, 90]. Even if the bilayer transport is absent, such a nonlinear Hall effect as artifact can be observed in a heterogeneous sample owing to a nonuniform current path. These possibilities have to be ruled out in the experiments.

3.2.4.2 Electric Field Effect on Ferromagnetism in Co-Doped TiO$_2$

Recent central issue in (Ti,Co)O$_2$ was the role of carriers in high-temperature ferromagnetism. Experimental results in Section 3.2.3 strongly suggest that the carrier-mediated exchange interaction plays a significant role. However, the systematic variation of the electron density in those samples was caused by that of the oxygen vacancy because the oxygen vacancy served as the electron donor. Thus, it was difficult to distinguish the role of the carriers from that of the oxygen vacancy to induce ferromagnetism. Several groups proposed the oxygen-vacancy-mediated mechanisms rather than the carrier-mediated one [91–93]. If it is the case, (Ti,Co)O$_2$ will not be useful for spintronic applications not only because the ferromagnetism cannot be controlled through the carriers but also because the carriers are not spin polarized. Therefore, the electric-field-effect study was strongly desired in order to demonstrate the key role of carriers without changing the amount of the oxygen vacancy.

For the electric-field-effect experiments, the application of high electric field is desired to accumulate high-density carriers. Although metal-oxide-semiconductor structure is usually used, the gate oxide insulator is often destroyed by dielectric breakdown. Recently, electric-double layer transistor was developed for the application of very high electric field, $\sim 50\,\mathrm{MV\,cm^{-1}}$, in which liquid electrolyte served as the gate insulator [94]. This transistor was adopted for the electric-field-effect study on anatase (Ti,Co)O$_2$ as follows [49].

The inset of Figure 3.13a shows a photograph of the Hall bar sample used for the electric-double layer transistor (EDLT). Pt planar electrode adjacent to the Hall bar was used as gate electrode. The gate electrode and the channel were dipped in the electrolyte: an ionic liquid, N,N-diethyl-N-(2-methoxyethyl)-N-methylammonium–bis(trifluoromethylsulfonyl)-imide (DEME–TFSI) (Figure 3.13 except device 2 in Figure 3.13d), or an electrolyte salt, CsClO$_4$ solved in polyethylene oxide (device 2 in Figure 3.13d). Figure 3.13a shows the temperature dependence of resistivity during cooling at different gate voltages. With increasing gate voltage, the (Ti,Co)O$_2$ channel transformed from an insulating state to a metallic state. Figure 3.13b shows temperature dependences of the conductivity, mobility, and electron density at different gate voltages evaluated from the ordinary Hall term, representing that the increase in conductivity was caused by the electrostatically increased electron density. Figure 3.13c shows the magnetic field dependence of anomalous Hall conductivity at different gate voltages. The anomalous Hall conductivity was almost negligible at zero gate voltage, representing a paramagnetic state. With increasing gate voltage, the anomalous Hall conductivity showed clear hysteresis, representing electric-field-induced ferromagnetic state, at 300 K. Figure 3.13d

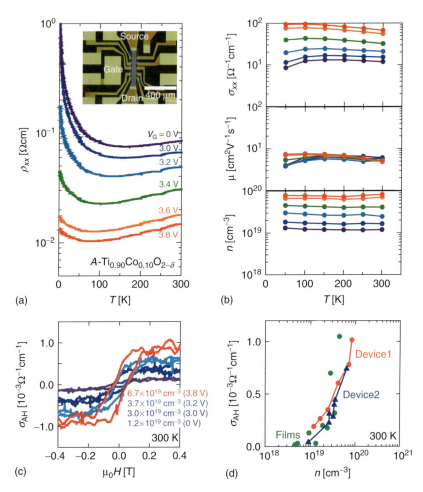

Figure 3.13 (a) Resistivity (ρ_{xx}) versus T curves in anatase $Ti_{0.90}Co_{0.10}O_2$ channel measured while cooling from 320 K with application of each gate voltage (V_G). The inset shows a top-view photograph of the Hall bar device used for the electric-double layer transistor (EDLT) [49]. (b) Conductivity (σ_{xx}), mobility (μ), and electron density (n) versus T plots at each V_G; gray scale correspond to those in (a) [49]. (c) Anomalous Hall conductivity (σ_{AH}) versus H curves at 300 K at different gate voltages V_G. n at each V_G are shown [49]. (d) The relation between σ_{AH} and n at 300 K. Interconnected circles and triangles denote the data for two different EDLTs (devices 1 and 2, respectively). V_G was from 0 to 3.8 V for device 1 and from 0 to 3.75 V for device 2. Isolated circles denote the data from nine $Ti_{0.90}Co_{0.10}O_2$ samples, in which n was varied by tuning the amount of the oxygen vacancy [49].

shows the relation between the anomalous Hall conductivity and the electron density at 300 K for the two EDLTs, together with the data for chemically doped (Ti,Co)O$_2$ with different amount of oxygen vacancy. For all of them, the anomalous Hall conductivity was steeply increased above $n \sim 1 \times 10^{19}$ cm^{-3}, corresponding to the paramagnetic-to-ferromagnetic transition. Extrinsic ferromagnetism such as magnetic segregation was undoubtedly ruled out because the only variable in the electric field effect was the electron density. The well-coincident relation between the electric field gating and chemically doped samples represents that the ferromagnetic exchange coupling in (Ti,Co)O$_2$ is mediated by the electron carriers.

3.2.5
A Principal Role of Electron Carriers on Ferromagnetism

By means of chemical doping, wider range of electron density was successfully covered than that by electric field effect for anatase (Ti,Co)O$_2$ (Figure 3.7d). As a result, the series of samples showed paramagnetic-to-ferromagnetic transition observed in the magnetization and the anomalous Hall effect (Figure 3.9b and Figure 3.11d) as well as insulator-to-metal transition (Figure 3.7c). Figure 3.14a,b summarizes the induced part of magnetization and the anomalous Hall conductivity at 300 K as a function of the electron density [59]. The magnetization increased from $n \sim 5 \times 10^{18}$ cm^{-3} (Figure 3.14a). The anomalous Hall conductivity was negligible for the insulating state and increased from $n \sim 1 \times 10^{19}$ cm^{-3} (Figure 3.14b). Both quantities were monotonically increasing functions of the electron density (except for a sample A), indicating a principal role of the electron carriers in the emergence of ferromagnetism. Taking into account the good coincidence of the data between the electric field effect and the chemical doping samples (Figure 3.13d), it is concluded that the role of oxygen vacancy on ferromagnetism is to provide electron carriers that mediate exchange interaction between localized spins at Co sites.

The sample A showed an inconsistent behavior: the reduced anomalous Hall conductivity and magnetization despite the highest electron density. It is noted that the coercive force of sample A was negligibly small, suggesting that the superparamagnetic feature, caused by the presence of Co nanoparticles, is due to excessively reductive growth conditions. This phenomenon is probably consistent with the cases for the observation of Co metal in samples annealed in vacuum [95] and the observation of the anomalous Hall effect with the presence of Co nanoparticles [96].

Around $n \sim 1 \times 10^{19}$ cm^{-3}, a finite magnetization was observed with negligible anomalous Hall conductivity. This result suggests an emergence of microscopic ferromagnetic embryos (bubbles or clusters) that are separated by the insulating matrix without contributing to macroscopic anomalous Hall effect. This state may correspond to ferromagnetic insulator state previously reported: the finite ferromagnetic magnetization without considerable electric conduction [92].

Figure 3.14c shows the anomalous Hall conductivity as a function of the electron density at 300 K for chemically doped rutile (Ti,Co)O$_2$. The rutile (Ti,Co)O$_2$ also showed a monotonically increasing anomalous Hall conductivity with the electron density, but the onset of the electron density was an order of magnitude larger

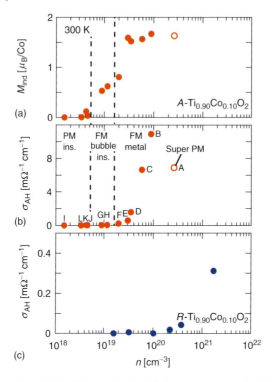

Figure 3.14 Electron density (n) dependence of (a) induced magnetization (M_{ind}) and (b) anomalous Hall conductivity (σ_{AH}) at 300 K for anatase $Ti_{0.90}Co_{0.10}O_2$. Open symbols denote a superparamagnetic sample. For the alphabet indices, see Figure 3.7d [59]. (Reprinted with permission. Copyright 2011, American Institute of Physics.) (c) n dependence of σ_{AH} at 300 K for rutile $Ti_{0.90}Co_{0.10}O_2$.

than the anatase (Ti,Co)O_2, representing that the rutile (Ti,Co)O_2 required higher electron density for the ferromagnetic ordering owing to the hopping conduction nature and the heavier electron mass ($m^* \sim 10m_0$ for rutile [97] and $m^* \sim m_0$ for anatase [98], where m_0 is the mass of free electron). The difference of those electrical conductions between rutile and anatase may have a significant consequence for ferromagnetic properties. The choice of anatase phase was important for field-effect study because fewer carriers are necessary to induce ferromagnetism.

3.2.6
Spectroscopic Properties

3.2.6.1 Photoemission Spectroscopy
Photoemission spectroscopy provides useful information of electronic structures such as valence band and ionic state of elements. Figure 3.15a shows valence band spectra for rutile (Ti,Co)O_2 with different Co contents x [99]. With increasing

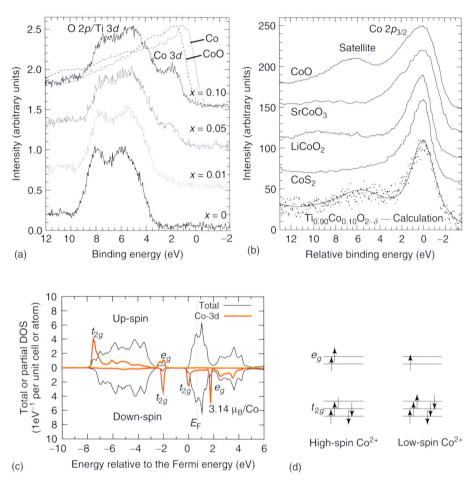

Figure 3.15 (a) Valence band X-ray photoemission spectra as a function of Co contents x for rutile $Ti_{1-x}Co_xO_2$. Valence band spectra of metallic Co (dotted line) and CoO (dashed line) are also shown for comparison [99]. (Reprinted with permission. Copyright 2006, American Physical Society.) (b) Comparison of the Co 2p spectra from CoO, $SrCoO_3$, $LiCoO_2$, and CoS_2 with that of rutile $Ti_{0.90}Co_{0.10}O_2$ thin film (dots). The solid line through the $Ti_{0.90}Co_{0.10}O_2$ spectrum shows the result of a cluster model calculation for CoO_6 [99]. (Reprinted with permission.

Copyright 2006, American Physical Society.) (c) Calculated density of states (DOS) in $Ti_{0.95}Co_{0.05}O_{1.95}$ with 2.5% oxygen vacancy within self-interaction-corrected local density approximation (SIC-LDA) in ferromagnetic state. Black and gray lines denote total DOS per unit cell and partial density of d states at Co site per Co atom, respectively. Calculated magnetic moment per Co atom was $3.14 \mu_B$ [108]. (Reprinted with permission. Copyright 2009, The Japan Society of Applied Physics.) (d) High- and low-spin configurations of Co^{2+}.

x, the valence band edge shifted to lower binding energy with an emergence of broad peak at $\sim$2 eV. It was suggested that the donor and acceptor levels of Co impurity are located just above the valence band maximum of TiO_2 in the dilute limit ($\sim$2 eV below the Fermi level) and in the conduction band minimum (degenerated), respectively, taking into account the charge transfer energy obtained from cluster model analysis of the Co 2p spectra. Figure 3.15b shows Co 2p spectrum (dots) for $x = 0.10$ sample together with that obtained by a cluster model calculation (line), with parameters of the charge-transfer energy $\Delta = 4.0$ eV, the on-site d–d Coulomb repulsion energy $U = 6.5$ eV, and ligand–metal transfer integral $(pd\sigma) = -1.1$ eV [99]. The satellite structure was significantly different from those of Co metal and nanoclusters [100, 101]. Among Co-containing compounds such as high-spin CoO [102], low- or intermediate-spin $SrCoO_3$ [103, 104], low-spin $LiCoO_2$ [102], and low-spin CoS_2 [105], the experimental spectrum of CoO showed the strongest resemblance in the Co 2p spectra, suggesting that Co ions in $(Ti,Co)O_2$ have divalent high-spin state at the film surface (Figure 3.15d). Ohtsuki *et al.* [106] performed independently soft and hard X-ray photoemission spectroscopy of anatase $(Ti,Co)O_2$. They observed metallic Ti^{3+} electrons at the Fermi level in the bulk and high-spin Co^{2+} below the Fermi level and proposed that Ti 3d carriers mediate the ferromagnetism. It is noted that the band bending at the surface due to carrier depletion was observed for both rutile and anatase $(Ti,Co)O_2$ [99, 106, 107]. The surface depletion could reduce the ferromagnetic coupling in the case of the carrier-mediated exchange interaction, as discussed in Section 3.2.6.2.

The band structure shows good coincidence with the electronic structure calculated by self-interaction-corrected local density approximation (SIC-LDA) (Figure 3.15c) [108]. In this model, both Co^{2+} high-spin state and the presence of oxygen vacancy were taken into account. The calculation reproduced well the valence band obtained experimentally, as seen in Figure 3.15a. A partial down-spin t_{2g} level was located at the conduction band minimum, which is expected to hybridize strongly with the electron carriers.

3.2.6.2 X-Ray Magnetic Circular Dichroism

X-ray magnetic circular dichroism (XMCD) is the difference in the absorption coefficients of X-ray photon helicity depending on the majority-spin direction of target elements, and hence is the element-selective magnetic measurement in contrast with standard magnetometer measurement. For 3d transition-metal elements, the L-edge absorption caused by the transition from 2p core level to 3d level is often used because it reflects the spin state of the final state of 3d level that plays a major role in the magnetism. For XMCD measurements of thin films, total electron yield (TEY) and total fluorescence yield (TFY) modes are available. The probing depths of the former and the latter are $\sim$5 and $\sim$100 nm, respectively; hence, magnetic state of the surface and the bulk can be evaluated separately (Figure 3.16a).

The XMCD spectrum of Co^{2+} in Co $L_{2,3}$-edge for rutile $(Ti,Co)O_2$ was observed with TEY mode, where surface treatments such as Ar ion sputtering and annealing were not applied (Figure 3.16b) [109]. Clear multiplet features at the Co $L_{2,3}$ edges

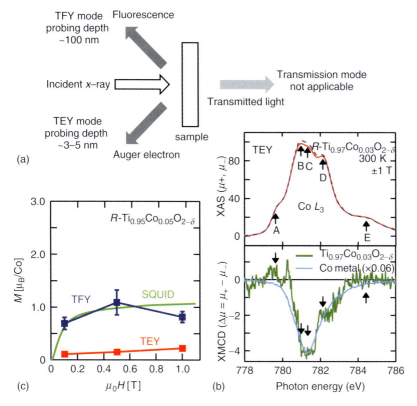

Figure 3.16 (a) Schematic illustration of X-ray magnetic circular dichroism (XMCD) measurement with total electron yield (TEY) and total fluorescence yield (TFY) modes. (b) Co L$_3$ region of X-ray absorption spectroscopy (XAS) (upper) and XMCD (bottom) spectra of rutile Ti$_{0.97}$Co$_{0.03}$O$_2$ thin film taken in the TEY mode at 300 K in 1 T [109]. (Reprinted with permission. Copyright 2006, American Institute of Physics.) (c) *M–H* relation measured with SQUID magnetometer and the magnetization estimated from the XMCD spectra for rutile Ti$_{0.95}$Co$_{0.05}$O$_2$ thin film [110]. (Reprinted with permission. Copyright 2011, IOP Publishing.)

were observed, ruling out the presence of Co metal. The spectral shapes of XMCD and X-ray absorption spectroscopy (XAS) were similar to those obtained by a full atomic-multiplet calculation for high-spin Co^{2+} ions under D_{2h}-symmetry crystal field of Ti site in rutile TiO$_2$. The orbital and spin magnetizations of Co ion were 0.013 μ_B and 0.12 μ_B per Co ion, respectively. The total magnetization was an order of magnitude smaller than that measured with magnetometer. Subsequent measurements were performed both in TEY and TFY modes [110]. The multiplet features of Co L$_{2,3}$ edge were observed in the XMCD and XAS spectra with TEY and TFY modes. The magnetization obtained with bulk-selective TFY mode using the optical sum rules [111] as well as with the cluster model calculation was similar to that measured with the magnetometer, whereas the magnetization obtained with surface-sensitive TEY mode was much smaller (Figure 3.16c). This result confirmed the presence of the magnetically dead layer at the sample surface. The possibility of CoO at the film surface was excluded by the significant deviations in the valence band and Co 2p line shapes (Figure 3.15a,b) [99]. Taking into account the presence of surface depletion observed with X-ray photoemission spectroscopy in (Ti,Co)O$_2$ [99, 106, 107] similar to nonmagnetic TiO$_2$ [112], it is plausible that the surface magnetization was reduced owing to the depleted carriers. This result is consistent with the carrier-mediated mechanism, and could explain the reduced magnetization in nanoparticle sample because of the high surface-to-volume ratio [113].

From the cluster model calculations of TFY spectrum, the estimated crystal symmetry of Co^{2+} ions was a mixture of D_{2h} low-spin, O_h low-spin, and O_h high-spin configurations (Figure 3.15d), contrasted with high-spin Co^{2+} ions under D_{2h}-symmetry crystal field of Ti site deduced from the TEY result [109]. This result suggested that the position of Co^{2+} ions in bulk lattice was significantly distorted from the pristine one owing to the adjacent oxygen vacancy. The local lattice distortion around Co ions in both rutile and anatase (Ti,Co)O$_2$ was also observed by using X-ray anomalous scattering measurement as discussed in Section 3.4.2.

3.3
Properties of Transition-Metal-Doped ZnO

Recently, there have been growing evidences of the high-temperature ferromagnetism in transition-metal-doped ZnO, although the origin of the ferromagnetism has still been under debate. In this section, we selected some recent studies on (Zn,Co)O in view of the success of (Ti,Co)O$_2$ studies described earlier. It should be noted that this section does not cover all the literature particularly at the initial stage of this topic. Reviews by Pan *et al.* and Ogale [44, 45] cover more complete literatures.

3.3.1
Structural Properties

Yang *et al.* [114, 115] observed no secondary phase in annealed Co- or Mn-implanted ZnO epitaxial thin films with X-ray diffraction, high-resolution transmission electron microscope, and selected area surface diffraction, where the sample showed anomalous Hall effect. On the other hand, Tay *et al.* [116] observed the presence of ZnCo$_2$O$_4$ or CoO in ferromagnetic (Zn,Co)O:Al thin films with high-resolution transmission electron microscope. Yan *et al.* [117] observed Co–Zn alloy segregation in ferromagnetic (Zn,Co)O bulk samples annealed in Zn vapor by X-ray diffraction. There is still controversy on the segregation of secondary phases among research groups and synthesis methods.

3.3.2
Magnetic and Electronic Properties

3.3.2.1 Magnetization
Recent studies reported a sufficiently large magnetization $>1\,\mu_B$ per transition metal [44, 45]. Such large magnetization could be attributed not to the uncompensated antiferromagnetic coupling proposed by Dietl *et al.* [118] but to the ferromagnetic magnetization. As discussed in Section 3.2, the magnetization measurement alone is not enough to evidence the intrinsic ferromagnetism. Not only magnetization measurements but various measurements such as MCD and anomalous Hall effect for systematic series of samples, are strongly desired to elucidate the relation between carrier conduction and ferromagnetism.

3.3.2.2 X-Ray Magnetic Circular Dichroism
The XMCD measurements are useful to rule out the possibility of ferromagnetism caused by unintentional phases. Recently, XMCD studies were performed for (Zn,V)O [119], (Zn,Mn)O [120–122], (Zn,Fe)O [123–125], and (Zn,Cu)O [126–128], in addition to (Zn,Co)O. Here, we review only the results of (Zn,Co)O.

Ferromagnetic XMCD signals were observed but the conclusions were different. Zhang *et al.* [125] reported intrinsic ferromagnetism in the nanoparticles from Co L-edge measurements, whereas Rode *et al.* [129] reported that the ferromagnetic phase coexisted with the paramagnetic phase in epitaxial thin films and attributed the former to the Co-metal clusters from Co L- and K-edge measurements.

Paramagnetic XMCD signals were also observed. In most of the studies, the multiplet feature was observed indicating Co^{2+} ionic state. Ney *et al.* [130, 131] reported paramagnetic epitaxial thin films from Co K-edge measurements. Farley *et al.* [120] reported paramagnetic Co L-edge for thin films in spite of the small ferromagnetic signal measured with magnetometer. Barla *et al.* and Tietze *et al.* [132, 133] observed paramagnetic TEY and TFY signals in epitaxial thin films; hence the samples were paramagnetic at the surface and in the bulk. Opel *et al.* [134] observed superparamagnetic TFY signal larger than TEY signal in epitaxial thin film claiming the metallic Co in the bulk and its oxidation at the surface, although the surface depletion discussed in Section 2.6.2 is possible from our viewpoint. Kobayashi *et al.* [135] observed the coexistence of paramagnetic and ferromagnetic components of Co^{2+} ions in homoepitaxial thin films, the magnetic field dependence of which was consistent with the magnetization measurement. They suggested a competition between antiferromagnetic Co-rich wurtzite nanoclusters and the electron-carrier-mediated ferromagnetism.

3.3.2.3 Magnetooptical Effect
Transition-metal-doped ZnO has the direct transition bandgap so that the magnetooptical spectrum should show sharper features, as shown in Figure 3.8d, than that of $(Ti,Co)O_2$. Pacuski *et al.* [136] reported magnetooptical spectroscopy of paramagnetic (Zn,Co)O epitaxial thin films. White *et al.* [137] reported paramagnetic

MCD in colloidal (Zn,Co)O nanocrystals excluding the possibilities of the spinodal decomposition and the uncompensated surface spins. Fukuma *et al.* [138] reported the presence of Co clusters in ferromagnetic (Zn,Co)O thin films. Kittilstved *et al.* [139] reported ferromagnetic MCD at 300 K for (Zn,Co)O polycrystalline thin film annealed in Zn vapor, which was coincident with the magnetization curve. Cho *et al.* [140] reported the reversible ferromagnetism in hydrogenated (Zn,Co)O, suggesting hydrogen-mediated ferromagnetism.

Neal *et al.* [141] observed ferromagnetic MCD or Kerr rotation of Ti-, V-, Mn-, and Co-doped ZnO thin films at the absorption edge of ZnO (Figure 3.17a), where the electron density was over 10^{20} cm^{-3} except (Zn,Mn)O. The spectra showed broad peaks at the absorption edge, suggesting the intrinsic nature of the ferromagnetism. Additional Al doping led to the enhanced MCD signal (Figure 3.17b) [142]. These results imply that sufficiently heavy electron doping will lead to the appearance of ferromagnetic MCD.

3.3.2.4 Anomalous Hall Effect

The observation of anomalous Hall effect, that is nonlinear magnetic field response of the Hall effect, was recently reported by many groups for (Zn,Mn)O

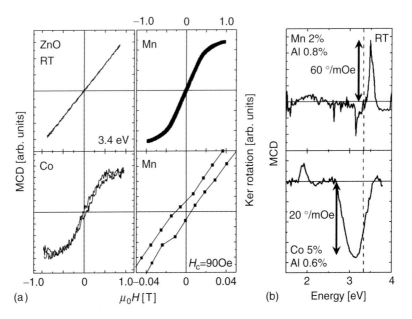

Figure 3.17 (a) Magnetic circular dichroism (MCD)/Kerr rotation versus *H* curves for the pure, Mn 2%-doped, and Co 2%-doped ZnO thin films at room temperature with photon energy of 3.4 eV. Bottom right panels show expanded figure for the Mn-doped thin film. The linear variation of the pure ZnO thin film (upper left) has been subtracted from the data [141]. (Reprinted with permission. Copyright 2006, American Physical Society.) (b) MCD spectra at room temperature for a $Zn_{0.972}Mn_{0.02}Al_{0.008}O$ and $Zn_{0.944}Co_{0.05}Al_{0.006}O$ thin films. The vertical dashed line indicates the ZnO band edge (3.35 eV) [142]. (Reprinted with permission. Copyright 2008, American Physical Society.)

[115, 142–146], (Zn,Co)O [114, 116, 143, 147–153], and (Zn,Cu)O [154, 155], while the absence of anomalous Hall effect was also reported for (Zn,Cr)O [156], (Zn,Mn)O [157–159], and (Zn,Co)O [157, 160]. Because the anomalous Hall effect without hysteresis can be seen even in superparamagnetic samples (e.g., sample A in Figure 3.14), the opened hysteresis may be required as an evidence of the ferromagnetic order. The opened hysteresis suggesting intrinsic ferromagnetism was reported by Behan *et al.* [142], Yang *et al.* [114], and Hsu *et al.* [151], where the electron density of the samples was within the range of $10^{19}-10^{20}$ cm^{-3}. Figure 3.18 shows the anomalous Hall effect in (Zn,Co)O thin films. In Figure 3.18a, a small hysteresis can be seen after the extraction of the anomalous Hall term [114]. In Figure 3.18b, the hysteresis curves of magnetization and anomalous Hall resistivity are almost identical [151], implying the intrinsic ferromagnetism.

The anomalous Hall conductivity was scaled approximately with the longitudinal conductivity (Figure 3.12). This scaling law was also followed by various ferromagnetic compounds including metals, metallic alloys, Mn-doped III–V compound

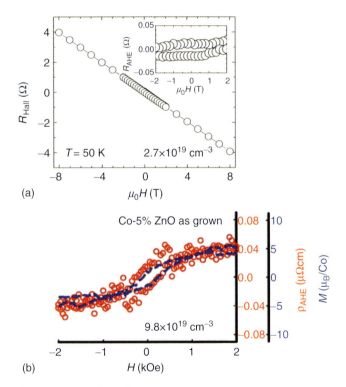

Figure 3.18 (a) The Hall resistance versus H curves for ZnO:Co at 50 K. The inset shows the anomalous Hall-effect data over an expanded field range [114]. (Reprinted with permission. Copyright 2008, American Institute of Physics.) (b) M versus H and AHE versus H curves for Zn$_{0.95}$Co$_{0.05}$O [151]. (Reprinted with permission. Copyright 2010, American Institute of Physics.)

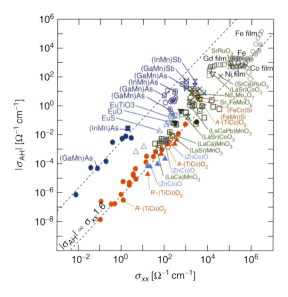

Figure 3.19 The relation between anomalous Hall conductivity (σ_{AH}) and longitudinal conductivity (σ_{xx}) for various ferromagnetic compounds. (Source: Modified from Refs. [161, 163].)

semiconductors, transition-metal complex oxides, silicides, and (Ti,Co)O$_2$ [161]. The theoretical background was discussed by Onoda *et al.* [162]. Figure 3.19 shows updated relation including the data of Eu-compounds [163] and (Zn,Co)O [114, 151]. The data of (Ti,Co)O$_2$ and (Zn,Co)O were located closely, possibly suggesting the similar mechanism of the anomalous Hall effect. Hsu *et al.* [151] suggested that the matching of the scaling law is an indication of the intrinsic ferromagnetism. However, further investigation of the underlying mechanism of the scaling law is required because a superparamagnetic sample could show a comparable magnitude of anomalous Hall conductivity (Figure 3.14).

3.3.2.5 A Role of Carriers on Ferromagnetism

Recently, quantitative comparisons between the magnetic and electric properties in (Zn,Co)O were started. Behan *et al.* [142] reported the carrier density dependence of the magnetization (Figure 3.20a). A finite magnetization for (Zn,Co)O was observed at an insulating regime $\leq 10^{18}$ cm^{-3} and at a metallic regime $\geq 10^{20}$ cm^{-3}, while the magnetization was negligible at the intermediate regime $10^{18}-10^{20}$ cm^{-3}, indicating a reentrant behavior. On the other hand, Yan *et al.* [114] reported that the magnetization monotonically increased as a function of the carrier density from $\sim 10^{18}$ cm^{-3} for (Zn,Co)O (Figure 3.20b). Park *et al.* [164] reported the ferromagnetism in (Zn,Mn)O being induced by improved mobility owing to the incorporation of hydrogen despite rather constant carrier density (Figure 3.20c). These results suggest a close relation between the ferromagnetism and the carrier

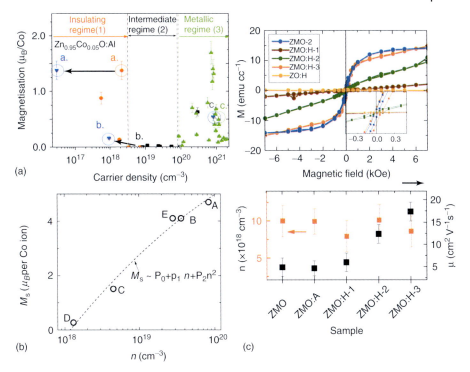

Figure 3.20 (a) Room-temperature magnetization for $Zn_{0.95}Co_{0.05}O$ thin films with varying Al doping, as a function of carrier density [142]. (Reprinted with permission. Copyright 2008, American Physical Society.) (b) The dependence of the saturated magnetization (M_S) on carrier density (n) in ZnO:Co thin films [114]. (Reprinted with permission. Copyright 2008, American Institute of Physics.) (c) Hysteresis loops of ZnO, $Zn_{0.96}Mn_{0.04}O$, and $Zn_{0.96}Mn_{0.04}O$:H thin films at room temperature (upper panel). The inset shows an expanded view of the hysteresis loops. Carrier mobility (μ) and n for each sample (bottom panel) [164]. (Reprinted with permission. Copyright 2007, Wiley.)

conduction, although further investigation is needed to reach a consensus on the mechanism by elucidating the role of carriers.

3.4
Discussion

3.4.1
Remarks on Sample Characterizations

Reproducible sample quality and properties are indispensable to conduct the research of magnetic oxide semiconductors. The importance of structural analyses was stressed in order to examine the presence of ferromagnetic segregation [165], although it is necessary to investigate the influence of such segregation on the macroscopic ferromagnetism observed by, for example, the anomalous Hall effect. It is equally important to investigate the relation between magnetic and electric/electronic properties, because the relation is the most important feature of the ferromagnetic semiconductor for semiconductor spintronics.

3.4.2
Origin of High-Temperature Ferromagnetism

The carrier-mediated exchange interaction takes a principal role in the high-temperature ferromagnetism, for example, for (Ti,Co)O$_2$. However, it is still a fundamental question why the T_C appears to be so high in these dilute magnetic systems, exemplified by magnetization and anomalous Hall-effect measurements for (Ti,Co)O$_2$ (Figure 3.21). The range of T_C was from 600 to 750 K [166–168]. For (Zn,Co)O, Song *et al.* [169] reported T_C of 790 K. For further investigation, it is necessary to evaluate T_C as a function of carrier density.

It has been discussed until recently as to which mechanism is dominant between the defect-mediated mechanisms such as bound magnetic polaron and oxygen vacancy-mediated superexchange interaction [91, 92, 170, 171], or the carrier-mediated mechanism [58, 65, 172]. As described in Section 3.2, recent studies on (Ti,Co)O$_2$ support the carrier-mediated mechanism, although more specification of the mechanism is needed such as double exchange interaction [173], Zener model [174–176], RKKY interaction [177, 178], and percolation of bound magnetic polaron [179, 180]. The observed ferromagnetic insulator state, a ferromagnetic embryo, possibly corresponds to magnetic/electronic phase separation [5], which is reminiscent of that reported in (Ga,Mn)As and perovskite manganese oxides [181–184].

Another important issue is the role of oxygen vacancy for ferromagnetism, as has been recognized in EuO for decades [185]. The oxygen vacancy may have several roles: the electron donor, the trapped center of bound magnetic polaron, and the source of lattice distortion. The first-principle calculation studies suggested that the oxygen vacancy enhances the ferromagnetic coupling [186–189]. Also, a structural correlation between the oxygen vacancy and the transition-metal ion was suggested

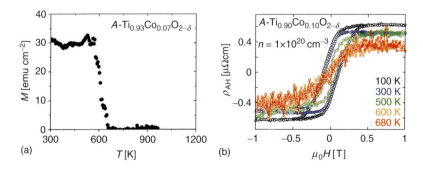

Figure 3.21 (a) *M* versus *T* data for anatase Ti$_{0.93}$Co$_{0.07}$O$_2$ [166]. (Reprinted with permission. Copyright 2003, American Physical Society.) (b) ρ_{AH} versus *H* curves for anatase Ti$_{0.90}$Co$_{0.10}$O$_2$ at high temperatures.

to lead to large exchange coupling [91]. Instead of the oxygen vacancy, the electron dopants such as Nb in $(Ti,Co)O_2$ [190] and Al, Ga, and H in $(Zn,Co)O$ [44, 131, 142, 164, 191–193] were used. At present, it is unclear whether such chemical dopants play the same role in ferromagnetism as oxygen vacancy or not.

For $(Ti,Co)O_2$, the charge neutrality in system requires the same amount of oxygen vacancy as that of Co^{2+} ions substituting Ti^{4+} ions. Such a significant amount of oxygen vacancy in conjunction with the aliovalent Co^{2+} could induce local lattice distortion as described in Section 3.2.6.2. By using X-ray anomalous scattering measurements, the substitution of doped element can be examined, in which the energy dependence of Bragg reflection intensity shows an anomaly at the absorption edge of the doped element. Figure 3.22 shows X-ray anomalous

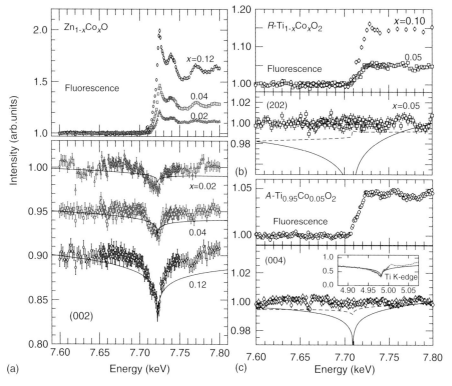

Figure 3.22 (a) Fluorescence spectra (upper panel) and X-ray energy dependences of 002 Bragg reflection intensity (bottom panel) for $Zn_{1-x}Co_xO$. Data are shifted vertically [194]. (b) Fluorescence spectra (upper panel) and X-ray energy dependence of 202 Bragg reflection intensity (bottom panel) for rutile $Ti_{1-x}Co_xO_2$ [194]. (c) Fluorescence spectrum (upper panel) and X-ray energy dependence of 004 Bragg reflection intensity (bottom panel) for anatase $Ti_{0.95}Co_{0.05}O_2$. Inset shows the result around the K edge of Ti. Solid and dashed lines denote calculations using the structure factor assuming random substitution of Zn or Ti with doped amount of Co ion and additionally assuming randomly created one oxygen vacancy in the octahedron at the Co site with certain shifts of the Co atom, respectively [194]. (Reprinted with permission. Copyright 2007, American Physical Society.)

scattering measurements of paramagnetic (Zn,Co)O and ferromagnetic rutile and anatase (Ti,Co)O$_2$ using a synchrotron radiation source [194]. The presence of Co atoms in the samples was confirmed from the fluorescence spectra without secondary phase confirmation by X-ray diffraction using a synchrotron radiation source. Zn sites were randomly substituted with Co atoms in paramagnetic (Zn,Co)O in good coincidence with the calculation of Bragg reflection intensity using the structure factor that assumes random substitution of Zn with a doped amount of Co ion (Figure 3.22a). However, Ti atoms were not exactly substituted with Co atoms in both rutile and anatase (Ti,Co)O$_2$ as the Bragg reflection intensity was negligible (Figure 3.22b,c). Additional randomness such as the presence of oxygen vacancy in CoO$_6$ octahedron could significantly reduce the scattering intensity (dashed lines in Figure 3.22b,c).

The coexistence of high- and low-spin states observed by XMCD [110] and X-ray anomalous scattering study (Figure 3.22) suggested that the random or glassy nature of the lattice structure around Co^{2+} ions in (Ti,Co)O$_2$ contrasted with the substitutional Co for Zn site in paramagnetic (Zn,Co)O$_2$. The electron microscopy combined with the first-principle calculation suggested a formation of Co$-$Ti^{3+}$-$V$_O$ complexes in (Ti,Co)O$_2$ [195]. In addition, the possibility of nanoscale chemical phase separation such as spinodal decomposition and spontaneously formed nanocrystals was proposed [5, 196, 197], because the thin film samples are often in metastable phase containing the transition metal beyond the thermodynamic solubility limit.

3.5
Summary and Outlooks

If the ferromagnetism originated from segregation or localized defects, the magnetic oxide semiconductors would not be useful for semiconductor spintronics. However, recent growing evidences support the scenario that high-temperature ferromagnetism is caused by the carrier-mediated mechanism, particularly for (Ti,Co)O$_2$. Accordingly, the magnetic oxide semiconductors would be promising materials for future semiconductor spintronics.

In order to elucidate the details of carrier-mediated ferromagnetism, further experimental and theoretical studies are necessary. It is important to study (i) fundamental parameters such as T_C and magnetization as a function of carrier density, (ii) microscopic features of magnetic/electronic phases from the aspects of phase separation and formation/percolation of bound magnetic polaron, and (iii) microscopic structures around the doped transition metal and oxygen vacancy. It would be noteworthy to point out that magnetic oxide semiconductors such as (Ti,Co)O$_2$ could be a counterpart of (Ga,Mn)As in terms of their carrier polarity. Further investigation would clarify the similarities and differences between these fascinating systems.

In addition to the mechanism of high-temperature ferromagnetism, the implementation of spintronic devices is a very important issue [6, 7, 198]. Progress has been scarcely seen in ferromagnetic oxide semiconductors in comparison with

(Ga,Mn)As [199, 200]. Despite low-temperature operation, tunneling magnetore-sistance effect was observed in magnetic tunneling junctions with $(Ti,Co)O_2$ and $(Zn,Co)O$ electrodes [201–203], probably because of the insufficient quality of the junctions. Nonetheless, the success of the electric field control of ferromagnetism at room temperature will directly lead to the electric field control of magnetooptical effect and the magnetization inversion at room temperature, for example. Further efforts will pave the way to semiconductor spintronics operating at room temperature.

Acknowledgments

This work was in collaboration with Y. Yamada, H. Toyosaki, T. Yamasaki, M. Nakano, K. Ueno, H. Kimura, Z. Jin, Y. Matsumoto, H. Koinuma, T. Hasegawa, T. Mizokawa, A. Fujimori, H. T. Yuan, H. Shimotani, Y. Iwasa, L. Gu, S. Tsukimoto, Y. Ikuhara, T. Matsumura, Y. Murakami, K. Ando, H. Saito, J. Okabayashi, H. Ofuchi, K. Ono, H. Fujioka, M. Oshima, K. Nakajima, and T. Chikyow. This work was supported in part by JSPS through NEXT Program initiated by CSTP, JST-PRESTO, NEDO Industrial Research Grant Program, and MEXT-KAKENHI.

References

1. Ohno, H. (1998) *Science*, **281**, 951.
2. Furdyna, J.K. (1988) *J. Appl. Phys.*, **64**, R20.
3. Willardson, R.K. and Beer, A.C. (eds) (1988) *Semiconductors and Semimetals*, Vol. 25, Academic Press, Boston, MA.
4. Onodera, K. *et al.* (1994) *Electron. Lett.*, **30**, 1954.
5. Dietl, T. (2010) *Nat. Mater.*, **9**, 965.
6. Wolf, S.A. *et al.* (2001) *Science*, **294**, 1488.
7. Žutić, I. *et al.* (2004) *Rev. Mod. Phys.*, **76**, 323.
8. Fukumura, T. *et al.* (1999) *Appl. Phys. Lett.*, **75**, 3366.
9. Fukumura, T. *et al.* (2001) *Appl. Phys. Lett.*, **78**, 958.
10. White, W.B. *et al.* (1965) *Trans. Br. Ceram. Soc.*, **64**, 521.
11. Dorain, P.B. (1958) *Phys. Rev.*, **112**, 1058.
12. Spałek, J. *et al.* (1986) *Phys. Rev. B*, **33**, 3407.
13. Mizokawa, T. *et al.* (2002) *Phys. Rev. B*, **65**, 085209.
14. Okabayashi, J. *et al.* (2004) *J. Appl. Phys.*, **95**, 3573.
15. Fukumura, T. *et al.* (2005) *Semicond. Sci. Technol.*, **20**, S103.
16. Goldstein, Y. *et al.* (1977) *Phys. Rev. Lett.*, **39**, 953.
17. Sawicki, M. *et al.* (1986) *Phys. Rev. Lett.*, **56**, 508.
18. Wojtowicz, T. *et al.* (1986) *Phys. Rev. Lett.*, **56**, 2419.
19. Koinuma, H. *et al.* (2004) *Nat. Mater.*, **3**, 429.
20. Matsumoto, Y. *et al.* (1999) *Jpn. J. Appl. Phys.*, **38**, L603.
21. Fukumura, T. *et al.* (2000) *Appl. Phys. Lett.*, **77**, 3426.
22. Shono, T. *et al.* (2000) *Appl. Phys. Lett.*, **77**, 1363.
23. Fukumura, T. *et al.* (2001) *Physica E*, **10**, 135.
24. Jin, Z. *et al.* (2001) *Appl. Phys. Lett.*, **78**, 3824.
25. Ando, K. *et al.* (2001) *J. Appl. Phys.*, **89**, 7284.
26. Ando, K. *et al.* (2001) *Appl. Phys. Lett.*, **78**, 2700.
27. Kimura, H. *et al.* (2002) *Appl. Phys. Lett.*, **80**, 94.

28. Momma, K. and Izumi, F. (2008) *J. Appl. Crystallogr.*, **41**, 653.

29. Matsumoto, Y. *et al.* (2001) *Science*, **291**, 854.

30. Matsumoto, Y. *et al.* (2001) *Jpn. J. Appl. Phys.*, **40**, L1204.

31. Tang, H. (1994) Electronic properties of anatase TiO_2 investigated by electrical and optical measurements on single crystals and thin films. PhD Thesis. Ecole Polytechnique Federale de Lausanne.

32. Diebold, U. (2003) *Surf. Sci. Rep.*, **48**, 53.

33. Abrahams, S.C. *et al.* (1971) *J. Chem. Phys.*, **55**, 3206.

34. Horn, M. *et al.* (1972) *Z. Kristallogr.*, **136**, 273.

35. Glassford, K.M. *et al.* (1992) *Phys. Rev. B*, **46**, 1284.

36. Burdett, J.K. (1987) *J. Am. Chem. Soc.*, **109**, 3639.

37. Dietl, T. *et al.* (2000) *Science*, **287**, 1019.

38. Sato, K. *et al.* (2000) *Jpn. J. Appl. Phys.*, **39**, L555.

39. Pearton, S.J. *et al.* (2003) *J. Appl. Phys.*, **93**, 1.

40. Prellier, W. *et al.* (2003) *J. Phys. Condens. Matter*, **15**, R1583.

41. Fukumura, T. *et al.* (2004) *Appl. Surf. Sci.*, **223**, 62.

42. Janisch, R. *et al.* (2005) *J. Phys. Condens. Matter*, **17**, R657.

43. Liu, C. *et al.* (2005) *J. Mater. Sci.: Mater. Electon.*, **16**, 555.

44. Pan, F. *et al.* (2008) *Mater. Sci. Eng. R*, **62**, 1.

45. Ogale, S.B. (2010) *Adv. Mater.*, **22**, 3125.

46. Elfimov, I.S. *et al.* (2002) *Phys. Rev. Lett.*, **89**, 216403.

47. Kenmochi, K. *et al.* (2004) *J. Phys. Soc. Jpn.*, **73**, 2952.

48. Osorio-Guillén, J. *et al.* (2006) *Phys. Rev. Lett.*, **96**, 107203.

49. Yamada, Y. *et al.* (2011) *Science*, **332**, 1065.

50. Saito, H. *et al.* (2003) *Phys. Rev. Lett.*, **90**, 207202.

51. Schallenberg, T. *et al.* (2006) *Appl. Phys. Lett.*, **89**, 042507.

52. Wang, M. *et al.* (2008) *Appl. Phys. Lett.*, **93**, 132103.

53. Brežný, B. *et al.* (1969) *J. Inorg. Nucl. Chem.*, **31**, 649.

54. Chen, S. *et al.* (1993) *J. Vac. Sci. Technol. A*, **11**, 2419.

55. Toyosaki, H. (2007) Spin polarized properties and devices of ferromagnetic oxide semiconductor, PhD thesis. Tohoku University (In Japanese).

56. Yamasaki, T. *et al.* (2008) *Appl. Phys. Express*, **1**, 111302.

57. Yamasaki, T. *et al.* (2009) *Appl. Phys. Lett.*, **94**, 102515.

58. Fukumura, T. *et al.* (2008) *New J. Phys.*, **8**, 055018.

59. Yamada, Y. *et al.* (2011) *Appl. Phys. Lett.*, **99**, 242502.

60. Khalid, M. *et al.* (2010) *Phys. Rev. B*, **81**, 214414.

61. Sugano, S. *et al.* (2000) *Magneto-Optics*, Springer-Verlag, New York.

62. Fukumura, T. *et al.* (2003) *Jpn. J. Appl. Phys.*, **42**, L105.

63. Chien, C.L. and Westgate, C.R. (eds) (1980) *The Hall Effect and its Applications*, Plenum Press, New York.

64. Pakhomov, A.B. *et al.* (1996) *J. Appl. Phys.*, **79**, 6140.

65. Toyosaki, H. *et al.* (2004) *Nat. Mater.*, **3**, 221.

66. Hosaka, N. *et al.* (1997) *J. Phys. Soc. Jpn.*, **66**, 877.

67. Inoue, M. *et al.* (2006) *J. Phys. D*, **39**, R151.

68. Ueno, K. *et al.* (2007) *Appl. Phys. Lett.*, **90**, 072103.

69. Hitosugi, T. *et al.* (2006) *J. Appl. Phys.*, **99**, 08M121.

70. Cho, J.H. *et al.* (2006) *J. Korean Phys. Soc.*, **48**, 1400.

71. Ramaneti, R. *et al.* (2007) *Appl. Phys. Lett.*, **91**, 012502.

72. Higgins, J.S. *et al.* (2004) *Phys. Rev. B*, **69**, 073201.

73. Weisheit, M. *et al.* (2007) *Science*, **315**, 349.

74. Maruyama, T. *et al.* (2009) *Nat. Nanotechnol.*, **4**, 158.

75. Chiba, D. *et al.* (2011) *Nat. Mater.*, **10**, 853.

76. Ohno, H. *et al.* (2000) *Nature*, **408**, 944.

77. Chiba, D. *et al.* (2003) *Science*, **301**, 943.

78. Chiba, D. *et al.* (2008) *Nature*, **455**, 515.

79. Nishitani, Y. *et al.* (2010) *Phys. Rev. B*, **81**, 045208.

80. Park, Y.D. *et al.* (2002) *Science*, **295**, 651.

81. Nazmul, A.M. *et al.* (2004) *Jpn. J. Appl. Phys.*, **43**, L233.
82. Zhao, T. *et al.* (2005) *Phys. Rev. Lett.*, **94**, 126601.
83. Kanki, T. *et al.* (2006) *Appl. Phys. Lett.*, **89**, 242506.
84. Stolichnov, I. *et al.* (2008) *Nat. Mater.*, **7**, 464.
85. Nepal, N. *et al.* (2009) *Appl. Phys. Lett.*, **94**, 132505.
86. Xiu, F. *et al.* (2010) *Nat. Mater.*, **9**, 337.
87. Sawicki, M. *et al.* (2010) *Nat. Phys.*, **6**, 22.
88. Bihler, C. *et al.* (2008) *Phys. Rev. B*, **78**, 45203.
89. Wieder, H.H. (1974) *Appl. Phys. Lett.*, **25**, 26.
90. Beck, W.A. *et al.* (1987) *J. Appl. Phys.*, **62**, 541.
91. Coey, J.M.D. *et al.* (2005) *Nat. Mater.*, **4**, 173.
92. Griffin, K.A. *et al.* (2005) *Phys. Rev. Lett.*, **94**, 157204.
93. Kaspar, T.C. *et al.* (2005) *Phys. Rev. Lett.*, **95**, 217203.
94. Ueno, K. *et al.* (2008) *Nat. Mater.*, **7**, 855.
95. Kim, J.-Y. *et al.* (2003) *Phys. Rev. Lett.*, **90**, 017401.
96. Shinde, S.R. *et al.* (2004) *Phys. Rev. Lett.*, **92**, 166601.
97. Frederikse, H.P.R. (1961) *J. Appl. Phys.*, **32**, 2211.
98. Tang, H. *et al.* (1994) *J. Appl. Phys.*, **75**, 2042.
99. Quilty, J.W. *et al.* (2006) *Phys. Rev. Lett.*, **96**, 027202.
100. Cattaruzza, E. *et al.* (1998) *Appl. Phys. Lett.*, **73**, 1176.
101. Wang, F.L. *et al.* (2004) *J. Appl. Phys.*, **95**, 5069.
102. van Elp, J. *et al.* (1991) *Phys. Rev. B*, **44**, 6090.
103. Saitoh, T. *et al.* (1997) *Phys. Rev. B*, **56**, 1290.
104. Potze, R.H. *et al.* (1995) *Phys. Rev. B*, **51**, 11501.
105. Bocquet, A.E. *et al.* (1992) *Phys. Rev. B*, **46**, 3771.
106. Ohtsuki, T. *et al.* (2011) *Phys. Rev. Lett.*, **106**, 047602.
107. Yamashita, N. *et al.* (2010) *Appl. Phys. Lett.*, **96**, 021907.
108. Kizaki, H. *et al.* (2009) *Appl. Phys. Express*, **2**, 053004.
109. Mamiya, K. *et al.* (2006) *Appl. Phys. Lett.*, **89**, 062506.
110. Singh, V.R. *et al.* (2011) *J. Phys. Condens. Matter*, **23**, 176001.
111. Chen, C.T. *et al.* (1995) *Phys. Rev. Lett.*, **75**, 152.
112. Lira, E. *et al.* (2011) *J. Am. Chem. Soc.*, **133**, 6529.
113. Bryan, J.D. *et al.* (2004) *J. Am. Chem. Soc.*, **126**, 11640.
114. Yang, Z. *et al.* (2008) *J. Appl. Phys.*, **104**, 113712.
115. Yang, Z. *et al.* (2009) *J. Appl. Phys.*, **105**, 053708.
116. Tay, M. *et al.* (2006) *J. Appl. Phys.*, **100**, 063910.
117. Yang, G. *et al.* (2011) *Appl. Phys. Lett.*, **99**, 082501.
118. Dietl, T. *et al.* (2007) *Phys. Rev. B*, **76**, 155312.
119. Ishida, Y. *et al.* (2007) *Appl. Phys. Lett.*, **90**, 022510.
120. Farley, N.R.S. *et al.* (2008) *New J. Phys.*, **10**, 055012.
121. Droubay, T.C. *et al.* (2009) *Phys. Rev. B*, **79**, 155203.
122. Thakur, P. *et al.* (2007) *Appl. Phys. Lett.*, **91**, 162503.
123. Kumar, R. *et al.* (2008) *J. Phys. D: Appl. Phys.*, **41**, 155002.
124. Kataoka, T. *et al.* (2010) *J. Appl. Phys.*, **107**, 033718.
125. Zhang, Z.H. *et al.* (2009) *Nat. Nanotech.*, **4**, 523.
126. Keavney, D.J. *et al.* (2007) *Appl. Phys. Lett.*, **91**, 012501.
127. Thakur, P. *et al.* (2010) *J. Appl. Phys.*, **107**, 103915.
128. Herng, T.S. *et al.* (2010) *Phys. Rev. Lett.*, **105**, 207201.
129. Rode, K. *et al.* (2008) *Appl. Phys. Lett.*, **92**, 012509.
130. Ney, A. *et al.* (2010) *Phys. Rev. B*, **81**, 054420.
131. Ney, A. *et al.* (2010) *Phys. Rev. B*, **82**, 041202(R).
132. Barla, A. *et al.* (2007) *Phys. Rev. B*, **76**, 125201.
133. Tietze, T. *et al.* (2008) *New J. Phys.*, **10**, 055009.
134. Opel, M. *et al.* (2008) *Eur. Phys. J. B*, **63**, 437.

135. Kobayashi, M. *et al.* (2010) *Phys. Rev. B*, **81**, 075204.

136. Pacuski, W. *et al.* (2006) *Phys. Rev. B*, **73**, 035214.

137. White, M.A. *et al.* (2008) *Chem. Mater.*, **20**, 7107.

138. Fukuma, Y. *et al.* (2008) *Appl. Phys. Lett.*, **93**, 142510.

139. Kittilstved, K.R. *et al.* (2006) *Appl. Phys. Lett.*, **89**, 062510.

140. Cho, Y.C. *et al.* (2009) *Appl. Phys. Lett.*, **95**, 172514.

141. Neal, J.R. *et al.* (2006) *Phys. Rev. Lett.*, **96**, 197208.

142. Behan, A.J. *et al.* (2008) *Phys. Rev. Lett.*, **100**, 047206.

143. Nielsen, K.-W. *et al.* (2005) *Superlattice. Microst.*, **37**, 327.

144. Theodoropoulou, N. *et al.* (2006) *J. Magn. Magn. Mat.*, **300**, 407.

145. Xu, Q. *et al.* (2007) *Appl. Phys. Lett.*, **91**, 092503.

146. Shim, W. *et al.* (2007) *J. Appl. Phys.*, **101**, 123908.

147. Xu, Q. *et al.* (2007) *J. Appl. Phys.*, **101**, 063918.

148. Potzger, K. *et al.* (2008) *Appl. Phys. Lett.*, **93**, 232504.

149. Akdogan, N. *et al.* (2009) *J. Appl. Phys.*, **105**, 043907.

150. Lu, Z. *et al.* (2009) *Appl. Phys. Lett.*, **94**, 152507.

151. Hsu, H.S. *et al.* (2010) *Appl. Phys. Lett.*, **96**, 242507.

152. Lee, Y.H. *et al.* (2011) *J. Magn. Magn. Mater.*, **323**, 1846.

153. Pearton, S.J. *et al.* (2007) *IEEE Trans. Electron. Dev.*, **54**, 1040.

154. Xu, Q. *et al.* (2008) *Appl. Phys. Lett.*, **92**, 082508.

155. Hou, D.L. *et al.* (2007) *Appl. Phys. Lett.*, **90**, 142502.

156. Roberts, B.K. *et al.* (2008) *J. Appl. Phys*, **103**, 07D133.

157. Xu, X.H. *et al.* (2006) *New J. Phys.*, **8**, 135.

158. Hong, N.H. *et al.* (2009) *Physica B*, **404**, 3978.

159. Yang, Z. *et al.* (2011) *J. Cryst. Growth*, **314**, 97.

160. Ye, S. *et al.* (2009) *Phys. Rev. B*, **80**, 245321.

161. Fukumura, T. *et al.* (2007) *Jpn. J. Appl. Phys.*, **46**, L642.

162. Onoda, S. *et al.* (2008) *Phys. Rev. B*, **77**, 165103.

163. Yamasaki, Y. *et al.* (2011) *Appl. Phys. Lett.*, **98**, 082116.

164. Park, S.Y. *et al.* (2007) *Adv. Mater.*, **19**, 3496.

165. Kaspar, T.C. *et al.* (2008) *Phys. Rev. B*, **77**, 201303(R).

166. Shinde, S.R. *et al.* (2003) *Phys. Rev. B*, **67**, 115211.

167. Khaibullin, R.I. *et al.* (2007) *Nucl. Instrum. Methods Phys. Res. B*, **257**, 369.

168. Ueno, K. *et al.* (2008) *J. Appl. Phys.*, **103**, 07D114.

169. Song, C. *et al.* (2006) *Phys. Rev. B*, **73**, 024405.

170. Kikoin, K. *et al.* (2006) *Phys. Rev. B*, **74**, 174407.

171. Yan, W. *et al.* (2009) *Appl. Phys. Lett.*, **94**, 042508.

172. Calderón, M.J. *et al.* (2007) *Annal. Phys.*, **322**, 2618.

173. Zener, C. (1951) *Phys. Rev.*, **82**, 403.

174. Zener, C. (1951) *Phys. Rev.*, **81**, 440.

175. Zener, C. (1951) *Phys. Rev.*, **83**, 299.

176. Dietl, T. *et al.* (2001) *Phys. Rev. B*, **63**, 195205.

177. Yosida, K. (1996) *Quantum Theory of Magnetism*, Springer, Berlin.

178. Priour, D.J. *et al.* (2004) *Phys. Rev. Lett.*, **92**, 117201.

179. Kübler, J. *et al.* (1975) *Phys. Rev. B*, **11**, 4440.

180. Kaminski, A. *et al.* (2002) *Phys. Rev. Lett.*, **88**, 247202.

181. Mayr, M. *et al.* (2002) *Phys. Rev. B*, **65**, 241202(R).

182. Alvarez, C. *et al.* (2002) *Phys. Rev. Lett.*, **89**, 277202.

183. Kaminski, A. *et al.* (2003) *Phys. Rev. B*, **68**, 235210.

184. Moreo, A. *et al.* (1999) *Science*, **283**, 2034.

185. Torrance, J.B. *et al.* (1972) *Phys. Rev. Lett.*, **29**, 1168.

186. Park, M.S. *et al.* (2002) *Phys. Rev. B*, **65**, 161201(R).

187. Weng, H. *et al.* (2004) *Phys. Rev. B*, **69**, 125219.

188. Errico, L.A. *et al.* (2005) *Phys. Rev. B*, **72**, 184425.

189. Wang, Q. *et al.* (2009) *Phys. Rev. B*, **79**, 115407.

190. Sakai, E. *et al.* (2010) *Appl. Phys. Express*, **3**, 043001.
191. Tiwari, A. *et al.* (2008) *Appl. Phys. Lett.*, **92**, 062509.
192. Lu, Z. *et al.* (2009) *Appl. Phys. Lett.*, **94**, 152507.
193. Lu, Z. *et al.* (2009) *Appl. Phys. Lett.*, **95**, 062509.
194. Matsumura, T. *et al.* (2007) *Phys. Rev. B*, **76**, 115320.
195. Griffin Roberts, K. *et al.* (2008) *Phys. Rev. B*, **78**, 014409.
196. Sato, K. *et al.* (2010) *Rev. Mod. Phys.*, **82**, 1633.
197. Meyerheim, H.L. *et al.* (2009) *Phys. Rev. Lett.*, **102**, 156102.
198. Fabian, J. *et al.* (2007) *Acta Phys. Slov.*, **57**, 565.
199. Jungwirth, T. *et al.* (2006) *Rev. Mod. Phys.*, **78**, 809.
200. Awschalom, D.D. *et al.* (2007) *Nat. Phys.*, **3**, 153.
201. Toyosaki, H. *et al.* (2005) *Jpn. J. Appl. Phys.*, **44**, L896.
202. Song, C. *et al.* (2007) *Appl. Phys. Lett.*, **91**, 042106.
203. Ramachandran, S. *et al.* (2008) *Solid State Commun.*, **145**, 18.

4
Effect of Ta Alloying on the Optical, Electronic, and Magnetic Properties of TiO$_2$ Thin Films

S. Dhar, A. Roy Barman, A. Rusydi, K. Gopinadhan, Ariando, Y. P. Feng, M. B. H. Breese, Hans Hilgenkamp, and T. Venkatesan

4.1
Introduction

Titanium dioxide (TiO$_2$) is one of the most intensively studied material systems in the field of surface science and metal oxides [1]. Highly conductive TiO$_2$ is a potential alternative to indium tin oxide (ITO) because of its various advantages, such as high refractive index (~2.8), nontoxicity, high surface photoactivity, and favorable overlap with the ultraviolet portion of the solar spectrum. It exhibits room-temperature (RT) ferromagnetic properties after doping with a dilute amount of magnetic elements [2]. Osorio-Guillen *et al.* [3] proposed that "dopants" such as tantalum (Ta) and niobium (Nb) will introduce donor states inside the conduction band of anatase TiO$_2$, producing free electrons as well as compensating defects. This substitution renders the system in a transparent conductive state [4–6], causes a blueshift of the bandgap [5, 6], and induces significant effects on the band structure [6] and dramatic effects on the photoluminescence (A. Roy Barman *et al.*, unpublished) [7], in addition to introducing magnetic phenomena such as Kondo scattering [8]. However, the most exciting new observation [9] in this system is the cationic-vacancy-induced RT ferromagnetism (FM) mediated by itinerant charge carriers.

In this chapter, we review the effect of Ta incorporation in TiO$_2$ on the band structure, and optical and electronic properties, in detail. This is followed by a brief review of the field of diluted magnetic semiconducting oxides (DMSOs), where a small amount of magnetic element is introduced into an oxide host to produce FM. Finally, we review the emerging field of cationic-vacancy-induced FM in anatase TiO$_2$ system without any magnetic element.

4.1.1
Properties of TiO$_2$

TiO$_2$ has three structural polymorphs, namely rutile, anatase, and brookite, which are shown in Figure 4.1, and their common material parameters are listed in

Functional Metal Oxides: New Science and Novel Applications, First Edition.
Edited by Satishchandra B. Ogale, Thirumalai V. Venkatesan, and Mark G. Blamire.

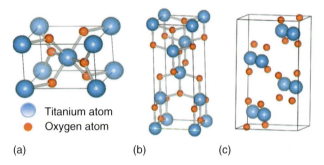

○ Titanium atom
● Oxygen atom

(a) (b) (c)

Figure 4.1 Structure of TiO$_2$ phases: (a) rutile, (b) anatase, and (c) brookite.

Table 4.1. In both rutile and anatase structures, the Ti atom is surrounded by six oxygen atoms in a distorted octahedral configuration. Both rutile and anatase phases are stable at RT and the anatase phase is easily converted into the rutile phase on heating at 550 °C.

Thin films of this material is generally used as white pigments, antireflection coating materials, waveguides, filters, gas sensors, and spacer materials in magnetic spin valve, and gate insulators. Both rutile and anatase have anisotropic optical properties because of their tetragonal crystal structure. They have similar electrical properties except that the electron mobility in the anatase phase is much higher (more than two orders of magnitude in the thin-film form) and this makes it more attractive than the rutile for optoelectronic applications. Many recent exciting phenomena such as the presence of magnetic centers leading to Kondo effect, transparent metallic conductivity, and so on, are seen in single-crystal anatase TiO$_2$ thin films, raising the importance of this phase further.

Table 4.1 Physical properties of bulk TiO$_2$ phases.

	Rutile	Anatase	Brookite
Crystal system	Tetrahedral	Tetrahedral	Orthorhombic
Point group	4/*mmm*	4/*mmm*	*Mmm*
Space group	P4$_2$/*mnm*	I4$_1$/*amd*	*Pbca*
Bandgap (eV)	3.00 (direct)	3.21 (indirect)	3.13
Electron mobility (cm^2 V^{-1} s^{-1})	1	10	—
Mass density (gm cc^{-1})	4.2743	3.895	4.123
a (Å)	4.578 (0.584)	3.7842	9.184
b (Å)	4.578	3.782	—
c (Å)	2.954	9.502	5.145
Volume	62.07	136.25	257.38
Molecule/unit cell	2	4	8

4.2
Ta Substitution in TiO$_2$: Doping or Alloying?

At a small percentage level of Ta or Nb, where significant modifications to the TiO$_2$ properties are seen, do these elements act as simple dopants or do they render the system into a new alloy [6]? In this section, this question is resolved in detail.

4.2.1
The PLD Phase Diagram

Since the TiO$_2$ film properties are phase dependent, it is important to know the phase boundary between anatase and rutile phases in the pulsed laser deposition (PLD) process, which is used for making the films. For making films, very high purity (99.999%) Ta$_2$O$_5$ and TiO$_2$ powders, either in pure form or a mixture of them, were ground for several hours and then sintered in a furnace at 1000–1100 °C for 20 h in air. Then, pellets were made and sintered again at 1100 °C in air for 24 h. Care was taken to avoid magnetic contaminants.

Figure 4.2 shows a phase diagram of Ti$_{1-x}$Ta$_x$O$_2$ ($x = 0.06$) as a function of oxygen partial pressure and growth temperature during the PLD process. The phase boundary (uncertainty on the order of ± 20 °C and $\pm 2 \times 10^{-5}$ Torr) is relatively sharp.

At low temperatures, no X-ray diffraction (XRD) peaks were seen, implying very small crystallites or near-amorphous films. But when the temperature was increased, by keeping oxygen partial pressure high, a direct transition from amorphous TiO$_2$ to the anatase phase was seen, whereas at lower oxygen partial pressures, there was a transition to the rutile phase followed by the anatase phase. No difference in the phase diagram was noticed for the pure TiO$_2$ films. This tells us that crystallographic phase-wise, there is not much effect of Ta on TiO$_2$.

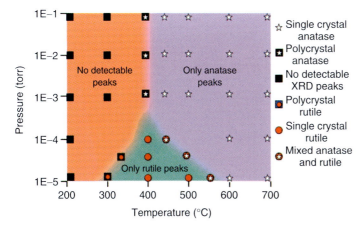

Figure 4.2 Phase diagram (for PLD process) of Ti$_{1-x}$Ta$_x$O$_2$ ($x \sim 0.06$) as a function of oxygen partial pressure and growth temperature.

4.2.2
Optical and Electronic Properties

Figure 4.3a,b shows a transmittance of 70–90% in the visible range for anatase and rutile films, respectively, with varying Ta concentrations. In order to extract the bandgap (E_g), $(\alpha)^{1/2}$ as a function of energy is plotted in Figure 4.3c,d. Both anatase and rutile thin films show a blueshift in E_g, with increasing Ta concentration. For anatase films, E_g varies from 3.3 (pure TiO$_2$) to 3.5 eV (8% Ta–TiO$_2$), while for rutile films, the variation is much steeper from 2.7 to 3.5 eV for the same Ta concentrations.

Figure 4.4 shows the variation in E_g with Ta concentration for the anatase and rutile films. The blue (anatase) and red (rutile) curves are the fit of Vegard's law to the data, which is an empirical rule of bandgap variation with the constituents [10, 11]. Although there exists a mixed view [12–15] about the nature of the optical bandgap of TiO$_2$ as many previous studies assign both anatase and rutile forms to have an indirect bandgap, for which $\alpha \propto (h\nu - E_g)^2$, where α is the inverse absorption length and E_g is the optical bandgap, a nice fit was obtained to the E_g versus Ta concentration for both the anatase and rutile films using appropriate bowing parameters (0.5 eV for rutile and 10.3 eV for anatase).

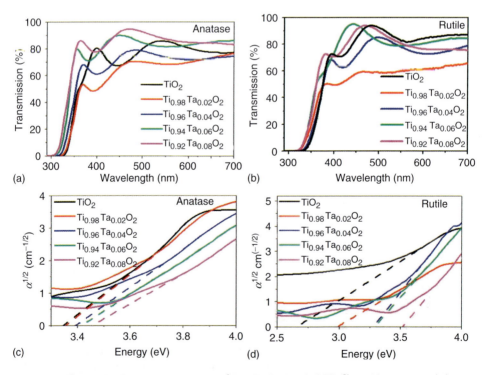

Figure 4.3 Transmittance spectra of Ti$_{1-x}$Ta$_x$O$_2$ ($x = 0$–0.08) films: (a) anatase and (b) rutile. The plot of $\alpha^{1/2}$ versus $h\nu$ for (c) anatase and (d) rutile phase showing increase in E_g.

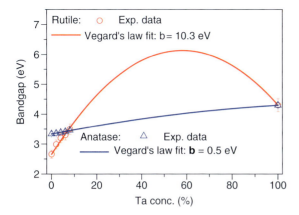

Figure 4.4 Blueshift in the bandgap for the anatase (blue color) and rutile (red color) $Ti_{1-x}Ta_xO_2$ films fitted with Vegard's law.

Alternatively, this large blueshift seen in the optical bandgap can also be due to the band-filling or the Moss–Burstein effect [16] in which free electrons in the conduction band (introduced by a dopant), shift the Fermi level higher than the conduction band minimum (CBM), resulting in an increase in the bandgap. To understand why the Moss–Burstein effect cannot explain the blueshift, Barman *et al.* [6] looked into the electronic transport properties of anatase and rutile TiO_2. Figure 4.5a shows the variation in carrier density and electron mobility with Ta concentration for the anatase films. As expected, there is an increase in the carrier density with the Ta concentration (Figure 4.5b). However, the mobility (Figure 4.5b) in the anatase phase is not Ta concentration (within 10%) dependent. While the RT resistivity of the rutile $Ti_{1-x}Ta_xO_2$ ($x = 0.08$) film is about two orders higher, its carrier density ($n = 2 \times 10^{21}$ cm^{-3}) was found to be of the same order as that of the anatase film. This suggests that the electron mobility in rutile thin films is significantly smaller (by a factor of 2–3 orders of magnitude) than the anatase films. As the calculated electron mobility of the rutile film was of the order of $\sim 10^{-2}$ cm^2 V^{-1} s^{-1}, it can be safely concluded that the effective mass of electron (m_e^*) is much higher for the rutile phase, resulting in a flatter band. Thus, the expected Moss–Burstein shift for rutile would be much smaller than that for the anatase phase, which was opposite to what was observed experimentally, ruling out the Moss–Burstein shift to be the primary mechanism for the blue shift in the bandgap with increasing Ta concentration.

Figure 4.5c shows an increase in the lattice constant with Ta concentration, which is explained by considering the larger ionic sizes of Ta^{5+} (0.064 nm) when compared to Ti^{4+} (0.061 nm). Figure 4.5c also shows the variation in RBS channeling (Rutherford backscattering) minimum yield of Ti (χ_{min}^{Ti}) and Ta (χ_{min}^{Ta}) with Ta concentration. The minimum yield for Ta is close to that of Ti, indicating a near-perfect substitution of Ta in Ti sites. The channeling minimum yield and the crystallinity have an inverse relationship.

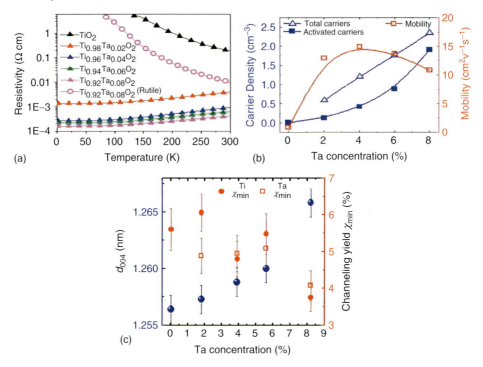

Figure 4.5 (a) Temperature dependence of resistivity of anatase Ti$_{1-x}$Ta$_x$O$_2$ ($x = 0$–0.08) (solid triangles) and rutile Ti$_{1-x}$Ta$_x$O$_2$ ($x = 0.08$) (open circles) films. (b) Variation in carrier density (left ordinate) and Hall mobility (right ordinate) with Ta concentration for anatase Ti$_{1-x}$Ta$_x$O$_2$ films. (c) Increase in lattice parameter (left) of anatase Ti$_{1-x}$Ta$_x$O$_2$ films and the decrease (right) in RBS-channeling minimum yield with Ta concentration.

It is important to note that both χ_{min}^{Ti} and χ_{min}^{Ta} is lower for the Ti$_{1-x}$Ta$_x$O$_2$ ($x = 0.08$) alloy. This tells us that Ta substitution in the TiO$_2$ lattice stabilizes the crystal structure. However, the nonlinear increase in carrier density with Ta addition, as shown in Figure 4.5b, requires an alternate explanation. Most likely, there are compensating defects forming (say Ti vacancy or Ti^{3+}) when Ta is added, in addition to the formation of donors. As the Ta concentration is increased, there is a nonlinear increase in activated carriers, implying a lesser rate of cationic defect formation than carriers. At a Ta concentration of 8%, the carrier activation is as high as 80%. This suggests that, as the cationic defects are created, the activation energy for further defect creation will increase owing to the stability of the crystal. In addition, oxygen vacancies are inhibited as a result of the large number of donors. Thus at high Ta concentration, oxygen vacancies are difficult to form and the number of cationic vacancies tends to saturate. This has, in fact, been verified using photoluminescence measurements [7]. This also explains why the crystal quality at the same deposition conditions is better for films containing higher Ta concentration.

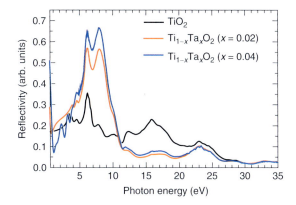

Figure 4.6 High-energy optical reflectivity for anatase $Ti_{1-x}Ta_xO_2$ ($x=0$, 0.02, and 0.04) films.

Figure 4.6 shows the high-energy optical reflectivity spectra obtained for an unalloyed and $Ti_{1-x}Ta_xO_2$ ($x=0$, 0.02 and 0.04) anatase films, showing the effect of Ta incorporation on the electronic band structure (A. Roy Barman *et al.*, unpublished). The sharp feature at $0\,eV$ noted only in $Ti_{1-x}Ta_xO_2$ ($x=0.02$ and 0.04) samples is due to their metallic nature and the Drude tail. The $Ti_{1-x}Ta_xO_2$ ($x=0.02$ and 0.04) samples have two intense peaks around 6 and 8 eV compared to pure TiO_2. Further, pure TiO_2 has a set of broad peaks of higher energy whose intensities are highly diminished for the $Ti_{1-x}Ta_xO_2$ ($x=0.02$ and 0.04) film. The significant shift of spectral weight from high energy to low energy with Ta addition attests to major changes in the band structure due to an alloying effect between TiO_2 and Ta_2O_5.

4.3
Diluted Magnetic Semiconductors (DMS)

The importance of diluted magnetic semiconductors (DMSs) materials in spintronics for manipulating the spin functionality in a controlled manner was recognized long ago [17–20]. Ferromagnetic properties are produced by incorporating a small amount of magnetic element in the semiconductor host material systems without compromising their basic semiconducting properties [18–28]. The quest for producing RT DMS materials [21] has accelerated after the discovery of the first DMS system in Mn-doped GaAs whose Curie temperature did not exceed 173 K [26, 27]. In the early 2000, functional wide-bandgap metal-oxide semiconductors became the leading host materials after the discovery of RT FM in Co-doped anatase [2] TiO_2- and Mn-doped [28] ZnO. Subsequently, many other wide-bandgap oxides were shown to be ferromagnetic at RT [19, 29–46]. Among these, TiO_2 and ZnO have been extensively studied, leading to some basic understanding in the development of this field. In most of these experiments, transition magnetic elements such as Co, Fe, Ni, and Mn were substituted at the cationic sites of the host material, which acts as magnetic centers. In addition to providing local moment, the magnetic impurity acts as an acceptor or donor for the host oxides depending on its final valence state, and hence the FM in DMS oxide is referred to as *charge mediated*. The observed FM in such systems is very much influenced by the concentration of magnetic elements, codoping, and deposition parameters such as oxygen partial pressure and substrate temperature [19].

Depending on the availability of itinerant charge carriers, two DMS oxide classes (i.e., conducting and insulating) are generally observed. Theoretical models [19–21, 47–51] such as Ruderman–Kittel–Kasuya–Yosida (RKKY)-type carrier-mediated FM for conducting system [21, 35, 47–49] and bound magnetic polaron as well as F-center models for insulating system [49–51] have been proposed. The Curie temperature (T_c) depends on the carrier density [21]. If the material is very resistive, T_c is expected to be lower than that observed in conducting materials.

In the carrier-mediated RKKY exchange model [21], a free carrier near a localized magnetic moment gets polarized. When this spin-polarized carrier moves through the lattice, it interacts with other magnetic centers, causing them to align in the same direction, giving rise to FM. It was found [34, 47] that the exchange is maximized when the ratio of carrier density (n_c) to the density of magnetic ions (n_i) is in between about 0.3 and 0.5. The situation is more complicated when the magnetic ions are not randomly substituted in the cation sites or the system is in a disordered or defective state. In such cases, theory [using coherent potential approximation (CPA)] predicts [7, 52] RT FM, for example, in GaMnAs when the magnetic ions form correlated pairs. Unlike the RKKY theory, the CPA model predicts that a threshold concentration of magnetic ions in the range 5–10% is needed for producing high-temperature FM.

On the other hand, in the bound magnetic polaron or F-center model, the incorporated magnetic dopants behave similar to a hydrogen atom with a large Bohr radius owing to the large dielectric constant of the host oxide. The interaction between the localized charge and the magnetic moment leads to its polarization, that is, the spins of the charges are aligned with the local magnetic moment. As the concentration of the magnetic moments increases, the large Bohr radii of the charges start to overlap, leading to exchange interaction between the charges and eventually the magnetic moments, producing a short-range ferromagnetic (or antiferromagnetic) state.

Despite all these developments, magnetic-impurity-based DMS oxides have become a subject of intense debate because of the many inconsistent results related to reproducibility, secondary phase formation, unknown magnetic impurities, substitutionality, solubility, clustering, and segregation-related issues of the magnetic impurities in the host matrix. Consequently, several leading groups [53, 54] in this field have published articles highlighting the sources of the problems and suggesting appropriate solutions. However, the skepticism in experimental data from such DMS materials have persisted, and the field is yet to overcome these perceptions of controversy. Hence, the traditional magnetic characterization of the sample via the highly sensitive superconducting quantum interference device (SQUID) is not sufficient, as it cannot distinguish between intrinsic and extrinsic source of the observed magnetic properties. Clearly, a variety of other magnetic measurements ranging from optical magnetic circular dichroism (OMCD), which directly probes the spin-polarized band, to element-specific soft X-ray magnetic circular dichroism (SXMCD) as well as various elemental analysis techniques such as time-of-flight secondary ion mass spectrometry (TOF-SIMS), RBS- channeling, proton-induced X-ray emission (PIXE), X-ray photoemission spectroscopy (XPS), X-ray absorption spectroscopy (XAS), and three-dimensional atomic profilometry (3DAP) have been deployed. Despite these negative perceptions, a new field [3, 8, 19, 20, 55–63] of defect-mediated FM in wide-bandgap oxide materials has emerged and will be reviewed in the rest of this chapter.

4.4
Defect-Mediated Ferromagnetism

It is well known that imperfections in materials (whether chemical or structural) modify intrinsic material properties, leading to many interesting effects such as luminescence, electrical conductivity, diffusion, and superconductivity. Recent studies involving various oxides such as TiO_2, ZnO, HfO_2, CeO_2, In_2O_3, SnO_2, Al_2O_3, SrO, and so on, reveal rather weak RT ferromagnetic properties in either pure or nonmagnetic element (such as Cr, Cu, C, N)-incorporated form grown under specific conditions [19, 58–62]. Coey *et al.* [50, 51] proposed the possible role of oxygen vacancy, that is, F-center defects, for the mechanism of FM in such material systems. Alternatively, FM should also be purely based on cationic defects without the intervention of magnetic elements [3, 19, 20, 55–57].

FM via cationic vacancy (with concomitant half-metallicity) in wide-bandgap semiconducting oxides was originally proposed on theoretical grounds by Elfimov *et al.* [55]. Later, Osorio-Guillén *et al.* [56, 57] pointed out that the cationic vacancies are readily formed in most of the wide-bandgap oxides because of either hole or electron doping. It was predicted that the wide-bandgap CaO can become a half-metallic ferromagnet with about 5% ($\sim 10^{21}$ cm^{-3}) cationic, that is, Ca vacancies. Such a large number of vacancies exceed the equilibrium concentration of vacancies in the solid by three orders of magnitude. Experimentally, such large numbers of cationic vacancies and hence the establishment of FM in CaO has never been realized. It is not clear whether this is due to the hygroscopic nature of this material or to the high vacancy concentration, leading to a rather unstable film. Interestingly, Zhang *et al.* [8] have recently observed signatures for the magnetic effect (Kondo scattering) below 100 K in their 5% niobium (Nb) incorporated anatase TiO_2 thin films grown by PLD.

Electrical measurements revealed the coupling between the localized carriers and magnetic moments. Figure 4.7a,b shows the resistivity and Hall coefficient (R_H) data, respectively, for different Nb concentrations [8]. The temperature-independent R_H indicates the degenerate state of Nb donors, and the low-temperature upturn in resistivity is attributed to the temperature-dependent scattering processes such

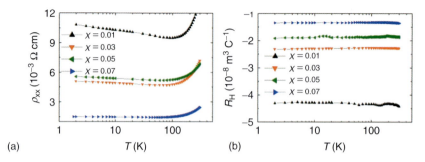

Figure 4.7 Temperature dependence of (a) the resistivity ρ and (b) the Hall coefficient of Ti$_{1-x}$Nb$_x$O$_2$ thin films ($x = 0.01$, 0.03, 0.05, and 0.07).

as weak localization and Kondo scattering. By using angle-dependent magnetoresistance, the predominant scattering process was identified as Kondo scattering.

Through XAS and XPS supported by the first-principles calculations, it was demonstrated [8] that the observed Kondo scattering is due to the presence of localized magnetic moments associated with cationic (Ti) vacancies (Figure 4.8) produced after Nb incorporation in anatase TiO$_2$ films, and not due to other defect states such as Nb^{4+} and Ti^{3+}. This experimental study indicated for the first time that magnetism could originate from cationic vacancies, supporting the basic idea of Elfimov et al. [55].

As a small number of defect-based magnetic entities can cause Kondo scattering, its presence and identification becomes a signature for the presence of such magnetic centers in a nonmagnetic material system. Once Kondo scattering is seen in a system, the natural question arises as to whether one can subsequently optimize the processing of the material to optimize the density of the defects (i.e., magnetic centers) to lead to FM, as predicted by Elfimov et al. [55] and supported by Osorio-Guillén et al. [3, 57]. Recent findings [9] show that under specific growth conditions, it is indeed possible to produce FM by generating a certain number of cationic vacancies and free carriers in Ta-incorporated anatase TiO$_2$ thin films. In the following sections, these results will be reviewed.

4.5
Magnetic Impurity Analysis in Ti$_{1-x}$Ta$_x$O$_2$ System

As discussed in Section 4.3, the field of DMS oxides has become controversial because of various artifacts influencing the magnetic properties. The most prominent cause has been identified as the presence of unintentional magnetic impurities coming from external sources during various thin-film processing steps. In this section, we demonstrate steps to rule out such possibilities by employing a variety of characterization techniques such as RBS, PIXE, TOF-SIMS, XPS, XAS, and 3DAP. All the samples were handled without any metallic substance.

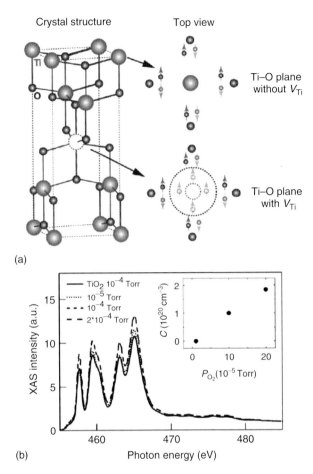

(a)

(b)

Figure 4.8 (a) Ti-vacancy-induced local magnetic moment: anatase TiO$_2$ structure (left); Ti–O plane (right) with and without a Ti vacancy (big dashed circles). The big and small solid spheres represent Ti and O, and small spheres and dash circles with arrows (spin direction) represent electrons and holes and V_Ti represents Ti vacancy. (b) XAS spectra of a pure TiO$_2$ film grown at 10^{-4} Torr, and Ti$_{0.95}$Nb$_{0.05}$O$_2$ films grown at various pressures at the Ti L edge.

Figure 4.9 shows typical XRD spectra of a pure and Ti$_{1-x}$Ta$_x$O$_2$ ($x \sim 0.05$) films deposited on single-crystal (001) LaAlO$_3$ (LAO) substrates by PLD at various temperatures. In all the cases, the anatase phase was formed. No peaks from common magnetic elements or related compounds are seen within two orders of magnitude (detection limit of XRD system).

The corresponding random RBS and ion-channeling spectra using 2 MeV He$^+$ ions are shown in Figure 4.10. The channeling yield of Ti is about 3–4%, indicating a good epitaxial thin film. The channeling of Ta is also about 3%, suggesting that more than 97% of the Ta is substituted into the Ti sites. The expanded view of

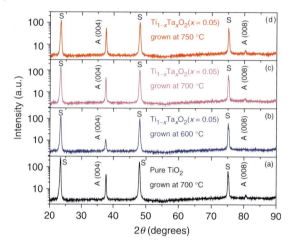

Figure 4.9 XRD spectra of a pure TiO$_2$ and Ti$_{1-x}$Ta$_x$O$_2$ ($x \sim 0.06$) films grown on an (001) LaAlO$_3$ substrate at various temperatures in 10^{-5} Torr of oxygen partial pressure.

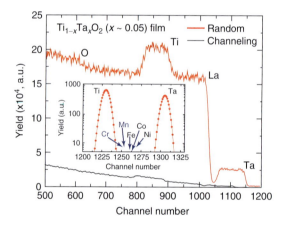

Figure 4.10 RBS-channeling spectra of Ti$_{1-x}$Ta$_x$O$_2$ ($x \sim 0.06$) thin film on an LAO substrate. Inset: the expanded view of the RBS spectrum of same film on an Si substrate, showing the positions of common magnetic elements generally seen in typical extrinsic DMSO materials.

the RBS spectrum of a similar film grown on a Si substrate is shown in the inset of Figure 4.10. If any of the common magnetic impurities (Cr, Fe, Co, Ni, and Mn) were present in these samples, they would have appeared between channel numbers 1250 and 1290, as indicated in Figure 4.10. The absence of signal from these elements demonstrates that there are no common magnetic impurities within the RBS detection limit of 50 ppm (limited by pulse pileup).

PIXE (a nondestructive multitrace elemental quantitative technique with sensitivity down to the parts per million) spectra from pure and Ti$_{1-x}$Ta$_x$O$_2$ ($x \sim 0.06$) target and film are shown in Figure 4.11a–d. From these figures, Ta corresponds

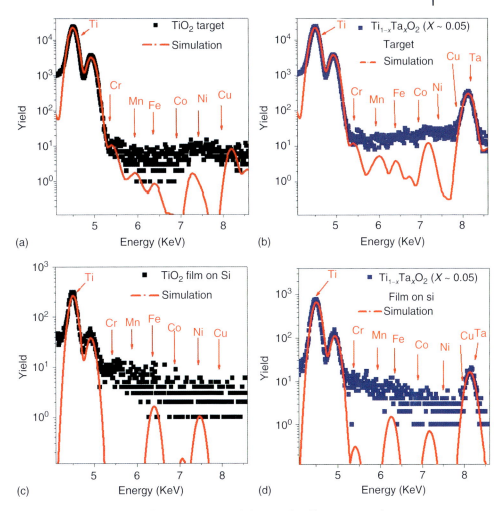

Figure 4.11 PIXE spectrum of (a) TiO_2 target and (b) TiO_2 thin film on an Si substrate. (c) $Ti_{1-x}Ta_xO_2$ ($x \sim 0.06$) target and (d) $Ti_{1-x}Ta_xO_2$ ($x \sim 0.05$) thin film on a silicon substrate.

to 55 000 ppm, showing contaminants at a level two orders of magnitude lower than that of Ta.

For elemental mapping, XAS technique (highly sensitive to electronic band structure, especially for unoccupied states) in the wide-scan mode is shown in Figure 4.12 along with the theoretical simulation. The XAS data from 440 to 1000 eV do not show any signal at energies where the absorptions of typical magnetic impurities Mn ($\sim$640 eV), Fe ($\sim$710 eV), Co ($\sim$780 eV), and Ni ($\sim$855 eV) are expected to occur at the positions shown by the theoretical calculations. It may be noted that the Ta $L_{3,2}$ edges are expected to occur at around 9881 and

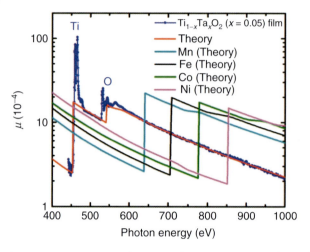

Figure 4.12 A comparison between wide-scan XAS spectra obtained from the $Ti_{1-x}Ta_xO_2$ ($x \sim 0.06$) film and theoretical calculations.

11136 eV, which are out of the soft X-ray range. It is very clear that the film is not contaminated by these magnetic impurities.

The spectra obtained from the most sensitive trace element analysis technique, namely TOF-SIMS (using 25 keV Bi-analysis ions), are shown in Figure 4.13. A reference target containing 1 at% each of Cr, Fe, Mn, Ni, and Co and 95% of Ti and the corresponding thin film were used for reference. It is clear from the data that all the magnetic elements in the pure and Ta-incorporated TiO_2 target, and pure and Ta-substituted TiO_2 films are at concentrations that are roughly five orders of magnitude lower than the Ti concentration, that is, substantially below 0.001%

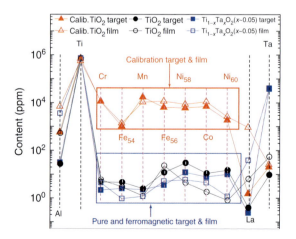

Figure 4.13 SIMS spectra of from pure TiO_2 and $Ti_{1-x}Ta_xO_2$ ($x \sim 0.06$) targets and their PLD-deposited films on silicon substrates. For comparison, the spectra from a similar target and film but with 1% magnetic impurities of Fe, Cr, Mn, Ni, and Co are also shown.

and about three orders of magnitude less than the Ta concentration. Therefore, we can safely conclude that there are no common magnetic impurities in these samples, as verified by all these different techniques.

4.6
Defect-Induced Ferromagnetism in Ti$_{1-x}$Ta$_x$O$_2$ Film

Having demonstrated that this new alloy (i.e., anatase Ti$_{1-x}$Ta$_x$O$_2$ film) is devoid of any magnetic impurities, in the rest of this chapter, we will discuss the observation of FM in this system and also rationalize the origin of this to cationic defects [9].

4.6.1
Ferromagnetic Properties

Magnetization of pure and Ti$_{1-x}$Ta$_x$O$_2$ ($x \sim 0.06$) thin films grown at 600–750 °C at 1×10^{-5} Torr are shown in Figure 4.14a. The magnetization of 4 emu g^{-1} implies 1.1 μ_B/Ta using 5.5% Ta ions obtained from the RBS analysis. The T_c of the sample was well above 100 °C. The LAO substrate treated under similar conditions shows only diamagnetic behavior. In a study involving more than 60 samples, the magnetization tended to peak at a deposition temperature of about 600 °C (repeated for about 20 samples) and an oxygen pressure of about 10^{-5} Torr.

All the samples prepared at the optimum condition showed (Figure 4.14b) FM with an average saturation magnetization of $\sim$1.1 emu g^{-1}, and with coercive fields ranging from 70 to 90 Oe. In order to obtain a better signal-to-noise ratio, the sample with maximum saturation magnetization value close to 4 emu g^{-1} was used for all the following characterizations.

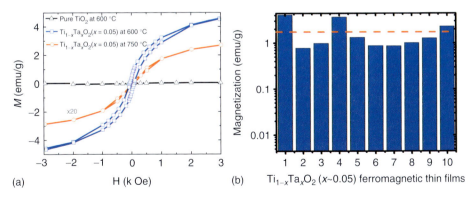

Figure 4.14 (a) Magnetic hysteresis loop for pure TiO$_2$ (black), Ti$_{1-x}$Ta$_x$O$_2$ ($x \sim 0.06$) thin films grown at 600 °C (blue) and at 750 °C (red) in 1×10^{-5} Torr. (b) Histogram for the magnetization data measured on samples grown at an optimized temperature of 600 °C and an optimized oxygen partial pressure of 1×10^{-5} Torr.

4.6.2
Intrinsic versus Extrinsic FM

Although the SQUID data clearly show the presence of RT FM in the Ta-substituted TiO$_2$ samples, they do not provide microscopic insight into the origin of the ferromagnetic ordering. Also, it cannot determine whether the magnetism is originating from an intrinsic or an extrinsic source. So, an elemental-specific technique such as SXMCD or OMCD is necessary.

4.6.2.1 SXMCD
SXMCD measurement is crucial, as this technique is not only element-specific but is also capable of estimating both the spin and orbital magnetic moments and their anisotropy. The SXMCD data were taken in the total yield mode without any external magnetic field. In order to compare the SXMCD results at various edges, that is, the Ti L$_{2,3}$ edges (2p $\rightarrow$ 3d transitions) and the O K edge (O 1s $\rightarrow$ 2p transition), they were normalized to an absolute scale by fitting to the Henke tables [64] far below and above the edges and absorption coefficients, μ^+ (parallel) and μ^- (antiparallel) were calculated. The μ^+ and μ^- at the Ti L$_{2,3}$ and O K edge of Ti$_{1-x}$Ta$_x$O$_2$ ($x \sim 0.06$) film grown at 600 °C are shown in Figure 4.15a,b, respectively. The μ^+ and μ^- at the Ti L$_{2,3}$ edge consists of two sets of peaks separated by 5–6 eV due to core hole spin–orbit coupling of Ti 2p$_j$ with $j = 1/2$ or 3/2. Moreover, owing to ligand-field splitting, the 3d bands can be identified as t_{2g}- and e_g-symmetry bands. As a result, the Ti 2p $\rightarrow$ 3d transitions consist of four dominant structures, and all the relevant transitions are shown in Table 4.2.

The SXMCD signals, which are defined as $\mu^- - \mu^+$ (Figure 4.15c,d), correspond to the remanent magnetization and are the most direct evidence for the intrinsic FM. One can see that the SXMCD signals are surprisingly robust despite the fact that there was no applied magnetic field during the measurement. The observation of SXMCD signal rules out the possibility of superparamagnetism. In contrast, the pure TiO$_2$ film, which is the reference sample, does not show any SXMCD signal at both resonant edges. Further, the higher crystallinity sample grown at 750 °C

Table 4.2 SXMCD peaks and corresponding allowed transitions.

Peak position of SXMCD signal (eV)	Elemental edge	Transition
458.2	Ti L$_{2,3}$	2p$_{3/2}$ $\rightarrow$ 3d(t_{2g})
460	Ti L$_{2,3}$	2p$_{3/2}$ $\rightarrow$ 3d(e_g)
463.5	Ti L$_{2,3}$	2p$_{1/2}$ $\rightarrow$ 3d(t_{2g})
465.5	Ti L$_{2,3}$	2p$_{1/2}$ $\rightarrow$ 3d(e_g)
530.6	O K	1s $\rightarrow$ Ti 3d(t_{2g})
533.2	O K	1s $\rightarrow$ Ti 3d(e_g)

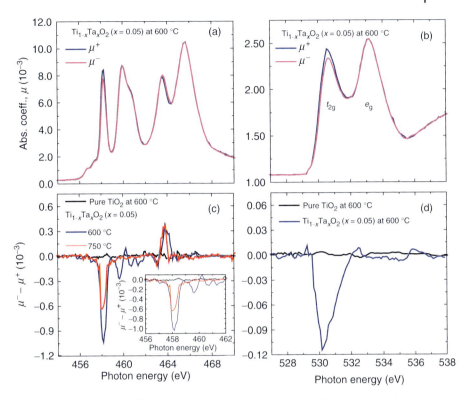

Figure 4.15 Absorption coefficient μ at (a) Ti L$_{2,3}$ edges and (b) O K edge of Ti$_{1-x}$Ta$_x$O$_2$ ($x \sim 0.06$) thin films grown at 600 °C. Here, right circular polarized light (μ^+, blue lines) and left circular polarized light (μ^-, magenta) and the corresponding SXMCD spectra (μ^- to μ^+) for pure TiO$_2$ (black), Ti$_{1-x}$Ta$_x$O$_2$ ($x \sim 0.06$) grown at 600 °C (blue) and at 750 °C (red) at (c) Ti L$_{2,3}$ edge and (d) O K edge.

shows (Figure 4.15c) an SXMCD signal $\sim$20 times smaller, which is consistent with the SQUID measurement.

The fact that both the Ti L and O K edges show an SXMCD signal dominantly at the t_{2g} state indicates strong p–d hybridization and suggests that the t_{2g}-derived state plays a dominant role in the observed FM in this system. By applying the X-ray MCD sum rule [65, 66], we estimated that the contribution to the orbital magnetic moment of Ti is 9 times stronger than that of O and that the spin alignment at Ti and O is parallel. These data confirm unambiguously that the FM is originating from Ti-related defects.

4.6.3
OMCD

The OMCD technique is used to study the spin-splitting energy between the up-spin (majority spin) and the down-spin (minority spin) and the spin polarization of

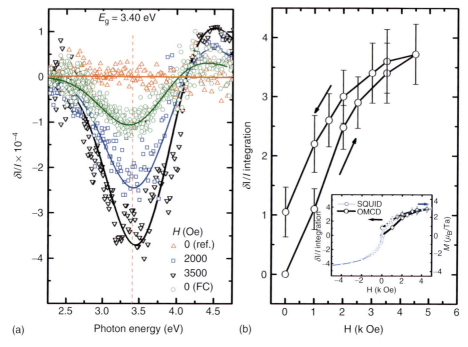

(a)

(b)

Figure 4.16 (a) OMCD signal of a Ti$_{1-x}$Ta$_x$O$_2$ ($x \sim 0.06$) film grown at 600 °C showing the dichroism and spin-polarized magnetization near the optical bandgap shown by the vertical dotted line.

(b) Magnetic hysteresis loop from the OMCD measurement showing ferromagnetic behavior. In the inset, this loop is overlapped with the SQUID data at 40 K.

carriers at the optical bandgap. Figure 4.16a shows the differential intensity change $\delta I/I$ as a function of photon energy from 2 to 5 eV at various magnetic fields.

The position of the measured optical bandgap of 3.42 eV is shown by a vertical dotted line. It is seen from the figure that the transitions around 3.5–4.5 eV are strongly influenced by the applied magnetic field. The change in sign of $\delta I/I$ at 4.5 eV is due to the optical transitions from non-spin-polarized occupied states to two possible unoccupied states: majority-spin states at E_F and the minority-spin states roughly 1 eV above E_F. Interestingly, this 1 eV spin-splitting energy is similar to those found in colossal magnetoresistive manganites [67], suggesting strong electron localization effects (in the energy or k space) in this system. Assuming that the bands close to E_F have low density of states (DOS), the width of the transition at 3.5 eV ($\sim$0.7 eV) corresponds to the width of the spin-majority band. This suggests that the occupied states correspond to O 2p states, whereas the majority and minority-spin unoccupied states correspond to Ti 3d states.

The integrated absolute value of the OMCD signal between 2.2 and 4.1 eV and between 4.2 and 4.75 eV is shown (Figure 4.16b) as a function of the applied magnetic field going from 0 to 4500 Oe and back to zero. Also, the hysteresis loop obtained from the SQUID measurement is compared (see inset of Figure 4.16b) with the OMCD data. This supports that the FM seen in both measurements has the same origin.

4.7
First-Principles Spin-Polarized GGA + U Calculations

In order to get an insight into these findings, first-principles calculations within a spin-polarized generalized-gradient approximation plus the onsite U parameter (GGA + U, generalized-gradient approximation) were performed. A 48-atom supercell, modeled by $2 \times 2 \times 1$ repetition of the 12-atom conventional unit cell of anatase TiO_2, which is proportional to 6.25% Ta doping, was employed to study the electronic structure of the $Ti_{1-x}Ta_xO_2$ alloy. The 6.25% Ta dopant was used in place of the actual 5.5% for the calculation because of computational difficulties, but this does not change the main conclusions. An effective onsite U parameter $(U_{eff} = U - J)$ of 5.8 eV and a scissor operator were used to make the calculated bandgap comparable to the experimental value.

The calculated electronic structures are shown in Figure 4.17a–d. Interestingly, the magnetic moment mainly resides at the Ti 3d (t_{2g}) bands that are hybridized

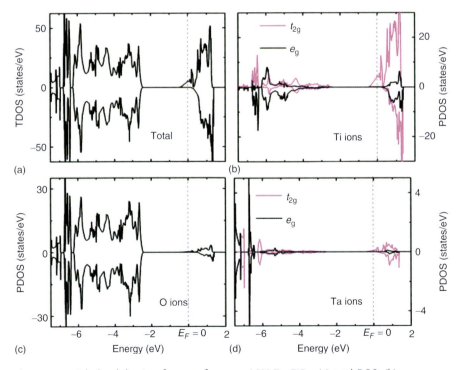

Figure 4.17 Calculated density of states of anatase 6.5% Ta–TiO_2: (a) total DOS, (b) partial DOS for O 2p states, (c) t_{2g} and e_g of Ti 3d, and (d) t_{2g} and e_g of Ta 5d.

with O 2p bands, and this is consistent with SXMCD results. These calculations also indicate that an isolated V_{Ti} produces a high-spin-polarization electronic state, which is mainly contributed by the O 2p orbital of the first-nearest O atoms around the V_{Ti}. Furthermore, the V_{Ti}–V_{Ti} interaction results in a stable ferromagnetic ground state.

It is worth mentioning that the spin-polarized density induced by the V_{Ti} extends very long to the third-nearest and even to the fifth-nearest O atoms. As a result, the spin-polarized V_{Ti} orbitals are delocalized. When the magnetic orbitals of two V_{Ti} overlap through the common spin-polarized, third- and fifth-nearest O atom, the overlapping spin density is nonzero, thereby leading to a long-range ferromagnetic alignment between the magnetic orbitals of two V_{Ti}. In this process, free electron carriers are expected to facilitate the spin-exchange coupling interaction between two V_{Ti} magnetic orbitals. In contrast, our calculations also show that the Ti^{3+}–Ti^{3+} interaction favors antiferromagnetism. This also fully supports our experimental observations in which the sample with high Ti^{3+}, that is, $Ti_{1-x}Ta_xO_2$ ($x \sim 0.06$) film grown at 750 °C has weak FM. Furthermore, the splitting energy between the up-spin (majority spin) and down-spin (minority spin) Ti 3d states is about 1.1 eV. This value is also consistent with the OMCD measurements of the Ta–TiO$_2$ (600 °C) sample shown in Figure 4.16a with a measured spin-splitting energy of about 1 eV.

The calculated OMCD based on the DOS is compared with the experimental results for the Ta–TiO$_2$ (600 °C) sample in Figure 4.18. Owing to the quantum mechanical selection rules for optical absorption, the spin-polarized majority and minority states are directly connected, with the appearance of a magnetooptical response. In the energy range of 2–5 eV, the optical transitions are dominated by

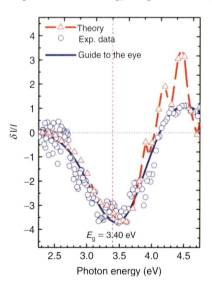

Figure 4.18 Comparison of calculated OMCD with the experimental data. The blue line through the OMCD experimental data points is a guide to the eye only.

charge-transfer excitations between O 2p and Ti 3d. The calculations consider the intersite transition from spin-up (spin-down)-occupied O 2p states to the spin-down (spin-up)-unoccupied Ti 3d states. Interestingly, the calculated OMCD spectrum tracks very well with the experimental data, including the peak position and width. This further supports that the FM truly derives from the intrinsic properties of the system and thus OMCD is important in revealing the intrinsic magnetism and spin-splitting band near the Fermi level.

4.7.1
XPS and XAS: Determination of Intrinsic Defects

In Figure 4.11a, the higher crystallinity $Ti_{1-x}Ta_xO_2$ $(x \sim 0.06)$ film grown at $750\,^{\circ}C$ shows less (factor of ~ 20) magnetization, suggesting the role of intrinsic defects in the observed magnetism. As mentioned earlier, the dominant defects tend to be cationic at this high Ta concentration [7]. Ti vacancy (V_{Ti}) and Ti^{3+} are the likely candidates, and they were identified via XPS (which can identify Ti^{3+}) and XAS (which can recognize the presence of V_{Ti}).

XPS analysis (Figure 4.19) confirmed that Ti is dominantly in the 4+ state for both pure TiO_2 and $Ti_{1-x}Ta_xO_2$ $(x \sim 0.06)$ films grown at 600 and $750\,^{\circ}C$. The Ta was found to be in the 5+ state. A detailed analysis suggested that the Ti^{3+}/Ti^{4+} ratios for pure TiO_2 (Figure 4.19a) and $Ti_{1-x}Ta_xO_2$ $(x \sim 0.06)$ films grown at $600\,^{\circ}C$ (Figure 4.19b) are comparable, which are ~ 0.65 and $\sim 0.59\%$, respectively, while the $Ti_{1-x}Ta_xO_2$ $(x \sim 0.06)$ film grown at $750\,^{\circ}C$ (Figure 4.19c) shows a ratio of 8.45%. Hence, at high crystallinity, weakly ferromagnetic $Ti_{1-x}Ta_xO_2$ $(x \sim 0.06)$ films grown at $600\,^{\circ}C$ exhibit Ti^{3+} signal 14 times higher than that of ferromagnetic films grown at $600\,^{\circ}C$. This clearly rules out the role of Ti^{3+} as the defect responsible for the observed FM in the $Ti_{1-x}Ta_xO_2$ $(x \sim 0.06)$ films grown at $600\,^{\circ}C$.

XAS measurements were performed to reveal the role of Ti vacancies created as a result of Ta incorporation. The XAS spectra taken at the Ti $L_{2,3}$ edges from the pure TiO_2 and $Ti_{1-x}Ta_xO_2$ $(x \sim 0.06)$ films grown at 600 and $750\,^{\circ}C$ are shown after background correction [9] in Figure 4.20a,b, respectively. The Ti $L_{2,3}$ edges $Ti_{1-x}Ta_xO_2$ $(x \sim 0.06)$ films grown at $600\,^{\circ}C$ increases dramatically in the t_{2g} bands when compared to pure TiO_2 film. The increasing spectral weight is a direct evidence [8] of the formation of V_{Ti} because a V_{Ti} creates four holes in the O 2p band, which is strongly hybridized with the Ti 3d band. The creation of these holes increases the number of unoccupied states near the Fermi level, for example, in the t_{2g} bands, and therefore increases the XAS signal. Interestingly, the SXMCD is strong also at t_{2g} bands.

From the ratio of t_{2g} to e_g bands for the pure and $Ti_{1-x}Ta_xO_2$ $(x \sim 0.06)$ films grown at $600\,^{\circ}C$ (Figure 4.20a), the estimated upper limit of V_{Ti} was about 3%, with the actual number likely to be lower by a factor of two or more owing to unaccounted defects that could increase the number of unoccupied states near the Fermi level. This is a factor of 5 larger than the number one would arrive for the vacancy concentration based on charge compensation. Altogether, one can say for sure that the actual vacancy concentration is somewhere between 0.6 and

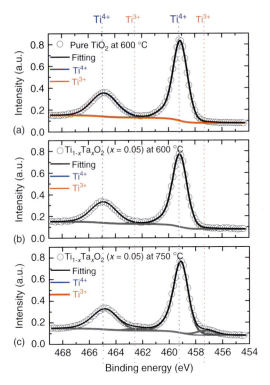

Figure 4.19 The XPS analysis (done at Ti 2p core levels) of (a) pure TiO$_2$ film grown at 600 °C and Ti$_{1-x}$Ta$_x$O$_2$ ($x \sim 0.06$) films grown at (b) 600 °C and at (c) 750 °C. The XPS shows that Ti is mainly in the 4+ state with negligible amount in 3+ state for both pure and Ti$_{1-x}$Ta$_x$O$_2$ ($x \sim 0.06$) films grown at 600 °C, while for 750 °C film, a significant amount of Ti^{3+} is seen.

3%. A much more detailed study would be needed to further narrow down these numbers. By the same analogy, when we compare the samples grown at 750 °C (Figure 4.20b), the t_{2g} peak decreases in height for the Ti$_{1-x}$Ta$_x$O$_2$ ($x \sim 0.06$) films grown at 750 °C with respect to the pure TiO$_2$ film (together with the support of XPS data), suggesting absence or reduced V_{Ti}.

In retrospect, it is important to note that the SXMCD signal in the present case is different from the one observed in LaMnO$_3$/SrTiO$_3$ originating from the Ti^{3+} states present at its interface, as recently reported by Garcia-Barriocanal et al. [68]. This also strongly suggests that the SXMCD signal in Ti$_{1-x}$Ta$_x$O$_2$ ($x \sim 0.06$) does not arise from Ti^{3+} defect and is most likely from the V_{Ti}. All these facts support the conclusion that the role of Ti^{3+}, if any, in the FM of the 600 °C sample is secondary while V_{Ti} is the key to the FM seen.

Now that V_{Ti} is established as the magnetic entity, we will try to develop a microscopic understanding of the FM. The four unpaired electrons in a Ti vacancy can align in three possible ways, which will yield 4, 2, or 0 μ_B. Statistically, we can assume a value of 2 μ_B per vacancy, which would mean that, to get the

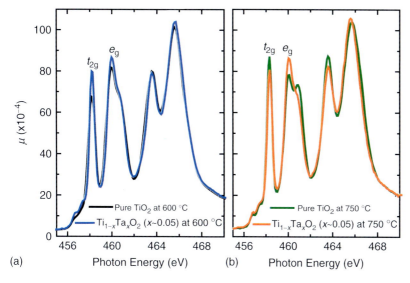

Figure 4.20 XAS of pure TiO_2 and $Ti_{1-x}Ta_xO_2$ ($x \sim 0.06$) films grown at (a) 600 °C and (b) 750 °C.

magnetization seen, $\sim 2.5\%$ vacancy would be needed. In addition, to compensate 50% of the electrons from the Ta, about 0.6% vacancies would be needed. So a total of about 3% V_{Ti} is adequate to explain the saturation magnetization as well as the electron compensation seen. The average distance between two Ti vacancies is about three to four unit cells. Unless the orbital magnetization of the Ti vacancy is extended over at least two unit cells, the direct exchange probability is very low. The fact that we lose FM (but instead see Kondo scattering [8]) in samples prepared at higher oxygen pressures, where the Ti vacancy concentration is higher but the carrier concentration is lower, strongly argues in favor of a carrier-mediated exchange.

4.8
Mechanism of Defect-Mediated FM

The RBS, XAS, SXMCD, XPS, and theoretical calculations altogether clearly show that the presence of Ta substitution, free carriers, and V_{Ti} are crucial for FM in this system. Figure 4.21 shows a scheme of the mechanism where the origin of FM in the nominal 5 at% Ta-substituted anatase TiO_2 thin films is related to magnetic centers associated with the cationic (Ti) vacancies. As the free electron carrier density of the $Ti_{1-x}Ta_xO_2$ ($x \sim 0.06$) film is about 7.6×10^{20} cm^{-3}, the mechanism of FM is most likely facilitated through itinerant electron-mediated RKKY interactions and is most likely not related to other exchange mechanisms.

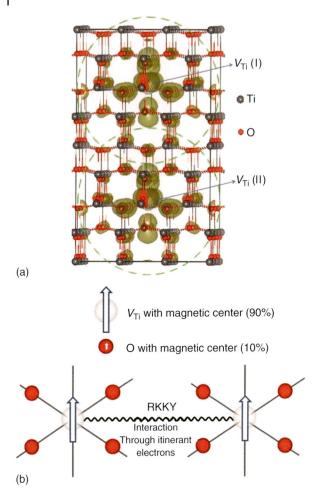

(a)

V_{Ti} with magnetic center (90%)

O with magnetic center (10%)

RKKY
Interaction
Through itinerant
electrons

(b)

Figure 4.21 (a) Three-dimensional spin density plot of anatase TiO$_2$ with two V_{Ti}. The yellow isosurface represents the spin density of V_{Ti}, and dashed green circles show the range of the delocalized magnetic orbitals of V_{Ti}. (b) Schematic of the maximum possible ferromagnetic ordering of magnetic centers (gray circle) at the sites of Ti vacancies coupled by itinerant electrons via a RKKY type of exchange mechanism.

4.9
Optimization of Ferromagnetism

4.9.1
Influence of Film Growth Temperature and Pressure

In PLD process, oxygen partial pressure is an extremely important control parameter for growing any oxide thin film. In Figure 4.22, saturation magnetizations and the carrier densities of the Ti$_{1-x}$Ta$_x$O$_2$ ($x \sim 0.06$) samples are plotted as a function of

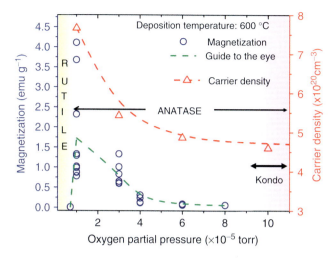

Figure 4.22 Variation in magnetization as a function of oxygen partial pressure (left ordinate). Blue open circles represent magnetization for multiple samples. Variation in carrier density as a function of oxygen partial pressure (right ordinate).

oxygen partial pressure, keeping the deposition temperature fixed at $600\,^\circ$C. It is seen that both the magnetization and the electron density for the films are highest at 1×10^{-5} Torr and gradually fall off as the pressure is increased. Structurally, $Ti_{1-x}Ta_xO_2$ ($x \sim 0.06$) thin films grown below 1×10^{-5} Torr will produce rutile films in which the electron mobility is orders of magnitude lower than in the anatase films. As such, it is understandable that, for the rutile films, the RKKY exchange mechanism is almost nullified and no perceptible magnetism is seen. At higher oxygen partial pressure ($\sim 1 \times 10^{-4}$ Torr), the thin films have fewer oxygen vacancies and the carrier density is found to be lower than that for the films grown at lower oxygen partial pressure (1×10^{-5} Torr). At this higher pressure, FM disappears and Kondo scattering is seen. As the films grown at 1×10^{-5} Torr oxygen partial pressures have a higher number of free carriers, the RKKY mechanism is enhanced. This explains the higher saturation magnetization for films grown at 1×10^{-5} Torr oxygen partial pressure.

Similarly, Figure 4.23 shows the saturation magnetization of the $Ti_{1-x}Ta_xO_2$ ($x \sim 0.06$) thin films as a function of the deposition temperature. The magnetization value peaks at $600\,^\circ$C and decreases considerably as the temperature is increased to $750\,^\circ$C. For temperatures below $600\,^\circ$C, the formation of rutile phase inhibits magnetism. RBS ion-channeling experiments show that both the Ti and the Ta minimum yields decrease with increasing deposition temperature of the films. This clearly tells us that films grown at higher temperatures have better crystallinity when compared to those grown at lower temperatures, implying that the cationic defects that provide the magnetic moments necessary for the FM in the $Ti_{1-x}Ta_xO_2$ ($x \sim 0.06$) thin films are getting annealed.

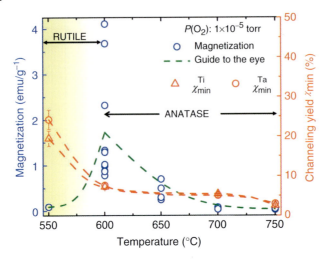

Figure 4.23 Variation in magnetization as a function of deposition temperature (left ordinate). Blue open circles represent magnetization for multiple samples. Decrease in minimum channeling yield for Ti and Ta with increasing deposition temperature (right ordinate).

4.9.2
Influence of Ta Concentration

Having discussed the importance of deposition parameters on the magnetization of Ti$_{1-x}$Ta$_x$O$_2$ thin films, we move on to explain the important role played by Ta incorporation in the TiO$_2$ crystal. As mentioned earlier, it has been calculated by theorists [3] that Ta^{5+} in TiO$_2$, being an electron donor, reduces the formation energies of such defects in the crystal, which act as electron acceptors (V_{Ti}, Ti^{3+}, or oxygen interstitials). On the one hand, the Ta incorporation favors magnetism by inducing defect centers with magnetic moments. On the other hand, it hinders the same by reducing the number of free carriers required for the RKKY mechanism as a result of compensation. Hence, there should be an optimum condition for Ta incorporation at which one would have simultaneously enough defect centers and charge carriers, and the experimental observation is shown in Figure 4.24.

In Figure 4.5b, it is shown that the carrier density of the thin films increases with Ta concentration. Although the increase in the carrier density with Ta concentration is quite expected, the nonlinearity of the increase is quite striking. This means that all the carriers from the Ta incorporation in the Ti$_{1-x}$Ta$_x$O$_2$ films are not activated, and the compensated carrier density as a function of Ta is shown in Figure 4.24. Defects such as V_{Ti} are known as *electron killers* and thus the amount of compensated carriers can be correlated to such defects. As such, it is quite clear from the transport data that the number of compensating defects and uncompensated carriers are optimum in the case of the sample that has about 6% Ta incorporation (Figure 4.5b). The saturation magnetization is the

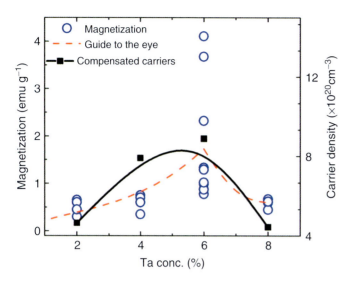

Figure 4.24 Variation in magnetization as a function of Ta incorporation in $Ti_{1-x}Ta_xO_2$ thin films. Blue open circles represent magnetization for multiple samples. The compensated (solid square) carrier densities as a function of Ta incorporation peaks around 6%.

highest for the samples with Ta incorporation of around 6%, which is consistent with the fact that the compensated carrier density scales with the measured magnetization.

If we assume that four charges are compensated by one V_{Ti}, then the vacancy concentration will be about 1.9×10^{20} cm^{-3} as at the optimum condition the free carrier density is about 7.6×10^{20} electrons cm^{-3}, which is about 50% of the activated Ta ($\sim$100% substitutionality). At this concentration, the average distance between the V_{Ti} is 1.74 nm. Clearly, there is a need for carrier-mediated exchange to cause ferromagnetic ordering. Samples with lower Ta concentrations have fewer carriers and lesser defects and hence do not favor the RKKY exchange mechanism. On the other hand, the samples with higher Ta incorporation having higher carriers but fewer defects (magnetic centers) also show lower magnetization values. Thus, it is clear that the amount of Ta incorporation in the $Ti_{1-x}Ta_xO_2$ thin films plays a pivotal role in the magnetization by inducing magnetic centers and introducing free carriers. This explains why the magnetization of the $Ti_{1-x}Ta_xO_2$ thin films is so sensitive to the PLD growth parameters, especially oxygen partial pressure, and this can be seen from Figure 4.18, in which the magnetic centers are created in the high-pressure regime where only Kondo effect is seen because of the lack of free carriers, while at lower pressure, when enough free carriers are generated, the RKKY interactions dominate, leading to FM. Thus, one needs to consider an intricate interplay of carrier, V_{Ti}, and Ta concentrations in order to understand the relationship between magnetic and electronic properties of this new ferromagnetic Ta–TiO$_2$ alloy.

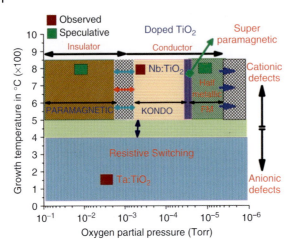

Figure 4.25 Phase diagram of the Nb- or Ta-incorporated TiO₂ system where different applications due to or properties arising from cationic or anionic defects are indicated.

4.10
Outlook for Defect-Mediated Properties of Ti$_{1-x}$Ta$_x$O$_2$

Future prospects of wide-bandgap DMS oxide materials rely on our knowledge in controlling and manipulating their intrinsic defects to realize exotic novel material properties similar to the ones demonstrated in the Ti$_{1-x}$Ta$_x$O$_2$ system. The recent discovery and advances in resistive switching in TiO$_2$ via anionic defects, along with this exciting breakthrough on FM originating from cationic defects, suggest a phase diagram based on defect-mediated properties in TiO$_2$ system, as shown in Figure 4.25. Thus, future experiments in optimizing oxides for various device applications must involve a rigorous study of the various defects, both anionic and cationic, in these systems.

References

1. Diebold, U. (2003) *Surf. Sci. Rep.*, **48**, 53.
2. Matsumoto, Y., Murakami, M., Shono, T., Hasegawa, T., Fukumura, T., Kawasaki, M., Ahmet, P., Chikyow, T., Koshihara, S., and Koinuma, H. (2001) *Science*, **291**, 854.
3. Osorio-Guillén, J., Lany, S., and Zunger, A. (2008) *Phys. Rev. Lett.*, **100**, 036601.
4. Hitosugi, T., Furubayashi, Y., Ueda, A., Itabashi, K., Inaba, K., Hirose, Y., Kinoda, G., Yamamoto, Y., Shimada, T., and Hasegawa, T. (2005) *Jpn. J. Appl. Phys.*, **244**, L1063.

5. Zhang, S.X., Kundaliya, D.C., Yu, W., Dhar, S., Young, S.Y., Salamanca-Riba, L.G., Ogale, S.B., Vispute, R.D., and Venkatesan, T. (2007) *J. Appl. Phys.*, **2102**, 013701.
6. Roy Barman, A., Motapothula, M., Annadi, A., Gopinadhan, K., Zhao, Y.L., Yong, Z., Santoso, I., Ariando , Breese, M., Rusydi, A., Dhar, S., and Venkatesan, T. (2011) *Appl. Phys. Lett.*, **98**, 072111.
7. Roy Barman, A. (2011) Defect mediated novel structural, optical, electrical and magnetic properties in Ti$_{1-x}$Ta$_x$O$_2$ thin

films. PhD thesis. National University of Singapore.

8. Zhang, S.X., Ogale, S.B., Yu, W., Gao, X., Liu, T., Ghosh, S., Das, G.P., Wee, A.T.S., Greene, R.L., and Venkatesan, T. (2009) *Adv. Mater.*, **21**, 2282.

9. Rusydi, A., Dhar, S., Roy Barman, A., Ariando, Qi, D.-C., Motapothula, M., Yi, J.B., Santoso, I., Feng, Y.P., Yang, K., Dai, Y., Yakovlev, N.L., Ding, J., Wee, A.T.S., Neuber, G., Breese, M.B.H., Ruebhausen, M., Hilgenkamp, H., and Venkatesan, T. (2012) *Phil. Trans. R. A*, **370**, 4927.

10. Denton, A.R. and Ashcroft, N.W. (1991) *Phys. Rev. A*, **43**, 3161.

11. Li, W., Pessa, M., and Likonen, J. (2001) *Appl. Phys. Lett.*, **78**, 2864.

12. Tang, H., Prasad, K., Sanjines, R., Schmid, P.E., and Levy, F. (1994) *J. Appl. Phys.*, **75**, 2042.

13. Domardzki, J., Kaczmarek, D., Prociow, E.L., Borkowska, A., Schmeisser, D., and Beukert, G. (2006) *Thin Solid Films*, **513**, 269.

14. Jimenez Gonzalez, A.E. and Gelover Santiago, S. (2007) *Semicond. Sci. Technol.*, **22**, 709.

15. Bao, D., Yao, X., Wakiya, N., Shinozaki, K., and Muzutani, N. (2001) *Appl. Phys. Lett.*, **79**, 3767.

16. Burstein, E. (1954) *Phys. Rev.*, **93**, 632.

17. Datta, S. and Das, B. (1990) *Appl. Phys. Lett.*, **56**, 665.

18. Wolf, S.A., Awschalom, D.D., Buhrman, R.A., Daughton, J.M., Von Molnár, S., Roukes, M.L., Chtchelkanova, A.Y., and Treger, D.M. (2001) *Science*, **294**, 1488.

19. Ogale, S.B. (2010) *Adv. Mater.*, **22**, 3125.

20. Wessels, B.W. (2008) *New J. Phys.*, **10**, 055008.

21. Dietl, T., Ohno, H., Matsukura, F., Cibert, J., and Ferrand, D. (2000) *Science*, **287**, 1019.

22. Korbecka, A. and Majewski, J.A. (2009) *Low Temp. Phys.*, **35**, 53.

23. Žutić, I., Fabian, J., and Das Sarma, S. (2004) *Rev. Mod. Phys.*, **76**, 323.

24. Ohno, Y., Young, D.K., Beschoten, B., Matsukura, F., Ohno, H., and Awschalom, D.D. (1999) *Nature*, **402**, 790.

25. Ohno, H. (1998) *Science*, **281**, 951.

26. MacDonald, A.H., Schiffer, P., and Samarth, N. (2005) *Nat. Mater.*, **4**, 195.

27. Oiwa, A., Mitsumori, Y., Moriya, R., Słupinski, T., and Munekata, H. (2002) *Phys. Rev. Lett.*, **88**, 137202.

28. Fukumura, T., Jin, Z., Ohtomo, A., Koinuma, H., and Kawasaki, M. (1999) *Appl. Phys. Lett.*, **75**, 3366.

29. Kundaliya, D.C., Ogale, S., Lofland, S., Dhar, S., Metting, C.J., Shinde, S.R., Ma, Z., Varughese, B., Ramanujachary, K.V., Salamanca-Riba, L., and Venketesan, T. (2004) *Nat. Mater.*, **3**, 709.

30. Shinde, S.R., Ogale, S., Das Sarma, S., Simpson, J.R., Drew, H.D., Lofland, S., Lanci, C., Buban, J.P., Browning, N.D., Kulkarni, V.N., Higgins, J., Sharma, R.P., Greene, R.L., and Venketesan, T. (2003) *Phys. Rev. B*, **67**, 115211.

31. Chambers, S.A., Thevuthasan, S., Farrow, R.F.C., Marks, R.F., Thiele, J.U., Samant, M.G., Kellock, A.J., Ruzycki, N., Ederer, D.L., and Diebold, U. (2006) *Surf. Sci. Rep.*, **61**, 345.

32. Kaspar, T.C., Droubay, T., Shutthanandan, V., Heald, S.M., Wang, C.M., McCready, D.E., Thevuthasan, S., Bryan, J.D., Gamelin, D.R., Kellock, A.J., Toney, M.F., Hong, X., Ahn, C.H., and Chambers, S.A. (2005) *Phys. Rev. Lett.*, **95**, 217203.

33. Coey, J.M.D., Douvalis, A.P., Fitzgerald, C.B., and Venkatesan, M. (2004) *Appl. Phys. Lett.*, **84**, 1332.

34. Behan, A.J., Mokhtari, A., Blythe, H.J., Score, D., Xu, X.-H., Neal, J.R., Fox, A.M., and Gehring, G.A. (2008) *Phys. Rev. Lett.*, **100**, 047206.

35. Fulumura, T., Toyosky, H., Ueno, K., Nakano, M., and Kawasaki, M. (2008) *New J. Phys.*, **10**, 055018.

36. Zhao, T., Shinde, S.R., Ogale, S.B., Zheng, H., Venkatesan, T., Ramesh, R., and Das Sarma, S. (2005) *Phys. Rev. Lett.*, **94**, 126601.

37. Ogale, S.B., Choudhary, R.J., Buban, J.P., Lofland, S.E., Shinde, S.R., Kale, S.N., Kulkarni, V.N., Higgins, J., Lanci, C., Simpson, J.R., Browning, N.D., Das Sarma, S., Drew, H.D., Greene, R.L., and Venkatesan, T. (2003) *Phys. Rev. Lett.*, **91**, 077205.

38. Zhang, S.X., Ogale, S.B., Yu, W., Shinde, S.R., Kundaliya, D.C., Tse, W.K., Young,

S.Y., Higgins, J.S., Salamanca-Riba, L.G., Herrera, M., Fu, L.F., Browning, N.D., Greene, R.L., and Venkatesan, T. (2007) *Phys. Rev. B*, **76**, 085323.

39. Shinde, S.R., Ogale, S.B., Higgins, J.S., Zheng, H., Millis, A.J., Kulkarni, V.N., Ramesh, R., Greene, R.L., and Venkatesan, T. (2004) *Phys. Rev. Lett.*, **92**, 166601.

40. Kale, S.N., Ogale, S.B., Shinde, S.R., Greene, R.L., and Venkatesan, T. (2003) *Appl. Phys. Lett.*, **82**, 2100.

41. Ueda, K., Tabata, H., and Kawai, T. (2001) *Appl. Phys. Lett.*, **79**, 988.

42. Ney, A., Ollefs, K., Ye, S., Kammermeier, T., Ney, V., Kaspar, T.C., Chambers, S.A., Wilhelm, F., and Rogalev, A. (2008) *Phys. Rev. Lett.*, **100**, 157201.

43. Venkatesan, M., Fitzgerald, C.B., Lunney, J.G., and Coey, J.M.D. (2004) *Phys. Rev. Lett.*, **93**, 177206.

44. Kaspar, T.C., Droubay, T., Heald, S.M., Nachimuthu, P., Wang, C.M., Shutthanandan, V., Johnson, C.A., Gamelin, D.R., and Chambers, S.A. (2008) *New J. Phys.*, **10**, 055010.

45. Fitzgerald, C.B., Venkatesan, M., Lunney, J.G., Dorneles, L.S., and Coey, J.M.D. (2005) *Appl. Surf. Sci.*, **247**, 493.

46. Schwartz, D.A. and Gamelin, D.R. (2004) *Adv. Mater.*, **16**, 2115.

47. Bouzerar, G., Ziman, T., and Kudrnovský, J. (2004) *Appl. Phys. Lett.*, **85**, 4941.

48. Xu, Q., Hartmann, L., Schmidt, H., Hochmuth, H., Lorenz, M., Schmidt-Grund, R., Strum, C., Psemann, D., and Grundmann, M. (2006) *Phys. Rev. B*, **73**, 205342.

49. Chattopadhyay, A., Das Sarma, S., and Millis, A.J. (2001) *Phys. Rev. Lett.*, **87**, 227202.

50. Coey, J.M.D., Venkatesan, M., and Fitzgerald, C.B. (2005) *Nat. Mater. Lett.*, **4**, 173.

51. Coey, J.M.D., Wongsaprom, K., Alaria, J., and Venkatesan, M. (2008) *J. Phys. D: Appl. Phys.*, **41**, 134012.

52. Kudrnovský, J., Drchal, V., Bouzerar, G., and Bouzerar, R. (2007) *Phase Transitions*, **80**, 333.

53. Ando, K. (2006) *Science*, **312**, 1883.

54. Abraham, D.W., Frank, M.M., and Guha, S. (2005) *Appl. Phys. Lett.*, **87**, 252502.

55. Elfimov, I.S., Yunoki, I.S., and Sawatzky, S. (2002) *Phys. Rev. Lett.*, **89**, 216403.

56. Venkatesan, M., Fitzgerald, C.B., and Coey, J.M.D. (2004) *Nature*, **430**, 630.

57. Osorio-Guillén, J., Lany, S., Barabash, S., and Zunger, A. (2006) *Phys. Rev. Lett.*, **96**, 107203.

58. Herng, T.S., Qi, D.-C., Berlijn, T., Yi, J.B., Yang, K.S., Dai, Y., Feng, Y.P., Santoso, I., Sanchez-Hanke, C., Gao, X.Y., Wee, A.T.S., Ku, W., Ding, J., and Rushydi, A. (2010) *Phys. Rev. Lett.*, **105**, 207201.

59. Pan, H., Yi, J.B., Shen, L., Wu, R.Q., Yang, J.H., Lin, J.Y., Feng, Y.P., Ding, J., Van, L.H., and Yin, J.H. (2007) *Phys. Rev. Lett.*, **99**, 127201.

60. Sundaresan, A., Bhargavi, R., Rangarajan, N., Siddesh, U., and Rao, C.N.R. (2006) *Phys. Rev. B*, **74**, 161306.

61. Hong, N.H., Sakai, J., Poirot, N., and Brizé, V. (2006) *Phys. Rev. B*, **73**, 132404.

62. Rusydi, A., Abbamonte, P., Eisaki, H., Fujimaki, Y., Smadici, S., Motoyama, N., Uchida, S., Kim, Y.-J., Rübhausen, M., and Sawatzky, G.A. (2008) *Phys. Rev. Lett.*, **100**, 036403.

63. Ariando , Wang, X., Baskaran, G., Liu, Z.Q., Huijben, J., Yi, J.B., Annadi, A., Roy Barman, A., Rusydi, A., Dhar, S., Feng, Y.P., Ding, J., Hilgenkamp, H., and Venkatesan, T. (2011) *Nat. Commun.*, **2**, 288.

64. Henke, B.L., Gullikso, F.M., and Davis, J.C. (1993) *At. Data Nucl. Data Tables*, **54**, 181.

65. Carra, P., Thole, B., Altarelli, M., and Wang, X. (1993) *Phys. Rev. Lett.*, **70**, 694.

66. Stöhr, J. (1999) *J. Magn. Magn. Mater.*, **200**, 470.

67. Rauer, R., Rübhausen, M., and Dörr, K. (2006) *Phys. Rev. B*, **73**, 692402.

68. Garcia-Barriocanal, J., Cezar, J.C., Bruno, F.Y., Thakur, P., Brookes, N.B., Utfeld, C., Rivera-Calzada, A., Giblin, S.R., Taylor, J.W., Duffy, J.A., Dugdale, S.B., Nakamura, T., Kodama, K., Leon, C., Okamoto, S., and Santamaria, J. (2010) *Nat. Commun.*, **1**, 82.

5
Defect-Induced Optical and Magnetic Properties of Colloidal Transparent Conducting Oxide Nanocrystals

Pavle V. Radovanovic

5.1
Introduction

Transparent conducting oxides (TCOs) possess many unique properties, including structural diversity (polymorphism) and a combination of transparency and conductivity [1, 2]. These properties are largely due to the presence of native defects or dopants [3], and are responsible for the technological importance of TCOs in optoelectronics [4], photovoltaics [5, 6], photocatalysis [7], and sensors [8, 9]. With the discoveries of the size-dependent properties of condensed matter [10], and the development of techniques for the preparation of materials in reduced dimensions [11–13], the attention has turned to the synthesis of TCO nanostructures and controlling their morphology, structure, and functional properties. The ability to impart and control new properties of TCOs in the nanostructured forms can be a basis for simultaneous control and manipulation of multiple functionally relevant properties. These phenomena can lead to potentially new applications of these materials in spintronics, optospintronics, or lighting. An integral part of these efforts includes the possibility of imparting various types of defects into TCO nanostructures during the synthesis or via postsynthetic processing [14]. This chapter summarizes some recent developments of the control of the crystal structure and defects of colloidal TCO nanocrystals (NCs), and discusses the resulting properties. The chapter is divided into two main parts. The first part of the chapter deals with transition-metal doping of TCO NCs as a way of imparting magnetic properties to these materials. The second part focuses on the native defects in colloidal TCO NCs and their role in defining and controlling optical emission. Owing to their functional importance, these defects are generally referred to as *functional defects*. Although specific case studies have been elaborated to illustrate certain phenomena, many conclusions have been shown to be general and widely applicable to TCO NCs. Richness of the properties associated with defects in TCO NCs will hopefully provide a basic background and stimulate new ideas for the future research in physics, chemistry, and materials science.

Functional Metal Oxides: New Science and Novel Applications, First Edition.
Edited by Satishchandra B. Ogale, Thirumalai V. Venkatesan, and Mark G. Blamire.
© 2013 Wiley-VCH Verlag GmbH & Co. KGaA. Published 2013 by Wiley-VCH Verlag GmbH & Co. KGaA.

5.2
Colloidal Transition-Metal-Doped Transparent Conducting Oxide Nanocrystals

Controlled doping of TCO NCs, such as ZnO, SnO_2, TiO_2, and In_2O_3, has attracted broad interest as a means of imparting new optical, electrical, and magnetic properties to these technologically important materials. Understanding the doping mechanism and its effect on the growth, structure, and properties of colloidal TCO NCs is important for a range of technologies, including electronics, optoelectronics, spintronics, and catalysis. This section discusses different aspects of colloidal transition-metal-doped TCO NCs, with an emphasis on introducing and controlling magnetic properties. First, the mechanism of dopant incorporation and the effect of dopant ions on the size, structure, and composition of TCO NCs are addressed. The discussion will then shift to the electronic structure of transition-metal dopant ions and the spectroscopic methods for its investigation. The final subsection of this part of the chapter is dedicated to ferromagnetic ordering of dopant ions in nanocrystalline TCOs, and the correlation between the electronic structure of dopant ions and the magnetic properties. The discussion in this section is not intended to be a comprehensive review of doped TCO NCs, but rather to underline some fundamental principles and observations common to different TCO NC systems.

5.2.1
Doping TCO NCs

A variety of colloidal transition-metal-doped TCO NCs, known as diluted magnetic semiconductor oxide (DMSO) NCs, has been synthesized to date by different solution methods. The growth of DMSO NCs has been studied by various structural and spectroscopic methods. These studies have provided important knowledge about the nucleation and growth of oxide NCs in the presence of impurities. A review of the synthesis and properties of these materials can be found in recent books and articles [15]. This section summarizes some of the general findings associated with the mechanism of dopant incorporation into nanocrystalline oxide lattices.

Early studies of DMSO NCs have largely focused on transition-metal-doped ZnO [16–19], driven by the promise of high-T_C ferromagnetic ordering of dopant ions in ZnO lattice [20]. Specifically, Co^{2+}-doped ZnO (Co^{2+}:ZnO) NCs served as an excellent model system for studies of the correlation between NC growth and dopant incorporation, owing to the characteristic ligand-field electronic absorption spectra of Co^{2+} in different coordination environments [16, 17]. The incorporation of Co^{2+} ions from the precursor has been investigated by simultaneously monitoring the electronic absorption spectrum of both ZnO NC bandgap and Co^{2+} ligand-field transitions during the reaction (Figure 5.1).

In the precursor form in solution, Co^{2+} ions reside in octahedral environment characterized by a broad, multiple-structured band centered at about $19\,000\,cm^{-1}$, which is assigned to the $^4T_{1g}(F) \rightarrow {}^4T_{1g}(P)$ transition. At the low-energy side,

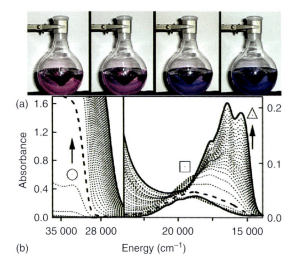

Figure 5.1 (a) Photographs of Co^{2+}:ZnO NCs taken at various times during NC synthesis (increasing time from left to right). The solution color changes from pink to blue, as a result of the conversion of octahedral Co^{2+} to tetrahedral Co^{2+}. (b) Electronic absorption spectra of the ZnO NC bandgap (left) and Co^{2+} ligand-field (right) energy regions collected during NC formation. The intensities of NC bandgap ($\bigcirc$), Co^{2+} intermediate ($\square$), and substitutionally incorporated Co^{2+} in ZnO NCs ($\triangle$) used for analysis of the dopant incorporation are indicated. (Source: From Ref. [17], with permission.)

this band overlaps with a weaker $^4T_{1g}(F) \rightarrow {}^4A_{2g}$ transition, which often appears as a shoulder. By substituting for Zn^{2+} ions in ZnO NCs having wurtzite crystal structure, Co^{2+} adopts a quasi-tetrahedral coordination. Tetrahedral Co^{2+} in ZnO exhibits a richly structured feature between about 14 500 and 19 000 cm^{-1} arising from spin-allowed $^4A_2 \rightarrow {}^4T_1(P)$ ligand-field transition. A comparison between the evolutions of ZnO NC bandgap and the tetrahedral Co^{2+} absorption spectra, monitored during the course of the reaction, revealed that Co^{2+} dopants inhibit the nucleation and growth of the NC host lattice. The dopants are quantitatively excluded from the nucleation process but are incorporated nearly isotropically during the NC growth [17]. In addition to octahedral Co^{2+} in the precursor and tetrahedral Co^{2+} substitutionally incorporated into the NCs, the presence of quasi-tetrahedral surface-bound Co^{2+} intermediate can also be identified from the absorption spectrum at about 19 500 cm^{-1}. Isosbestic points suggest that this surface-bound, distorted, tetrahedral Co^{2+} species is the direct intermediate of the substitutional Co^{2+}, implying that doping occurs by internalization of the surface-bound Co^{2+} ions during NC growth rather than, for example, by their kinetic trapping from the solution.

Similar conclusions about the doping mechanism and the influence of dopant ions on the NC growth have subsequently been made for different dopant ions and colloidal TCO NC host lattices [21–25]. Further discussion in this section will be limited to Cr^{3+}-doped TCO NCs having six-coordinate substitutional sites. Owing to its kinetic stability and high affinity for six-coordinate sites associated with large ligand-field stabilization energy (LFSE) of the d^3 configuration, Cr(III) is an excellent model system for investigation of doping TCO NCs having six-coordinate substitutional sites (TiO_2, SnO_2, and In_2O_3) [22, 23, 26]. Hence, Cr^{3+} is, in a way, a dopant complementary to Co^{2+}, which has a tendency to be incorporated in a tetrahedral coordination environment. Furthermore, similarly to tetrahedral Co^{2+}, octahedral Cr^{3+} has characteristic ligand-field transitions in the visible spectral region, which allow for its reliable spectroscopic characterization during NC synthesis and in the final product. The Tanabe–Sugano energy state diagram of Cr^{3+} (d^3) ion in octahedral coordination is shown in Figure 5.2. The ground state of the free ion is described by ^4F term. In an octahedral ligand field, this term splits, giving rise to two spin-allowed transitions $^4A_{2g} \rightarrow {}^4T_{2g}(F)$ and $^4A_{2g} \rightarrow {}^4T_{1g}(F)$. Using ligand exchange on NC surfaces, ligand-field spectroscopy can effectively probe dopant ion binding to the NC surfaces and help elucidate the mechanism of dopant incorporation [22, 23, 26]. The transitions of substitutional Cr^{3+} in SnO_2 NCs, prepared by n-trioctylphosphine oxide (TOPO) treatment and capping, in the visible range are indicated with a solid line in Figure 5.3 [23]. These transitions appear at about 16 250 cm^{-1} [$^4A_2 \rightarrow {}^4T_2$ (F)] and 23 100 cm^{-1} [$^4A_2 \rightarrow {}^4T_1$(F)]. In addition, a small shoulder is observed at about 14 400 cm^{-1}, and is assigned to formally spin-forbidden $^4A_2 \rightarrow {}^2E$ and $^4A_2 \rightarrow {}^2T_1$ transitions. The absorption spectrum of Cr^{3+}:SnO_2 NCs removed from the reaction mixture during the synthesis and capped with 1-dodecylamine (DDA) is shown with the dashed line in Figure 5.3. The spectrum of these DDA-capped NCs shows two peaks centered at about 17 000 and 23 850 cm^{-1}. The blueshift of the $^4A_2 \rightarrow {}^4T_2$(F) and

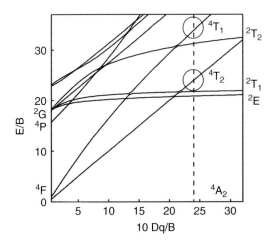

Figure 5.2 Energy diagram of a d^3 transition-metal ion in an octahedral coordination. The dashed line denotes the ligand-field strength characteristic for SnO_2 lattice. The energies expected for the spin-allowed Cr^{3+} ligand-field transitions are labeled with circles. (Source: From Ref. [23], with permission.)

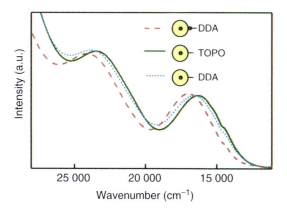

Figure 5.3 Ligand-field electronic absorption spectra of Cr^{3+} doped in SnO_2 NCs. As-synthesized DDA-capped NCs (dashed line) were TOPO-treated and capped (solid line) followed by reverse ligand exchange to DDA (dotted line). (Source: From Ref. [23], with permission.)

$^4A_2 \rightarrow {}^4T_1(F)$ transitions relative to the same transitions in internally doped NCs arises from nitrogen coordination of Cr^{3+} ions exposed on NC surfaces, which has higher ligand-field strength than the oxygen ions of the NC host lattice. On thermal treatment of NCs in TOPO, which has been shown to remove surface-bound dopant ions by reversible coordination, the $^4A_2 \rightarrow {}^4T_2(F)$ and $^4A_2 \rightarrow {}^4T_1(F)$ bands redshift to about 16 250 and 23 100 cm^{-1}, respectively, and remain largely unchanged following subsequent ligand exchange with DDA (dotted line in Figure 5.3). These results indicate that dopant ions remaining upon TOPO treatment are internally incorporated into NC host lattice. The findings of these ligand-exchange studies further imply that doping mechanism involves internalization of surface-bound dopant ions during NC growth. Importantly, this work suggests that special care must be taken to remove NC-surface-bound dopant ions, which could lead to misinterpretation of the observed physical properties of doped NCs.

The mechanism of dopant incorporation has been a subject of significant debate in recent scientific literature, and several different concepts have been proposed to describe the dopant incorporation. Specifically, it has been proposed that NC shape determines the binding of dopant ions to NC surfaces, and their incorporation into colloidal NCs [27]. On the other hand, Lewis acid/base chemistry has been invoked as a key factor in determining dopant ion incorporation from the solution phase [28]. The ability to stabilize different crystal structures of the same material and achieve the incorporation of dopant ions into different phases of colloidal NCs provides an opportunity to understand the correlation between NC structure and dopant incorporation [22]. Metal oxides are known for their polymorphism, and the variety of possible crystal structures [1], enabling the stabilization of high-energy phases with unique properties. Recent work on transition-metal doping of complex polymorphic TCO NCs, such as In_2O_3, proves that NC growth and dopant incorporation follow the same fundamental principles for different dopants, and different crystal and electronic structures of the same host material [22]. The next section describes some results on *in situ* phase control of colloidal TCO NCs using size–structure correlation, and the effect of dopant ions on the growth and structure of colloidal In_2O_3 NCs.

5.2.2
Phase transformation of TCO NCs

The development of the colloidal synthesis methods has enabled a high degree of control of the NC size, morphology, surface area, and surface structure [13, 29, 30]. The control of these parameters has provided a path to high-energy crystal structures, which are often accessible in bulk only under extreme temperature and pressure [31]. The ability to prepare metastable crystal structures in solution can significantly expand the range of available materials and their properties. In order to facilitate solution-phase synthesis of metastable colloidal NC, it is essential to develop a fundamental understanding of the phase transformation mechanism at the nanoscale. Furthermore, it is necessary to understand the correlation between doping process and the phase stabilization and transformation to effectively manipulate the properties of different phases of the same material. As indicated earlier, TCOs are polymorphic, and as such TCO NCs represent not only good model systems for the studies of phase transformation but also technologically important materials whose properties may be modulated by structural modification.

Traditionally, it has been widely accepted that the kinetics of phase transition in NCs is mechanistically simpler than that in bulk, owing to their large surface areas, small volumes, and, presumably, low concentration of crystal lattice defects [31, 32]. This section focuses on In_2O_3 as a model system for the study of phase transformation of NCs in solution. In_2O_3 exists in two known allotropic modifications: cubic bixbyite and corundum phases [33]. Cubic bixbyite phase (often labeled as bcc-In_2O_3) has a body-centered cubic crystal structure in which In^{3+} ions reside in two distinct six-coordinate sites, known as b and d sites. The unit cell of bcc-In_2O_3, which is stable at ambient conditions, is shown in Figure 5.4a. The b sites adopt the C_{3i} point group symmetry, while d sites are strongly distorted (C_2 point group), as shown in Figure 5.4c. One-fourth of In^{3+} cations reside in b sites, while three-fourths of In^{3+} cations are located in d sites (the unit cell consists of 8 b sites and 24 d sites). Corundum phase is the metastable form of In_2O_3 [34]. It has a rhombohedral crystal structure (rh-In_2O_3) with In^{3+} ions residing in C_{3v} symmetry sites (Figure 5.4b,d). The density of rh-In_2O_3 is higher than that of bcc-In_2O_3 by about 2.6%.

The cohesive energy as a function of volume per unit formula for the two phases is shown in Figure 5.5 [35]. The rh-In_2O_3 phase has a higher energy minimum, which corresponds to lower volume per unit cell (higher density). The potential energy curves for the two structures intersect, indicating the possibility of pressure-induced phase transformation. Figure 5.6 shows the calculated unit cell volume as a function of an applied pressure [36]. The abrupt decrease in the unit cell volume by 1.72 $Å^3$ at 3.83 GPa indicates the occurrence of the phase transformation. The

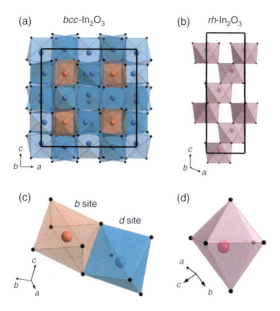

Figure 5.4 (a,b) Crystal unit cell of bcc-In$_2$O$_3$ (a) and rh-In$_2$O$_3$ (b). Unit cells are viewed along *b*-axes. (c,d) The In^{3+} sites in (c) bcc-In$_2$O$_3$ and (d) rh-In$_2$O$_3$. Indium ions are shown as large spheres, and oxygen ions as small spheres. (Source: From Ref. [22], with permission.)

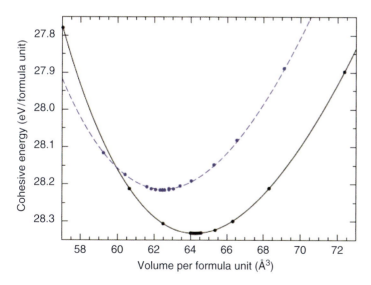

Figure 5.5 Cohesive energy as a function of volume per formula unit of In$_2$O$_3$ for bcc-In$_2$O$_3$ (solid line) and rh-In$_2$O$_3$ (dashed line). Dots indicate volumes sampled by the calculations. (Source: From Ref. [35], with permission.)

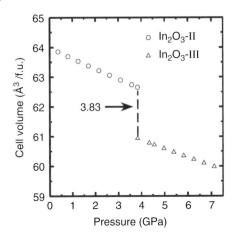

Figure 5.6 Calculated volume per formula unit of In_2O_3 as a function of pressure for bcc-In_2O_3 (In_2O_3-II) and rh-In_2O_3 (In_2O_3-III). (Source: From Ref. [36], with permission.)

phase transition from bcc-In_2O_3 to metastable rh-In_2O_3 is associated with breaking of the chemical bonds between atoms and rearrangement of anion and cation packing.

Early reports on the synthesis and properties of In_2O_3 NCs focus on the stable bcc-In_2O_3 phase [37–39]. With the development of the new synthetic approaches, the direct preparations of rh-In_2O_3 NCs have also been achieved [40–44]. The reported examples include preparations of rh-In_2O_3 powders under ambient pressure by calcination of $In(OH)_3$ or InOOH, indicating that In_2O_3 synthesis involves dehydration of initially formed hydroxide precursors [40–42]. Nearly monodispersed colloidal rh-In_2O_3 nanocubes have been prepared in an elaborate multistep synthesis, from indium isopropoxide [43]. This method also involves rapid dehydration of initially stabilized InOOH nanoparticles. Recent results have shown that rh-In_2O_3 phase is an intermediate in the formation of bcc-In_2O_3 NCs [44]. Detailed studies of the kinetics of this phase transformation have revealed the role of an interface nucleation mechanism, in which a new bcc-In_2O_3 phase is formed at the interface of contacting rh-In_2O_3 NCs [45]. The nucleation of the new phase is likely facilitated by the defect sites formed at the interface between contacting NCs. On nucleation, the new phase propagates rapidly throughout the NC volume owing to their small sizes [46]. The described process of phase transformation is kinetically controlled, and the mechanism is dependent on the probability of NC contact formation. Hence, the mechanism depends on the concentration of NCs in solution and on the reaction temperature. An increase in the synthesis temperature, as well as a decrease in the concentration of NCs can favor the surface nucleation mechanism, for which the stable phase is nucleated at the surface of individual NCs. These mechanistic results have provided a guideline for the solution-phase preparation of metastable high-energy structures in the nanocrystalline form. Specifically, rh-In_2O_3 NCs were synthesized by a simple procedure from inexpensive inorganic salts by controlling the reaction kinetics and NC size via reaction temperature

(Figure 5.7) [44]. The stabilization of metastable rh-In_2O_3 NCs below the critical size of about 5 nm is enabled by the surface energy and/or surface stress contribution. For sufficiently small NCs having high surface-to-volume ratio, lower surface energy of a metastable phase could overcome the Gibbs free energy of the stabilization of the thermodynamically stable phase (ΔG), leading to the spontaneous metastability. Because of the small ΔG between bulk phases, rh-In_2O_3 can also be stabilized in NCs below the critical size through an increase in surface stress.

An interesting question that arises in the context of colloidal NC doping, discussed in the Section 5.2.1 of this chapter, is how the presence of dopant ions influences the phase stabilization and phase transformation of colloidal NCs. Substitutional doping of bulk In_2O_3 with ions smaller than In^{3+} (i.e., Sn^{4+} and Fe^{3+}) has been shown to stabilize rh-In_2O_3 phase [47, 48]. This phenomenon has been explained by shrinking of the host lattice on incorporation of dopants, which then favors the stabilization of the higher density rh-In_2O_3 phase. In NCs, however, the dopant ions also lead to the inhibition of NC growth, which by itself causes the stabilization of rh-In_2O_3, as described earlier. Systematic investigation of the NC size and structure in the presence of different concentrations of dopant ions in the reaction mixture has revealed that correlation [22].

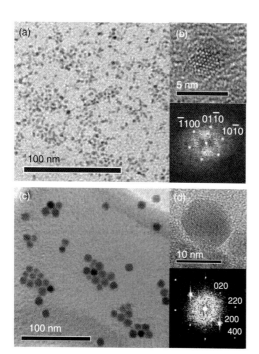

Figure 5.7 (a,c) Transmission electron microscopy (TEM) images of (a) rh-In_2O_3 NCs and (c) bcc-In_2O_3 NCs. (b,d) Lattice-resolved TEM images (top) and the corresponding fast Fourier transform images (bottom) of a single rh-In_2O_3 NC (b) and bcc-In_2O_3 NC (d). (Source: From Ref. [44], with permission.)

Figure 5.8 shows the transmission electron microscopy (TEM) images of single In_2O_3 NCs doped with chromium and manganese ions [22]. With increasing concentration of the dopant precursor, smaller NCs are formed for both dopants, which indeed have a corundum crystal structure. Although both dopant ions have the same effect on NC size and phase, the final doping concentrations are different under identical conditions. This study has also shown that Cr^{3+} is more readily doped in rh-In_2O_3, while higher doping levels of Mn^{3+} can be achieved in bcc-In_2O_3. It should be noted here that both Mn^{3+} and Cr^{3+} ions have similar ionic radii in octahedral coordination ($r(Mn^{3+}) = 0.65$ Å, $r(Cr^{3+}) = (0.62$ Å$)$), and that average In–O distance is nearly identical in both In_2O_3 phases (about 2.17–2.18 Å). Higher doping concentrations of Mn^{3+} in bcc-In_2O_3 than in rh-In_2O_3 NCs suggest that dopant incorporation is not the main factor for the stabilization of rh-In_2O_3 NCs. The dependence of the doping concentration on the NC type and structure has been a point of intense debate in the literature. This phenomenon can be considered from both thermodynamic and kinetic points of view. Thermodynamically, the propensity of dopant incorporation can be described by the excess enthalpy of isovalent dopant ion mixing (ΔH_m), which is dependent on the difference between In–O and dopant–O bond distances [49]. A large bond length difference could lead to destabilization of the doped NCs, resulting in low doping concentrations or, in extreme cases, no doping at all [50]. The highest doping concentrations of

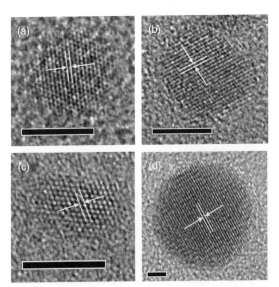

Figure 5.8 High-resolution TEM images of Mn^{3+}:In_2O_3 NCs synthesized with [Mn]/[In] = 0.10 (a,b) and Cr^{3+}:In_2O_3 NCs synthesized with [Cr]/[In] = 0.10 (c,d). The d spacings of about 2.74 Å (a,c) correspond to {110} lattice plane of rh-In_2O_3, and the d spacings of 2.91 Å (b) and 2.54 Å (d) correspond, respectively, to {222} and {400} lattice planes of bcc-In_2O_3. The critical size for nanocrystal transformation from rh-In_2O_3 (a,c) to bcc-In_2O_3 (b,d) is about 5 nm. Scale bars: 5 nm (a–c) and 2 nm (d). (Source: From Ref. [22], with permission.)

Mn^{2+} doped in II–VI NCs have been found to coincide with the minimum value of ΔH_m [49]. On the basis of the ionic radius and bond length considerations alone, a significant role of thermodynamics in determining doping concentrations in different phases of In$_2$O$_3$ NCs is not likely. The NC doping can also be explained using kinetic considerations. Specifically, the competition between dopant and host cations for the surface binding sites, as well as the competition between surface anions and the coordinating ligands for the cations in solution can also determine NC doping concentrations. Doping different phases of the same material under identical conditions can be particularly instructive for elucidating the mechanism of dopant incorporation. For instance, in the case of In$_2$O$_3$, syntheses of both cubic bixbyite and corundum phases have been performed under the same conditions, eliminating the influence of solution ligands on the incorporation of dopant ions [22]. This leaves the competition between dopant and host cations for the available surface sites as the most probable source of the difference in doping concentration of the same ions in bcc- and rh-In$_2$O$_3$ NCs. The probability of dopant incorporation into NCs can be associated with their average residence time (τ) on the NC surfaces, which is related to their heat of adsorption (ΔH_{ads}):

$$\tau = \tau_0 \exp\left(\frac{\Delta H_{ads}}{RT}\right) \tag{5.1}$$

Interestingly, there is a notable structural similarity between α-Mn$_2$O$_3$ and bcc-In$_2$O$_3$, as both materials have a cubic bixbyite crystal structure, and between Cr$_2$O$_3$ and rh-In$_2$O$_3$ (both have a corundum crystal structure). This structural similarity correlates well with the higher doping concentration of Cr^{3+} and Mn^{3+} ions in the corresponding In$_2$O$_3$ phases. For example, Cr^{3+} is found to exhibit higher doping concentrations in rh-In$_2$O$_3$ than in bcc-In$_2$O$_3$, while the opposite is true for Mn^{3+}. These findings suggest that binding of dopant ions to the host NC surface sites, as a necessary step for their subsequent incorporation, depends on the relationship between the NC structure and the nature of the dopant ion. Since NC surfaces are defined by the lattice structure, it is likely that the structural similarity enables competitive binding of dopant ions to the available surface sites relative to the host cations, leading to high doping concentrations. As discussed in the literature, in addition to surface binding, incorporation of dopant ions into NCs is influenced by other, less-investigated, microscopic phenomena, including surface strain and defects [49, 50]. These phenomena will likely continue to generate interest in the NC doping community, and the availability of NCs having different structures may be particularly useful in that regard.

5.2.3
Electrical and Magnetic Properties of Doped TCO Nanocrystals

As stated at the beginning of this part of this chapter, doping NCs is an effective way of imparting new properties or manipulating the native properties of these nanostructures. Various dopant ions have been studied in different oxide NCs, and there are some excellent review articles covering specific aspects of this topic

[15, 51]. The readers are referred to those sources for specific details. Here, the focus is on the electrical and magnetic properties of doped TCO NCs, which are a subject of a flurry of the current research activity. Owing to the wide scope of the topic and the space limitation, this discussion will still mostly be limited to In_2O_3 and ZnO NC host lattices.

5.2.3.1 Electrical Properties of TCO Nanocrystals

The combination of optical transparency and electrical conductivity in TCOs has for decades been an intriguing area of research for both experimentalists and theorists, driven in part by the technological significance of these materials in optoelectronics, photovoltaics, and, more recently, spintronics and photocatalysis [5–7, 52, 53]. TCOs are generally n-type semiconductors, with charge carriers originating from the shallow donor states. The donor states are introduced by the presence of defects [3, 54], and the concentration of carriers in TCO NCs is dependent on the ability to incorporate dopant ions into NC host lattices. Perhaps the most investigated and widely applied TCO is indiumtin oxide (ITO), in which tin ions substitute for In^{3+} ions. ITO NCs having different compositions have been synthesized by a variety of methods ranging from relatively simple colloidal syntheses [55–58] to microwave and solvothermal methods [59, 60]. These methods have resulted in NCs having different sizes and size distributions, as well as different dispersability in organic solvents. Figure 5.9 shows an example of colloidal ITO NCs having different Sn doping concentrations [55]. The doping concentration is correlated with a strong absorption band in the near-infrared region, giving these samples a characteristic blue color, which is absent in undoped In_2O_3 NCs. This transition arises from the collective oscillations of the free electrons in the conduction band, and is referred to as the surface plasmon resonance (SPR), in analogy to the same phenomenon in metal nanoparticles [55, 58]. Supporting evidence that this band is a signature of delocalized free electrons is that its energy and intensity for samples having different doping concentrations follow the behavior predicted by the Drude–Lorentz theory of free-electron gas [58]. This is illustrated in Figure 5.10, which shows the absorption spectra of ITO NCs having different Sn^{4+} concentrations and therefore different free-electron densities. With increasing electron density, the peak energy shifts to higher values and its intensity increases. This behavior can be described through the equation for plasma frequency (ω_p):

$$\omega_p = \sqrt{\frac{Ne^2}{m^*\varepsilon_{opt}\varepsilon_0}} \tag{5.2}$$

where N is the change carrier density, e is the electron charge, m^* is the effective mass of an electron, ε_0 is the vacuum permittivity, and ε_{opt} is the dielectric constant measured in the transparent region of the spectrum of an undoped semiconductor. The inset in Figure 5.10 shows the Drude model fit to the experimental data for absorption band energy and intensity. It should be noted that, above a certain doping concentration (in this case, about 12%), the effective concentration of free electrons decreases, most likely owing to electron trapping at the dopant sites. The ability to manipulate the NC phase in solution, described in Section 5.2.2, has enabled the

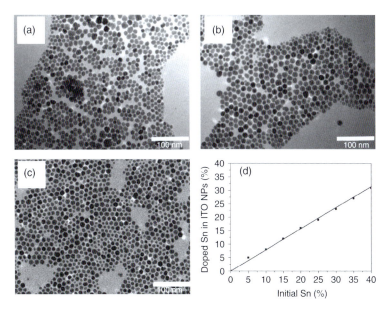

Figure 5.9 (a–c) TEM images of 11.3 ± 1.9 nm In_2O_3 NCs (a), 12.8 ± 1.8 nm ITO (15% Sn) NCs (b), and 11.2 ± 1.8 nm (30% Sn) ITO NCs (c). Scale bars: 100 nm. (d) Relationship between the Sn concentration (at. %) in the initial reaction mixture and ITO NCs. (Source: From Ref. [55], with permission.)

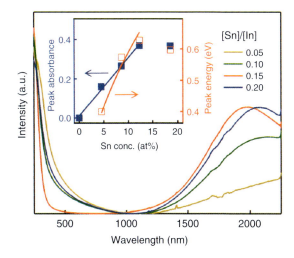

Figure 5.10 Absorption spectra of bcc-ITO NCs isolated from the samples synthesized with different precursor concentration ratios ([Sn]/[In]), as indicated in the graph. Inset: maximum absorbance (solid squares) and peak energy position (open squares) of the plasmon band as a function of Sn^{4+} doping concentration. (Source: From Ref. [58], with permission).

preparation of ITO NCs with different crystal structures. In contrast to bcc-ITO, for which high free-electron concentration can be introduced, the optical properties of rh-ITO in the near-infrared spectrum remain unchanged even with high Sn^{4+} doping concentration [58]. The reason for this difference is the larger bandgap energy of rh-In_2O_3 NCs, which leads to significantly larger donor activation energy than in bcc-ITO NCs. The larger bandgap energy is associated with the electronic structure of corundum In_2O_3 phase and/or quantum confinement due to smaller sizes of rh-ITO NCs.

The phenomena described for ITO NCs are general and broadly applicable for other TCO NCs, judged by the qualitatively similar results for aluminum zinc oxide (AZO) [61] and antimony tin oxide (ATO) NCs [62]. This is a very promising finding because of the lower cost and greater abundance of ZnO relative to In_2O_3, which could lead to broader technological implementations of this class of materials.

5.2.3.2 Magnetic Properties of Doped TCO NCs

Diluted magnetic semiconductors (DMSs), prepared by doping semiconductors with paramagnetic transition-metal ions, are often seen as a key component of the emerging semiconductor spintronics technology, which relies on the interaction of spin and charge degrees of freedom for information manipulation and storage [63]. The presence of paramagnetic dopant centers allows for the exchange interactions between unpaired d electrons on the dopant ions and the host lattice charge carriers (electrons or holes). These interactions are termed *sp–d exchange interactions*, which could enable simultaneous control and manipulation of spin and charge degrees of freedom (i.e., polarizations of semiconductor charge carriers). The ideal candidate material for spintronic applications should operate at room temperature [exhibit ferromagnetic phase transition temperature (T_C) above 300 K], have high charge-carrier density and mobility, and be transparent to the visible light. Research on introducing magnetic dopants into TCO host lattices has been revived in the past decade on the report of long-range magnetic ordering in Co-doped TiO_2 [64]. Despite the opposing opinions about the conclusions of this paper, it remains a seminal contribution, which has caused an avalanche of research activity in the ensuing time [52, 65]. An Achilles' heel of the DMSOs, and more generally DMSs, is the lack of consistency of the results reported for the nominally same samples. In some cases, ferromagnetism has been reported even in undoped TCOs [66, 67]. All these observations clearly point to the fact that sample preparation critically determines the magnetic ordering of dopant ions. The sensitivity of the magnetic properties to the seemingly small differences in the sample preparation procedure requires the answers to the following questions: what is the origin of ferromagnetism in transition-metal-doped TCOs? Is it an inherently intrinsic or an extrinsic phenomenon? Can magnetic ordering be controlled by material properties such as size, doping concentration, conductivity, transparency, or other properties? This chapter does not attempt to give comprehensive answers to these questions (after all, most of them are still a subject of intense debate in the literature), but rather points to some critical results from the author's perspective, in light of colloidal DMSO NCs.

The discussion about the origin of the magnetic properties of DMSOs should start with the fact that doping concentrations of typical DMSOs exhibiting ferromagnetic behavior are well below the expected percolation limit. This consideration excludes traditional descriptions of the magnetic interactions in oxides, such as double-exchange or superexchange mechanisms [68]. The conflict with the established magnetic theories has raised serious concerns that observed high-temperature magnetization in DMSOs is caused by the formation of secondary phases involving dopant ions. On the other hand, a growing number of research papers have brought no evidence of the secondary phase formation in spite of detailed structural characterization, supporting the hypothesis of intrinsic dilute ferromagnetic ordering [69, 70]. In analogy to bulk III–V DMSs (particularly Mn-doped GaAs), it has been suggested that the ferromagnetism originated from charge-carrier-mediated ordering of dopant centers [71]. A breakthrough came from the studies of colloidal DMSO NCs [18]. As indicated in the earlier-mentioned discussion, doped NCs can be prepared in the solution under mild conditions, and as such can serve as building blocks for macroscopic materials and devices. Ligand-field electronic absorption spectroscopic investigations of colloidal free-standing DMSO NCs have confirmed internal substitutional incorporation of dopant ions into TCO NC host lattices [16–18]. Magnetic circular dichroism (MCD) spectroscopy has played a particularly important role in these studies. MCD measures the difference in absorption of left- and right-circularly polarized light as a function of magnetic field, and enables optical detection magnetization at the molecular level. One of the very first systems developed and studied was Co^{2+}-doped ZnO (Co^{2+}:ZnO) NCs [17]. Figure 5.11a,b shows the MCD spectrum of Co^{2+}:ZnO NCs. Quasi-tetrahedral Co^{2+} ligand-field

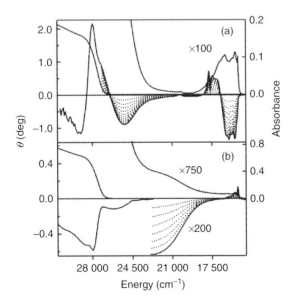

Figure 5.11 MCD spectra of 7 K absorption and 5 K variable-field (0–6.5 T) of (a) 1.7% Co^{2+}:ZnO NCs and (b) 1.5% Ni^{2+}:ZnO NCs. (Source: From Ref. [17], with permission.)

transition ($^4A_2 \rightarrow {}^4T_1(P)$) is observed at about 17 000 cm^{-1}. This richly structured transition is in excellent agreement with the analogous transition in bulk Co^{2+}:ZnO [72], confirming homogeneous internal substitutional incorporation of Co^{2+} with negligible quantum confinement effect on the dopant ion. The intensity of this $^4A_2 \rightarrow {}^4T_1(P)$ transition increases with increasing magnetic field according to the spin state $S = 3/2$ characteristic for Co^{2+} in tetrahedral coordination, as determined by fitting to the Brillouin function [17]. Similarly, a strong MCD feature at about 28 000 cm^{-1} follows the same behavior as a function of the magnetic field. This transition arises from ZnO bandgap transition. The same saturation magnetization behavior indicates that NC band structure undergoes a strong interaction with the dopant ion, leading to a giant Zeeman splitting of the semiconductor band states. Such exchange interactions are the basis of the coexistence of and the interactions between spin and charge degrees of freedom in semiconductor lattices. These studies have clearly shown that dopant ions are paramagnetic in colloidal DMSO NCs, a result that is hardly surprising considering the small sizes of NCs and the dilute magnetic nature of DMSOs. In this scenario, the ferromagnetic phase cannot be stabilized in individual NCs even at very low temperatures.

However, a seminal paper by Radovanovic and Gamelin [18] has shown the possibility of ferromagnetism in nanocrystalline aggregates prepared from colloidal Ni^{2+}:ZnO NCs by slow aggregation at room temperature (Figure 5.12). Under these conditions, the dopant ions are stable and cannot be expelled from NCs or diffuse within NCs to form secondary phases. How can then NC aggregation turn on a long-range ordering in NCs? First, aggregation can lead to the formation of the stable ferromagnetic domains, and the domain walls as the origin of the magnetic hysteresis through the domain wall movement. The more difficult question is, what could be the mechanism of the long-range ordering of the dopant ions?

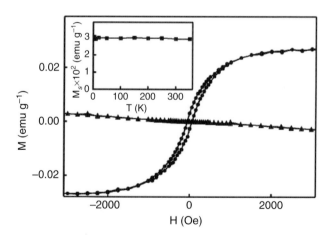

Figure 5.12 Magnetic hysteresis loop measurements of rapidly (▲) and slowly (·) aggregated 0.93% Ni^{2+}:ZnO NCs, at 350 K. Inset: Temperature dependence of the saturation magnetization (M_s) for the slowly aggregated NCs. (Source: From Ref. [18], with permission.)

In free-standing NCs, such a mechanism is obviously lacking; otherwise, at least superparamagnetism as a form of parallel magnetic exchange interactions would be observed. The most likely explanation, proposed initially by Radovanovic and Gamelin [18], is that structural defects, such as oxygen vacancies, formed at the interface between contacting NCs, mediate magnetic ordering of the dopant ions.

Similar results were later reported for other DMSO NCs, including Co^{2+}- and Cr^{3+}-doped TiO_2 [26, 73], Co^{2+}- and Mn^{2+}-doped ZnO [17, 74], Cr^{3+}-doped SnO_2 [23], In_2O_3 [75], and even $BaTiO_3$ [76], suggesting the generality of this aggregation-induced dilute magnetic ordering in oxide NCs. The experiments involving mild annealing of nanocrystalline films prepared from colloidal Ni^{2+}:SnO_2 NCs in the oxygen-rich environment lead to a decrease in the magnitude of the observed ferromagnetism, implying the role of interfacial oxygen vacancies in mediating ferromagnetic ordering of dopant ions [21]. Furthermore, it has subsequently been shown that extended defects (i.e., grain boundaries) are necessary for activating the ferromagnetism in bulk DMSO films [77] prepared by traditional gas-phase deposition techniques. The exact nature of this defect-mediated magnetic ordering and the role of defects in such a process are, however, still relatively poorly understood. On the basis of a systematic analysis of a large collection of the experimental magnetization data for DMSOs, Coey *et al.* [78] have proposed that the ferromagnetic exchange of dopant ions is mediated by shallow donor electrons in the impurity band (arising from oxygen vacancies) that form bound magnetic polarons. This hypothesis is schematically shown in Figure 5.13. According to this model, high Curie temperature results when empty minority or majority spin d states overlap with the Fermi level in the impurity band. This proposal is in general agreement with the observations for nanocrystalline films prepared from colloidal NCs and with those for structurally imperfect thin films prepared by vacuum deposition techniques. The illustration in Figure 5.13 may, however, be a somewhat idealistic picture, given that extended structural defects appear to be an integral component of ferromagnetic exchange involving dopant ions. For example, it is likely that defect–dopant ion exchange interactions are limited only to dopant ions in the immediate vicinity of grain boundaries, which could explain the larger effective magnetic moments of the films prepared from smaller NCs [76].

The mechanism of diluted magnetic ordering and the possibility of its manipulation in oxides will certainly continue to fascinate researchers in physics, chemistry, and materials science. We have only begun to assemble the pieces of the complex mosaic that is dilute magnetism in doped TCOs. At this point, there are strong and growing evidences that magnetic ordering of dilute magnetic centers in TCOs is associated with internally incorporated dopant ions, but the mechanism (mediation) of their ordering appears to be somewhat an external effect, determined by the presence of extended structural defects. The ability to obtain NCs with controlled crystal and electronic structures, and electrical, optical, and magnetic properties as building blocks ensures exciting future developments with significant fundamental and technological implications.

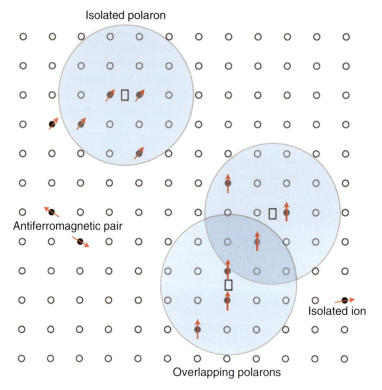

Figure 5.13 Schematic representation of the mechanism of magnetic ordering and magnetic polaron formation. A donor electron in its hydrogenic orbit couples with its spin antiparallel to impurities with an open 3d shell. Cation sites are represented by small circles (dopants are shown as filled circles), oxygen sites are omitted, and the oxygen vacancy sites are represented by squares. (Source: From Ref. [78], with permission.)

5.3
Native Defects in Colloidal Transparent Conducting Oxide Nanocrystals

5.3.1
Type of Native Defects in TCO NCs and Photoluminescence Involving Oxygen Vacancy: ZnO Nanocrystals

Extrinsic or external defects are foreign atoms, or impurities, in a host lattice. A class of external defects are dopants, which are intentionally introduced impurities, as described in the previous part of this chapter. Unlike external defects, native defects (or intrinsic defects) refer to the defects associated with the atoms of the host lattice. The presence or absence of these atoms leads to deviation from the ideal crystal structure. For a comprehensive review on defects in semiconductors, the reader is referred to a recent book [79]. Native defects are often described using the Kröger-Vink notation (A_S^C), where:

- *A* denotes the host lattice species, such as an atom (Zn, In, Ga, O) or vacancy (V);
- *S* stands for a lattice site that the species A occupies. For example, "i" indicates an interstitial site, and an atom symbol indicates a substitutional site (if Al occupies a Zn site in ZnO, *A* should be written as Al, and *S* should be replaced by Zn); and
- *C* indicates the charge of the species with respect to the site it occupies (e.g., "×" indicates null charge, "·" indicates a single positive charge, and "'" indicates a single negative charge).

The most common TCO native defects relevant for the discussion here are

- oxygen vacancy (V_O) with different charges, for example, $V_O^{\times}$, $V_O^{\cdot}$, and $V_O^{\cdot\cdot}$,
- cation vacancy (V_M) with the corresponding charge, depending on the type of the cation,
- interstitial cation site (e.g., $M_i^{\times}$), and
- cation-oxygen vacancy pair (V_O, V_M) with the corresponding charge.

The discussion of native defects in this chapter will mostly focus on oxygen vacancies and their influence on the functional properties of colloidal TCO NCs, particularly photoluminescence (PL). Perhaps the most studied case of PL involving native defect sites in NCs is visible emission in ZnO NCs. Zinc oxide is an efficient blue-green emitting phosphor, which has found important applications in vacuum fluorescent displays and field-emission displays. The origin and the mechanism of this emission have been a theme of much investigation in bulk ZnO [80, 81], and, more recently, in ZnO NCs prepared by colloidal methods [82–85]. Different defect species have been invoked as the origin of visible PL of ZnO, including interstitial zinc ions and oxygen vacancies.

In recent years, oxygen vacancies have been increasingly proposed as the most likely candidates for the recombination centers responsible for the visible luminescence of ZnO [81–83, 85]. The development of the colloidal syntheses and other methods for preparation of NCs has allowed for studies of ZnO emission as a function of NC size, surface structure, and composition, as well as synthesis conditions, such as temperature, growth time, ripening, and so on. One of the first systematic studies [82] of PL in ZnO NCs revealed two distinct emission peaks when NCs were excited in the bandgap (Figure 5.14). A weak and narrow emission band with the maximum at about 3.6 eV has been assigned to the exciton emission or recombination of shallowly trapped charge carriers. A much stronger and broad PL band is observed in the visible region, with a maximum at about 2.2 eV. Given the large Stokes shift, this visible emission must involve a recombination of deeply trapped electrons or holes. Both ultraviolet and visible emission bands shift to higher energy with decreasing NC size owing to the quantum size effect. A detailed analysis of this visible emission energy for different NC sizes has allowed for the conclusion that this feature is due to the transition of an electron from the conduction band to a deep trap in the ZnO NCs. Furthermore, temperature-dependent studies of the ratio of the visible to exciton luminescence led to the formulation of the model in which the photogenerated hole is transferred from the valence band to the $V_O^{\cdot}$ trap level within NC in a two-step process [83]. The first step of this process

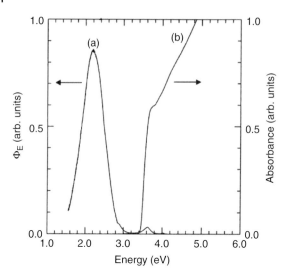

Figure 5.14 (a) Room-temperature PL spectrum of a suspension of ZnO NCs on excitation with 4.4 eV light (Φ_E is photon flux per constant energy interval). (b) Absorption spectrum of the same suspension of ZnO NCs. (Source: From Ref. [82], with permission.)

was proposed to be surface trapping, which these authors associated with the O^{2-} site. Norberg and Gamelin [84], however, have shown that visible trap emission intensities are directly correlated with the concentration of the surface hydroxides. On the basis of this correlation, it was suggested that the identity of surface traps is OH^- dangling bonds. Different mechanisms involving defect states have also been proposed. Among them, it has been suggested that visible emission is determined by singly charged oxygen vacancies ($V_O^{\bullet}$), and corresponds to a transition of holes from the valence band to the deep donor energy level [85]. Other reports claim that visible PL in ZnO nanostructures is altogether a surface phenomenon, arising from the structural near-surface defects [86] or ZnO-capping ligand complexes [87]. It is expected that future research will continue to shine new light on the origin of PL of ZnO NCs, fueled by additional opportunities for technological applications. As an integral part of this effort, understanding the nature of native defects as well as surface traps should enable the optimization and modulation of ZnO NC emission.

5.3.2
Donor–Acceptor Pair Recombination: Ga$_2$O$_3$ Nanocrystals

One of the well-known radiative recombination mechanisms involving native defects is donor–acceptor pair (DAP) [74] recombination [88]. This mechanism involves radiative recombination of an electron trapped on a donor with a hole trapped on an acceptor, where donor and acceptor are relatively shallow defect sites. This chapter describes a case study of the DAP luminescence in Ga$_2$O$_3$ NCs. Gallium oxide is a native n-type semiconductor with a large bandgap energy (about

4.9 eV), relatively high conductivity, and strong luminescence in the visible region. The conductivity of bulk β-Ga_2O_3 (stable monoclinic phase of Ga_2O_3) comes from oxygen vacancies, which act as shallow donors with the activation energy of about 0.03 eV [89, 90]. At the same time, acceptor states are also formed in Ga_2O_3. It has been suggested that an electron trapped on a donor site can recombine with a hole created on an acceptor, which is proposed to be a gallium vacancy (V_{Ga}''') or a gallium-oxygen vacancy pair $(V_O, V_{Ga})'$, to produce blue emission [89, 91]. This process is the reverse of the creation of an electron on a donor and a hole on an acceptor by the acceptor excitation, and occurs according to the following equation:

$$(V_O, V_{Ga})^\times + V_O^\times \rightarrow (V_O, V_{Ga})' + V_O^\bullet + h\nu \tag{5.3}$$

The blue PL produced as a result of DAP formation, involves a transfer of an electron trapped on a donor cluster to a neutral acceptor, as illustrated in Figure 5.15a [92]. This intersite electron transfer is known as a *tunnel transfer*, and in spite of the term *tunneling* is the rate-determining step in the emission process. Consequently, the blue emission of Ga_2O_3 is characterized by a long lifetime, measured to be as long as a few milliseconds [89, 91]. The electron transfer is followed by the radiative recombination of the exciton at the acceptor site. The resulting luminescence has a localized character with a large Stokes shift, indicating that the acceptor state involved in the DAP recombination is much deeper than the donor state. Another notable feature of this emission is a significant spectral broadening. This feature indicates a strong electron–phonon coupling (Figure 5.15b), as well as a broad distribution of donor-acceptor pairs [88, 92]. The energy of the blue luminescence can expressed as

$$E = E_g - (E_D + E_A) + E_C \pm nE_{phonon} \tag{5.4}$$

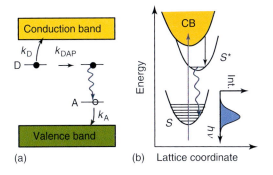

(a) (b) Lattice coordinate

Figure 5.15 (a) Schematic representation of the DAP recombination mechanism, involving a transfer of an electron trapped on a donor (D) to an acceptor (A) site (k_{DAP}). De-trapping of an electron from a donor to the conduction band (k_D) and a hole from an acceptor to the valence band (k_A) are also indicated. (b) The configuration coordinate diagram showing the exciton recombination on an acceptor site, involving strong coupling to lattice vibrations. The equilibrium value of a lattice coordinate changes between the excited (S^*) and ground states (S), leading to a large wave function overlap for a transition to a high vibronic state and contributing to broadening of the PL band. (Source: From Ref. [92], with permission.)

where E_g is the bandgap energy, E_D and E_A are the donor and acceptor-binding energies, respectively, $E_C = e^2/4\pi\varepsilon r$ is the Coulomb interaction between a donor and an acceptor, and E_{phonon} describes the phonon energies involved in the radiative transitions.

Until recently, almost nothing was known about the DAP behavior in reduced dimensions. The main reason for this was the lack of controlled synthesis of NCs with sufficiently small sizes and uniform size distribution. The previous documented preparations of Ga_2O_3 nanostructures include the syntheses of nanocrystalline powder by solvolysis of $GaCl_3$ in *N,N*-dimethylformamide (DMF) [93], and nanorods by the chemical vapor deposition method [94]. The recent development of the colloidal synthesis of Ga_2O_3 NCs opened the opportunities not only to manipulate visible PL in this highly transparent material but also to learn more about the DAP mechanism in the solid state and the defect formation in reduced dimensions [95]. The NCs were prepared from gallium acetylacetonate ($Ga(acac)_3$) in oleylamine (OA) as a coordinating solvent in the temperature range from 200 to 310 °C. Figure 5.16 shows the TEM image and XRD pattern of a typical sample prepared at 310 °C. These colloidal NCs have an average size of 6.0 ± 1.1 nm and exhibit a metastable cubic crystal structure (γ-Ga_2O_3), in analogy to metastable NC formation described in Section 5.2.2 of this chapter. The PL studies of the NC samples prepared at different temperatures revealed a strong blueshift of the characteristic PL band with decreasing NC size (Figure 5.17). This unique size-dependent PL shift is not a consequence of quantum confinement, as Ga_2O_3 is characterized by the formation of tightly bound excitons (Frenkel excitons), and its bandgap energy is insensitive to size in the range of NC sizes prepared in the described study. On the basis of Eq. (5.4), two factors can determine the energy of the DAP emission: (i) the binding energies of donor and acceptor sites (E_D and E_A, respectively) and (ii) the Coulomb interaction between them.

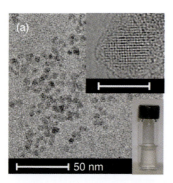
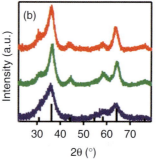

Figure 5.16 (a) TEM image of 6.0 ± 1.1 nm Ga_2O_3 NCs. Insets: (top) High-resolution TEM image of a single NC (scale bar: 5 nm). Lattice spacings match {311} d spacing of cubic Ga_2O_3. (bottom) Photograph of an NC colloidal suspension. (b) X-ray diffraction (XRD) patterns of Ga_2O_3 NCs prepared at different temperatures (top to bottom: 310, 245, and 220 °C). Black lines: XRD of bulk cubic Ga_2O_3. (Source: From Ref. [95], with permission.)

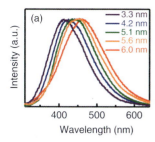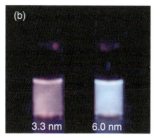

Figure 5.17 (a) 300 K PL spectra of γ-Ga2O3NCs having different sizes, excited at 250 nm. (b) Photograph of colloidal 3.3 and 6.0 nm γ-Ga$_2$O$_3$NCs illuminated with 250 nm light. (Source: From Ref. [95], with permission.)

The influence of the donor- and acceptor-binding energies on the DAP emission energy can be elucidated using temperature-dependent steady-state PL measurements. These experiments were performed in the temperature range 5–300 K, which revealed a notable change in the PL intensity behavior with decreasing NC size (Figure 5.18) [92]. For large NCs, the PL intensity remains fairly steady with increasing temperature until about 150 K, followed by a drop at higher temperatures. The PL intensity of small NCs, on the other hand, increases initially with raising temperatures, followed by a decrease above about 200 K. Analysis of these results using the expression for thermal quenching of the DAP emission reveals that the activation energy increases by about 30 meV with decreasing NC size from 6.0 to 3.3 nm. This slight increase in the activation energy in the studied temperature range suggests an increase in the donor trap depth with decreasing NC size. In the framework of the Bohr model, these results suggest that the Bohr radius of the donor decreases with decreasing NC size. This phenomenon has mostly been associated with an increase in the concentration of induced charges at the surfaces of colloidal NCs [96]. Supporting evidence for a decrease in the Bohr radius with decreasing NC size has been obtained using time-resolved PL studies

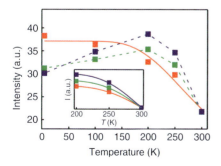

Figure 5.18 Temperature dependence of the maximum PL band intensity of colloidal γ-Ga$_2$O$_3$ NCs having the average sizes of 6.0 ± 1.0 nm (red), 4.2 ± 0.9 nm (green), and 3.3 ± 0.5 nm (blue). The solid line is the fit to Eq. (4) in Ref. [92]. Inset shows the same fits in the range 200–300 K for all three NC sizes. (Source: From Ref. [92], with permission.)

(see below). Importantly, these results indicate that binding energies of the defect centers show the opposite trend from the DAP emission energy as a function of NC size suggesting that the Coulomb interactions are a dominant component in determining the blueshift of the DAP emission with decreasing NC size. The dependence of E_D and Coulomb interactions on the NC size demonstrates the size tunability of the electronic structure and optical properties of complex oxide NCs that are not subject to quantum confinement.

In addition to controlling the PL properties via NC size, other ways of manipulating DAP emission in Ga_2O_3 NCs have also been demonstrated. Specifically, quantum yield manipulation has been demonstrated by *in situ* control of the native defect formation via adjustment of the chemical environment during NC synthesis [97]. Figure 5.19 illustrates the effect of hydrogen (as a reducing agent) and oxygen (as an oxidizing agent) on the PL intensity. The oxidative environment is found to decrease the PL intensity, while the presence of a reducing agent enhances the emission of Ga_2O_3 NCs. This difference is directly determined by the concentration of oxygen vacancies in NCs. The reducing (oxygen-poor) environment leads to an increase in the oxygen vacancy concentration in NCs, allowing for an enhancement of the PL intensity. Using hydrogen as a reducing agent in a proof-of-principle

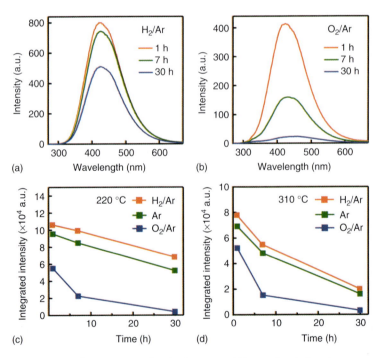

Figure 5.19 (a,b) PL spectra of γ-Ga_2O_3 NCs prepared for different durations in H_2/Ar (a) and O_2/Ar (b) atmosphere. (c,d) Integrated PL intensity versus synthesis time for colloidal Ga_2O_3 NCs synthesized in different chemical environments at 220 °C (c) and 310 °C (d). The data correspond to the same NC concentrations. (Source: From Ref. [97], with permission.)

study, quantum yield of up to about 30% has been reported. It is expected that stronger reducing agents can further enhance the PL efficiency, making Ga_2O_3 NCs technologically applicable as building blocks for light-emitting devices. The PL energy and efficiency can also be tuned by the NC composition. Alloying Ga_2O_3 NCs with indium has allowed for the formation of gallium indium oxide (GIO) NCs with variable structure and continuously tunable composition [24]. Alloying throughout the entire composition range has enabled wide tunability of NC PL in the visible range. The increasing concentration of In^{3+} in GIO NCs reduces the NC bandgap energy, concurrently changing the energy states of the donors and acceptors, and their interactions. This change in the electronic structure of donors and acceptors causes a systematic redshift of the PL.

Another interesting and important feature of PL of Ga_2O_3 NCs is the size dependence of the PL decay dynamics. With decreasing NC size, the half-life of the DAP emission decreases from $5.67\,\mu s$ for $6.0\,nm$ diameter NCs to $4.66\,\mu s$ for $3.3\,nm$ diameter NCs. An increase in the PL decay rate with decreasing NC size is largely a consequence of the reduced average separation between donors and acceptors. The investigation of the PL decay dynamics therefore offers an opportunity to study the mechanism of DAP recombination and defect formation in NCs [98]. The intensity of the emitted light at time t on excitation is dependent on the average probability of an electron existing at a donor site, which is expressed by the following equation:

$$\langle Q(t) \rangle = \exp \left[4\pi n \int_0^{R_c} \left\{ \exp\left[-W(r)\,t \right] - 1 \right\} r^2 dr \right] \tag{5.5}$$

where $W(r)$ is the recombination rate, or recombination probability per unit time, for a DAP with the separation r, and is given by

$$W(r) = W_{max} \exp\left(-\frac{2r}{R_d} \right) \tag{5.6}$$

R_d and R_c are the Bohr radius of the donor and an average radius of NCs, respectively, n is the number density of donors per unit volume, and W_{max} is a constant. Starting from this basic theoretical model, two models have been developed for simulating the PL decay dynamics of colloidal Ga_2O_3 NCs [99], on the basis of the relative positioning of donor and acceptor sites within a NC:

1) 3D model, in which a single acceptor resides in the center of a spherical NC, and donor sites are randomly distributed throughout the NC volume (Figure 5.20a);
2) 2D model, in which both donors and acceptors are randomly distributed over the NC surface area (Figure 5.20c).

The mathematical approach related to the development of these two models has been described elsewhere [99], but one of the main differences between these two models is the expression for density of the donors as majority defects in NCs. In the 3D model, the density of donors is expressed as a number density per unit volume,

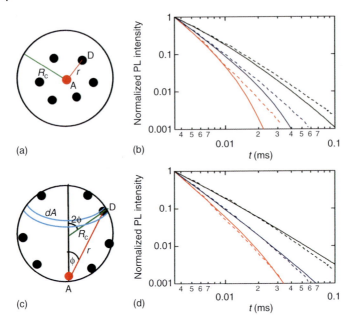

Figure 5.20 (a,c) Schematic representation of the 3-D DAP model (a) and 2-D DAP model (b). (b,d) Normalized PL decay rate (dashed lines) and the corresponding 3-D DAP (b) and 2-D DAP (d) model best fits (solid lines) for Ga_2O_3 NCs with $R_c = 1.6$ nm (red), 2.1 nm (blue), and 3.0 nm (black); $W_{max} = 10^7$ s^{-1}, $R_d = 0.9$ nm (b), and $R_d = 1.1$ nm (d). (From Ref. [99], with permission.)

while in the 2D model, the number density per unit area. In the fitting procedure, the value of W_{max} was fixed (10^7 s^{-1}), while the number of donors per NC and the Bohr radius of the donor (R_d) were treated as adjustable parameters. The results of the fitting (solid lines) to the experimental data (dashed lines) for three different NC sizes using the 3D model are shown in Figure 5.20b. The best overall fit was obtained for $R_d = 0.9$ nm. This model reproduces the trend of increasing decay rate with decreasing NC size, although there is a notable difference between the functional form of best fits and the experimental data. The experimental data is more linear (or less curved) than the functional form associated with the 3D model. Similar fitting was also performed using the 2D model [99]. The results of this fitting are shown in Figure 5.20d. In this case, the functional form is evidently more linear than that of the 3D model, allowing for a better match with the experimental data. A better fit of the 2D relative to 3D model to the experimental data suggests the preferential formation of native defect sites on NC surfaces. Another important result obtained from this theoretical modeling is that the number of defects per NC is significantly larger for the 2D relative to the 3D model. This result also supports the hypothesis that native defect formation in oxide NCs is inherently associated with NC surfaces. This work also led to some other, more subtle conclusions. As the second-order correction to the 2D model, the fitting was also performed by fixing W_{max} and treating the Bohr radius as a free parameter. This simulation suggested that the Bohr radius of the donor decreases with decreasing NC size, which is in agreement with the results of the temperature-dependent PL studies (see earlier text). The PL studies of Ga_2O_3 NCs presented here provide another instructive example of how TCO NC size, structure, and composition control can allow for the manipulation of the native defect formation and interaction. Thermodynamically, it is expected that the density of defects in NCs is lower than that in bulk, because the materials in reduced dimensions generally tend to expel the impurities because of their small volume. However, the results described in this chapter indicate that surfaces could play an important role in fostering native defect formation, allowing for the preparation of defect-rich nanostructures. The ability to control native defects in NCs provides an opportunity to design and control electrical and PL properties for photonics and optoelectronics.

5.4
Summary and Outlook

The intent of this chapter was to provide an overview of some recent research activities and developments in the area of colloidal TCO NCs. The content of the chapter is, to some degree, based on the personal research interests of the author and focuses predominantly on controlling the properties of colloidal TCO NCs by controlling their structures and defects. Examples of introducing and controlling functional properties via phase manipulation, dopants (external defects), and native defects were provided, as an illustration of some general fundamental principles. An important point derived from this body of research is that structural transformation and defect control could represent a viable path to multifunctionality of nanostructured materials. For example, the oxygen vacancies, which are responsible for the conductivity of TCOs, have also been invoked in explaining the PL properties of oxide NCs described here. This possibility renders TCO NCs promising materials for a number of currently existing technologies, or even some yet-unforeseen applications. It is the author's hope that the examples provided in this chapter will spur the imagination of interested researchers about the other ways in which defects can be used to manipulate functional properties of metal oxide NCs. Successful developments will require an interdisciplinary approach and application of a variety of structural and spectroscopic techniques, but promise to bring advancements in solid-state physics and chemistry of fundamental and practical importance.

Acknowledgments

The author is truly grateful to all of his coworkers and collaborators who have contributed over the years to the research work described in this chapter. The author acknowledges generous financial support from Natural Sciences and Engineering Research Council of Canada, Canada Foundation for Innovation, Canada Research Chairs Program, and the University of Waterloo.

References

1. Rao, C.N.R. and Raveau, B. (1998) *Transition Metal Oxides: Structure, Properties, and Synthesis of Ceramic Oxides*, 2nd edn, John Wiley & Sons, Inc., New York.
2. Cox, P.A. (1992) *Transition Metal Oxides: An Introduction to Their Electronic Structure and Properties*, Oxford University Press, Oxford.
3. Lany, S. and Zunger, A. (2007) Dopability, intrinsic conductivity, and nonstoichiometry of transparent conducting oxides. *Phys. Rev. Lett.*, **98**, 045501.
4. Falk, G. (2012) Sintering of transparent conductive oxides: from oxide ceramic powders to advanced optoelectronic materials, in *Sintering of Ceramics—New Emerging Techniques* (ed A. Lakshmanan), InTech.
5. Delahoy, A.E. and Guo, S. (2011) Transparent conducting oxides for photovoltaics, in *Handbook of Photovoltaic Science and Engineering* (eds A. Luque and S. Hegedus), John Wiley & Sons, Ltd, Chichester.
6. Fortunato, E., Ginley, D., Hosono, H., and Paine, D.C. (2007) Transparent conducting oxides for photovoltaics. *MRS Bull.*, **32**, 242–247.
7. Walter, M.G., Warren, E.L., McKone, J.R., Boettcher, S.W., Mi, Q., Santori, E.A., and Lewis, N.S. (2010) Solar water splitting cells. *Chem. Rev.*, **110**, 6446–6473.
8. Kolmakov, A., Zhang, Y., Cheng, G., and Moskovits, M. (2003) Detection of CO and O_2 using tin oxide nanowire sensors. *Adv. Mater.*, **15**, 997–1000.
9. Xu, X., Zhuang, J., and Wang, X. (2008) SnO_2 quantum dots and quantum wires: controllable synthesis, self-assembled 2D architectures, and gas-sensing properties. *J. Am. Chem. Soc.*, **130**, 12527–12535.
10. Bawendi, M.G., Steigerwald, M.L., and Brus, L.E. (1990) The quantum mechanics of larger semiconductor clusters ("quantum dots"). *Annu. Rev. Phys. Chem.*, **41**, 477–496.
11. Nozik, A.J. and Micic, O.I. (1998) Colloidal quantum dots of III-V semiconductors. *MRS Bull.*, **23**, 24–30.
12. Hu, J., Odom, T.W., and Lieber, C.M. (1999) Chemistry and physics in one-dimension: synthesis and properties of nanowires and nanotubes. *Acc. Chem. Res.*, **32**, 435–445.
13. Murray, C.B., Norris, D.J., and Bawendi, M.G. (1993) Synthesis and characterization of nearly monodisperse CdE (E = S, Se, Te) semiconductor nanocrystallites. *J. Am. Chem. Soc.*, **115**, 8706–8715.
14. Beaulac, R., Archer, P.I., Ochsenbein, S.T., and Gamelin, D.R. (2008) Mn^{2+}-doped CdSe quantum dots: new inorganic materials for spin-electronics and spin-photonics. *Adv. Funct. Mater.*, **18**, 3873–3891.
15. Bryan, J.D. and Gamelin, D.R. (2005) Doped semiconductor nanocrystals: synthesis, characterization, physical properties, and applications. *Prog. Inorg. Chem.*, **54**, 47–126.
16. Radovanovic, P.V., Norberg, N.S., McNally, K.E., and Gamelin, D.R. (2002) Colloidal transition-metal-doped ZnO quantum dots. *J. Am. Chem. Soc.*, **124**, 15192–15193.
17. Schwartz, D.A., Norberg, N.S., Nguyen, Q.P., Parker, J.M., and Gamelin, D.R. (2003) Magnetic quantum dots: synthesis, spectroscopy, and magnetism of Co^{2+} and Ni^{2+}-doped ZnO nanocrystals. *J. Am. Chem. Soc.*, **125**, 13205–13218.
18. Radovanovic, P.V. and Gamelin, D.R. (2003) High-temperature ferromagnetism in Ni^{2+}-doped ZnO aggregates prepared from colloidal diluted magnetic semiconductor quantum dots. *Phys. Rev. Lett.*, **91**, 157202.
19. Karmakar, D., Mandal, S.K., Kadam, R.M., Paulose, P.L., Rajarajan, A.K., Nath, T.K., Das, A.K., Dasgupta, I., and Das, G.P. (2007) Ferromagnetism in Fe-doped ZnO nanocrystals: experiment and theory. *Phys. Rev. B*, **75**, 144404.
20. Dietl, T., Ohno, H., Matsukura, F., Cibet, J., and Ferrand, D. (2000) Zener model description of ferromagnetism in zinc-blende magnetic semiconductors. *Science*, **287**, 1019–1022.
21. Archer, P.I., Radovanovic, P.V., Heald, S.M., and Gamelin, D.R. (2005) Low-temperature activation and deactivation

of high-curie-temperature ferromag-netism in a new diluted magnetic semiconductor: Ni^{2+}-doped SnO_2. *J. Am. Chem. Soc.*, **127**, 14479–14487.

22. Farvid, S.S., Dave, N., Wang, T., and Radovanovic, P.V. (2009) Dopant-induced manipulation of the growth and struc-tural metastability of colloidal indium oxide nanocrystals. *J. Phys. Chem. C*, **113**, 15928–15933.

23. Dave, N., Pautler, B.G., Farvid, S.S., and Radovanovic, P.V. (2010) Synthesis and surface control of colloidal Cr^{3+}-doped SnO_2 transparent magnetic semicon-ductor nanocrystals. *Nanotechnology*, **21**, 134023.

24. Farvid, S.S., Wang, T., and Radovanovic, P.V. (2011) Colloidal gallium indium oxide nanocrystals: a multifunctional light-emitting phosphor broadly tunable by alloy composition. *J. Am. Chem. Soc.*, **133**, 6711–6719.

25. Singhal, A., Achary, S.N., Manjanna, J., Jayakumar, O.D., Kadam, R.M., and Tyagi, A.K. (2009) Colloidal Fe-doped indium oxide nanoparticles: facile synthesis, structural, and mag-netic properties. *J. Phys. Chem. C*, **113**, 3600–3606.

26. Bryan, J.D., Santangelo, S.A., Keveren, S.C., and Gamelin, D.R. (2005) Acti-vation of high-T_C ferromagnetism in Co^{2+}:TiO_2 and Cr^{3+}:TiO_2 nanorods and nanocrystals by grain boundary defects. *J. Am. Chem. Soc.*, **127**, 15568–15574.

27. Erwin, S.C., Zu, L., Haftel, M.I., Efros, A.L., Kennedy, T.A., and Norris, D.J. (2005) Doping semiconductor nanocrys-tals. *Nature*, **436**, 91–94.

28. Archer, P.I., Santangelo, S.A., and Gamelin, D.R. (2007) Inorganic cluster synthesis of TM^{2+}-doped quantum dots (CdSe, CdS, CdSe/CdS): physical prop-erty dependence on dopant locale. *J. Am. Chem. Soc.*, **129**, 9808–9818.

29. Bowen Katari, J.E., Colvin, V.L., and Alivisatos, A.P. (1994) X-ray photoelec-tron spectroscopy of CdSe nanocrystals with applications to studies of the nanocrystal surface. *J. Phys. Chem.*, **98**, 4109–4117.

30. Zhuang, Z., Peng, Q., Zhang, B., and Li, Y. (2008) Controllable synthesis of Cu_2S nanocrystals and their assembly into a superlattice. *J. Am. Chem. Soc.*, **130**, 10482–10483.

31. Tolbert, S.H. and Alivisatos, A.P. (1995) High-pressure structural transformations in semiconductor nanocrystals. *Annu. Rev. Phys. Chem.*, **46**, 595–625.

32. Chen, C.-C., Herhold, A.B., Johnson, C.S., and Alivisatos, A.P. (1997) Size dependence of structural metastability in semiconductor nanocrystals. *Science*, **276**, 398–401.

33. Prewitt, C.T., Shannon, R.D., Rogers, D.B., and Sleight, A.W. (1969) C rare earth oxide-corundum transition and crystal chemistry of oxides having the corundum structure. *Inorg. Chem.*, **8**, 1985–1993.

34. Gurlo, A., Kroll, P., and Riedel, R. (2008) Metastability of corundum-type In_2O_3. *Chem. Eur. J.*, **14**, 3306–3310.

35. Fuchs, F. and Bechstedt, F. (2008) Indium-oxide polymorphs from first principles: quasiparticle electronic states. *Phys. Rev. B*, **77**, 155107.

36. Karazhanov, S.Z., Ravindran, P., Vajeeston, P., Ulyashin, A., Finstad, T.G., and Fjellvåg, H. (2007) Phase sta-bility, electronic structure, and optical properties of indium oxide polytypes. *Phys. Rev. B*, **76**, 075129–075113.

37. Chen, C., Chen, D., Jiao, X., and Chen, S. (2007) In_2O_3 Nanocrystals with a tunable size in the range of 4–10 nm: one-step synthesis, characterization, and optical properties. *J. Phys. Chem. C*, **111**, 18039–18043.

38. Franzman, M.A., Perez, V., and Brutchey, R.L. (2009) Peroxide-mediated synthesis of indium oxide nanocrystals at low temperatures. *J. Phys. Chem. C*, **113**, 630–636.

39. Narayanaswamy, A., Xu, H., Pradhan, N., Kim, M., and Peng, X. (2006) Formation of nearly monodisperse In_2O_3 nanodots and oriented-attached nanoflowers: hydrolysis and alcoholy-sis vs pyrolysis. *J. Am. Chem. Soc.*, **128**, 10310–10319.

40. Epifani, M., Siciliano, P., Gurlo, A., Barsan, N., and Weimar, U. (2004) Am-bient pressure synthesis of corundum-type In_2O_3. *J. Am. Chem. Soc.*, **126**, 4078–4079.

41. Yu, D., Wang, D., and Qian, Y. (2004) Synthesis of metastable hexagonal In_2O_3 nanocrystals by a precursor-dehydration route under ambient pressure. *J. Solid State Chem.*, **177**, 1230–1234.

42. Yu, D., Yu, S.-H., Zhang, S., Zuo, J., Wang, D., and Qian, Y. (2003) Metastable hexagonal In_2O_3 nanofibers templated from InOOH nanofibers under ambient pressure. *Adv. Funct. Mater.*, **13**, 497–501.

43. Lee, C.H., Kim, M., Kim, T., Kim, A., Paek, J., Lee, J.W., Choi, S.-Y., Kim, K., Park, J.-B., and Lee, K. (2006) Ambient pressure synthesis of size-controlled corundum-type In_2O_3 nanocubes. *J. Am. Chem. Soc.*, **128**, 9326–9327.

44. Farvid, S.S., Dave, N., and Radovanovic, P.V. (2010) Phase-controlled synthesis of colloidal In_2O_3 nanocrystals via size-structure correlation. *Chem. Mater.*, **22**, 9–11.

45. Farvid, S.S. and Radovanovic, P.V. (2012) Phase transformation of colloidal In_2O_3 nanocrystals driven by the interface nucleation mechanism: a kinetic study. *J. Am. Chem. Soc.*, **134**, 7015–7024.

46. Grunwald, M. and Dellago, C. (2009) Nucleation and growth in structural transformations of nanocrystals. *Nano Lett.*, **9**, 2099–2102.

47. Frank, G., Olazcuaga, R., and Rabenau, A. (1977) Occurrence of corundum-type indium(III) oxide under ambient conditions. *Inorg. Chem.*, **16**, 1251–1253.

48. Li, X., Xia, C., Pei, G., and He, X. (2007) Synthesis and characterization of room-temperature ferromagnetism in Fe- and Ni-co-doped In_2O_3. *J. Phys. Chem. Solids*, **68**, 1836–1840.

49. Beaulac, R., Ochsenbein, S.T., and Gamelin, D.R. (2010) in *Nanocrystal Quantum Dots* (ed V.I. Klimov), CRC Press, Boca Raton, FL, pp. 397–453.

50. Nag, A., Chakraborty, S., and Sarma, D.D. (2008) To dope Mn^{2+} in a semiconducting nanocrystal. *J. Am. Chem. Soc.*, **130**, 10605–10611.

51. Archer, P.I. and Gamelin, D.R. (2007) in *Magnetism in Semiconducting Oxides* (ed N.H. Hong), Research Signpost Press, pp. 23–52.

52. Chambers, S.A. (2006) Ferromagnetism in doped thin-film oxide and nitride semiconductors and dielectrics. *Surf. Sci. Rep.*, **61**, 345–381.

53. Hamberg, I. and Granqvist, C.G. (1986) Evaporated Sn-doped In_2O_3 films: basic optical properties and applications to energy-efficient windows. *J. Appl. Phys.*, **60**, R123–R159.

54. Kilic, C. and Zunger, A. (2002) Origins of coexistence of conductivity and transparency in SnO_2. *Phys. Rev. Lett.*, **88**, 095501.

55. Kanehara, M., Koike, H., Yoshinaga, T., and Teranishi, T. (2009) Indium tin oxide nanoparticles with compositionally tunable surface plasmon resonance frequencies in the near-IR region. *J. Am. Chem. Soc.*, **131**, 17736–17737.

56. Gilstrap, R.A., Capozzi, C.J., Carson, C.G., Gerhardt, R.A., and Summers, C.J. (2008) Synthesis of a nonagglomerated indium tin oxide nanoparticle dispersion. *Adv. Mater.*, **20**, 4163–4166.

57. Choi, S.-I., Nam, K.M., Park, B.K., Seo, W.S., and Park, J.T. (2008) Preparation and optical properties of colloidal, monodisperse, and highly crystalline ITO nanoparticles. *Chem. Mater.*, **20**, 2609–2611.

58. Wang, T. and Radovanovic, P.V. (2011) Free electron concentration in colloidal indium tin oxide nanocrystals determined by their size and structure. *J. Phys. Chem. C*, **115**, 406–413.

59. Bühler, G., Thölmann, D., and Feldmann, C. (2007) One-pot synthesis of highly conductive indium tin oxide nanocrystals. *Adv. Mater.*, **19**, 2224–2227.

60. Ba, J., Fattakhova Rohlfing, D., Feldhoff, A., Brezesinski, T., Djerdj, I., Wark, M., and Niederberger, M. (2006) Nonaqueous synthesis of uniform indium tin oxide nanocrystals and their electrical conductivity in dependence of the tin oxide concentration. *Chem. Mater.*, **18**, 2848–2854.

61. Buonsanti, R., Llordes, A., Aloni, S., Helms, B.A., and Milliron, D.J. (2011) Tunable infrared absorption and visible transparency of colloidal aluminum-doped zinc oxide nanocrystals. *Nano Lett.*, **11**, 4706–4710.

62. Nutz, T., Zum Felde, U., and Haase, M. (1999) Wet-chemical synthesis of doped nanoparticles: blue-colored colloids of

n-doped SnO_2: Sb. *J. Chem. Phys.*, **110**, 12142–12150.

63. Zutic, I., Fabian, J., and Das Sarma, S. (2004) Spintronics: fundamentals and applications. *Rev. Mod. Phys.*, **76**, 323–410.

64. Matsumoto, Y., Murakami, M., Shono, T., Hasegawa, T., Fukumura, T., Kawasaki, M., Ahmet, P., Chikyow, T., Koshihara, S., and Koinuma, H. (2001) Room-temperature ferromagnetism in transparent transition metal-doped titanium dioxide. *Science*, **291**, 854–856.

65. Pearton, S.J., Abernathy, C.R., Overberg, M.E., Thaler, G.T., Norton, D.P., Theodoropoulou, N., Hebard, A.F., Park, Y.D., Ren, F., Kim, J., and Boatner, L.A. (2003) Wide band gap ferromagnetic semiconductors and oxides. *J. Appl. Phys.*, **93**, 1–13.

66. Sundaresan, A., Bhargavi, R., Rangarajan, N., Siddesh, U., and Rao, C.N.R. (2006) Ferromagnetism as a universal feature of nanoparticles of the otherwise nonmagnetic oxides. *Phys. Rev. B*, **74**, 161306.

67. Coey, J.M.D., Venkatesan, M., Stamenov, P., Fitzgerald, C.B., and Dorneles, L.S. (2005) Magnetism in hafnium dioxide. *Phys. Rev. B*, **72**, 024450.

68. Goodenough, J.B. (1963) *Magnetism and the Chemical Bond*, Interscience Publishers, New York.

69. Fitzgerald, C.B., Venkatesan, M., Douvalis, A.P., Huber, S., and Coey, J.M.D. (2004) SnO_2 doped with Mn, Fe or Co: room temperature dilute magnetic semiconductors. *J. Appl. Phys.*, **95**, 7390–7392.

70. Philip, J., Punnoose, A., Kim, B.I., Reddy, K.M., Layne, S., Holmes, J.O., Satpati, B., Leclair, P.R., Santos, T.S., and Moodera, J.S. (2006) Carrier-controlled ferromagnetism in transparent oxide semiconductor. *Nat. Mater.*, **5**, 298–304.

71. Ohno, H., Shen, A., Matsukura, F., Oiwa, A., Endo, A., Katsumoto, S., and Iye, Y. (1996) (Ga, Mn) as: a new diluted magnetic semiconductor based on GaAs. *Appl. Phys. Lett.*, **69**, 363–365.

72. Weakliem, H.A. (1962) Optical spectra of Ni^{2+}, Co^{2+}, and Cu^{2+} in tetrahedral sites in crystals. *J. Chem. Phys.*, **36**, 2117–2140.

73. Bryan, J.D., Heald, S.M., Chambers, S.A., and Gamelin, D.R. (2004) Strong room-temperature ferromagnetism in Co^{2+}-doped TiO_2 made from colloidal nanocrystals. *J. Am. Chem. Soc.*, **126**, 11640–11647.

74. Norberg, N.S., Kittilstved, K.R., Amonette, J.E., Kukkadapu, R.K., Schwartz, D.A., and Gamelin, D.R. (2004) Synthesis of colloidal Mn^{2+}:ZnO quantum dots and high-T_C ferromagnetic nanocrystalline thin films. *J. Am. Chem. Soc.*, **126**, 9387–9398.

75. Farvid, S.S., Ju, L., Worden, M., and Radovanovic, P.V. (2008) Colloidal chromium-doped In_2O_3 nanocrystals as building blocks for high-T_C ferromagnetic transparent conducting oxides structures. *J. Phys. Chem. C*, **112**, 17755–17759.

76. Ju, L., Sabergharesou, T., Stamplecoskie, K.G., Hegde, M., Wang, T., Combe, N.A., Wu, H., and Radovanovic, P.V. (2012) Interplay between size, composition and phase transition of nanocrystalline Cr^{3+}-doped $BaTiO_3$ as a path to multiferroism in perovskite-type oxides. *J. Am. Chem. Soc.*, **134**, 1136–1146.

77. Kaspar, T.C., Heald, S.M., Wang, C.M., Bryan, J.D., Droubay, T., Shutthanandan, V., Thevuthasan, S., McCready, D.E., Kellock, A.J., Gamelin, D.R., and Chambers, S.A. (2005) Negligible magnetism in excellent structural quality $Cr_xTi_{1-x}O_2$ anatase: contrast with high-T_C ferromagnetism in structurally defective $Cr_xTi_{1-x}O_2$. *Phys. Rev. Lett.*, **95**, 217203.

78. Coey, J.M.D., Venkatesan, M., and Fitzgerald, C.B. (2005) Donor impurity band exchange in dilute ferromagnetic oxides. *Nat. Mater.*, **4**, 173–179.

79. McCluskey, M.D. and Haller, E.E. (2012) *Dopants and Defects in Semiconductors*, Taylor & Francis, Boca Raton, FL.

80. Özgür, Ü., Alivov, Y.I., Liu, C., Teke, A., Reshchikov, M.A., Doğan, S., Avrutin, V., Cho, S.-J., and Morkoç, H. (2005) A comprehensive review of ZnO materials and devices. *J. Appl. Phys.*, **98**, 041301.

81. Vanheusden, K., Warren, W.L., Seager, C.H., Tallant, D.R., Voigt, J.A., and Gnade, B.E. (1996) Mechanisms behind green photoluminescence in ZnO

phosphor powders. *J. Appl. Phys.*, **79**, 7983–7990.

82. van Dijken, A., Meulenkamp, E.A., Vanmaekelbergh, D., and Meijerink, A. (2000) Identification of the transition responsible for the visible emission in ZnO using quantum size effects. *J. Lumin.*, **90**, 123–128.

83. van Dijken, A., Meulenkamp, E.A., Vanmaekelbergh, D., and Meijerink, A. (2000) The kinetics of the radiative and nonradiative processes in nanocrystalline ZnO particles upon photoexcitation. *J. Phys. Chem. B*, **104**, 1715–1723.

84. Norberg, N.S. and Gamelin, D.R. (2005) Influence of surface modification on the luminescence of colloidal ZnO nanocrystals. *J. Phys. Chem. B*, **109**, 20810–20816.

85. Zhang, L., Yin, L., Wang, C., Iun, N., Qi, Y., and Xiang, D. (2010) Origin of visible photoluminescence of ZnO quantum dots: defect-dependent and size-dependent. *J. Phys. Chem. C*, **114**, 9651–9658.

86. Zhou, X., Kuang, Q., Jiang, Z.-Y., Xie, Z.-X., Xu, T., Huang, R.-B., and Zheng, L.-S. (2007) The origin of green emission of ZnO microcrystallites: surface-dependent light emission studied by cathodoluminescence. *J. Phys. Chem. C*, **111**, 12091–12093.

87. Fu, Y.-S., Du, X.-W., Kulinich, S.A., Qiu, J.-S., Qin, W.-J., Li, R., Sun, J., and Liu, J. (2007) Stable aqueous dispersion of ZnO quantum dots with strong blue emission via simple solution route. *J. Am. Chem. Soc.*, **129**, 16029–16033.

88. Dean, P.J. (1973) Inter-impurity recombinations in semiconductors. *Prog. Solid State Chem.*, **8**, 1–126.

89. Binet, L. and Gourier, D. (1998) Origin of the blue luminescence of β-Ga$_2$O$_3$. *J. Phys. Chem. Solids*, **59**, 1241–1249.

90. Lorenz, M.R., Woods, J.F., and Gambino, R.J. (1967) Some electrical properties of the semiconductor β-Ga$_2$O$_3$. *J. Phys. Chem. Solids*, **28**, 403–404.

91. Harwig, T. and Kellendonk, F. (1978) Some observations on the photoluminescence of doped β-galliumsesquioxide. *J. Solid State Chem.*, **24**, 255–263.

92. Wang, T. and Radovanovic, P.V. (2011) Size-dependent electron transfer and trapping in strongly luminescent colloidal gallium oxide nanocrystals. *J. Phys. Chem. C*, **115**, 18473–18478.

93. Chen, T. and Tang, K. (2007) γ-Ga$_2$O$_3$ quantum dots with visible blue-green light emission property. *Appl. Phys. Lett.*, **90**, 053104.

94. Vanithakumari, S.C. and Nanda, K.K. (2009) A one-step method for the growth of Ga$_2$O$_3$-nanorod-based white-light-emitting phosphors. *Adv. Mater.*, **21**, 3581–3584.

95. Wang, T., Farvid, S.S., Abulikemu, M., and Radovanovic, P.V. (2010) Size-tunable phosphorescence in colloidal metastable γ-Ga$_2$O$_3$ nanocrystals. *J. Am. Chem. Soc.*, **132**, 9250–9252.

96. Tsu, R. and Babic, D. (1994) Doping of a quantum dot. *Appl. Phys. Lett.*, **64**, 1806–1808.

97. Wang, T. and Radovanovic, P.V. (2011) In-situ enhancement of the blue photoluminescence of colloidal Ga$_2$O$_3$ nanocrystals by promotion of defect formation in reducing conditions. *Chem. Commun.*, **47**, 7161–7163.

98. Thomas, D.G., Hopfield, J.J., and Augustyniak, W.M. (1965) Kinetics of radiative recombination at randomly distributed donors and acceptors. *Phys. Rev.*, **140**, A202–A220.

99. Hegde, M., Wang, T., Miskovic, Z.L., and Radovanovic, P.V. (2012) Origin of size-dependent photoluminescence decay dynamics in colloidal γ-Ga$_2$O$_3$ nanocrystals. *Appl. Phys. Lett.*, **100**, 141903.

Part III
Ferroelectrics

Functional Metal Oxides: New Science and Novel Applications, First Edition.
Edited by Satishchandra B. Ogale, Thirumalai V. Venkatesan, and Mark G. Blamire.
© 2013 Wiley-VCH Verlag GmbH & Co. KGaA. Published 2013 by Wiley-VCH Verlag GmbH & Co. KGaA.

6
Structure–Property Correlations in Rare-Earth-Substituted BiFeO₃ Epitaxial Thin Films at the Morphotropic Phase Boundary

Daisuke Kan, Ching-Jung Cheng, Valanoor Nagarajan, and Ichiro Takeuchi

6.1
Introduction

A material compositionally tuned to be at a phase boundary can be viewed as being *"on the brink."* Materials on the brink often display a colossal susceptibility response in reaction to weak external stimuli, such as temperature, electric field, or magnetic field. For instance, giant piezoelectric responses are achieved in some ferroelectrics by tuning their compositions to their morphotropic phase boundaries (MPBs) [1–5], where two different structural phases with very similar energy levels can coexist. Composition tuning through A- or B-site substitution in perovskite ferroelectrics is a well-known route to achieve high-performance MPB behaviors in traditional ferroelectrics.

Perovskite-structured $BiFeO_3$ (BFO) [6–8] with a rhombohedral distortion along the [111] direction (space group $R3c$) has recently attracted enormous interest owing to its multiferroic nature (i.e., coexistence of ferroelectricity and antiferromagnetism) as well as superior ferroelectric properties at room temperature. Although BFO has been identified as a possible environmentally friendly alternative to current Pb-based piezoelectrics, it suffers from a high leakage current, large coercive fields as well as the possession of electromechanical coefficients much smaller than those of traditional Pb-based piezoelectrics. In order to overcome those shortcomings, chemical substitution into either A- or B-site (of the BFO perovskite cell) has been attempted so as to improve both the electromechanical response and the leakage resistance [9–13]. The rationale behind such experiments is the fact that the lattice, and hence the functional behavior of a parent ferroelectric perovskite can be easily tuned through chemical substitution.

For the past 6 years, a collaborative program between the University of Maryland and the University of New South Wales has led to significant developments in the understanding of the structure–property correlations in rare-earth (RE) substituted BFO thin films investigated as a Pb-free candidate material. Our effort has also substantially benefited from theoretical insight provided by Karin Rabe and Lucia Pálová of Rutgers University, who performed first principles calculations surveying the energy landscape of different possible structures of rare-earth substituted

Functional Metal Oxides: New Science and Novel Applications, First Edition.
Edited by Satishchandra B. Ogale, Thirumalai V. Venkatesan, and Mark G. Blamire.
© 2013 Wiley-VCH Verlag GmbH & Co. KGaA. Published 2013 by Wiley-VCH Verlag GmbH & Co. KGaA.

BFO. In this chapter, we summarize our research efforts over the duration of the aforementioned program to review our findings and take readers in a systematic manner over the progress of the research program.

The outline of the chapter is as follows: we begin with the combinatorial discovery of a MPB in Sm-substituted $BiFeO_3$ (Sm-BFO) thin films. This is followed by detailed structural investigations on the Sm-BFO system with high-resolution transmission electron microscopy (HRTEM), electron diffraction, and aberration-corrected transmission electron microscope (TEM). Then we extend our findings to a variety of RE dopants and identify a universal parameter that governs the structural and ferroelectric properties across the MPB, independent of the RE dopant type. Furthermore, we reveal a universal structural fingerprint that is essential for the enhancement in the electromechanical response at the MPB. The demonstration of polarization rotation using chemical substitution is then discussed, and we close the chapter with a discussion of the experiments currently planned within the program.

6.2
Combinatorial Discovery of a MPB in Sm-substituted $BiFeO_3$ (Sm-BFO)

As discussed earlier, the central motivation for our research program was the discovery of novel Pb-free MPB compositions in substituted BFO. This requires synthesis and characterization of an enormously large number of individual samples. To overcome this challenge, we first implemented the combinatorial technique to fabricate pseudobinary thin film composition spreads of $Bi_{1-x}(RE)_xFeO_3$ or $BiFe_{1-y}(TM)_yO_3$ [14], where RE and TM denote RE and transition metal cations, respectively. The technique allows us to track precisely substitution-induced changes in crystal structures as well as physical properties as a function of composition in epitaxial thin-film forms [15]. In our studies, the 200 nm thick composition spreads of $(Bi_{1-x}RE_x)FeO_3$ (RE-BFO) were epitaxially grown on $SrRuO_3$(SRO)-buffered (001) $SrTiO_3$ (STO) substrates by pulsed laser deposition equipped with moving masks and a multitarget system (Pascal Inc.).

A number of interesting compounds were identified from our initial screening, where we prepared composition spreads of $Bi_{1-x}La_xFeO_3$, $Bi_{1-x}Pr_xFeO_3$, and $Bi_{1-x}Sm_xFeO_3$ systems. Initial piezoelectric force microscopy (PFM) scans revealed a significant enhancement of the piezoresponse only in the Sm-substituted system near $x = 0.15–0.2$. We discuss later in Section 6.6 why such an enhancement is not observed for La or other dopants of similar ionic size. A two-dimensional (2D) X-ray diffraction (XRD) image also showed that the Sm substitution results in a substantial structural change at this composition, as denoted by the change in the out-of-plane lattice constant. In Figure 6.1a, we plot the zero-bias out-of-plane dielectric constant ε_{33} and the loss tangent tan δ measured at 10 kHz as a function of Sm concentration. The dielectric constant reaches a maximum at $x = 0.14$, which is in agreement with the observation that a structural transition from the rhombohedral phase to an orthorhombic one takes place at the composition.

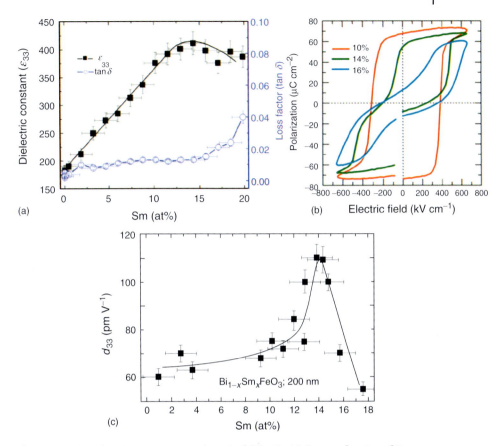

Figure 6.1 (a) Dielectric constant ε_{33} and tan δ of $(Bi_{1-x}Sm_x)FeO_3$ as a function of Sm composition. The data were measured at 1 MHz (zero bias). (b) PE hysteresis loops at 25 kHz at compositions of $x = 0.1$, 0.14, and 0.16. (c) High-field d_{33} determined by PFM as a function of Sm composition. In (a) and (c), the curve is a guide to the eye.

The loss tangent at this composition is relatively low, namely, 0.01. In addition, we found that the polarization vs electric field (P-E) hysteresis loop shows concomitant changes across the structural transition at $x = 0.14$, as shown in Figure 6.1b. With increasing Sm, the square-shaped ferroelectric hysteresis loops begin to get distorted, and, at the transition boundary, the loops show a transition from single to double hysteresis behaviors.

Figure 6.1c plots the high-field piezoelectric coefficient d_{33} evaluated by quantitative PFM as a function of Sm concentration. Around $x = 0.11–0.13$, the effective d_{33} displays a rapid increase in peaking with 110 pm V^{-1} at $x = 0.14$, where the structural transition occurs. Beyond this d_{33} value, it rapidly decreases to ~55 pm V^{-1} for $x = 0.17$. The measured remanent and high-field d_{33} here are comparable to the values previously reported for epitaxial thin films of Pb-based compounds such as Pb(Zr,Ti)O$_3$ (PZT) at the MPB [16]. Compared with nominally similar thin film samples of the same thickness, the MPB discovered here exhibit the best intrinsic piezoelectric properties known. The added advantage of the present system is that it is much simpler in terms of the solid-state crystal chemistry than some of the other reported Pb-free compounds [17], and it has a relatively easy processing method.

6.3
Structural Evolution across the MPB in Sm-BFO

Having discovered the presence of an MPB in the Sm-BFO system, the next step was to investigate the interplay between variations in the underlying polar orders and corresponding crystal structures [2, 18–22]. This is because, in ferroelectrics, there is a very close link between their electromechanical response and the underlying crystal structures. Therefore, a systematic study aimed at deciphering the subtle changes in local crystal structure and microstructure across the MPB in Sm-BFO was deemed necessary. Indeed, the extensive use of electron diffraction analyzes coupled with HRTEM yielded considerable insight into the underlying mechanisms that governed the structure–property correlations in this system.

Figure 6.2a is a cross-sectional TEM image of the 10% Sm-BFO grown on an SRO-buffered STO substrate. The inset in Figure 6.2a shows a zone-axis selected-area diffraction pattern (ZADP) from the bulk of the Sm-BFO film showing the fundamental reflections along the [100] zone axis. However, we see that, in local

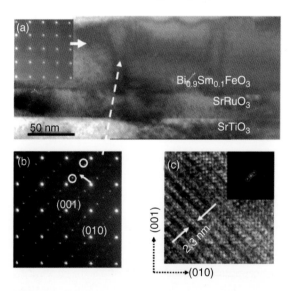

Figure 6.2 (a) Cross-sectional TEM image of 10% Sm-BFO where the inset shows only fundamental reflections. (b) [100] ZADP with circled 1/4{011} and arrowed 1/2{011} superstructure reflections from antiparallel cation displacement. (c) HRTEM image showing periodic ordering quadrupled along the (011) plane. The inset shows the corresponding FFT.

pockets that have a distinct contrast (shown by the dotted arrow), in addition to the fundamental reflections, clear 1/4{011} superstructure reflections (circled) are present (Figure 6.2b). The corresponding HRTEM image acquired for such local regions (Figure 6.2c) displays distinct commensurate ordering along the [011] direction, and the optical diffraction pattern obtained via fast Fourier transform (FFT) is in close agreement with Figure 6.2b.

The diffraction pattern from the local pockets (Figure 6.2b) shows a remarkable resemblance to [100] ZADPs from the well-established antiferroelectric (AFE) system of $PbZrO_3$ (PZO) [23, 24], which has the antiparallel displacement of the A-site cations and, as a result, have the quadrupled unit cell along the [011] direction with dimensions of $\sqrt{2}a_{pc} \times 2\sqrt{2}a_{pc} \times 2a_{pc}$, where a_{pc} is the lattice parameter of the pseudocubic perovskite unit. Indeed, the ordering periodicity of $\sim$2.3 nm (seen in Figure 6.2c) has a very close match with the above-suggested quadrupled structure.

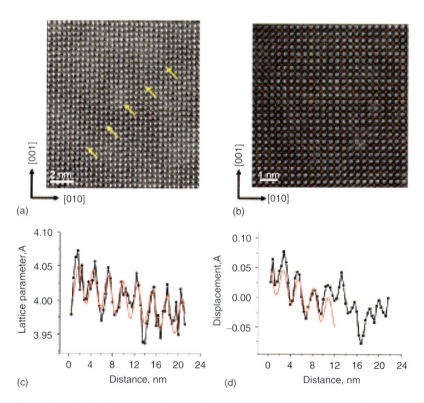

Figure 6.3 (a) [100] Zone axis STEM bright-field image of 10% Sm-BFO. The 1/4-related superstructure domain boundaries are marked by arrows. (b) Aberration-corrected Z-contrast image of the antipolar cluster. This image was used to obtain atomic displacements of Bi (marked with blue plus markers) and Fe (marked with red plus markers) columns. (c) Quantification of displacements of Bi atoms along the super-structure. (d) Quantification of displacements of Fe atoms along the superstructure.

Figure 6.3a is the [100] zone axis cross-sectional aberration-corrected scanning transmission electron microscope (STEM) bright-field image of a region where domain boundaries of the 1/4-related superstructure are marked by arrows. It is clear that chemical ordering between Bi^{3+} and Sm^{3+} ions at the atomic scale does not occur at all, because there is virtually no contrast difference between the A-site atomic positions. Therefore, this provides direct evidence that the superstructure spots observed for the local pockets originate from distortions to the parent lattice rather than from chemical ordering at the cationic sites of the perovskite cell.

Furthermore, we have quantified atomic displacements of the cations within the superstructure, as shown in the dark-field image (Figure 6.3b), which is a magnified portion of Figure 6.3a, with Bi columns marked with blue plus markers and Fe columns marked with red plus markers. The observed sets of atomic positions in the xy plane are used to obtain 2D maps of the lattice parameters (difference in positions between neighboring Bi atoms) and cation displacements (Δx and Δy) (the difference between the Fe position and the average of the neighboring Bi positions) (for full details, see Ref. [19]). Figure 6.3c plots the profile of the Bi–Bi distance (out-of-plane lattice parameter) along the superstructure direction, calculated from the full 2D map. The profile exhibits modulation with a consistent periodicity, which is estimated as 2.745 ± 0.026 nm peak to peak. It is also notable that the sine fit underestimates the peak values of the modulation profile, which appears to be close to a zigzag. Figure 6.3d displays the profile of the Fe off-center displacement map taken in the same manner, revealing the clear oscillation. The sine fit on the entire range was not successful because of the existence of an apparent stacking fault in the middle of the profile; however, the fit on half of the data yields a consistent periodicity value of 2.773 ± 0.064 nm peak to peak, suggesting that the lattice parameter modulation and the displacement modulation have the same origin. It is especially noteworthy that, after a correction for sloping background likely arising from drift, the absolute values of the displacements change sign for every neighboring two-unit-cell domain, indicating that the apparent antipolar order in this system is actually a lamellar array of highly dense ferroelectric domains with alternating polarizations.

As the Sm^{3+} content is increased further toward the composition of $x = 0.14$, the diffraction data become highly complex. Figure 6.4a depicts a typical [100] ZADP, which shows not only circled 1/4{011} and 1/2{011} reflections but also arrowed 1/2{010} reflections. It is to be noted though, that the intensity as well as the shape of these 1/4{011} spots is qualitatively different from those in the shown in Figure 6.2b. To understand the origin of this complex diffraction pattern at $x = 0.14$, we performed detailed TEM investigations of the microstructure. Figure 6.4b is a plan-view HRTEM micrograph that reveals the nanoscale phase mixture in this composition. Optical diffraction (FFT) was employed to analyze the four principal representative regions (phases), and the corresponding patterns are shown in Figure 6.4c–f. First, at the top left corner of Figure 6.4b (region I), it has only the fundamental perovskite reflections from the parent BFO lattice. Figure 6.4d indicates that region II has antiparallel cation displacements, and is structurally similar to that of 10% Sm-BFO. In region III, the cell-doubled orthorhombic phase

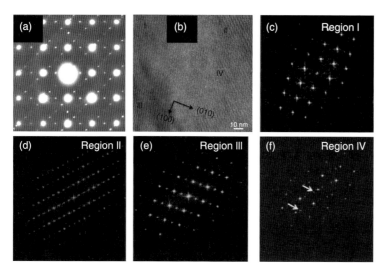

Figure 6.4 (a) [100] ZADP of 14% Sm-BFO showing both circled 1/4{011} and 1/2{011} superstructure reflections and arrowed 1/2{010} reflections. (b) Plan-view zoom-in HRTEM image of 10% Sm-BFO indicating a complex nanoscale mixture consisting of regions I, II, III, and IV. (c) FFT of region I with fundamental reflections. (d) FFT of region II with cell-quadrupled antiparallel cation shift reflections. (e) FFT of region III with cell-doubled orthorhombic reflections. (f) FFT of region IV with incommensurate reflections.

exists (Figure 6.4e). This structure is linked to the macroscopic double hysteresis $P-E$ loop behavior discussed later. We have also looked at areas in between regions I, II, and III, and it was found that in the nanoscale regions on the right of Figure 6.4b (region IV), the FFT yields incommensurate reflections (arrowed) along the [1−10] direction (Figure 6.4f).

When more Sm is added beyond $x = 0.14$, that is, for $x = 0.2$, the 1/4{011} spots due to the antiparallel cation displacement component totally vanishes. Instead, one finds strengthening of two other distinctive superstructure spots highlighted by

x = 0.2

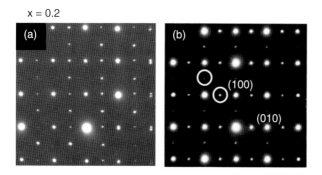

Figure 6.5 (a) [100] ZADP of 20% Sm-BFO. (b) Plan-view diffraction pattern with 1/2{010} and 1/2{100} weak reflections in addition to the fundamental spots.

the arrowed and boxed positions of 1/2{010} and 1/2{011} reflections, respectively, as shown in Figure 6.5a. It is important to note that patterns such as the ones in Figure 6.5a are observed in entire regions of the 20% Sm-BFO sample, implying that this is not a cluster effect. Figure 6.5b displays the plan-view diffraction pattern viewed down the [001] zone axis. It shows main perovskite reflections along with circled 1/2{010} and 1/2{100} spots, that is, absence of weak 1/2{011} reflections. These observations are consistent with the fact that the orthorhombic structure has unit cell dimensions of $\sqrt{2}a_{pc} \times \sqrt{2}a_{pc} \times 2a_{pc}$, and grows with the (110)$_{ortho}$ orientation. (The subscript "ortho" denotes the orthorhombic notation.) The film forms twin structure, where one of the epitaxial relationships between the film and substrate is [1–10]$_{ortho\text{-}Sm\text{-}BFO}$//[100]$_{STO}$ and [001]$_{ortho\text{-}Sm\text{-}BFO}$//[010]$_{STO}$, and the other is [1–10]$_{ortho\text{-}Sm\text{-}BFO}$//[010]$_{STO}$ and [001]$_{ortho\text{-}Sm\text{-}BFO}$//[100]$_{STO}$.

Figure 6.6 schematically illustrates the Sm-substitution-induced structural evolution. At low Sm concentration before the structural transition (MPB), Sm substitution induces antiparallel cation displacements in the local pockets. This can be best viewed as AFE clusters, where the substituted Sm^{3+} dramatically modifies the local dipole alignments, although ferroelectric PE hysteresis loops are macroscopically observed. Further increase in the Sm^{3+} concentration leads to concomitant changes in local chemical environment, which in turn affects the perovskite unit cell via cation–cation interactions as well as distortions to the oxygen octahedral network. The clusters of AFE regions first undergo nucleation and then grow within a ferroelectric (FE) matrix. At the critical concentration of $x = 0.14$, FE regions are so frustrated by the surrounding AFE matrix that it results in an incommensuration phenomenon, in which the system adopts a complex nanoscale mixture bridging the Bi-rich rhombohedral and Sm-rich orthorhombic phases,

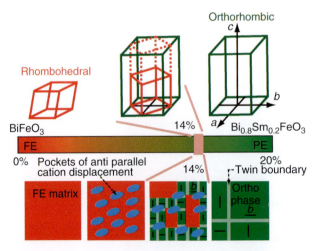

Figure 6.6 Schematic illustration describing structural evolution across the rhombohedral ferroelectric to orthorhombic paraelectric phase boundary in Sm-BFO, which ultimately is linked to its functional behavior.

which probably play a key role in the enhancement of piezoelectric properties in Sm-BFO at the MPB [14].

6.4
Universal Behavior in RE-Substituted BFO

Having understood the substitution-induced evolution of the structure and the concomitant changes in the electromechanical response of Sm-BFO, we next considered a set of fundamental key questions: (i) are there other compositions with similar behaviors and (ii) what is the mechanism behind the structural transition and the concomitant enhancement of piezoelectric properties? To this end, we have systematically investigated substitution of various trivalent REs into the A-site of BFO [25]. Here, we focus on results with Sm, Gd, and Dy as the RE^{3+} dopants to substitute for Bi^{3+} in the A-site of the perovskite structure. The ionic radii of these trivalent ions with a coordination number of 12 are Bi^{3+} (1.36 Å) $>$ Sm^{3+} (1.28 Å) $>$ Gd^{3+} (1.27 Å) $>$ Dy^{3+} (1.24 Å) [26]. For each composition spread with a different RE^{3+} dopant, the dopant concentration is continuously varied from $x = 0$ (pure BFO) to a fixed value (typically $x = 0.3$) along the direction of the spread sample. We used scanning XRD extensively as the main tool to track and map systemic changes in film structure as the phase transition takes place across each composition spread.

Figure 6.7a−c shows typical 2D XRD images taken for a Dy-substituted BFO (Dy-BFO) composition spread sample. They capture the essential and common characteristics of the substitution-induced structural evolution we observed for all RE dopants studied. The key findings from the XRD patterns discussed later have also been corroborated by TEM images and ZADPs observed for individual samples [18, 19, 27]. For pure BFO ($x_{Dy} = 0$, Figure 6.7a), only (001) and (002) reflections from the film and the substrate are clearly observed, confirming the epitaxial growth of the film on the substrate. With increasing Dy concentration, extra diffraction spots begin to appear in addition to the fundamental spots. In Figure 6.7b, where $x_{Dy} = 0.07$, the spots (with arrows in red) that are assigned as 1/4{011} appeared. This X-ray superstructure spot results from antiparallel cation displacements along the [011] direction in the local regions. This structure closely resembles that of the well-established AFE structure in PZO [23, 24, 28, 29], as discussed in the earlier sections. It is important to point out that, in the composition range where the 1/4 spots are observed, macroscopic properties of the films are those consistent with an FE phase and not with an AFE phase, as the structure associated with the 1/4 spots is a minority phase.

On further increasing the Dy concentration (Figure 6.7c), the 1/4{011} superstructure spots disappear and new superstructure spots (marked with yellow arrows) begin to emerge. These new spots correspond to the 1/2{010} reflection spots, indicating that unit-cell doubling has taken place along the in-plane direction

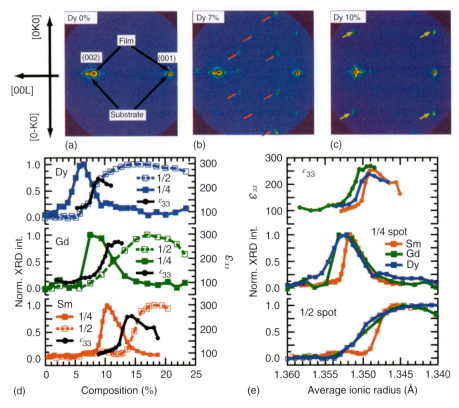

Figure 6.7 (a–c) 2D XRD images for the Dy-BFO composition spread film on (001) SrTiO₃ substrate. (b) $x_{Dy} = 0$. (c) $x_{Dy} = 0.07$. (d) $x_{Dy} = 0.1$. The 2D XRD images are recorded in the $2\theta - \chi$ plane with the incident X-ray beam parallel to the $[100]_{STO}$ direction. For (b, c), the spots with arrows belong to 1/4{011} and 1/2{010}, respectively. All spots are indexed by the pseudocubic-unit-cell notation. (d) Normalized XRD intensities of (0, 1/4, 7/4) (filled square, denoted as 1/4, or 1/4 spot) and (0, 1/2, 2) (open square, denoted as 1/2, or 1/2 spot) XRD superstructure spots and ε_{33} at zero bias (black dot) as a function of RE dopant composition. (e) All data in (d) are plotted against r_{ave}. The ionic radii in 12 coordination are 1.36 Å for Bi^{3+}, 1.28 for Sm^{3+}, 1.27 for Gd^{3+}, and 1.24 for Dy^{3+} [26]. This change in r_{ave} corresponds to the change in the Goldschmidt tolerance factor from 0.954 ($r_{ave} = 1.36$ Å) to 0.947 ($r_{ave} = 1.34$ Å) with 0.645 and 0.14 Å for the ionic radii of Fe^{3+} and O^{2-}, respectively.

when compared to the rhombohedral BFO phase. From X-ray reciprocal space mapping, we see that a film showing the 1/2{010} spots is in a (pseudo-)orthorhombic structural phase with the unit cell having the dimensions of $\sqrt{2}a_{pc} \times \sqrt{2}a_{pc} \times 2a_{pc}$.

To further elucidate the sequence of structural evolution as a function of the substituting concentration, we plot the normalized intensities of the (0, 1/4, 7/4) XRD spot (a 1/4{011} spot) arising from the PZO-type minority phase and the (0, 1/2, 2) spot (a 1/2{010} spot) due to cell-doubling in the majority phase for

Dy-BFO, in the upper panel of Figure 6.7d. The 1/4{011} spot intensity peaks at 6% substitution. Beyond this concentration, the 1/4{011} spot decreases in intensity, while the 1/2{010} spot gets enhanced, as an indication of the structural phase transition taking place from the rhombohedral to the orthorhombic phase. Also plotted in the upper panel of Figure 6.7d are the zero-bias ε_{33}, which show the peaking behavior at the structural boundary. The maximum value of ε_{33} is $\sim$250. It is noteworthy that the maximum d_{33} value at the peak is almost double that for pure BFO thin films. The observed structural evolution and the enhanced properties are found to be common to all three dopants as shown in Figure 6.7d. A key observation here is that the composition of the structural transition shifts toward the larger composition values as the ionic radius of the dopant becomes larger. For RE = Dy, which has the smallest ionic radius of the three elements, 8% substitution is sufficient to induce the transition. On the other hand, for RE = Sm, which has the largest radius of the three, 14% substitution is necessary. This observation is attributed to a hydrostatic pressure effect caused by the smaller radii of the isovalent RE ions. Figure 6.7e captures the behavior universal to all the RE dopants, where the same data shown in Figure 6.7d are plotted together against the average A-site ionic radius r_{ave}. The curves collapse together, revealing an underlying common behavior as a function of the average ionic radius.

In Figure 6.8a, we show contour plots of the (0, 1/4, 7/4) and the (0, 1/2, 2) XRD spot intensities as functions of temperature and the average ionic radius r_{ave} for the case of the Sm-BFO thin films, in order to closely monitor the occurrence of the PZO-type structure and the orthorhombic structure, respectively. The intensities are color-coded by the side bar. As the temperature is raised from room temperature, the 1/4 spot gradually decreases in intensity, while the 1/2 spot intensity gets enhanced. This indicates that, on the higher temperature side, the orthorhombic phase is the stable phase. This behavior is also confirmed by synchrotron XRD measurements of 11% Sm-BFO films [30]. More importantly, the same rhombohedral-to-orthorhombic structural phase transition induced by RE substitution occurs with increasing temperature. Essentially the same contour plots of the (0, 1/4, 7/4) and the (0, 1/2, 2) XRD spot intensities are observed for Dy-BFO, indicating again that r_{ave} is the critical parameter that governs the structural properties of RE-BFO. On the basis of this, we arrive at a universal phase diagram for RE-BFO, as illustrated in Figure 6.8b. On the lower temperature side, the composition region (in light blue) where the local region gives rise to the 1/4 spot is seen to "bridge" the ferroelectric rhombohedral phase regions (in blue) and the orthorhombic phase (in green). This bridging phase has been previously shown to result in lattice incommensuration at the rhombohedral–orthorhombic phase boundary [18]. As the temperature is elevated, the composition region with the 1/4 spot phase disappears, and the orthorhombic phase lies right adjacent to the FE region.

The ferroelectric properties display concomitant evolution when the material undergoes the structural transition, and hence the ferroelectric properties of RE-BFO also exhibit a universal behavior. This is shown in Figure 6.9 where the room-temperature polarization hysteresis loops are plotted for various RE

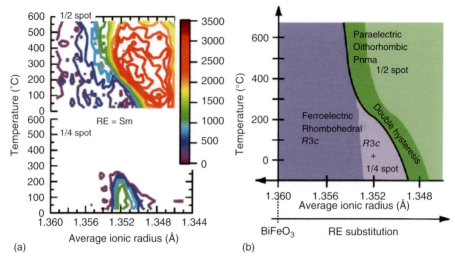

(a)   (b)

Figure 6.8 Phase diagram for RE-substituted BiFeO$_3$ (RE = Sm, Gd, and Dy). (a) Contour plots of intensities (counts/s) of (0, 1/2, 2) (1/2 spot, upper panel) and (0, 1/4, 7/4) spots (1/4 spot, bottom panel) as functions of temperature and r_{ave} for the Sm-BFO composition spread thin films. (b) Proposed phase diagram for RE-BFO. The black line represents the structural phase boundary between the rhombohedral (in blue) and the orthorhombic (in green) structural phases. At the lower temperature side, the 1/4 spot is observed in the region in light blue. The double hysteresis loop behavior (see text and Figure 6.9) emerges in the region in dark green.

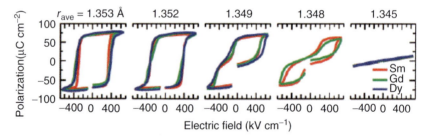

Figure 6.9 Universal behavior in polarization hysteresis loops across the structural transition for RE-substituted BiFeO$_3$ (RE-BFO) thin films (RE = Sm, Gd, and Dy). Room-temperature PE hysteresis loops for 200 nm thick films are plotted for r_{ave}. For each r_{ave}, the hysteresis loops collapse into a single curve. This clearly demonstrates the universal behavior in ferroelectric properties of RE-BFO.

substituting concentrations such that r_{ave} is the same for each group of curves. For each r_{ave}, the hysteresis loops collapse together, and the general shape and FE character of the hysteresis loop are constant. This indicates that the average ionic radius is the universal parameter to describe not only the structural but also the ferroelectric properties. As we cross the structural boundary (drawn with a black line in Figure 6.8b), which is located at $r_{ave} = 1.35$ Å at room temperature, the polarization hysteresis loop consequently exhibits a transition from single to double hysteresis behaviors with decreasing r_{ave}, as shown in Figure 6.9. The corresponding Goldschmidt tolerance factor for this r_{ave} is 0.9509. (For pure BFO, the tolerance factor is 0.954.) For the FE composition region immediately adjacent to the transition ($r_{ave} > 1.35$ Å), the FE square-shaped hysteresis loops with saturation polarizations of $70\,\mu C\,cm^{-2}$ are observed. As one approaches the transition ($r_{ave} \sim 1.35$ Å), first the hysteresis loop becomes distorted, and then it develops into a fully developed double hysteresis loop beyond the transition ($r_{ave} < 1.35$ Å).

The first-principles calculation [25], in combination with our experimental results, suggested that the origin of the double hysteresis behavior and the concomitant enhancement in the piezoelectric coefficient is an electric-field-induced transformation from a paraelectric orthorhombic phase to the polar rhombohedral phase. An important consequence of this electric-field-induced phase transformation events is the enhancement of d_{33} at the boundary. Similar electric-field-induced transformation and the resulting enhancement of piezoelectric response have been observed in Pb-based ferroelectric materials at the MPBs [1–5], for example, $Pb(Zr,Nb)O_3$–$PbTiO_3$, PZT, and $Pb(Mg,Nb)O_3$–$PbTiO_3$. We believe that the enhancement in the electromechanical properties we see in the present RE-BFO thin films [14] is of similar nature. At the structural boundary of the substituted BFO, the lattice strain caused by the structural transformation is estimated to be ~1%, based on the difference in the lattice parameter of 0.04 Å along the $[001]_{STO}$ direction between the rhombohedral and orthorhombic phases [31]. Taking into account that a small spread in E_c present in samples owing to local variation in dopant concentration (estimated to be ±1% or better), the 1% strain corresponds to $750–1000\,pm\,V^{-1}$. This value is up to an order of magnitude larger than our observed d_{33} value of $110\,pm\,V^{-1}$ at the boundary [14]. Given that some degree of hysteresis is expected at the electric-field-induced transition, and that the clamping effect is known to significantly reduce d_{33} in epitaxial thin films compared to bulk [32], we believe the agreement is not unreasonable.

6.5
Structural Fingerprint of MPB in RE-Substituted BFO

While the origin of the enhanced electromechanical response was theoretically identified to be an electric-field-induced transformation from the orthorhombic phase to the polar rhombohedral phase, the picture of the MPB in terms of its microstructural makeup that facilitates such a transformation was still unclear. Critically, it was of fundamental interest to investigate whether there is a common "structural fingerprint" irrespective of the type of the RE dopant, whose occurrence would substantially enhance the electric-field-induced electromechanical susceptibilities [27, 31].

To answer that question, we investigated substitution-induced evolution in the dielectric and piezoelectric structural properties RE-BFO (RE = Dy, Gd, Sm, and

La). We began by examining dielectric properties of RE = Dy, Gd, and Sm cases where all the RE^{3+} dopants have the smaller ionic radius than the Bi^{3+} ion. ε_{33} versus E hysteresis loop measurements revealed that all RE cases (RE = Dy, Gd, and Sm) show the clear universal evolution in the loop shape as a function of the dopant composition: from the typical ferroelectric double-humped curve before the MPB through the MPB composition to a quadruple-humped curve beyond the MPB. The presence of the four humps in the ε_{33}–E loop in compositions for the paraelectric orthorhombic phase is consistent with the observation of the double-hysteresis P–E loops for these compositions [25]. There is a continuous enhancement of the dielectric constant, reaching maximum values of 250 at the phase boundary compositions of 8% for RE = Dy^{3+}, 12% for Gd^{3+}, and 14% for Sm^{3+}.

The remnant and high-field d_{33} values evaluated with quantitative PFM also show an enhancement at the MPB compositions, which is observed for all RE (= Dy, Gd, and Sm) cases, with the value reaching 110 pm V^{-1} for RE = Sm [14]. For compositions beyond the MPB, there is not only a drastic drop in the d_{33} response but also a change in the shape of the loop.

Comprehensive ZADP studies of RE-BFO films (RE = Dy, Gd, and Sm) with various substituting concentrations before, at, and beyond the MPB reveal the universal trend in structural evolution. In low-concentration-substituted compositions before the MPB, the substitution results in the 1/4 spots, which are due to antipolar PZO-like clusters (10–20 nm in size) having antiparallel A-site cation displacements along the [011] direction. At the MPB composition, the picture is more complex, with the coexistence of the parent rhombohedral ferroelectric phase, the antipolar PZO-like phase (the 1/4 spots), and the cell-doubled orthorhombic paraelectric phase (the 1/2 spots). We note that this complex pattern is found for the entire film, and not restricted to local regions, unlike in the PZO-like phase. The strong lattice interactions between ferroelectric rhombohedral, cell-doubled orthorhombic, and an intermediate PZO-like phase can lead to lattice incommensuration at the MPB [18], which provides a "facile platform" for an electric-field-induced phase transformation. A similar scenario has been proposed for the well-known Pb-based MPB piezoelectrics [2, 33, 34], where such a nanoscale phase mixture has been associated with a low-energy pathway for a polarization transition between the different polar axes, which results in enhanced piezoelectric properties [35]. Therefore, we can conclude that the occurrence of such complex phase coexistence is the underlying structural fingerprint for the enhanced electromechanical properties observed for the smaller RE-substituted BFO. In compositions beyond the MPB, cell-doubled orthorhombic phases signified by 1/2{010} and 1/2{011} reflections are observed throughout the entire sample. The occurrence of both reflections is consistent with the fact that the orthorhombic phase is similar to the one in the Sm-BFO with the unit cell having the dimensions of $\sqrt{2}a_{pc} \times \sqrt{2}a_{pc} \times 2a_{pc}$.

We have also investigated the La-substituted BFO thin films (La-BFO), where the ionic radius of La is comparable to that of Bi^{3+} ion. Figure 6.10 summarizes our findings for the RE = La^{3+} case. We find that the rhombohedral-to-orthorhombic structural phase transition occurs at 23% La^{3+} substitution, which is a much higher

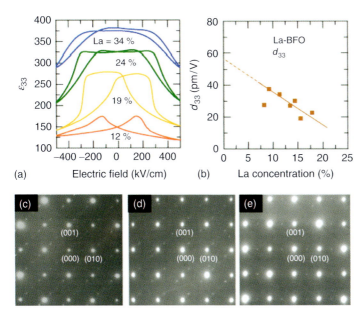

Figure 6.10 (a) $\varepsilon_{33}-E$ loops for La^{3+}-substituted BFO (La-BFO) across the structural boundary. (b) Piezoelectric coefficient d_{33} for La-BFO at 400 kV cm^{-1} as a function of the La^{3+} concentration. (c) [100] ZADP of 14% La-BFO showing 1/4{011} spots; (d) [100] ZADP 23% La-BFO showing the coexistence of weak 1/4{011} and 1/2{010} spots. (e) [100] ZADP 26% La-BFO showing 1/2{010} spots only.

concentration compared to the smaller RE dopants [25, 27]. Figure 6.10a shows the $\varepsilon_{33}-E$ loops for a series of La-BFO films across the structural boundary. No sharp quadruply humped loop is observed as one crosses the boundary, which is in contrast to the smaller RE cases. In Figure 6.10b, we plot the d_{33} at 400 kV cm^{-1} for the La-BFO films as a function of La^{3+} concentration. The d_{33} value shows a monotonous decrease (down to almost zero) as one approaches the phase boundary, showing that there is no distinct increase in the electromechanical properties in contrast to the other smaller RE cases [27] (Figure 6.1c), which is in agreement with previous reports [11, 36]. We have also examined the [100] ZADP of the three La-BFO compositions across the phase boundary, that is, before (Figure 6.10c; 14% La substitution), at the phase boundary (Figure 6.10d; 23%), and for the orthorhombic phase (Figure 6.10e; 26%). The diffraction patterns at the two ends of our composition series display certain features (1/4 ordering → 1/2 ordering) similar to those of the smaller RE cases [27] (Figures 6.2b, 6.4a, and 6.5a). The presence of 1/4 superstructure spots indicates that a phase isostructural to PZO certainly forms initially. ZADPs similar to our findings have also been observed in 20% La-BFO thin films on an entirely different Si substrate [11], and thus any strain effects from the underlying substrate as a cause can be excluded. The key difference between the La^{3+}-substituted system and the smaller RE^{3+}-substituted systems is found in the ZADP for the phase boundary composition (Figure 6.10d). Here, the phase coexistence is distinctly weak, as revealed by much fainter superstructure spots. The 1/2{010} spots are barely visible, which implies that the microstructural platform that would facilitate direct lattice interactions between the antipolar PZO-like phase and the rhombohedral/orthorhombic phase is absent here. Thus, the required domain mixture in order to facilitate electric-field-induced transformation does not occur and, as a result, no peak in the electromechanical properties is observed. Beyond the structural transition (Figure 6.10e), superstructure reflections have almost disappeared, with much weaker intensity of 1/2{010} reflections only. The absence of the 1/2{011} reflections beyond the structural transition implies that the structure of the La-BFO has a dimension of $a_{pc} \times a_{pc} \times 2a_{pc}$, instead of $\sqrt{2}a_{pc} \times \sqrt{2}a_{pc} \times 2a_{pc}$ as observed for the smaller RE^{3+}-substituted case.

Our results demonstrate the strong correlation between the occurrence of a nanoscale phase coexistence and the enhanced electromechanical properties observed for (smaller) RE-substituted BFO systems. We believe that the unique microstructure provides a facile platform that mediates the theoretically proposed electric-field-induced phase transition from the orthorhombic phase to the ferroelectric rhombohedral phase. In contrast, for the RE = La case (a same sized RE^{3+} ion as Bi^{3+}), there is a distinct lack of this phase coexistence and, consequently, the absence of enhancement in the electromechanical response.

6.6
Chemical-Substitution-Induced Polarization Rotation in BFO

One remaining question to be addressed is that RE-BFO has almost constant remnant polarization before the structural boundary although RE substitution reduces Bi concentration [37], which is responsible for the ferroelectric polarization [37, 38]. The constant remnant polarization with the RE substitution concentration may indicate that the substitution modifies the direction of the ferroelectric polarization.

In order to examine this possibility, we systematically studied P-E hysteresis loops in different orientation directions and at different substitution compositions. Probing the direction and the absolute value of the ferroelectric polarization can reveal how the polarization vector evolves with chemical substitution. The room-temperature $P-E$ hysteresis loop measurements [37] show that, for the film grown on the (001) substrate, the square-shaped ferroelectric hysteresis loop undergoes a clear transition to a double hysteresis loop as we cross the structural phase boundary at Sm 14% (Figure 6.9). In contrast, the films grown on the (110) and (111) substrates undergo different evolutions in $P-E$ loops. For pure BFO (Sm 0%), square-shaped ferroelectric $P-E$ loops with saturated polarization values of 85 and $110 \, \mu C \, cm^{-2}$ are observed for the (110) and (111) film, respectively. These values are in good agreement with a previous report [39] and obey the simple projection rule. As we increase the Sm composition, the saturated and remnant polarization values for the (110) film continuously decrease. Beyond the structural transition at Sm $\sim$14%, the $P-E$ loops display curves characteristic of a paraelectric phase. No

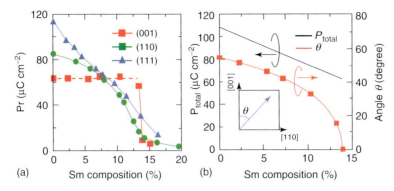

Figure 6.11 (a) Remnant polarization as a function of Sm composition for Sm-BFO films grown on (001), (110), and (111) STO substrates. (b) Extracted rotation angle θ as a function of Sm composition (red). In (b), the red solid line, which is drawn based on the relationship of $P_{total} \times \cos\theta = 60\,\mu\text{C cm}^{-2}$, is a guide to the eye. The *rotation angle* θ is defined as the angle of the polarization vector with respect to the [001] axis as shown in the inset. The calculation was performed by assuming that the total polarization P_{total} decreases linearly with substitution (the black line on the graph).

double hysteresis behavior is observed at the structural boundary. A similar trend is seen for the film on the (111) substrate.

In Figure 6.11a, we plot the remnant polarization (P_r) values as a function of the Sm composition for the three orientations. Since the orthorhombic phase is paraelectric [25], P_r becomes zero for all orientations beyond Sm 14%. For the (001) film, P_r appears to show a roughly constant value of ~60 μC cm^{-2} until Sm 14%, while P_r steadily decreases in value to zero for the (110) and (111) films. Since the ferroelectric polarization in BFO mainly arises from Bi 6s lone-pair electrons [38], Sm substitution for Bi necessarily reduces the absolute value of polarization. Hence, this cannot explain the fact that P_r along the [001] direction appears to remain unchanged in the rhombohedral structural phase.

It is also important to point out that the structural phase associated with the 1/4 spots displaying the antiparallel cation displacements is not the origin of the observed evolution in P_r in Figure 6.11a, either. Since the structural phase associated with the 1/4 spots appears only in spatially localized pockets [18, 19], it is unlikely that this structural phase affects the macroscopic physical properties such as P_r. If this phase were to influence the overall properties, we would expect to see a variation in P_r with pronounced changes at around Sm 10% for all orientations, since the 1/4 diffraction spots show a maximum intensity at this composition (Figure 6.7d). This is not the case in our observed evolution in Figure 6.11a. Instead, our observation points to the fact that Sm substitution changes the direction of the polarization from the [111] direction for pure BFO to the [001] direction.

To better understand this, we extracted the rotation angle by assuming that the total amount of the polarization P_{total} linearly decreases from 110 μC cm^{-2} (at Sm

0%) to 60 μC cm^{-2} (at Sm 14%) owing to the Sm substitution. The polarization projected along the [001] direction $P_{[001]}$ is calculated to be $P_{total} \times \cos\theta$, where we define θ as the angle of the polarization vector with respect to the [001] axis, as shown in the inset of Figure 6.11b. On the basis of our observations, $P_{[001]}$ ($= P_{total} \times \cos\theta$) remains approximately constant at ~60 μC cm^{-2} until Sm 14%. The calculated angle based on this is plotted in Figure 6.11b. The angle clearly changes from 54.7° (the [111] direction) for pure BFO to 0° (the [001] direction) at the structural boundary.

We attribute the rotation of the ferroelectric polarization vector to the structural modifications caused by the chemical substitution. Figure 6.12a displays changes in lattice parameters of the 400 nm thick (001) Sm-BFO thin films on the STO substrate as a function of the Sm composition. This also confirms the structural transition from the rhombohedral to the orthorhomic phase occurring at Sm ~14%. A careful inspection of the difference between the out-of-plane and in-plane lattice parameters suggests that the structure before the structural transition (Sm composition slightly <14%) can be described as a monoclinic phase with

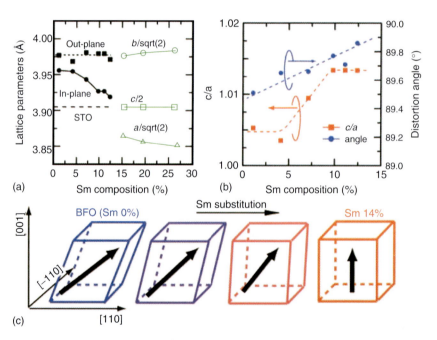

(a)

(b)

(c)

Figure 6.12 Structural properties of Sm-BFO film grown on (001) STO substrate. (a) Evolution in lattice parameters with Sm substitution. (b) Changes in tetragonality c/a and rhombohedral distortion angle as a function of Sm composition. The value c/a is calculated as the ratio of the out-of-plane lattice parameter to the in-plane lattice parameter. (c) Schematic describing the rotation of the ferroelectric polarization vector associated with the structural changes induced by chemical substitution. The black arrow denotes the ferroelectric polarization vector. Note that the unit cells are drawn in terms of the monoclinic lattice with a dimension of $\sqrt{2}a_{pc} \times \sqrt{2}a_{pc} \times 2a_{pc}$.

the distortion along the [110] direction. Hereafter, we call this as a monoclinic structure. We see that the monoclinic structure is maintained until Sm $\approx$14%, which is confirmed by the appearance of the multiple (203) reflections due to the crystallographic twin formation observed in the X-ray reciprocal space mappings. As we substitute Sm for Bi and increase its concentration, the in-plane lattice parameters decrease, while the out-of-plane lattice parameters remain constant. This increases the c/a ratio slightly from 1.005 for pure BFO to 1.013 at Sm $\sim$13%, as displayed in Figure 6.12b.

We note that, as we approach the structural boundary, the monoclinic distortion angle, which is determined from the separation between the {203} reflections [11], also continuously increases and reaches 90° at the boundary. This indicates that the monoclinic structure changes to a tetragonal-like one, with the c/a ratio of 1.013 at Sm 13%. This evolution is schematically depicted in Figure 6.12c. For pure BFO, the polarization points to the [111] direction. As we increase the Sm substitution, the lattice becomes compressed and less distorted, resulting in the tetragonal-like structure, which allows the polarization to point toward the [001] direction. We also found a similar trend in the structural evolution for the films on (001) LSAT ($(LaAlO_3)_{0.3-}(Sr_2AlTaO_6)_{0.7}$) substrates. Therefore, we conclude that it is the chemical pressure effect induced by the Sm substitution rather than epitaxial strain that results in the observed structural evolution that leads to the rotation of the polarization vector from the [111] to the [001] direction.

It has been previously reported that a similar change in the ferroelectric polarization vector can be induced by epitaxial strain in BFO thin films [40, 41]. A recent report [41] has shown that the rhombohedral BFO suffering from a large compressive epitaxial strain undergoes a structural transition to a polymorph tetragonal-like structure with the large c/a ratio of $\sim$1.27, where the polarization vector is primarily parallel to the [001] direction. For the chemical-substitution-stabilized tetragonal-like structural phase in our case, the c/a ratio 1.013 (Figure 6.12b) is not as large as the one induced by the strain, while the distortion angle is almost 90° in contrast with the angle of $\sim$88.7° in the strain-driven tetragonal-like phase. The implication is that there are two approaches to stabilize the ferroelectric polarization along the [001] direction in BFO: one is the application of compressive epitaxial strain, resulting in the very large c/a ratio; the other is chemical substitution, making the distortion angle $\sim$90°. An important consequence of the chemical substitution is that one can dial in on an arbitrary polarization vector angle along any direction between the [111] and [001] directions by tuning the substitution composition.

The impact of the rotation in the polarization vector can be seen from the evolution of out-of-plane piezoelectric constant d_{33} for Sm-BFO films grown on (100), (110), and (111) oriented substrates. As we showed in a previous report [14], the (001) film exhibits the enhancement in d_{33} at $\sim$14% substitution where the structural transition takes place, and the effective d_{33} value reaches 110 pm V^{-1}. For the (110) and (111) films, d_{33} shows a decrease in value and reaches zero as the Sm composition is increased up to 14% [37]. It should be noted that the rhombohedral-to-orthorhombic phase transition takes place at Sm

~14% independent of the substrate orientation. This implies that the structural transition is not playing a direct role in the evolution in d_{33} for the (110) and (111) films. Instead, it is the rotation of the polarization vector that acts as a determining factor for the observed d_{33} behavior. As one can see in Figure 6.11a, the polarization components along the [110] and [111] directions diminish as a result of the rotation of polarization vector toward the [001] direction. In the composition region adjacent to structural boundary (Sm composition slightly <14%), the polarization vector lies along the [001] direction. This dictates the direction along which the dominant piezoelectric coefficient exists and is the main reason why only the (001) film shows the enhancement in d_{33} at the structural boundary.

As mentioned earlier, on the basis of first-principles calculations [25], the origin of the double hysteresis loops at the structural boundary is the electric-field-induced transformation from the paraelectric orthorhombic phase to the polar rhombohedral one. Accordingly, the polar phase induced by the applied electric field is energetically favorable owing to the nonzero polarization P through the coupling energy, $-P \times E$. Because the polarization vector is pointing along the [001] direction at the boundary, maximum energy stability is achieved by applying the electric field E along this direction through $-P \times E$. As a result, the double hysteresis behavior is most prominent only when E is applied along the [001] direction [37]. This is also consistent with the fact that the enhancement in d_{33} is observed for the (001) oriented film [14].

6.7
Concluding Remarks and Future Perspectives

In conclusion, we have reviewed our effort over the past 6 years aimed at understanding the structure–property correlations in RE-substituted BFO epitaxial thin films at MPB. At the boundary, a structural phase transition from the original rhombohedral phase to an orthorhombic phase takes place, and consequently, ferroelectric polarization hysteresis loops undergo a transition into double hysteresis behavior, associated with substantial enhancement in dielectric and piezoelectric properties. We also discover a universal behavior at the MPB, independent of the RE dopant species, whereby we find that the universal parameter for structural and ferroelectric properties in the substituted BFO is the average A-site ionic radius. First-principles calculations propose that the origin of the double hysteresis loop and the concomitant enhancement in the piezoelectric coefficient is an electric-field-induced transformation from a paraelectric orthorhombic phase to the polar rhombohedral phase. Extensive electron diffraction analysis reveals that the complex phase coexistence at the MPB acts as a facile platform to facilitate the theoretically proposed orthorhombic-to-rhombohedral structural transformation. We also find that chemical substitution results in the continuous rotation in the polarization vector from the [111] direction for pure BFO to the [001] direction for

RE-BFO at the MPB. This polarization rotation can explain why the double hysteresis behavior with enhanced piezoelectric coefficient d_{33} is prominently observed only for the (001) oriented film. While epitaxial film samples explored here are of significant interest for thin film device applications, in order to fully harness the enhanced piezoelectric properties of these MPB compounds, it is highly desirable to fabricate bulk samples. Unfortunately to date, bulk BFO (polycrystalline and single crystals) in general display poor insulating properties which preclude them from their use in bulk actuator applications. To this end, researchers are working to increase their electrical insulation through chemical modification and improved synthesis methods [42, 43]. Fully insulating bulk BFO at MPB free of substrate clamping are expected to display large piezoelectric coefficients due to the electric-field induced structural transition discussed here which can potentially rival that of other common bulk actuator materials at MPBs. Breakthroughs in this front may be just around the corner, and the community anxiously awaits the development of bulk polycrystalline BFO with robust insulation and high breakdown voltage.

In this chapter, we did not discuss the effect of the systematic substitution on magnetic properties in BFO. Magnetic properties of pure thin film BFO have been extensively investigated [44]. Because of the direct coupling between ferroelectric polarization and the G-type antiferromagnetic moments [44], the continuous polarization rotation [37] discussed in this chapter has immediate implications for tuning of magnetic properties. Indeed, there has already been some evidence of a magnetic transition at the rhombohedral-orthorhombic structural boundary at the MPB [45]. Taken together with the discussed electric-field induced structural transition, this may lead to an extremely large magnetoelectric (ME) susceptibility at the MPB. Such a phenomenon can be highly unique, and it can be the basis of an entirely new type of ME devices which is yet to be explored.

Acknowledgments

We would like to acknowledge the work of Dr A.Y. Borisevich (Oak Ridge National Laboratories) on TEM observations, Dr S.B. Emery, Prof. B.O. Wells, and Prof. S.P. Alpay (University of Connecticuts) on synchrotron XRD measurements, and Dr L. Pálová and K.M. Rabe (Rutgurs University) on first-principles calculations. This work is supported by UMD-NSF-MRSEC, ARO W.M. Keck Foundation. The work at UNSW was supported by ARC Discovery and Linkage LP0991794 projects, DEST-ISL, and NEDO.

References

1. Park, S.-E. and Shrout, T.R. (1997) *J. Appl. Phys.*, **82**, 1804–1811.
2. Guo, R., Cross, L.E., Park, S.-E., Noheda, B., Cox, D.E., and Shirane, G. (2000) *Phys. Rev. Lett.*, **84**, 5423.
3. Kutnjak, Z., Petzelt, J., and Blinc, R. (2006) *Nature*, **441**, 956–959.
4. Fu, H. and Cohen, R.E. (2000) *Nature*, **403**, 281–283.

5. Bellaiche, L., García, A., and Vanderbilt, D. (2001) *Phys. Rev. B*, **64**, 060103.

6. Wang, J., Neaton, J.B., Zheng, H., Nagarajan, V., Ogale, S.B., Liu, B., Viehland, D., Vaithyanathan, V., Schlom, D.G., Waghmare, U.V., Spaldin, N.A., Rabe, K.M., Wuttig, M., and Ramesh, R. (2003) *Science*, **299**, 1719–1722.

7. Catalan, G. and Scott, J.F. (2009) *Adv. Mater.*, **21**, 2463–2485.

8. Lebeugle, D., Colson, D., Forget, A., and Viret, M. (2007) *Appl. Phys. Lett.*, **91**, 022907.

9. Cheng, Z., Wang, X., Dou, S., Kimura, H., and Ozawa, K. (2008) *Phys. Rev. B*, **77**, 092101.

10. Yuan, G.L., Or, S.W., Liu, J.M., and Liu, Z.G. (2006) *Appl. Phys. Lett.*, **89**, 052905.

11. Chu, Y.H., Zhan, Q., Yang, C.-H., Cruz, M.P., Martin, L.W., Zhao, T., Yu, P., Ramesh, R., Joseph, P.T., Lin, I.N., Tian, W., and Schlom, D.G. (2008) *Appl. Phys. Lett.*, **92**, 102909.

12. Khomchenko, V.A., Kiselev, D.A., Bdikin, I.K., Shvartsman, V.V., Borisov, P., Kleemann, W., Vieira, J.M., and Kholkin, A.L. (2008) *Appl. Phys. Lett.*, **93**, 262905.

13. Zhu, W.-M., Su, L.W., Ye, Z.-G., and Ren, W. (2009) *Appl. Phys. Lett.*, **94**, 142908.

14. Fujino, S., Murakami, M., Anbusathaiah, V., Lim, S.-H., Nagarajan, V., Fennie, C.J., Wuttig, M., Salamanca-Riba, L., and Takeuchi, I. (2008) *Appl. Phys. Lett.*, **92**, 202904.

15. Koinuma, H. and Takeuchi, I. (2004) *Nat. Mater.*, **3**, 429–438.

16. Nagarajan, V., Stanishevsky, A., Chen, L., Zhao, T., Liu, B.-T., Melngailis, J., Roytburd, A.L., Ramesh, R., Finder, J., Yu, Z., Droopad, R., and Eisenbeiser, K. (2002) *Appl. Phys. Lett.*, **81**, 4215–4217.

17. Saito, Y., Takao, H., Tani, T., Nonoyama, T., Takatori, K., Homma, T., Nagaya, T., and Nakamura, M. (2004) *Nature*, **432**, 84–87.

18. Cheng, C.-J., Kan, D., Lim, S.-H., McKenzie, W.R., Munroe, P.R., Salamanca-Riba, L.G., Withers, R.L., Takeuchi, I., and Nagarajan, V. (2009) *Phys. Rev. B*, **80**, 014109.

19. Cheng, C.-J., Borisevich, A.Y., Kan, D., Takeuchi, I., and Nagarajan, V. (2010) *Chem. Mater.*, **22**, 2588–2596.

20. Liu, Y., Withers, R.L., and Wei, X.Y. (2005) *Phys. Rev. B*, **72**, 134104.

21. Munkholm, A., Streiffer, S.K., Ramana Murty, M.V., Eastman, J.A., Thompson, C., Auciello, O., Thompson, L., Moore, J.F., and Stephenson, G.B. (2001) *Phys. Rev. Lett.*, **88**, 016101.

22. He, F., Wells, B.O., Ban, Z.-G., Alpay, S.P., Grenier, S., Shapiro, S.M., Si, W., Clark, A., and Xi, X.X. (2004) *Phys. Rev. B*, **70**, 235405.

23. Woodward, D.I., Knudsen, J., and Reaney, I.M. (2005) *Phys. Rev. B*, **72**, 104110.

24. Sawaguchi, E., Maniwa, H., and Hoshino, S. (1951) *Phys. Rev.*, **83**, 1078.

25. Kan, D., Pálová, L., Anbusathaiah, V., Cheng, C.J., Fujino, S., Nagarajan, V., Rabe, K.M., and Takeuchi, I. (2010) *Adv. Funct. Mater.*, **20**, 1108–1115.

26. Jia, Y.Q. (1991) *J. Solid State Chem.*, **95**, 184.

27. Cheng, C.-J., Kan, D., Anbusathaiah, V., Takeuchi, I., and Nagarajan, V. (2010) *Appl. Phys. Lett.*, **97**, 212905.

28. Karimi, S., Reaney, I.M., Levin, I., and Sterianou, I. (2009) *Appl. Phys. Lett.*, **94**, 112903.

29. Karimi, S., Reaney, I.M., Han, Y., Pokorny, J., and Sterianou, I. (2009) *J. Mater. Sci.*, **44**, 5102–5112.

30. Emery, S.B., Cheng, C.-J., Kan, D., Rueckert, F.J., Alpay, S.P., Nagarajan, V., Takeuchi, I., and Wells, B.O. (2010) *Appl. Phys. Lett.*, **97**, 152902.

31. Kan, D., Cheng, C.-J., Nagarajan, V., and Takeuchi, I. (2011) *J. Appl. Phys.*, **110**, 014106.

32. Nagarajan, V., Roytburd, A., Stanishevsky, A., Prasertchoung, S., Zhao, T., Chen, L., Melngailis, J., Auciello, O., and Ramesh, R. (2003) *Nat. Mater.*, **2**, 43–47.

33. Schönau, K.A., Schmitt, L.A., Knapp, M., Fuess, H., Eichel, R.-A., Kungl, H., and Hoffmann, M.J. (2007) *Phys. Rev. B*, **75**, 184117.

34. Jin, Y.M., Wang, Y.U., Khachaturyan, A.G., Li, J.F., and Viehland, D. (2003) *Phys. Rev. Lett.*, **91**, 197601.

35. Rao, W.-F. and Wang, Y.U. (2007) *Appl. Phys. Lett.*, **90**, 182906.

36. Uchida, H., Ueno, R., Funakubo, H., and Koda, S. (2006) *J. Appl. Phys.*, **100**, 014106.

37. Kan, D., Anbusathaiah, V., and Takeuchi, I. (2011) *Adv. Mater.*, **23**, 1765–1769.

38. Baettig, P., Schelle, C.F., LeSar, R., Waghmare, U.V., and Spaldin, N.A. (2005) *Chem. Mater.*, **17**, 1376–1380.

39. Li, J., Wang, J., Wuttig, M., Ramesh, R., Wang, N., Ruette, B., Pyatakov, A.P., Zvezdin, A.K., and Viehland, D. (2004) *Appl. Phys. Lett.*, **84**, 5261–5263.

40. Jang, H.W., Baek, S.H., Ortiz, D., Folkman, C.M., Das, R.R., Chu, Y.H., Shafer, P., Zhang, J.X., Choudhury, S., Vaithyanathan, V., Chen, Y.B., Felker, D.A., Biegalski, M.D., Rzchowski, M.S., Pan, X.Q., Schlom, D.G., Chen, L.Q., Ramesh, R., and Eom, C.B. (2008) *Phys. Rev. Lett.*, **101**, 107602.

41. Zeches, R.J., Rossell, M.D., Zhang, J.X., Hatt, A.J., He, Q., Yang, C.-H., Kumar, A., Wang, C.H., Melville, A., Adamo, C., Sheng, G., Chu, Y.-H., Ihlefeld, J.F., Erni, R., Ederer, C., Gopalan, V., Chen, L.Q., Schlom, D.G., Spaldin, N.A., Martin, L.W., and Ramesh, R. (2009) *Science*, **326**, 977–980.

42. Kalantari, K., Sterianou, I., Karimi, S., Ferrarelli, M.C., Miao, S., Sinclair, D.C., Reaney, I.M. (2011) *Adv. Funct. Mater.*, **21**, 3737–3743.

43. Walker, J., Sorrel, C., Nagarajan, V., 2012 (Submitted).

44. Zhao, T., Scholl, A., Zavaliche, F., Lee, K., Barry, M., Doran, A., Cruz, M.P., Chu, Y.H., Ederer, C., Spaldin, N.A., Das, R.R., Kim, D.M., Baek, S.H., Eom, C.B., Ramesh, R. (2006) *Nat. Mater.*, **5**, 823–829.

45. Levin, I., Karimi, S., Provenzano, V., Dennis, C.L., Wu, H., Comyn, T.P., Stevenson, T.J., Smith, R.I., Reaney, I.M. (2010) *Phys. Rev. B*, **81**, 020103.

7
Antiferroelectricity in Oxides: a Reexamination

Karin M. Rabe

7.1
Introduction

As a result of progress both in synthesis and in first-principles studies, there has been a lot of interest and progress in the investigation and rational design of functional materials with enhanced or novel properties [1–8]. Insofar as functional properties arise from coupling to macroscopic fields and stresses, attention has focused on ferroic materials: ferromagnets, ferroelectrics, piezoelectrics, and multiferroic combinations, with order parameters that have macroscopic conjugate fields. The potential of antiferroic materials for functional properties is less obvious, because an antiferroic order parameter does not couple directly to a macroscopic field, but rather to a microscopic staggered field. However, such materials can also exhibit characteristic functional behavior in applied macroscopic fields. Antiferromagnets have been well studied [9]. Here, we consider the electric analog: antiferroelectrics. The definition of antiferroelectricity is in general more subtle than for antiferromagnets, as only certain classes of materials have well-defined localized electric dipoles that can be regarded as analogs of the local magnetic moments in antiferromagnets. In this chapter, these materials, which include hydrogen-bonded antiferroelectrics and antiferroelectric liquid crystals, are not discussed. Rather, the focus is on antiferroelectricity in oxides, which require a more broadly applicable definition of antiferroelectricity and criteria for establishing a material as antiferroelectric.

As discussed more fully in the following section, antiferroelectrics are characterized by an antipolar crystal structure with a related ferroelectric polar structure at low free energy. As recognized by Shirane in the earliest studies of antiferroelectric perovskite oxides [10, 11], the competition between ferroelectric and antiferroelectric phases is an intrinsic feature of antiferroelectrics (Figure 7.1).

The free energy difference is in general tunable in a variety of ways, including composition substitution, epitaxial strain, and size effects. Therefore, proximity to ferroelectric phases through a first-order phase boundary in the phase diagram is an important indication of antiferroelectricity. The most characteristic property resulting from the low-free-energy difference is the electric-field-induced transition

Functional Metal Oxides: New Science and Novel Applications, First Edition.
Edited by Satishchandra B. Ogale, Thirumalai V. Venkatesan, and Mark G. Blamire.
© 2013 Wiley-VCH Verlag GmbH & Co. KGaA. Published 2013 by Wiley-VCH Verlag GmbH & Co. KGaA.

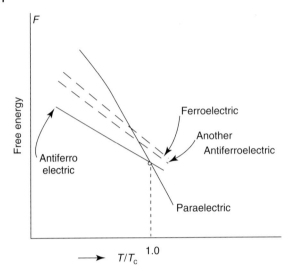

Figure 7.1 Schematic free energy curves for antiferroelectric $PbZrO_3$, showing the presence of a low-free-energy alternative ferroelectric phase. (Source: From Ref. [11].)

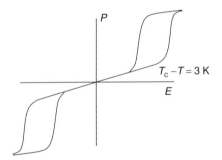

Figure 7.2 Double hysteresis loop for $PbZrO_3$, as redrawn in Refs. [10, 12].

from the antiferroelectric to the related ferroelectric state in a double hysteresis loop (Figure 7.2).

In systems where the two phases have different cell volumes or shapes, this transition is accompanied by large, nonlinear strain responses. Other characteristic properties include dielectric anomalies at the antiferroelectric–paraelectric transition in zero field, large electrostriction coefficients [13], and giant electrocaloric effects [14, 15]. The types of applications that this functionality would enable [16–18] include high-energy storage capacitors, electrocaloric refrigerators, high-strain actuators, and transducers. For realization of these applications, a number of materials properties, including operating temperature, critical electric fields, strain change at the field-induced transition, and switching time and reversibility of the field-induced transition, need to be optimized. This tuning is accomplished largely through compositional substitution, leading to multicomponent materials with complex stoichiometries.

In this chapter, aspects of the fundamental physics of antiferroelectric materials and their properties are reviewed. First, the difficulties in formulating a precise definition of antiferroelectricity are discussed, drawing on previous discussion in the literature. For analysis of the characteristic properties of antiferroelectrics, we use Landau theory functionals. The microscopic origins of the macroscopic behavior are examined. This is fairly straightforward for systems with clearly defined reorientable local-dipolar entities, such as antiferroelectric liquid crystals and hydrogen-bonded antiferroelecrics, and these systems are not discussed in any detail in this chapter. The main focus is on the subtler case of antiferroelectric oxides, for which models based on symmetry-adapted lattice modes prove to be useful. Specific examples of antiferroelectric oxides are presented with available first-principles results and discussion of the role of compositional tuning in producing antiferroelectricity, and the nature of the electric-field-induced ferroelectric phases. In thin films, additional tuning is possible through the effects of strain and finite size. This chapter concludes with some remarks on materials design, including the optimization of properties relevant to technological applications of antiferroelectrics.

7.2
Definition and Characteristic Properties

According to the standard reference book by Lines and Glass [19], an antiferroelectric is a phase obtained by condensation of a nonpolar lattice mode that "exhibits large dielectric anomalies near the transition temperature and that can be transformed to an induced ferroelectric phase by application of an electric field. " A similar definition, emphasizing that the concept of antiferroelectricity is "based not only on the crystal structure but also on the dielectric behavior of the crystal," appears in a recent overview article in Landolt Bornstein [20]. There is a close analogy to the definition for a ferroelectric as a phase obtained by condensation of a polar lattice mode, which can be switched between two or more symmetry-related variants by an applied electric field. These definitions of both ferroelectrics and antiferroelectrics are often extended to include the possibility that the transition temperature is above the decomposition temperature of the material, or that the critical electric field is above (but not too far above) the breakdown field of the material. The definition clearly distinguishes antiferroelectrics from the much larger group of materials with centrosymmetric structures, which can be described by unit cells with oppositely directed dipoles generated by ionic displacements from a higher symmetry reference structure, but without any distinctive behavior in applied electric field.

Other formulations of the definition of antiferroelectricity are more closely based on the analogy to antiferromagnetism, with a staggered electric polarization playing the role of the staggered magnetization. In the simplest cases, the formulation of the staggered polarization involves identification in the crystal structure of two (or more) symmetry-related sublattices of polarized ions or molecular units with opposite polarization; this is, in essence, the definition introduced by Kittel [21]. An energetic criterion was subsequently added to this structure-based definition by Shirane [10, 11], yielding a definition of an antiferroelectric [22] as an antipolar crystal whose free energy is comparable to that of the polar crystal obtained by aligning the sublattice polarizations. This extends the analogy to antiferromagnetism, with a transition to a ferroelectric phase being induced by an applied electric field.

The importance of both structural and energetic aspects of the definition of antiferroelectrics can be made clear using a Landau energy functional formulation. Using the concept of a well-defined sublattice polarization as a starting point, Kittel [21] established some characteristic properties of antiferroelectrics by introducing the free energy expression:

$$A(P_a, P_b, T) = A_0 + f(P_a^2 + P_b^2) + g P_a P_b + h(P_a^4 + P_b^4) + j(P_a^6 + P_b^6) - (P_a + P_b)E, \tag{7.1}$$

where P_a and P_b are the polarizations of two sublattices, and E is the electric field, with $g > 0$ favoring the antipolar alignment of the two sublattices. This can be transformed into a Landau functional [23–25]

$$G_1(P_F, P_A, T) = (1/2)(f + g/2)P_F^2 + (1/2)(f - g/2)P_A^2$$
$$+ (h/8)(P_F^4 + 6P_A^2 P_F^2 + P_A^4) + \mathcal{O}(P^6) - P_F E \tag{7.2}$$

where $P_F = P_a + P_b$ is the macroscopic polarization, and $P_A = P_a - P_b$ is the staggered polarization. Temperature dependence is introduced into the quadratic coefficients as $f(T) = g/2 + \lambda(T - T_c)$, so that the quadratic coefficient of P_A vanishes at T_c, and $\lambda > 0$ corresponds to a low-temperature antiferroelectric phase. The temperature dependences of the quadratic terms in P_F and P_A are linked, as are the quartic coefficients, by the original two-sublattice model. Depending on the choice of parameters, the transition can be first or second order. For a second-order transition, $h > 0$, and the sixth-order terms can be neglected. There is a phase transition from the high-temperature phase to the antiferroelectric phase at T_c, below which the spontaneous polarization of the sublattices and thus P_A become nonzero. The temperature dependence of the dielectric constant is such that it is a maximum at the Curie temperature and changes slope slightly. For a first-order transition, $h < 0$ and $j > 0$, and there will be a discontinuity in the sublattice polarization and the dielectric constant at the paraelectric–antiferroelectric transition.

Using this function, the analysis of the behavior in applied electric field is straightforward [23–25]. For values of the parameters giving antiferroelectric states in zero field, double hysteresis loops are obtained, corresponding to the loss of local stability of the higher energy state. This assumption yields a hysteresis loop that will, in general, be wider than would be expected in a real system, where more complex switching mechanisms may allow the system to overcome the barrier to the lower free-energy phase before the loss of local stability of the higher energy state. Antiferroelectrics also exhibit an electrocaloric effect, with a change in temperature with applied field under adiabatic conditions [14], which can be described with a thermodynamic analysis.

In addition to characteristic electrical properties, antiferroelectrics exhibit characteristic properties related to coupling of the sublattice polarizations to strain; polarization–strain coupling is significant in many perovskite ferroelectrics [26]. Most nonpolar space groups forbid piezoelectricity (one exception being $P4\overline{3}m$) and therefore most antiferroelectrics are not piezoelectric. Symmetry does not forbid electrostriction. A change in the temperature dependence of the lattice parameters, with an accompanying effect on dielectric properties, is generally observed at the transition from the antiferroelectric to the paraelectric phase. There is also generally a lattice parameter change at the electric-field-induced antiferroelectric–ferroelectric transition; in $PbZrO_3$, this takes the form of a large volume expansion [27]. In this case, the transition back to the antiferroelectric state can also be driven by compressive stress at fields close to the critical field [28, 29]. This corresponds to the observation that an antiferroelectric phase is generally stabilized under pressure [30, 31].

An important refinement of the model involves introducing the coupling of the antiferroelectric and ferroelectric order parameters to strain [32]. The addition of strain terms is generally at lowest order. A simple model incorporating hydrostatic pressure and volume change was introduced in Ref. [33] via a transformation from a Kittel model generalized to include strain:

$$G_1 = \frac{1}{2}\alpha(P_F^2 + P_A^2) + \frac{\beta}{8}(P_F^4 + 6P_A^2 P_F^2 + P_A^4) + \frac{\gamma}{24}(P_F^6 + 15P_A^4 P_F^2 + 15P_A^2 P_F^4 + P_A^6)$$
$$+ \frac{1}{2}\eta(P_F^2 - P_A^2) - \frac{1}{2}\chi_T p^2 + Qp[P_F^2 + P_A^2 + \Omega(P_F^2 - P_A^2)] \tag{7.3}$$

where p is the hydrostatic pressure, and Q and Ω are electrostrictive coefficients. Choice of $\beta < 0$, $\gamma > 0$, and $\eta > 0$ gives a first-order antiferroelectric transition. α is the only temperature-dependent parameter, with an assumed linear dependence $\frac{d\alpha}{dT} = \frac{1}{\varepsilon_0 C}$, where ε_0 is the permittivity of free space, and C is the Curie–Weiss constant. Analysis of the model shows that the antiferroelectric transition temperature $T_N(p)$ depends on pressure, according to $T_N(p) = T_N - 2Q\varepsilon_0 C(1 - \Omega)p$. If $\Omega < 0$, there is the possibility of an antiferroelectric-to-ferroelectric phase transition with increasing pressure at a critical pressure $(\eta/2Q)|\Omega|$; conversely, if $\Omega > 0$, a material can be driven from a ferroelectric to an antiferroelectric phase with applied pressure. In the latter case, a poled ferroelectric ceramic releases all polarization charges and can supply very high instantaneous current [34]. Finally, the volume change and entropy change at the antiferroelectric–paraelectric transition are related to the sublattice polarization P_A by $\Delta V/V = Q(1 - \Omega)P_A^2$ and $\Delta S = (1/2\varepsilon_0 C)P_A^2$. By fitting to experimental data, the Q value for $PbZrO_3$ ($2.03 \times 10^{-2} m^4 C^{-2}$) was found to be similar to that of perovskite ferroelectrics such as $BaTiO_3$ and $PbTiO_3$, while that for $Pb(Mg_{1/2}W_{1/2})O_3$ is several times larger; the difference was associated with the degree of cation disorder.

Models are valuable for a unified description of the properties of antiferroelectrics. They are also useful for the study of individual materials, such as $PbZrO_3$, for which a detailed phenomenological model has been obtained [35], and can be used as well to predict the behavior of epitaxially strained thin films and thin layers in superlattices.

7.3
Microscopic Origins of Macroscopic Behavior

The question now becomes that of identifying antiferroelectric materials and relating their macroscopic behavior to microscopic aspects of the crystal structure and energetics. To make this macroscopic–microscopic connection, we begin by considering simple microscopic models in which the degrees of freedom are reorientable localized electric dipoles on a bipartite lattice, so that the lattice sites can be divided into two sublattices, with each site being neighbored by sites in the other sublattice. Early theoretical studies [21, 36, 37] took the interactions between dipoles to be given simply by the electrostatic dipole–dipole interaction. However, it has been well established in first-principles studies of ferroelectricity that short-range interactions deviate strongly from this asymptotic form, and that the form of the short-range interactions dominates the phase energetics [38, 39]. Furthermore, a single nearest-neighbor interaction favoring antialignment is not enough to give the characteristic double hysteresis loop with a jump in polarization from the antiferroelectric state to the field-induced ferroelectric state (S.E. Reyes-Lillo and K.M. Rabe, unpublished.). Next-nearest-neighbor intrasublattice interactions to stabilize sublattice polarization formation have been shown to produce a double hysteresis loop in an Ising model (fixed-length dipole with two opposite orientations) [40]. This model yields the correct stabilization of an antiferroelectric phase and a ferroelectric phase with similar energy because a substantial energy is associated with the formation of the sublattice polarization, with a smaller energy associated with the difference between alignment and antialignment of the polarizations of the two sublattices.

To connect this genre of models directly to real antiferroelectric materials requires the identification of localized reorientable dipoles occupying well-defined sites in the crystal, in analogy to the case of antiferromagnets. However, isolation of dipoles on two (or more) sublattices is much less common in the case of antiferroelectrics than in antiferromagnets, in which many systems have easily identified localized spins arising from open d or f shells in constituent atoms. Indeed, this has been the subject of much attention in the investigation of ferroelectric materials, where most are not describable as a lattice of separable reorientable dipoles or polarizable entities, and definitions of the polarization incorporate not only ion positions but also the quantum mechanical wave functions of the electrons [41, 42]. Similarly, in the present case, a definition of antiferroelectrics that required localized dipoles would be too restrictive, including only hydrogen-bonded antiferroelectrics (in which the dipole is created by the two positions for a hydrogen in the bond), crystals incorporating small reorientable dipolar molecules [36], and, if the requirement for crystallinity is loosened, antiferroelectric liquid crystals, where the dipole is associated with a single molecule or molecular subunit.

Another approach to making the macroscopic–microscopic connection is to follow the analogy to the soft mode theory of ferroelectricity. The first step is to identify the lattice modes of definite symmetry in the structure of the antiferroelectric phase relative to a high-symmetry reference structure (generally the paraelectric phase), and then to expand the energy in symmetry invariants to obtain the energy in the form of a Landau functional [43]. A simple Kittel-type antiferroelectric would be characterized by a single lattice mode in which the ions involved are divided into two groups with equal and opposite displacements. This mode would be expected to have a wave vector at the zone boundary, doubling the unit cell; however, if the high-symmetry unit cell contains an even number of symmetry-equivalent ions involved in the mode, then the antiferroelectric phase need not have an enlarged unit cell [43].

In the more general case, other types of modes can generate the structure of the antiferroelectric phase. The distortion produced by the mode might involve more than two sublattices, with either collinear or noncollinear sublattice polarizations [44, 45], the only requirement being that the net polarization is zero. The mode might have its wave vector in the interior of the Brillouin zone, and it could be commensurate or incommensurate. Moreover, as we will see later in the discussion of individual ferroelectric materials, the antiferroelectric phase can be described by a group of coupled modes [46], with a primary unstable antipolar mode accompanied by other distortions such as oxygen octahedron rotations to which coupling is symmetry-allowed, lowering the energy of the phase.

The other key ingredient for an antiferroelectric is the presence of a low-energy alternative ferroelectric phase, which we presume to be obtained as a distortion of the same high-symmetry reference structure. The latter condition promotes a small energy difference and ease of transformation between the two phases. For the simple Kittel-type antiferroelectric, this is described by including in the model of the single polar mode, which is obtained from the antipolar mode by reversing the polarization of one of the two sublattices. In the more general case, the ferroelectric phase is generated by a more general zone-center polar mode and may be accompanied by other modes that are observed to be present in the low-energy alternative ferroelectric phase. The relation of this polar mode to the primary unstable antipolar mode might be more complex than simple alignment of antipolar ionic displacements. In three dimensions, there is also the possibility that the direction of the polarization of the ferroelectric phase could be different from the direction of the sublattice polarizations of the antiferroelectric phase and could depend on the direction and magnitude of the applied electric field; so the zone-center polar mode included should be more than one dimensional.

Levanyuk and Sannikov [47] have pointed out that the form of the Landau functional obtained for antiferroelectrics from a Kittel-type model is not specific to an antipolar phase with well-defined sublattices but describes any system with a nonpolar structural transition. Specifically, any system with a nonpolar structural transition, described by order parameter η, will have a generic $\eta^2 P^2$ coupling that will give the double hysteresis loops and dielectric anomalies of the Kittel model, which we consider to be the defining macroscopic properties of an antiferroelectric, and thus, according to Levanyuk, "the concept of an antiferroelectric state turns out to be superfluous."

However, this statement is too strong: it should not be inferred from this analysis that any system with a nonpolar structural transition should be considered an antiferroelectric. Particular values of the parameters are required to get a structural transition with an alternative ferroelectric phase at low energy, and thus to produce the characteristic behavior in accessible applied electric fields. The Kittel two-sublattice model, when mapped to the Landau form, leads to relations between the coupling coefficients that satisfy these requirements, although this is a sufficient but not necessary condition for obtaining parameters that give antiferroelectric behavior. So, we propose here to recognize that the "antipolar" character of materials to be regarded as antiferroelectrics in fact varies on a continuum. Classic Kittel two-sublattice systems lie at one extreme. Close to this extreme is the subclass of antiferroelectrics in which the antipolar and polar instabilities are clearly related: the eigenvectors transform directly one into the other with change in wave vector, or belong to the same isolated unstable phonon branch in the high-symmetry reference structure, as computed from first principles. At the other extreme are cases where the relation between the nonpolar and the ferroelectric structures is weakened to the level that the two structures are merely distortions of the same high-symmetry reference structure with a small energy difference separating the ferroelectric phase from the nonpolar phase. As there is no clear demarcation point between the two extremes, we include all these as antiferroelectrics; this is further justified as the materials at this opposite end of the spectrum still exhibit the macroscopic properties considered characteristic of antiferroelectrics.

As an application of this formulation of the definition of antiferroelectricity, we consider orthorhombic *Pnma* GdFeO$_3$-type perovskites. This structure is generated from the cubic ideal perovskite structure by an M-point oxygen octahedron rotation around [001] combined with an R-point oxygen octahedron rotation around [110]. In this space group, additional lattice modes are symmetry allowed, including a mode at the X-point $\frac{\pi}{a}$(001) that involves antipolar displacements of the A cations along the orthorhombic *a* direction (approximately along the primitive perovskite [110] direction); cations in the same *xy*-plane move together and the displacements alternate from plane to plane. However, this antipolar distortion alone is not sufficient to establish the material as antiferroelectric. In the general case, these displacements are induced by the primary oxygen octahedron rotation instabilities and cannot be aligned by an applied field, so that there is no related ferroelectric phase at low energy. A *Pnma* perovskite can be antiferroelectric in the Kittel limit if the antipolar X-mode belongs to a unstable phonon branch that includes a related polar instability and is in fact the primary instability, with the oxygen octahedron rotations resulting in a relatively small further lowering of the energy. First-principles calculations of the phonon dispersion can assist in identifying a material of this type, with BiCrO$_3$, to be discussed in Section 7.4, as a possible example [48]. With the broader definition of antiferroelectricity that we are using here, the high-symmetry reference structure needs only have a zone-center instability that produces a competing ferroelectric phase at a sufficiently low energy that an electric field can drive a first-order transition to this phase. In this sense, $(Bi,Sm)FeO_3$ satisfies this condition near the rhomobohedral–orthorhombic compositional phase boundary, with a first-order electric-field-induced phase transition from the orthorhombic *Pnma* phase to the polar *R3c* phase [49].

Here, we have seen that reexamination of the definition and the generally accepted properties of antiferroelectrics leads to a rather inclusive definition of antiferroelectricity. The definition hinges not on a structural criterion, as almost any nonpolar distortion of a high-symmetry reference structure could satisfy the requirements, but on energetic criteria. The essential point is that the candidate antiferroelectric phase be close in energy to a competing ferroelectric phase, to which it can be driven by an applied electric field through a first-order transition. This criterion can be evaluated in several ways. If the candidate antiferroelectric phase is generated by a single antipolar mode with a clear relation to a zone-center polar mode, it can be inferred that the zone-center polar mode will also be unstable and generate the necessary competing ferroelectric phase. The presence of a low-energy competing ferroelectric can also be inferred from proximity, across a first-order phase boundary, of the candidate antiferroelectric phase to the ferroelectric phases in the generalized phase diagram for the system, which includes control parameters such as pressure, epitaxial strain, and compositional substitution. First-principles calculations can be used to establish the relation between antipolar and zone-center polar instabilities and the low energy of a competing ferroelectric phase. In the end, direct experimental observation of an electric-field-induced transition to a ferroelectric phase is the most definitive indication of antiferroelectricity. In general, even if the phase satisfies this energetic criterion even for some range of control parameters, the entire phase will be referred to as antiferroelectric; this corresponds to the analogous case for ferroelectrics, in which a phase is referred to as ferroelectric even if it is not switchable by applied fields throughout its region of stability in the phase diagram.

7.4
Antiferroelectric Materials: Structure and Properties

In this section, we consider the structure and properties of individual antiferroelectric materials as determined from experimental observation and theoretical analysis. The main focus is on $PbZrO_3$ and related systems; two excellent recent reviews [16, 18] provide additional details and references. Other well-established antiferroelectric materials include certain niobate perovskites, vanadates, and complex perovskite oxides; in each of these cases, we consider the extent to which the system satisfies the structural and energetic criteria for antiferroelectricity. The possibility of antiferroelectricity has also been raised for a number of additional systems, including sulfide perovskites and spinels, and the evidence relating to these systems will be discussed.

$PbZrO_3$ is, by far, the most thoroughly studied antiferroelectric oxide. Reference [16] provides a recent detailed review; here, we highlight the most relevant aspects of the structure and properties. At high temperatures, $PbZrO_3$ has a paraelectric cubic ideal perovskite structure, and at temperatures below $T_c = 505$ K, the structure is an antiferroelectric orthorhombic distorted perovskite structure [10]. In a very narrow temperature range separating these two phases, an intermediate ferroelectric rhombohedral phase with polarization along [111] has been reported in some studies [50].

The details of the structure of the antiferroelectric phase were controversial for some time, with even the space group assignment differing between nonpolar orthorhombic *Pbam* and polar orthorhombic *Pba2*. The main features of the structure can be described in terms of modes of the primitive perovskite structure. In addition to the Σ_2 mode [51] at $q = \frac{2\pi}{a}(1/4,1/4,0)$, producing an antipolar arrangement with shifts of the lead ions along [110], as shown in Figure 7.3, the other dominant distortion in the structure is an oxygen octahedron rotation mode R_5^- with rotation around [110]. Symmetry analysis shows that these two modes generate the space group *Pbam* with a $\sqrt{2}a_0 \times 2\sqrt{2}a_0 \times 2a_0$ unit cell (eight formula units), as shown in Figure 7.4. In this structure, additional modes R_4^-, M_5^-, and X_1^+ and a mode with $q = \frac{2\pi}{a}(1/4,1/4,1/2)$ are allowed without breaking further symmetry. These are found to be relatively small, and indeed are set to zero in some structure determinations to reduce the number of free parameters in the refinement (see, for example, Ref. [53]). It has been suggested that the structure also includes a small zone-center polar distortion, putting it into the *Pba2* space group [54]. First-principles calculations [55–57] played a critical role in resolving this ambiguity in favor of the nonpolar *Pbam* structure. These calculations showed

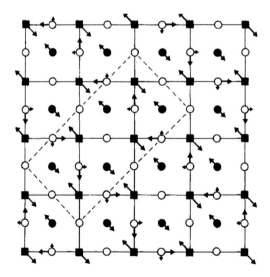

Figure 7.3 Projection of atomic displacements associated with the Σ_2 mode at $q = \frac{2\pi}{a}(1/4,1/4,0)$ onto the *ab*-plane. Squares and circles indicate Pb and oxygen atoms, respectively. Filled and open circles show atoms on the Pb atomic layer and atoms on the Zr atomic layer, respectively. (Source: From Ref. [52].)

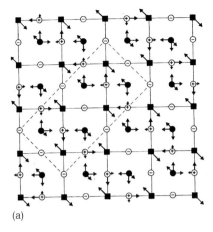

(a)

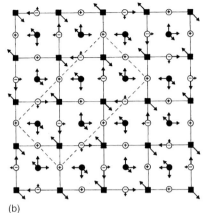

(b)

Figure 7.4 Projection of atomic displacements associated with the Σ_2 and R_5^- modes onto the ab-plane. Squares and circles indicate Pb and oxygen atoms, respectively. Filled and open circles show atoms on the Pb atomic layer and atoms on the Zr atomic layer, respectively. (a) Atoms in the planes $z = 0$ and $z = 1/2c_p$ (open symbols) and (b) atoms in the planes $z = c_p$ and $z = 3/2c_p$ (open symbols), where c_p is the pseudocubic lattice parameter in the z-direction. (Source: From Ref. [52].)

that optimized positions of the O atoms in the *Pbam* structure were in good agreement with available experiments, including subsequent determinations [53]. In the initial study [55], the energy of the antiferroelectric *Pbam* structure was found to be slightly lower (20 meV fu^{-1}) than that of the optimized rhombohedral *R3m* ferroelectric structure; in Ref. [57], this comparison was extended to the *Pbam* structure with full optimization of oxygen positions and the rhombohedral *R3c* ferroelectric structure observed to be stable with 7% Ti substitution [58], with the antiferroelectric structure found to be 31 meV fu^{-1} lower. This small energy difference is a key signature of an antiferroelectric material and is also consistent with the transformation to the rhombohedral FE phase at small Ti doping.

As shown in Figure 7.2, double hysteresis loops were observed in the earliest work on PbZrO$_3$, identifying it as an antiferroelectric [10]. The critical field is rather high and can be lowered by doping as discussed in the following paragraphs. The change in volume at the field-induced transition is relatively large, which is favorable for strain-related properties but may reduce the reversibility of the transition. A more detailed discussion of the field-induced ferroelectric phase can be found in Ref. [59].

Given that the antiferroelectric phase of PbZrO$_3$ is not described by two oppositely polarized sublattices, and the additional complication of the intermediate ferroelectric phase, it should be asked to what degree PbZrO$_3$ behaves according to the models described earlier. Analysis of the amplitudes of Pb displacements, oxygen octahedron rotations and lattice strains as a function of temperature are reported to be well described by a simple Landau theory for a first-order transition out of the paraelectric phase [60]. This theory was developed in more detail in

Ref. [35] to describe the dielectric response and elastic properties. In these theories, the antiferroelectric distortion is described by an order parameter that does not specifically require a two-sublattice character for the distortion.

The compositional substitution phase diagrams of $PbZrO_3$ provide further evidence of its antiferroelectric character. The fact that, in many cases, a small fraction of compositional substitution induces a first-order transition to a ferroelectric phase confirms that there is an alternative ferroelectric phase at low energy in pure $PbZrO_3$. By lowering the energy difference in favor of the ferroelectric phase, these substitutions also lower the critical field for the electric-field-induced transition [61]. Some examples include $Pb_{1-x}Ba_xZrO_3$ [62] in which Ba substitution reduces the relative free energy of the ferroelectric phase, thus lowering the critical field and widening the range of stability of the intermediate ferroelectric phase. This system illustrates the challenge that compositional tuning to improve one materials property (here, the critical field) may worsen others: in Ref. [63], it is shown that $Pb_{1-x}Ba_xZrO_3$ exhibits temperature irreversibility (i.e., the antiferroelectric phase is not recovered from the ferroelectric phase on cooling) and other references with similar observations are discussed. The tuning of critical electric field, structural parameters, and associated properties such as reversibility of the transitions has prompted extensive experimentation with complex substitutions, such as $Pb_{0.89}Nb_{0.02}[(Zr_{0.57}Sn_{0.43})_{0.94}Ti_{0.06}]_{0.98}O_3$ (also referred to as PNZST43/100y/2) [27, 64–67].

A study of compositional substitution phase diagrams also suggests that there are low-energy alternative antiferroelectric phases of distinct structure as well; this was noted already in the early work by Shirane [11]. Substitution of Pb by Sr is shown to result in an antiferroelectric intermediate phase separating the high-temperature paraelectric phase from the low-temperature $PbZrO_3$-like antiferroelectric phase [11]. In $PbZr_{1-x}Sn_xO_3$, a distinct antiferroelectric $P222_1$ phase is observed for $x > 0.05$; only for $x > 0.3$ does an $R3m$ ferroelectric phase become stable [68].

$PbHfO_3$ was first reported as antiferroelectric soon after the identification of antiferroelectricity in $PbZrO_3$ [69]. Despite their chemical similarity, the phase sequence is different from that of $PbZrO_3$ [70–72]: phase I (up to 163° C) is tetragonal with $c/a < 1$, phase II (163-215° C) is tetragonal with a larger c/a (still <1), and phase III (above 215° C) is cubic. Phases I and II have both been characterized as antiferroelectric. Phase I appears to be isomorphic to the orthorhombic AFE structure of $PbZrO_3$. Less is known about phase II. It appears to have a very large unit cell [71] and to be nonpolar, consistent with Raman scattering observations [73], and it has been suggested that it should be assigned to one of the orthorhombic space groups $P222_1$, $Pmm2$, $Pmmm$, or $P222$ [74, 75].

$NaNbO_3$ and $AgNbO_3$ are perovskite systems that both show antiferroelectric phases as part of a complex sequence of phases with temperature [76–79]. The room temperature antiferroelectric phases in the two materials are isostructural in orthorhombic space group $Pbcm$ with eight formula units per $\sqrt{2}a_0 \times \sqrt{2}a_0 \times 4a_0$ unit cell. The structure is characterized by oxygen octahedron rotations as well as a $\vec{q}=(0,0,\pi/2)$ mode, involving antipolar Nb displacements, that is responsible for the designation as an antiferroelectric [79]. In $NaNbO_3$, there is a transition to a low-temperature ferroelectric $R3c$ phase at 173 K and a sequence of phase transitions starting at 633 K and ending with a transition to a cubic paraelectric phase at 913 K [80]. In $AgNbO_3$, the room temperature antiferroelectric phase appears stable down to low temperature, and there is a different transition sequence with increasing temperature: two transitions have been tentatively identified at 340 and 540 K but without any apparent change in symmetry, and starting at 626 K, there is a sequence of phase transitions through the orthorhombic and tetragonal phases that end with a transition to a cubic paraelectric phase at 852 K [80]. First-principles calculations have focused mainly on the zone-center polar modes and oxygen octahedron rotations [81–86]. The lattice instabilities of the cubic perovskite structure are identified for $AgNbO_3$ and $NaNbO_3$. Both show unstable modes for oxygen octahedron rotations at R and M, as well as an unstable ferroelectric mode. This can produce a set of competing low-energy states consistent with the observed sequence of phases. The electric-field-induced transition to a ferroelectric phase in $NaNbO_3$ has been well studied [87–89] and occurs at high field perpendicular to the orthorhombic c-axis. The dependence of the transition on the direction of the field has been studied in detail; a clear distinction can be drawn between the transitions in which the polarization of the ferroelectric phase is along essentially the same direction as the sublattice polarizations of the antiferroelectric phase, and those in which it is along a different direction [89]. In addition, a small amount of K substitution for Na produces a ferroelectric phase at room temperature [90].

While, as discussed earlier, the orthorhombic *Pnma* perovskites are generally not antiferroelectric, there can be some exceptions to this rule. The clearest would be in cases where the local-dipolar distortions are driven by the A-site Pb or Bi displacements. The first-principles phonon dispersion for $BiCrO_3$ shows a clear antipolar instability at the X-point related to a polar instability at Γ, and characteristic antiferroelectric hysteresis loops have been observed in films [91]. However, it should be noted that the measurements were made for a lower symmetry phase, not for the *Pnma* phase. $BiFeO_3$ is a ferroelectric but, with RE doping on the Bi site, undergoes a transition to an orthorhombic *Pnma* phase. In this phase near the critical doping, double hysteresis loops are observed with the electric field inducing a transition back to the polar $R3c$ ferroelectric phase [49]. With our inclusive definition of antiferroelectricity, the doped system (and even pure $SmFeO_3$, as it is the same phase) can be considered antiferroelectric. Other nonpolar Bi perovskites, such as $BiMnO_3$, might also be antiferroelectric, given proximity to a ferroelectric phase; indeed $BiMnO_3$ was thought for a while to be ferroelectric before additional experiments and first-principles calculations established that the ground-state structure is nonpolar [92, 93]. Antiferroelectricity has also been discussed for $Sr_{1-x}Ca_xTiO_3$ [94].

Antiferroelectric phases have also been identified in double perovskites. The degree of cation disorder is a key feature in characterizing double perovskite systems. Cation disorder generally yields relaxor behavior, while a number of well-ordered systems show phases isostructural to $PbZrO_3$, which are thus characterized as antiferroelectric. A theoretical analysis is given in Ref. [95]. One example is $Pb(In_{1/2}Nb_{1/2})O_3$ [96–98] which is seen to have an antiferroelectric phase isostructural with $PbZrO_3$. Study of the $Pb(Nb_{1/2}B_{1/2})O_3$ series [99] shows a transition from ferroelectric (B = Cr, Fe, Mn, Sc, In) to nonpolar (B = Lu, Yb, Tm) low-temperature phases depending on the B ionic radius; the case of B = In is at the crossover and thus the close competition between ferroelectric and nonpolar phases characteristic of antiferroelectricity is expected. A phase sequence from paraelectric to antiferroelectric to low-temperature ferroelectric has been identified for a number of lead-based double perovskite systems, including $Pb(Yb_{1/2}Nb_{1/2})O_3$ [100], PYN-PT [101], $Pb(Yb_{1/2}Ta_{1/2})O_3$ [102], $Pb(Co_{1/2}W_{1/2})O_3$ [103], and $Pb(Sc_{1/2}Ta_{1/2})O_3$ [103, 104]. No detailed structural determination has been made for the antiferroelectric phases. Additional observations of antiferroelectric phases have been reported for non-lead perovskites, including $Na(Bi_{1/2}Ti_{1/2})O_3$ [105], $Bi(Mg_{1/2}Ti_{1/2})O_3$ (BMT) [106], and BNT-BT [107]. More information and references about these systems can be found in Ref. [18].

Antiferroelectricity has been noted in a number of other compounds, including $NaVO_3$ (mentioned in Ref. [108]) and WO_3 [109]. $BiNbO_4$ is found to be paraelectric above 570° C, ferroelectric between 360 and 570° C, and nonpolar below 360° C [110]. The latter phase can be presumed to be antiferroelectric owing to its proximity to the intermediate ferroelectric phase. A dielectric anomaly observed at the Jahn-Teller transition of $DyVO_4$ (T_c = 15.2 K) was attributed to a strain-induced staggered polarization on two sublattices within the zircon unit cell [111]. The electric field needed to produce alignment of the two sublattices was estimated from a model analysis to be two orders of magnitude greater than that for $PbZrO_3$. To the extent that this meets the energetic criterion for antiferroelectricity, $DyVO_4$ would be a rare example of an antiferroelectric in which the unit cell does not change in the paraelectric–antiferroelectric transition. Finally, some exotic examples include early reports of antiferroelectricity in spinels [112], betaine phosphate [113], hydroxyapatite, and thiourea, and, more recently, in hybrid organic–inorganic metal-organic-framework (MOF) structures [114].

7.5
Relation to Alternative Ferroelectric Phases

As previously noted, none of the known antiferroelectric compounds conform to the collinear two-sublattice Kittel model. Therefore, in all known compounds, the electric-field-induced ferroelectric phase is not obtained from the antiferroelectric structure simply by flipping one of the sublattice polarizations. Even in a hypothetical Kittel-type compound, application of an electric field noncollinear with the sublattice polarization could lead to a nontrivial induced phase. The relation between the structure of the antiferroelectric phase and the structure of the alternative ferroelectric phases induced by electric fields or by changes in control parameters such as compositional substitution and epitaxial strain is therefore of considerable interest.

First-principles studies of the electric-field-induced transitions for noncollinear fields (polarization rotation) have proved illuminating for ferroelectrics [115–117]. Modeling of electric-field-induced transitions in antiferroelectrics includes studies of $PbZrO_3$ and $NaNbO_3$ [89].

7.6
Antiferroelectricity in Thin Films

The behavior of materials in a thin film form is modified by strain, size effects, and structural and electronic relaxation and reconstruction at the surfaces and interfaces. These factors can strongly affect the energy difference between the antipolar and polar phases. In particular, an antiferroelectric material, in which this energy difference is small, can, in some cases, be tuned to the phase boundary with the related ferroelectric phase. Similarly, a bulk ferroelectric material with a low-energy alternative antipolar phase (which may not be manifest in observations of the bulk) could be driven antiferroelectric in a thin film form; in both cases, the depolarization field associated with any uncompensated polarization along the normal would tend to suppress the formation of a single-domain ferroelectric film. This tuning needs to be taken into consideration in the application of these materials in thin film devices: it can be beneficial as well as detrimental that the material in thin film form behaves differently from the material in bulk.

Thin films of $PbZrO_3$ have received the most attention. The electric-field-induced transition in a film on a Pt-coated Si substrate was reported to occur at 140 kV cm^{-1} for increasing field and 71 kV cm^{-1} for decreasing field, with a maximum polarization of 40 $\mu C\,cm^{-2}$ [118]. In another study, thin films of $PbZrO_3$ were grown on crystalline Si(100) with thicknesses ranging from 100 to 900 nm. At thicknesses below 500 nm, ferroelectric hysteresis loops were observed with spontaneous polarization of 31 $\mu C\,cm^{-2}$ and coercive field of 100 kV cm^{-1} at thickness of 260 nm; similar behavior was observed for thin films of antiferroelectric $BiNbO_4$ [119]. In this work, the appearance of ferroelectricity was attributed to the electric field produced at the interface, rather than to strain or size effects, because an earlier experiment reported double hysteresis loops in a 380 nm thick film of $PbZrO_3$ on a Pt-coated Si substrate [120]. In contrast, in $PbZrO_3$ grown on $SrTiO_3$ with an $SrRuO_3$ buffer layer, corresponding to over 5% compressive strain, 8 nm thick films showed a ferroelectric rhombohedral phase with a high density of misfit dislocations, while thicker films showed the rhombohedral phase for the first 8 nm and then a bulklike orthorhombic phase; this was attributed to the higher strain near the interface [17].

Superlattices of antiferroelectrics with ferroelectrics can show rich behavior owing to the competition between the two phases. In single-unit-cell layers, it can be argued that the antiferroelectric phase cannot develop fully, and alternative structures are stabilized [121]. In thicker layers, the behavior evolves in a systematic way [122].

7.7
Properties for Applications

There has been increasing interest in the technological applications of anti-ferroelectric materials [17]. Proposed applications include sensors, actuators, energy and charge storage devices, voltage regulators, and electrooptic devices [123–128].

The relevance to energy storage applications comes from the enhancement of $\int D dE$ by the shape of the double hysteresis loop (Figure 7.5). The energy stored is roughly twice what is stored by a linear dielectric with the same polarization at a field just above the transition field. Patented devices include a cardiac defibrillator [129].

The strain change at the electric-field-induced transition has received particular attention, with proposals of strain and force generators [131] and electrostrictors with high-strain response and small temperature dependence of strain [107]. In an applied electric field, the strain difference between the antiferroelectric and ferroelectric phases can result in a large nonlinear response. Specifically, the electric-field-induced transition from the antiferroelectric to the ferroelectric phase, with a large jump in electrical polarization as seen in the double hysteresis loop, can be accompanied by a large strain change (in $PbZrO_3$, a volume expansion). Observations in $(Pb_{0.97}La_{0.02})(Sn,Ti,Zr)O_3$ (PLZT) [34] and lead-free perovskites [132] suggest the application of this behavior for large strain actuators. In PLZT, it is observed that the field-induced polarization is almost temperature independent while the field-induced strain depends strongly on temperature. An intriguing twist has been noted in Ref. [27]. In a system in which applied pressure drives a transition from the ferroelectric phase to an antiferroelectric phase, the decrease in volume of a poled ferroelectric ceramic when a reversed polarity electric field is applied can induce a transition into the antiferroelectric phase, with the counter-intuitive effect that an electric field drives a system from ferroelectric to antiferro-electric.

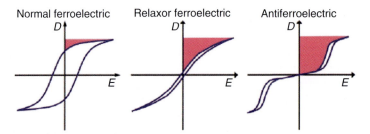

Figure 7.5 Schematic hysteresis loops for ferroelectrics, relax or ferroelectrics, and anti-ferroelectrics. The shaded area corresponds to the stored energy density. (Source: From Ref. [130].)

7.8
Prospects

This reexamination of antiferroelectricity in oxides raises many questions. Antiferroelectric materials overall have received much less attention than ferroelectrics, and only a few materials have been thoroughly studied. Much of the work has been on compositional tuning of $PbZrO_3$ and the lead-free perovskites to control essential features such as transition temperature, critical field, strain response, and reversibility of the electric-field-induced transition, leading to complex multi-component systems. On the other hand, many nonpolar phases have been given the designation antiferroelectric without verification of the energetic criterion of a distinct low-energy ferroelectric phase.

A systematic approach to design and discovery of new antiferroelectric materials is therefore indicated. For example, it is striking that there is no oxide antiferroelectric that conforms to the simple two-sublattice Kittel model. Also, it would be a curiosity to find a definitive example of an antiferroelectric driven by a $q = 0$ lattice mode. Although the perovskite structure could in principle support such a phase, driven by the zone-center silent mode in which two oxygens move in equal and opposite directions, no compound in this structure is reported in ICSD. A more promising route to new antiferroelectrics is to follow the paradigm set by the canonical antiferroelectric oxide $PbZrO_3$. As we have seen earlier, the antipolar structure of $PbZrO_3$ is generated by a non-zone-boundary phonon, with auxiliary oxygen octahedron rotation distortions stabilizing the antiferroelectric phase; a number of perovskite systems have been shown to have this same structure. It is possible that the reason this particular structure appears so prevalent is that, given the familiarity of $PbZrO_3$, it is easy to recognize. In fact, it would seem that there should be other, perhaps many, choices for antipolar phonons that could yield low-energy structures when coupled to other modes. First-principles calculations of full phonon dispersion of high-symmetry reference structures, including but not limited to perovskites, can be very valuable in identifying such instabilities and structures. Another promising approach, on the basis of combination of high-throughput first-principles calculations with crystallographic database mining, is described in Ref. [133].

The functional properties of antiferroelectrics arise for the most part from the changes in properties of the material at the electric-field-induced transition, the most well-studied examples being the electric polarization, the strain, and optical properties. As more classes of antiferroelectric materials are developed, there are opportunities to improve functional behavior by targeting particular properties for contrast in the two phases. For example, the two phases could have different magnetic ordering, leading to spin-lattice effects and magnetoelectric properties that are the subject of great current interest multiferroics. It has already been demonstrated that antiferroelectric lattice instabilities can be favored by magnetic ordering, specifically, in perovskite $LuCrO_3$, in which the silent mode becomes unstable for the higher energy ferromagnetic ordering [134]. Yet other functionalities could be accessed by consideration of ferrielectrics, in which the sublattice polarizations of the "antiferroelectric" phase do not completely cancel, resulting in a net spontaneous polarization. These would show the characteristic behavior of antiferroelectrics at the electric-field-induced transition, but also could have a switchable polarization at lower fields [135].

In conclusion, this brief review and reexamination of the fundamental physics of antiferroelectricity suggests that these are a rich and interesting class of functional materials whose potential has yet to be fully tapped. Further study, especially in materials design, synthesis, and characterization, should lead to significant progress, with the possibility of a variety of novel device technologies.

Acknowledgments

I acknowledge the support of ONR grant N00014-09-1-0302, MURI-ARO grant W911-NF-07-1-0410, and NSF MRSEC DMR-0820404. Part of the preparation of this chapter was carried out at the Aspen Center for Physics. I thank J. W. Bennett, C. J. Fennie, D. R. Hamann, K. Garrity, S. E. Reyes-Lillo, and D. Vanderbilt for useful discussions.

References

1. Fennie, C.J. and Rabe, K.M. (2006) Magnetic and electric phase control in epitaxial $EuTiO_3$ from first principles. *Phys. Rev. Lett.*, **97**, 267602.

2. Fennie, C.J. (2008) Ferroelectrically-induced weak ferromagnetism by design. *Phys. Rev. Lett.*, **100**, 167203.

3. Benedek, N.A. and Fennie, C.J. (2011) Hybrid improper ferroelectricity: a mechanism for controllable polarization-magnetization coupling. *Phys. Rev. Lett.*, **106**, 107204.

4. Schlom, D.G., Chen, L.Q., Eom, C.B., Rabe, K.M., Streiffer, S.K., and Triscone, J.M. (2007) Strain tuning of ferroelectric thin films. *Annu. Rev. Mater. Res.*, **37**, 589.

5. Armiento, R., Kozinsky, B., Fornari, M., and Ceder, G. (2011) Screening for high-performance piezoelectrics using high-throughput density functional theory. *Phys. Rev. B*, **84**, 014103.

6. Zhang, X., Yu, L., Zakutayev, A., and Zunger, A. (2012) Sorting stable versus unstable hypothetical compounds: the case of multi-functional ABX half-Heusler filled tetrahedral structures. *Adv. Funct. Mater.*, **22**, 1425.

7. Roy, A., Bennett, J.W., Rabe, K.M., and Vanderbilt, D. (2012) Half-Heusler semiconductors as piezoelectrics. *Phys. Rev. Lett.*, **109**, 037602.

8. Bennett, J.W., Garrity, K., Rabe, K.M., and Vanderbilt, D. (2012) Hexagonal ABC compounds as semiconducting ferroelectrics. *Phys. Rev. Lett.*, **109**, 167602.

9. Spaldin, N.A. (2010) *Magnetic Materials: Fundamentals and Applications*, 2nd edn, Cambridge University Press.

10. Shirane, G., Sawaguchi, E., and Takagi, Y. (1951) Dielectric properties of lead zirconate. *Phys. Rev.*, **84**, 476.

11. Shirane, G. (1952) Ferroelectricity and antiferroelectricity in ceramic $PbZrO_3$ containing Ba or Sr. *Phys. Rev.*, **86**, 219.

12. Strukov, B.A. and Levanyuk, A.P. (1998) *Ferroelectric Phenomena in Crystals: Physical Foundations*, Springer-Verlag.

13. Uchino, K., Nomura, S., Cross, L.E., Newnham, R.E., and Jang, S.J. (1981) Electrostrictive effect in perovskites and its transducer applications. *J. Mater. Sci.*, **16**, 569.

14. Mischenko, A.S., Zhang, Q., Scott, J.F., Whatmore, R.W., and Mathur, N.D.

(2006) Giant electrocaloric effect in thin film $PbZr_{0.95}Ti_{0.05}O_3$. *Science*, **311**, 1270.

15. Parui, J. and Krupanidhi, S.B. (2008) Electrocaloric effect in antiferroelectric $PbZrO_3$ thin films. *Physica Status Solidi*, **2**, 230.

16. Liu, H. and Dkhil, B. (2011) A brief review on the model antiferroelectric $PbZrO_3$. *J. Kristallogr.*, **226**, 163.

17. Chaudhuri, A.R., Arredondo, M., Hahnel, A., Morelli, A., Becker, M., Alexe, M., and Vrejoiu, I. (2011) Epitaxial strain stabilization of a ferroelectric phase in $PbZrO_3$ thin films. *Phys. Rev. B*, **84**, 054112.

18. Tan, X., Ma, C., Fredrick, J., Beckman, S., and Webber, K.G. (2011) The antiferroelectric-ferroelectric phase transition in lead-containing and lead-free perovskite ceramics. *J. Am. Ceram. Soc.*, **94**, 4091.

19. Lines, M.E. and Glass, A.M. (1977) *Principles and Applications of Ferroelectrics and Related Materials*, Cambridge University Press.

20. Mitsui, T. (2005) Ferroelectrics and antiferroelectrics, in *Springer Handbook of Condensed Matter and Materials Data*, *Part 4*, Springer-Verlag, pp. 903–938.

21. Kittel, C. (1951) Theory of antiferroelectric crystals. *Phys. Rev.*, **82**, 729.

22. Jona, F. and Shirane, G. (1962) *Ferroelectric Crystals*, Pergamon Press.

23. Cross, L.E. (1967) Antiferroelectric-ferroelectric switching in a simple Kittel antiferroelectric. *J. Phys. Soc. Jpn.*, **23**, 77.

24. Okada, K. (1969) Phenomenological theory of antiferroelectric transition. I. Second-order transition. *J. Phys. Soc. Jpn.*, **27**, 420.

25. Okada, K. (1974) Phenomenological theory of antiferroelectric transition. III. Phase diagram and bias effects of first-order transition. *J. Phys. Soc. Jpn.*, **37**, 1226.

26. Cohen, R.E. (1992) Origin of ferroelectricity in perovskite oxides. *Nature*, **358**, 136.

27. Tan, X., Frederick, J., Ma, C., Jo, W., and Rodel, J. (2010) Can an electric field induce an antiferroelectric phase out of a ferroelectric phase?. *Phys. Rev. Lett.*, **105**, 255702.

28. Zeuch, D.H., Montgomery, S.T., and Holcomb, D.J. (2000) Uniaxial compression experiments on lead zirconate titanate 95/5-2Nb ceramic: evidence for an orientation-dependent, maximum compressive stress criterion for onset of the ferroelectric to antiferroelectric polymorphic transformation. *J. Mater. Res.*, **15**, 689.

29. Avdeev, M., Jorgensen, J.D., Short, S., Samara, G.A., Venturini, E.L., Morosin, P., and Yang, B. (2006) Pressure-induced ferroelectric to antiferroelectric phase transition in $Pb_{0.99}(Zr_{0.95}Ti_{0.05})_{0.98}Nb_{0.02}O_3$. *Phys. Rev. B*, **73**, 064105.

30. Samara, G., Sakudo, T., and Yoshimitsu, K. (1975) Important generalization concerning the role of competing forces in displacive phase transitions. *Phys. Rev. Lett.*, **35**, 1767.

31. Yang, P. and Payne, D.A. (1996) The effect of external field symmetry on the antiferroelectric-ferroelectric phase transformation. *J. Appl. Phys.*, **80**, 4001.

32. Mason, W.P. (1952) Properties of a tetragonal antiferroelectric crystal. *Phys. Rev.*, **88**, 480.

33. Uchino, K., Cross, L.E., and Newnham, R.E. (1981) Electrostrictive effects in antiferroelectric perovskites. *J. Appl. Phys.*, **52**, 1455.

34. Pan, W.Y., Dam, C.Q., Zhang, Q.M., and Cross, L.E. (1989) Large displacement transducers based on electric field forced phase transitions in the tetragonal $(Pb_{0.97}La_{0.02})(Ti,Zr,Sn)O_3$ family of ceramics. *J. Appl. Phys.*, **66**, 6014.

35. Haun, M.J., Harvin, T.J., Lanagen, M.T., Zhuang, Z.Q., Jang, S.J., and Cross, L.E. (1989) Thermodynamic theory of $PbZrO_3$. *J. Appl. Phys.*, **65**, 3173.

36. Takagi, Y. (1952) Ferroelectricity and antiferroelectricity of a crystal containing rotatable polar molecules. *Phys. Rev.*, **85**, 315.

37. Cohen, M.H. (1952) Ferroelectricity versus antiferroelectricity in barium titanate. *Phys. Rev.*, **84**, 369.

38. Zhong, W., Vanderbilt, D., and Rabe, K.M. (1995) First-principles theory of

ferroelectric phase transitions for perovskite: the case of BaTiO$_3$. *Phys. Rev. B*, **52**, 6301.

39. Waghmare, U.V. and Rabe, K.M. (1997) Ab initio statistical mechanics of the ferroelectric phase transition in PbTiO$_3$. *Phys. Rev. B*, **55**, 6161.

40. Misirlioglu, I.B., Pintilie, L., Boldyreva, K., Alexe, M., and Hesse, D. (2007) Antiferroelectric hysteresis loops with two exchange constants using the two dimensional Ising model. *Appl. Phys. Lett.*, **91**, 202905.

41. King-Smith, R.D., and Vanderbilt, D. (1993) Theory of polarization of crystalline solids,. *Phys. Rev. B*, **47**, 1651.

42. Resta, R. (1994) Macroscopic polarization in crystalline dielectrics: the geometric phase approach. *Rev. Mod. Phys.*, **66**, 899.

43. Blinc, R. and Zeks, B. (1974) *Soft Modes in Ferroelectrics and Antiferroelectrics*, North-Holland.

44. Shuvalov, L.A. and Sonin, A.S. (1961) *Sov. Phys. Crystallogr.*, **6**, 258.

45. Hatt, R.A. and Cao, W. (2000) Landau-Ginzburg model for antiferroelectric phase transitions based on microscopic symmetry. *Phys. Rev. B*, **62**, 818.

46. Waghmare, U.V. and Rabe, K.M. (1997) Lattice instabilities, anharmonicity and phase transitions in PbZrO$_3$ from first principles. *Ferroelectrics*, **194**, 135.

47. Levanyuk, A.P. and Sannikov, D.G. (1968) Anomalies in dielectric properties in phase transitions. *Sov. Phys. JETP*, **28**, 134.

48. Hill, N.A., Battig, P., and Daul, C. (2002) First principles search for multiferroism in BiCrO$_3$. *J. Phys. Chem. B*, **106**, 3383.

49. Kan, D., Palova, L., Anbusathaiah, V., Cheng, C.J., Fujino, S., Nagarajan, V., Rabe, K.M., and Takeuchi, I. (2010) Universal behavior and electric-field-induced structural transition in rare-earth-substituted BiFeO$_3$. *Adv. Funct. Mater.*, **20**, 1108.

50. Roleder, K., Kugel, G.E., Handerek, J., Fontana, M.D., Carabatos, C., Hafin, M., and Kania, A. (1988) The first evidence of two phase transitions

in PbZrO$_3$ crystals derived from simultaneous Raman and dielectric measurements. *Ferroelectrics*, **80**, 161.

51. Stokes, H.T., Kisi, E.H., Hatch, D.M., and Howard, C.J. (2002) Group theoretical analysis of octahedral tilting in ferroelectric perovskites. *Acta Crystallogr., Sect. B*, **58**, 934.

52. Fujishita, H. and Hoshino, S. (1984) A study of structural phase transitions in antiferroelectric PbZrO$_3$ by neutron diffraction. *J. Phys. Soc. Jpn.*, **53**, 226.

53. Fujishita, H. and Katano, S. (1997) Reexamination of the antiferroelectric structure of PbZrO$_3$. *J. Phys. Soc. Jpn.*, **66**, 3484.

54. Jona, F., Shirane, G., Mazzi, F., and Pepinsky, R. (1957) X-ray and neutron diffraction study of antiferroelectric lead zirconate, PbZrO$_3$. *Phys. Rev.*, **105**, 849.

55. Singh, D.J. (1995) Structure and energetic of antiferroelectric PbZrO$_3$. *Phys. Rev. B*, **52**, 12559.

56. Johannes, M.D. and Singh, D.J. (2005) Crystal structure and electric field gradients of PbZrO$_3$ from density functional calculations. *Phys. Rev. B*, **71**, 212101.

57. Kagimura, R. and Singh, D.J. (2008) First-principles investigation of elastic properties and energetics of antiferroelectric and ferroelectric phases of PbZrO$_3$. *Phys. Rev. B*, **77**, 104113.

58. Corker, D.L., Glazer, A.M., Whatmore, R.W., Stallard, A., and Fauth, F. (1998) A neutron diffraction investigation into the rhombohedral phases of the perovskite series PbZr$_{1-x}$Ti$_x$O$_3$. *J. Phys.: Condens. Matter*, **10**, 6251.

59. Leyderman, A.V., Leontev, I.N., Fesenko, O.E., and Leontev, N.G. (1998) Dipole order and stability of the ferroelectric and antiferroelectric states in lead zirconate. *Phys. Solid State*, **40**, 1204.

60. Whatmore, R.W. and Glazer, A.M. (1979) Structural phase transitions in lead zirconate. *J. Phys. C*, **12**, 1505.

61. Jankowska-Sumara, I. (2007) Calorimetric study and Landau analysis of the phase transitions in PbZr$_{1-x}$Sn$_x$O$_3$ single crystals with $0 \leq x \leq 4$. *Phys. Status solidi B*, **244**, 1887.

62. Yoon, K.H., Hwang, S.C., and Kang, D.H. (1997) Dielectric and field-induced

strain behavior of $(Pb_{1-x}Ba_x)ZrO_3$ ceramics. *J. Mater. Sci.*, **32**, 17.

63. Pokharei, B.P. and Pandey, D. (1999) Irreversibility of the antiferroelectric to ferroelectric phase transition in $Pb_{0.90}Ba_{0.10}ZrO_3$ ceramics. *J. Appl. Phys.*, **86**, 3327.

64. Viehland, D., Forst, D., Li, Z., and Xu, J.F. (1995) Incommensurately modulated polar structures in antiferroelectric Sn-modified lead zirconate titanate: the modulated structure and its influences on electrically induced polarizations and strains. *J. Am. Ceram. Soc.*, **78**, 2101.

65. Tan, X., Jo, W., Granzow, T., Frederick, J., Aulbach, E., and Rödel, J. (2009) Auxetic behavior under electrical loads in an induced ferroelectric phase. *Appl. Phys. Lett.*, **94**, 042909.

66. He, H. and Tan, X. (2007) Raman spectroscopy study of the phase transitions in $Pb_{0.99}Nb_{0.02}[(Zr_{0.57}Sn_{0.43})_{1-y}Ti_y]_{0.98}O_3$ ceramics. *J. Phys.: Condens. Matter*, **19**, 136003.

67. Tan, X., Fredrick, J., Ma, C., Aulbach, E., Marsilius, M., Hong, W., Granzow, T., Jo, W., and Rodel, J. (2010) Electric-field-induced antiferroelectric to ferroelectric phase transition in mechanically confined $Pb_{0.99}Nb_{0.02}[(Zr_{0.57}Sn_{0.43})_{0.94}Ti_{0.06}]_{0.98}O_3$. *Phys. Rev. B*, **81**, 014103.

68. Jankowska-Sumara, I. (2004) Phase transitions in $PbZr_{1-x}Sn_xO_3$ single crystals. *Ferroelectrics*, **313**, 81.

69. Shirane, G. and Pepinsky, R. (1953) Phase transitions in antiferroelectric $PbHfO_3$. *Phys. Rev.*, **91**, 812.

70. Madigou, V., Baudour, J.L., Bouree, F., Favotto, Cl., Rubin, M., and Nihoul, G. (1999) Crystallographic structure of lead hafnate ($PbHfO_3$) from neutron powder diffraction and electron microscopy. *Philos. Mag. A*, **79**, 847.

71. Fujishita, H., Ishikawa, Y., Ogawaguchi, A., Kato, K., Nishibori, E., Takata, M., and Sakata, M. (2005) A study of structures and order parameters in antiferroelectric $PbHfO_3$ by synchrotron radiation. *J. Phys. Soc. Jpn.*, **74**, 2743.

72. Fujishita, H. Ogawaguchi, A., Katano, S. (2008) A study of structures and order parameters in antiferroelectric

$PbHfO_3$ using neutron diffraction. *J. Phys. Soc. Jpn.*, **77**, 064601.

73. Jankowska-Sumara, I., Kugel, G.E., Roleder, K., and Dec, J. (1995) Raman scattering in pure and Ti-doped $PbHfO_3$ antiferroelectric crystals. *J. Phys.: Condens. Matter*, **7**, 3957.

74. Zaitsev, S.M., Zhavoronko, G.P., Tatarenko, A.A., Kuprianow, M.F., Filipiev, V.S., and Fesenko, G.E. (1979) *Kristallografica*, **24**, 826.

75. Leontiev, N.G., Kolesova, R.V., Eremkin, V.V., Fesenko, O.E., and Smotriakow, V.G. (1984) Space group of high-temperature orthorhombic phase of lead hafnate. *Sov. Phys. Crystallogr.*, **29**, 238.

76. Sakowski-Cowley, A.C., Lukaszewicz, K., and Megaw, H.D. (1969) The structure of sodium niobate at room temperature, and the problem of reliability in pseudosymmetric structures. *Acta Crystallogr., Sect. B*, **25**, 851.

77. Darlington, C.N.W. and Knight, K.S. (1999) High-temperature phases of $NaNbO_3$ and $NaTaO_3$. *Acta Crystallogr., Sect. B*, **55**, 24.

78. Mishra, S.K., Choudhury, N., Chaplot, S.L., Krishna, P.S.R., and Mittal, R. (2007) Competing antiferroelectric and ferroelectric interactions in $NaNbO_3$: Neutron diffraction and theoretical studies. *Phys. Rev. B*, **76**, 024110.

79. Sciau, Ph., Kania, A., Dkhil, B., Suard, E., and Ratuszna, A. (2004) Structural investigation of $AgNbO_3$ phases using x-ray and neutron diffraction. *J. Phys.: Condens. Matter*, **16**, 2795.

80. Kania, A. and Kwapulinski, J. (1999) $Ag_{1-x}Na_xNbO_3$ (ANN) solid solutions: from disordered antiferroelectric $AgNbO_3$ to normal antiferroelectric $NaNbO_3$. *J. Phys.: Condens. Matter*, **11**, 8933.

81. King-Smith, R.D. and Vanderbilt, David. (1994) First principles investigation of ferroelectricity in perovskite compounds. *Phys. Rev. B*, **49**, 5828.

82. Zhong, W., King-Smith, R.D., and Vanderbilt, D. (1994) Giant LO-TO splittings in perovskite ferroelectrics. *Phys. Rev. Lett.*, **72**, 3618.

83. Zhong, W. and Vanderbilt, D. (1995) Competing structural instabilities in

cubic perovskites. *Phys. Rev. Lett.*, **74**, 2587.

84. Vanderbilt, D. and Zhong, W. (1998) First-principles theory of structural phase transitions for perovskites: Competing instabilities. *Ferroelectrics*, **181**, 206.

85. Dieguez, O., Rabe, K.M., and Vanderbilt, D. (2005) First-principles study of epitaxial strain in perovskites. *Phys. Rev. B*, **72**, 144101.

86. Prosandeev, S.A. (2005) Comparative analysis of the phonon modes in $AgNbO_3$ and $NaNbO_3$. *Phys. Solid State*, **47**, 2130.

87. Cross, L.E. and Nicholson, B.J. (1955) The optical and electrical properties of single crystals of sodium niobate. *Philos. Mag.*, **46**, 212.

88. Cross, L.E. (1956) *Philos. Mag.*, **1**, 76.

89. Ulinzheev, A.V., Leiderman, A.V., Smotrakov, V.G., Topolov, V.Yu., and Fesenko, O.E. (1997) Phase transitions induced in $NaNbO_3$ crystals by varying the direction of an external field. *Phys. Solid State*, **39**, 972.

90. Cross, L.E. (1958) Electric double hysteresis in $(K_xNa_{1-x})NbO_3$ single crystals. *Nature*, **181**, 178.

91. Kim, D.H., Lee, H.N., Varela, M., and Christen, H.M. (2006) Antiferroelectricity in multiferroic $BiCrO_3$ epitaxial films. *Appl. Phys. Lett.*, **89**, 162904.

92. Hill, N.A. and Rabe, K.M. (1999) First-principles investigation of ferromagnetism and ferroelectricity in bismuth manganite. *Phys. Rev. B*, **59**, 8759.

93. Baettig, P., Seshadri, R., and Spaldin, N.A. (2007) Anti-polarity in ideal $BiMnO_3$. *J. Am. Chem. Soc.*, **129**, 9854.

94. Ranjan, R., Pandey, D., and Lalla, N.P. (2000) Novel features of $Sr_{1-x}Ca_xTiO_3$ phase diagram: Evidence for competing antiferroelectric and ferroelectric interactions. *Phys. Rev. Lett.*, **84**, 3726.

95. Bokov, A.A., Raevskii, I.P., and Smotrakov, V.G. (1983) *Sov. Phys. Solid State*, **25**, 1168.

96. Turik, A.V., Kupriyanov, M.F., Zhestkov, V.F., Shevchenko, N.B., and Kogan, V.A. (1985) *Sov. Phys. Solid State*, **27**, 1686.

97. (a) Groves, P. (1986) Structural phase transitions and long-range order in ferroelectric perovskite lead indium niobate. *J. Phys. C*, **19**, 111; (b) Groves, P. (1986) The influence of B-site cation order on the phase transition behaviour of antiferroelectric lead indium niobate. *J. Phys. C*, **19**, 5103.

98. Randall, C.A., Barber, D.J., Groves, P., and Whatmore, R.W. (1988) TEM study of the disorder-order perovskite, $Pb(In_{1/2}Nb_{1/2}O_3$. *J. Mater. Sci.*, **23**, 3678.

99. Kuprianova, M.F., Turika, A.V., Zaitseva, S.M., and Fesenkoa, E.G. (1983) Phase transitions in $PbNb_{0.5}B_{0.5}O_3$ (B=Sc, In). *Phase Transit.*, **4**, 65.

100. Yasuda, N. and Inagaki, H. (1991) Large dielectric dispersion in ordered perovskite $Pb(Yb^{1/2}Nb^{1/2})O_3$. *Jpn. J. Appl. Phys.*, **30**, L2050.

101. Kwon, R. and Choo, W.K. (1991) The antiferroelectric crystal structure of the highly ordered complex perovskite $Pb(Yb^{1/2}Nb^{1/2})O_3$. *J. Phys.: Condens. Matter*, **3**, 2147.

102. Yasuda, N. and Konda, J. (1993) Successive paraelectric-antiferroelectric-ferroelectric phase transitions in highly ordered perovskite lead ytterbium tantalate. *Appl. Phys. Lett.*, **62**, 535.

103. Randall, C.A., Markgraf, S.A., Bhalla, A.S., and Baba-Kishi, K. (1989) Incommensurate structures in highly ordered complex perovskites $Pb(Co_{1/2}W_{1/2}O_3$ and $Pb(Sc_{1/2}Ta_{1/2})O_3$. *Phys. Rev. B*, **40**, 413.

104. Baba-Kishi, K. and Barber, D.J. (1990) Transmission electron microscope studies of phase transitions in single crystals and ceramics of ferroelectric $Pb(Sc_{1/2}Ta_{1/2})O_3$. *Appl. Crystallogr.*, **23**, 43.

105. Yi, J.Y. and Lee, J.K. (2011) Stabilized antiferroelectric phase in lanthanum-doped $Na(Bi_{1/2}Ti_{1/2})O_3$. *J. Phys. D*, **44**, 415302.

106. Khalyavin, D.D., Salak, A.N., Vyshatko, N.P., Lopes, A.B., Olekhnovich, N.M., Pushkarev, A.V., Maroz, I.I., and Radyush, Y.V. (2006) Crystal structure of metastable perovskite $Bi(Mg_{1/2}Ti_{1/2})O_3$: Bi-based structural

analogue of antiferroelectric $PbZrO_3$. *Chem. Mater.*, **18**, 5104.

107. Zhang, S.T., Kounga, A.B., Jo, W., Jamin, C., Seifert, K., Granzow, T., Rodel, J., and Damjanovic, D. (2009) High-strain lead-free antiferroelectric electrostrictors. *Adv. Mater.*, **21**, 4716.

108. Miller, R.C., Wood, E.A., Remeika, J.P., and Savage, A. (1962) $Na(Nb_{1-x}V_x)O_x$ system and ferrielectricity. *J. Appl. Phys.*, **33**, 1623.

109. Ueda, K. and Kobayashi, J. (1953) Antiparallel dipole arrangement in tungsten trioxide. *Phys. Rev.*, **91**, 1565.

110. Popolitov, V.I., Lobachev, A.N., and Peskin, V.F. (1982) Antiferroelectrics, ferroelectrics and pyroelectrics of a stibiotantalite structure. *Ferroelectrics*, **40**, 9.

111. Unoki, H. and Sakudo, T. (1977) Dielectric anomaly and improper antiferroelectricity at the Jahn-Teller transitions in rare-earth vanadates. *Phys. Rev. Lett.*, **38**, 137.

112. Schmid, H. and Ascher, E. (1974) Are antiferroelectricity and other physical properties hidden in spinel compounds?. *J. Phys. C*, **7**, 2697.

113. Albers, J., Klopperpieper, A., Rother, H.J., and Ehses, K.H. (1982) Antiferroelectricity in betaine phosphate. *Physica Status Solidi*, **74**, 553.

114. Zhang, W. (2010) Exceptional dielectric phase transitions in a perovskite-type cage compound. *Angew. Chem. Int. Ed.*, **49**, 6608.

115. Fu, H. and Cohen, R.E. (2000) Polarization rotation mechanism for ultrahigh electromechanical response in single crystal piezoelectrics. *Nature*, **403**, 281.

116. Bellaiche, L., Garcia, A., and Vanderbilt, D. (2001) Electric-field induced polarization paths in $Pb(Zr_{1-x}Ti_x)O_3$ alloys. *Phys. Rev. B*, **64**, 060103.

117. Ye, Z.G. (1996) Relaxor ferroelectric $Pb(Mg_{1/3}Nb_{2/3})O_3$: Properties and present understanding. *Ferroelectrics*, **184**, 193.

118. Bharadwaja, S.S.N. and Krupanidhi, S.B. (1999) Growth and study of antiferroelectric lead zirconate thin films by pulsed laser ablation. *J. Appl. Phys.*, **86**, 5862.

119. Ayyub, P., Chattopadhyay, S., Pinto, R., and Multani, M.S. (1998) Ferroelectric behavior in thin films of antiferroelectric materials. *Phys. Rev. B*, **57**, R5559.

120. Yamakawa, K., Trolier-McKinstry, S., Dougherty, J.P., and Krupanidhi, S.B. (1995) Reactive magnetro co-sputtered antiferroelectric lead zirconate thin films. *Appl. Phys. Lett.*, **67**, 2014.

121. Blok, J., Rabe, K.M., and Vanderbilt, D. (2011) Interplay of epitaxial strain and rotations in $PbTiO_3/PbZrO_3$ superlattices from first principles. *Phys. Rev. B*, **84**, 205413.

122. Boldyreva, K., Pintilie, L., Lotnyk, A., Misirlioglu, I.B., Alexe, M., and Hesse, D. (2007) Thickness-driven antiferroelectric-to-ferroelectric phase transition of thin $PbZrO_3$ layers in epitaxial $PbZrO_3/Pb(Zr_{0.8}Ti_{0.2})O_3$ multilayers. *Appl. Phys. Lett.*, **91**, 122915.

123. Jaffe, B., Cook, W.R., and Jaffe, H. (1971) *Piezoelectric Ceramics*, Academic Press, p. 174.

124. Yamakawa, K., Gachigi, K.W., Trolier-McKinstry, S., and Dougherty, J.P. (1997) Structural and electrical properties of antiferroelectric lead zirconate thin films prepared by reactive magnetron co-sputtering. *J. Mater. Sci.*, **32**, 5169.

125. Xu, B., Pai, N.G., and Cross, L.E. (1998) Lanthanum doped lead zirconate titanate stannate antiferroelectric thin films from acetic acid-based sol-gel method. *Mater. Lett.*, **34**, 157.

126. Xu, B., Ye, Y., and Cross, L.E. (2000) Dielectric properties and field-induced phase switching of lead zirconate titanate stannate antiferroelectric thick films on silicon substrates. *J. Appl. Phys.*, **87**, 2507.

127. Li, X., Zhai, J., and Chen, H. (2005) $(Pb,La)(Zr,Sn,Ti)O_3$ antiferroelectric thin films grown on $LaNiO_3$-buffered and Pt-buffered silicon substrates by sol-gel processing. *J. Appl. Phys.*, **97**, 024102.

128. Yoshiro, O. (1976) Electrooptic ceramic material, US Patent 3998523, issue date.

129. Dougherty, J.P. (1996) Cardiac defibrillator with high energy storage

antiferroelectric capacitor, US Patent 5545184, issue date.

130. Zhu, L. and Wang, Q. (2012) Novel ferroelectric polymers for high energy density and low loss dielectrics. *Macromolecules*, **45**, 2937.

131. Pan, W., Zhang, Q., Bhalla, A., and Cross, L.E. (1989) Field-forced antiferroelectric-to-ferroelectric switching in modified lead zirconate titanate stannate ceramics. *J. Am. Ceram. Soc.*, **72**, 1455.

132. Zhang, S.T., Kounga, A.B., Aulbach, E., Ehrenberg, H., and Rodel, J. (2007) Giant strain in lead-free piezoceramics $Bi_{0.5}Na_{0.5}TiO_3$-$BaTiO_3$-$K_{0.5}Na_{0.5}NbO_3$ system. *Appl. Phys. Lett.*, **91**, 112906.

133. Bennett, J.W., Garrity, K.F., Rabe, K.M., and Vanderbilt, D. (2013) Orthorhombic ABC semiconductors as antiferroelectrics. *Phys. Rev. Lett.*, **110**, 017603.

134. Ray, N. and Waghmare, U.V. (2008) Coupling between magnetic ordering and structural instabilities in perovskite biferroics: a first-principles study. *Phys. Rev. B*, **77**, 134112.

135. Suzuki, I. and Okada, K. (1978) Phenomenological theory of antiferroelectric transition. IV. Ferrielectric. *J. Phys. Soc. Jpn.*, **45**, 1302.

Part IV
Multiferroics

Functional Metal Oxides: New Science and Novel Applications, First Edition.
Edited by Satishchandra B. Ogale, Thirumalai V. Venkatesan, and Mark G. Blamire.
© 2013 Wiley-VCH Verlag GmbH & Co. KGaA. Published 2013 by Wiley-VCH Verlag GmbH & Co. KGaA.

8
Probing Nanoscale Electronic Conduction in Complex Oxides

Jan Seidel and Ramamoorthy Ramesh

The rich physics of transition metal oxides results in a wide variety of properties that are related to a delicate balance between charge, spin, and orbital degrees of freedom [1]. There has been a large body of work on these materials, mainly on members of the perovskite family, which offer the advantage of very high structural quality growth. Using modern synthesis methods, such as pulsed laser deposition (PLD), molecular beam epitaxy (MBE), and metal-organic chemical vapor deposition (MOCVD), among others, it is now possible to engineer interfaces between complex transition metal oxides with atomic-scale precision. Such interfaces locally break the symmetry, induce stress, and vary the bonding between ions. This, in turn, gives rise to changes in bandwidth, orbital interactions, and level degeneracy, opening venues for modifying the electronic structure of these strongly correlated materials. Charge transfer can induce carrier densities that are different at the interface than in the bulk, resulting in physical properties at the interface, which may be completely different from those of the parent bulk materials [2–6]. The recent development of local scanning-probe- based measurements now allows for nanoscale probing of these interfaces between strongly correlated oxides. Tuning and controlling the physical properties of these interfaces between different oxide materials provides a new playground for research and offers a new nanoelectronics characterization platform for future nanotechnology [7]. Recent developments in the field of electronic and transport properties of such materials are the subject of this chapter.

8.1
Scanning-Probe-Based Transport Measurements

Atomic force microscopy (AFM) and its variations (e.g., conductive-atomic force microscopy (c-AFM) and piezoresponse force microscopy (PFM)) are very well suited for the direct characterization of prototype oxide structures, including domain walls [8]. With c-AFM, one can probe local conductivity at interfaces and topological defects, which are of concern to ferroelectric capacitor-based-applications. PFM is also under continuous development, and is currently undergoing a shift in

Functional Metal Oxides: New Science and Novel Applications, First Edition.
Edited by Satishchandra B. Ogale, Thirumalai V. Venkatesan, and Mark G. Blamire.
© 2013 Wiley-VCH Verlag GmbH & Co. KGaA. Published 2013 by Wiley-VCH Verlag GmbH & Co. KGaA.

focus from imaging static domains to dynamic characterization of the switching process (with developments such as stroboscopic PFM and PFM spectroscopy) [9–12]. Ferroelectric thin films typically contain various structural defects, such as cationic and/or anionic point defects, dislocations, and grain boundaries. Since the electric and stress fields around such defects in a ferroelectric thin film are likely to be inhomogeneous, it is expected that the switching behavior near a structural defect will be different from the one found in a single domain state. The role of a single ferroelastic twin boundary has been studied, for example, in tetragonal $PbZr_{0.2}Ti_{0.8}O_3$ ferroelectric thin films [13]. It was shown that the potential required to nucleate a $180°$ domain is lower near ferroelastic twin walls. An increased interest in local conductivity measurements has recently arisen from the nonvolatile memory application perspective and because of the potential for electroresitive memory devices [14], which are also interesting for the characterization of oxygen vacancy movement and ionic battery materials on the nanoscale [15, 16]. In many cases, the presence of extended defects and oxygen vacancy accumulation makes the identification of polarization-mediated transport mechanisms difficult, although direct probing of polarization-controlled tunneling into ferroelectric surfaces has been demonstrated [17, 18]. The combination of local electromechanical and conductivity measurements has revealed a connection between local current and pinning at bicrystal grain boundaries in bismuth ferrite [19]. Electroresistance in ferroelectric structures has recently been reviewed by Watanabe [20]. The presence of extended defects and oxygen vacancy accumulation has been shown to influence transport mechanisms at domain walls [21, 22]. Recently, direct probing of polarization-controlled tunneling into ferroelectric surface has been shown. Near-field scanning optical microscopy has been used to observe pinning and bowing of a single ferroelectric domain wall under a uniform applied electric field [23, 24]. Scanning microwave microscopy has been applied to the study of mesoscopic metal–insulator transition at ferroelastic domain walls in VO_2 [25, 26].

scanning tunneling spectroscopy (STS) can be used to probe directly the superconducting order parameter at nanometer length scales. Scanning tunneling microscopy (STM) and scanning tunneling spectroscopy (STS) have been used to investigate the electronic structure of ferroelastic twin walls in $YBa_2Cu_3O_{7-\delta}$ [27]. Twin boundaries play an important role in pinning the vortices and thereby enhance the currents that oxide superconductors can support while remaining superconducting. An unexpectedly large pinning strength for perpendicular vortex flux across such boundaries was found, which implies that the critical current at the boundary approaches the theoretical "depairing" limit. In the case of insulators, STM/STS are by definition a lot more difficult to implement, primarily because a reliable tunneling current cannot be used to establish proximal contact. The emergence of ferroelectrics with smaller bandgaps and the possibility of conduction at domain walls (see later text) have stimulated renewed interest in exploring STM as a probe of the local electronic structure. The emergence of combined AFM/STM or scanning electron microscopy (SEM)/STM systems should be a boon in terms of exploring the electronic properties of domain walls in such insulating materials. Research using such combined tools is in its infancy [28–31].

8.2
Domain Wall Conductivity

Since ferroic phases can arise in two or more distinct orientations, they can form domains, separated by domain walls. Domains are a representation of long-range ordering with respect to at least one macroscopic tensor property of the material (order parameter). When orientation states are changed, the interfaces (domain walls) move; thus the domain structure can be manipulated by external fields, which is a central feature of ferroic materials. In general, in the vicinity of such a transition, one or more macroscopic properties of the material associated with the order parameter can become large and very susceptible to external fields. Field-induced phase transitions around the transition temperature are a common feature. Subsets of ferroic phase transitions are ferroelectric, ferroelastic, and ferromagnetic transitions. These involve the emergence of spontaneous polarization, spontaneous strain, or spontaneous magnetization, and are commonly referred to as "*primary*" or "first-order ferroics." Many so-called smart materials and structures have at least one of these properties and are designed to change them in a preconceived manner through the application of external fields. Thus, given their intrinsic structure, ferroic materials are ideally suited for this purpose (Figure 8.1).

The changes in structure (and as a consequence electronic structure) that occur at ferroelectric (multiferroic) domain walls can thus lead to changes in transport behavior. Indeed, domain wall conductivity has been shown in different ferroic materials, although with different transport behavior: the domain walls of $BiFeO_3$

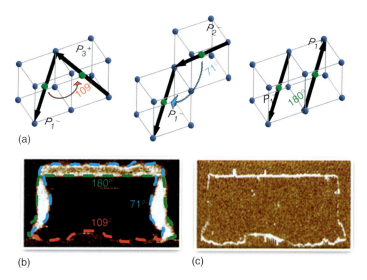

(a)

(b)　　　　　　　(c)

Figure 8.1 (a) The three different types of domain walls in rhombohedral bismuth ferrite. Arrows indicate polarization directions in adjacent domains. (b) In-plane PFM image of a written domain pattern in a mono- domain BFO (110) film showing all three types of domain wall. (c) The corresponding c-AFM image showing conduction at both 109° and 180° domain walls.

were found to be more conductive than the domains [19], while those of $YMnO_3$ were found to be more insulating or conductive depending on their orientation [32, 33]. Multiferroic $YMnO_3$, a so-called improper ferroelectric multiferroic, in which ferroelectricity is induced by structural trimerization coexisting with magnetism, domain walls are found to be charged and stable. This material exhibits a conductive "cloverleaf" pattern of six domains emerging from one point, and the ferroelectric state has been reported to be more conducting than the paraelectric state. The increase in the Y–O bond distance at domain walls may be responsible for the reduction of local conduction. The observed conduction suppression at domain walls at high voltages (still much less than the electric coercivity) is in striking contrast with what has been reported on $BiFeO_3$.

A useful clue for interpreting these results is perhaps the analysis of the paraelectric phase. The high-temperature, high-symmetry phase of $BiFeO_3$ is more conducting than the ferroelectric phase [34], but the converse is true for $YMnO_3$ [30]. This illustrates an important point: in some respects the internal structure of the walls can be considered to be in the paraelectric state; by way of a trivial example, the 180° domain walls of a ferroic are nonpolar, just the same as its paraphase. The examples of $BiFeO_3$ and $YMnO_3$ suggest that the paraphase-like behavior can be applied to domain wall properties other than just the polarization: the insulating paraelectric state of $YMnO_3$ is consistent with the insulating nature of its domain walls, and, conversely, the conducting state of the paraphase of $BiFeO_3$ is consistent with its domain wall conductivity. Nevertheless, in the conductivity of the $BiFeO_3$ walls at least, there are several other considerations: Octahedral rotations, electrostatic steps arising from rigid rotation of the polar vector, and increased carrier density at the wall are all thought to play a role in the domain wall conductivity of $BiFeO_3$ and potentially also of other perovskites.

Detailed electronic properties of domain walls in bismuth ferrite have been investigated by Lubk *et al.* [35]. Using density functional theory, the layer-by-layer densities of states were calculated. It was found that the domain walls have a significantly reduced bandgap compared to the *R3c* bulk structure. Structural changes at the wall thus lead to scenarios that approach the ideal cubic structure, in which the 180° Fe–O–Fe bond angle maximize the Fe 3d–O 2p hybridization and hence the bandwidth of the material. Figure 8.2 shows the local bandgap extracted from the layer-by-layer densities of states across the three wall types. In all cases, a reduction in the bandgap in the wall can be seen, with the 180° wall showing the largest effect. In no case, however, does the bandgap approach zero in the wall region. The same first-principles calculation supporting the experimental work of Seidel *et al.* also give an insight into the changes in Fe–O–Fe bond angle in $BiFeO_3$, in addition to the fact that walls in which the rotations of the oxygen octahedra do not change their phase when the polarization reorients are significantly more favorable than those with rotation discontinuities; that is, antiphase octahedral rotations are energetically very costly.

The analysis of the local polarization and electronic properties also revealed steps in the electrostatic potential for all wall types, and these must also contribute to the conductivity. Steps in the electrostatic potential at domain walls are correlated

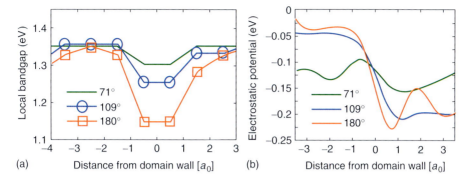

Figure 8.2 (a) Local band gap at domain walls in bismuth ferrite extracted from the layer-by-layer densities of states. (b) Macroscopically and planar-averaged electrostatic potential for the domain boundaries with continuous oxygen octahedral rotations. (Source: Adapted from Lubk *et al.* [35].)

with (and caused by) small changes in the component of the polarization normal to the wall [21]. These changes in normal polarization are a consequence of the fairly large rotation of the polar vector across the domain wall, and are not exclusive of BiFeO$_3$. Tetragonal PbTiO$_3$, for example, shows a similar effect for a 90° wall [36]. Extended phase-field calculations for tetragonal BaTiO$_3$ also allow the calculation of the intrinsic electrostatic potential drop across the 90° domain wall, regardless of the consideration of the ferroelectric as an n-type semiconductor or a dielectric [37]. This potential change creates a large electric field that promotes an asymmetric charge distribution around the walls, where electrons and oxygen vacancies concentrate on the opposite sides. The increased charge density presumably promotes increased conductivity.

As mentioned, the semirigid rotation of the polar vector across a ferroelectric–ferroelastic wall leads to an electrostatic potential that is screened by free charges, which enhance the local charge density and thus, presumably, the conductivity. Since this polar rotation [38] is not exclusive of BiFeO$_3$, other perovskite ferroelectrics should also be expected to display enhanced conductivity. In BiFeO$_3$, several other factors might be further helping the enhancement of conductivity. First, the magnetoelectric coupling between polarization and spin lattice is such that the magnetic sublattice rotates with the polarization [39, 40]. Since spins rotate rigidly, they might favor a more rigid rotation of the polarization and hence a bigger electrostatic step at the wall (and, of course, the polarization of BiFeO$_3$ is itself bigger than that of other known perovskite ferroelectrics, which means that all other things being equal, a rigid polar rotation in BiFeO$_3$ will cause a bigger electrostatic step). But perhaps the most obvious consideration is the fact that BiFeO$_3$ has intrinsically smaller bandgap than other prototypical perovskite ferroelectrics (~2.7 eV instead of 3.5–4 eV). This means that the screening charges accumulated at the wall will be closer to the bottom of the conduction band and hence will more easily contribute to the conductivity. It would be interesting to see whether highly insulating single-crystal samples such as those studied by Chishima *et al.* [41] display the same domain-wall conductivity as do the thin-film samples studied so far. The current density of these single-crystal samples can be as low as 10^{-9} A cm^{-2} even at electric fields in excess of 50 kV cm^{-1}, while typical resistivities of thin films are in the region of 10^6–10^8 Ω cm, comparable to the resistivity of good-quality BiMnO$_3$ [42].

The role of defect accumulation at the walls also deserves close scrutiny. Localized states are found in the spectrum of ferroelectric semiconductors, and states localized at the walls and inside the domain but close to the wall split off from the bulk continuum. These nondegenerate states have a high dispersion, in contrast with the "heavy-fermion" states at an isolated domain wall [43]. Charged double layers can be formed as a result of coupling between polarization and space charges at ferroelectric/ferroelastic domain walls [44]. Charged domain wall energies are about one order larger than the uncharged domain wall energies [45], and phenomenological calculations show decoration of walls by defects such as oxygen vacancies. The presence of charge and defect layers at the walls means that such walls promote electrical failure by providing a high-conductivity pathway from electrode to electrode [44].

The control of the electronic structure at walls by doping and strain in ferroelectric and ferroelastic oxides opens a way to effectively engineer nanoscale functionality in such materials. For the case of BiFeO$_3$, A-site doping with Ca, and magnetic B-site substitution, such as Co or Ni, might prove to be a viable way to achieve new domain wall properties by manipulating the electronic structure, spin structure, and dipolar moment in this material [14]. Of obvious future interest is the question of what sets the limits to the current transport behavior at walls: can one "design" the topological structure of the domain wall to controllably induce electronic phase transitions within the wall arising from the correlated electron nature? Is it possible to trigger an Anderson transition by doping the domain walls or straining them? (Figure 8.3).

Recently, the observation of tunable electronic conductivity at domain walls in La-doped BFO linked to oxygen vacancy concentration has been reported [22]. Specific

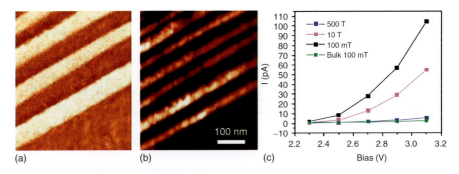

(a)   (b)   (c)

Figure 8.3 (a) PFM phase images of a BFO sample with 109° stripe domains. (b) Simultaneously acquired c-AFM image of the same area showing that each 109° domain wall is electrically conductive. (c) Current levels for samples with different oxygen cooling pressures and thus varying density of oxygen vacancies.

growth conditions have been used to introduce varying amounts of vacancies in thin film samples [46]. The conductivity at 109° walls in such samples is thermally activated with activation energies of 0.24–0.5 eV. From a broader perspective, these results are a first step toward realizing the tantalizing possibility of inducing an insulator–metal transition locally within the confines of the domain wall through careful design of the electronic structure, the state of strain, and chemical effects at the domain wall. For actual device applications, the magnitude of the wall current needs to be increased. The choice of the right shallow-level dopant and host material might prove to be key factors in this respect. Further study of correlations between local polarization and conductivity is an exciting approach to understanding the conduction dynamics and associated ferroelectric properties in the presence of strong coupling between electronic conduction and polarization in complex oxides.

8.3
Photovoltaic Effects at Domain Walls

Recently, it was reported that an anomalous photovoltaic (PV) effect in BFO thin films arises from a unique, new mechanism, namely structurally driven steps of the electrostatic potential at nanometer-scale domain walls [47–49]. In conventional solid-state PVs, electron–hole pairs are created by light absorption in a semiconductor and separated by the electric field spanning a micrometer-thick depletion region. The maximum voltage these devices can produce is equal to the semiconductor electronic bandgap. Interestingly, domain walls can give rise to a fundamentally different mechanism for PV charge separation, which operates over a distance of 1–2 nm and produces voltages that are significantly higher than the bandgap. Recent investigations using c-AFM under light illumination reveal these high photovoltages at 71° and 109° domain walls in BFO (Figure 8.4) [49].

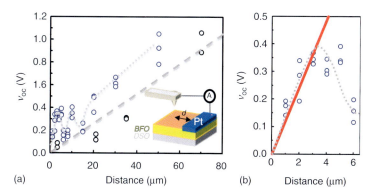

Figure 8.4 Local measurements of open circuit voltage (V_{OC}). (a) 109° domain walls (blue) show a different oscillating behavior of V_{OC} with distance when compared to 71° domain walls (black). (Inset) Schematic nanoscale PV measurement setup using a c-AFM tip as a variable-distance counter electrode for I–V characterization. (b) Initial large slope (red line) indicating a large PV effect at 109° walls.

The charge separation happens at previously unobserved nanoscale steps of the electrostatic potential, which naturally occur at ferroelectric domain walls in the complex oxide $BiFeO_3$. Electric-field control over domain structure allows the PV effect to be reversed in polarity or turned off.

Currently, the overall efficiency of those PV devices is limited by the conductivity of the bulk bismuth ferrite material. Methods to increase the carrier mobility, as well as to induce the spatially periodic potential in an adjacent material with a lower gap than BFO are possible routes to achieve larger current densities under white-light illumination, and more generally, they would demonstrate what the source of periodic potential and the PV current flow can be in different materials. Low-bandgap semiconductors with asymmetric electron and hole mobilities are possible candidates to show such an effect. In addition, photoelectrochemical effects at domain walls are a possible further interesting route, for example, for applications in water splitting [50].

8.4
Local Characterization of Doped Oxides and Defects

Defect–domain wall interaction is a very important area of research that deserves increased attention [51]. Point defects, for example, can broaden domain walls [52, 53]. The width of twin walls in $PbTiO_3$, for example, can be strongly modified by the presence of point defects within the wall. The intrinsic wall width of $PbTiO_3$ is about 0.5 nm, but clusters of point defects can increase the size of the twin wall up to 15 nm [54]. Trapped defects at the domain boundary play a significant role in the spatial variation of the antiparallel polarization width in the $BaMgF_4$ single crystal as seen by PFM [55]. Asymmetric charge distribution around $90°$ domain walls in $BaTiO_3$ has also been reported, where electrons and oxygen vacancies concentrate on the opposite sides [37].

Interaction between the order parameter and the point defect concentration causes point defects to accumulate within twin walls [52]; conversely, such defects contribute to the twin-wall kinetics and hysteresis [56] as they tend to clamp the walls. Oxygen vacancies, in particular, have been shown to have a lower formation energy in the domain wall than in the bulk, thereby confirming the tendency of these defects to migrate to and pin the domain walls [57]. This leads to a mechanism for the domain wall to have a memory of its location during annealing [43].

Domain wall (super) conductivity was studied by Aird and Salje [58]. Exposing WO_3 to sodium vapor, they observed preferential doping along the ferroelastic domain walls. Transport measurements showed superconductivity with a critical temperature of 3 K, while magnetic measurements did not, suggesting that super-conductivity was confined to the domain walls only, which provided a percolating superconductive path while occupying a very small volume fraction of the crystal. Later, Bartels *et al.* [59] used c-AFM to show the converse behavior. The domain walls of a calcium-doped lead orthophosphate crystal were found to be more resistive than the domains.

The concept of doping has also been applied to bismuth ferrite in an attempt to modify the electronic and magnetic properties and to reduce leakage currents [14, 60]. B-site doping of BFO with Ti^{4+} has been shown to reduce leakage by over three orders of magnitude, while doping with Ni^{2+} leads to a higher conductivity by over two orders of magnitude [61]. Likewise, doping with Cr or Mn has also been shown to greatly reduce leakage currents in BFO films [62].

Yang *et al.* (Figure 8.5) investigated Ca-doping of BFO with results showing strong similarity to phase diagrams of high-T_C superconductors and colossal magnetoresistive manganites where a competition between energetically similar ground states is introduced by hole doping. Control of the conductive properties by band-filling was observed in Ca-doped BFO. Application of an electric field enabled this control to the extent that a p−n junction can be formed, erased, and inverted in this material. A "dome-like" feature in the phase diagram is observed around a Ca concentration of 1/8, where a new pseudo-tetragonal phase appears and the electric modulation of conduction is found to be largest. c-AFM measurements reveal that the material exhibits resistive switching and that subsequent application of electric fields can reverse the effect. The observed reversible modulation of electric conduction accompanied by the modulation of the ferroelectric state is a consequence of the spatial movement of naturally produced oxygen vacancies under an electric

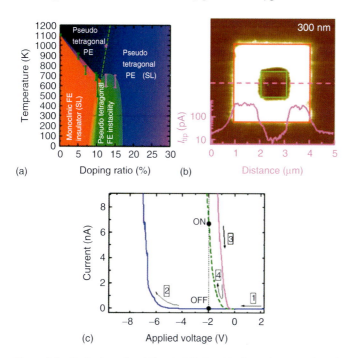

Figure 8.5 Ca-doping of multiferroic $BiFeO_3$. (a) Phase diagram of Ca-doped $BiFeO_3$. (b) c-AFM image of an electrically poled and re-poled area of the doped $BiFeO_3$ film. The electrically poled area (white) has become conducting. (c) *I*–*V* curve acquired during the switching process. (Source: Adapted from Yang *et al.* [14].)

field that act as donor impurities to compensate Ca acceptors and maintain a highly stable Fe^{3+} valence state. This observation might lead to new concepts for merging magnetoelectrics and magnetoelectronics at room temperature by combining electronic conduction with electric and magnetic degrees of freedom [63–65].

8.5
Local Electronic Probing of Oxide Interfaces

Over the last decade, many groups have explored the field of complex oxide interfaces. Many interesting phenomena have been observed in epitaxial heterostructures, such as metallic and superconducting behavior at the interface between two insulators [66, 67], metals and insulators [68], or gated oxide surfaces [69], which are promoted by electronic reconstruction as a result of charge-transfer or carrier-redistribution effects. Charge transfer has been found to trigger novel magnetic phases at the manganite–ruthenate interfaces [70, 71]. Unusual phenomena such as inducing magnetism in a superconductor and the rearrangement of the magnetic domains in superconductor/ferromagnet heterostructures due to orbital rearrangement and strong hybridization [72] as well as other ways of combining superconductors with metallic ferromagnets have been reported [73].

One of the most studied systems is the paradigmatic interface between $LaAlO_3$ and $SrTiO_3$ (LAO/STO). The electronic properties of this special interface can be controlled by an electric field for three to four unit-cell thick LAO films, which is just below the thickness needed to trigger an insulator-to-metal transition [74]. Transport measurements as a function of back-gate voltage show very large changes

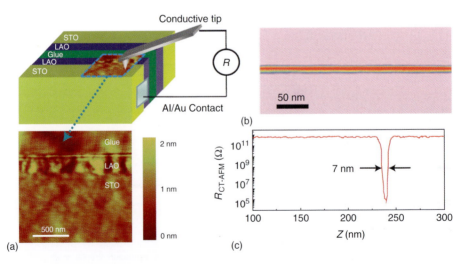

Figure 8.6 (a) Mapping the spatial distribution of charge carriers at $LaAlO_3$/$SrTiO_3$ interfaces. (b) Schematic diagram of the c-AFM experiment. (c) CT-AFM resistance image and resistance profile across the interface. (Source: Adapted from Ref. [75].)

at room temperature. The metallic state at the interface with a width of ~7 nm has been shown directly by c-AFM (Figure 8.6) [75]. Even more exciting is that this voltage-induced insulator-to-metal transition can be exploited at the nanoscale, as demonstrated by Cen *et al.* (Figure 8.7) [76, 77]. c-AFM was used to apply a local voltage to an LAO/STO film on the verge of the insulator-to-metal transition, which changes the local resistance on the nanoscale. This technique allows the writing of narrow lines or dots as small as 2 nm. Functional arrangements of lines and dots can be used to create nanoscale oxide electronic devices.

Another interesting aspect of nanoscale oxide materials is their potential use in tunneling devices. Control of the growth of these materials at the unit-cell level has led to research on the application of high-quality ultrathin ferroelectric films as switchable tunnel barriers for various applications [78]. The polarization reversal of a ferroelectric barrier has been predicted to show large changes in tunneling resistance [79] owing to the control of polarization charges at the interface. Using this effect may lead to the development of novel resistive memory devices.

Garcia *et al.* [17] used c-AFM to measure the tunneling resistance in ultra-thin BaTiO$_3$ thin films and demonstrated resistive readout of the polarization states. Another study investigated tunneling into ferroelectric thin film surfaces of Pb(Zr$_{0.2}$Ti$_{0.8}$)O$_3$. In this study, the tunneling current is found to exhibit a pronounced hysteresis with abrupt switching events that coincide with the local switching of ferroelectric polarization (Figure 8.8). A 500-fold amplification of the tunneling current on ferroelectric switching has been demonstrated.

STM and STS measurements in cross-sectional samples have also been used to directly investigate the nature of the unusual local electronic conductivity at the ferroelectric domain walls in multiferroic BFO [31]. In situ cleaved samples with ordered stripe arrays show decreases in the bandgap at the domain boundaries. In addition, a shift toward the Fermi level in the band edges of 109° and 71° domain walls has been measured (Figure 8.9). The demonstrated approach in this work serves as a model technique to investigate and understand electronic structure at oxide interfaces.

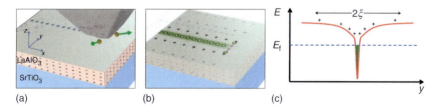

(a) (b) (c)

Figure 8.7 Oxide nanoelectronics on demand. (a) AFM tip moving left to right above LaAlO$_3$–SrTiO$_3$ heterostructure, removing oxygen-containing ions, and locally changing the charge state of the surface. (b) View of the same structure revealing the conducting nanowire formed at the interface. Electrons screen the surface charges by ionizing nearby states in the SrTiO$_3$ (lateral modulation doping) as well as from the top surface. (c) Illustration of potential profile across the nanowire. Modulation doping occurs over a screening length x on the order of micrometers. (Source: Adapted from Ref. [77].)

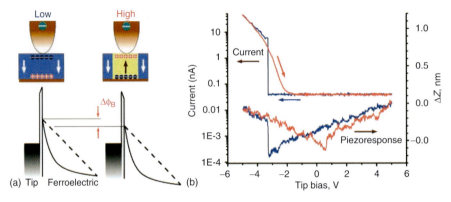

Figure 8.8 Polarization control of electron tunneling into ferroelectric surfaces. (a) Schematic polarization domain structure and interfacial band alignment for low- and high-conducting states of a thin PZT film. (b) Simultaneous measurements of local conductance and piezoresponse on the surface of a 30-nm PZT film. (Source: Adapted from Ref. [18].)

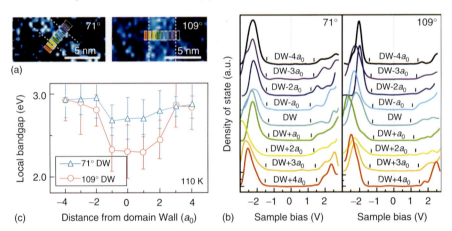

Figure 8.9 Layer-by-layer dI/dV measurements across 71° and 109° domain walls in BiFeO$_3$ acquired at 110 K. Bars in (a) denote positions where the electronic spectra are probed, and (b) shows the corresponding STS spectra. The band edges are indicated by black tick marks in (b). (c) Extracted local bandgap across the domain walls. (Source: Adapted from Ref. [31].)

8.6
Nanoscale Electronic Properties of CMR Manganites

Colossal magnetoresistance (CMR) materials such as doped manganites have been the focus of research for the past two decades [80, 81]. Such materials are especially interesting because of their large electronic correlations. Manganese-based perovskite oxides exhibit half-metallic character, and their CMR response is interesting for oxide nano(magneto)electronics. Small perturbations caused by

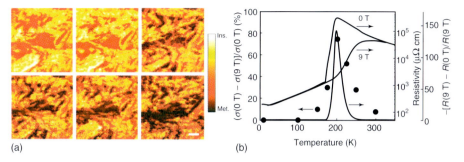

Figure 8.10 Spatially inhomogeneous metal–insulator transition in a doped manganite. (a) Generic STS images (scale bar 100 nm) of the local electronic structure of (La, Ca) MnO₃ taken just below T_C in magnetic fields of 0, 0.3, 1, 3, 5, and 9 T (from top left to bottom right). Parts of the surface are insulating (light colors), whereas others are metallic (dark colors) or in an intermediate state. (b) Temperature dependence of resistivity in magnetic fields of 0 and 9 T and the corresponding magnetoresistance ratio of a La₀.₇₅Ca₀.₂₅MnO₃ single crystal (solid lines, right axes). The circles denote the relative change to the metallic state of the image-averaged tunneling conductance induced by a magnetic field of 9 T. (Source: Adapted from Ref. [84].)

changes in temperature, magnetic fields, electric fields, or strain can drastically modify the magnetic and transport properties of these materials.

Using magnetic force microscopy, Zhang *et al.* [82] investigated magnetoresistance in these CMR materials by imaging the percolation of ferromagnetic, metallic domains, showing that such doped manganites were inhomogeneous and that phase separation was common. Ionic ordering on the A-site has also been observed by STM/STS, showing that these materials can be electronically homogeneous down to the 1-nm scale [83]. STM/STS was also used by Fäth *et al.* to investigate the spatially inhomogeneous metal–insulator transition in a doped manganite (Figure 8.10).

Orbital control in manganite materials is another interesting idea [85]. A modified electronic structure of manganese was demonstrated in La₀.₆₇Ca₀.₃₃MnO₃ thin films, that is, splitting of the e_g and t_{2g} levels by biaxial strain. Here, strain enhances selective orbital occupancy and charge localization. Artificial CMR heterostructures have been used to modulate the metal–insulator transition by over 50 K with the application of an electric field [86].

8.7
Future Directions

Nanoscale conduction phenomena in complex oxides form an exciting and growing field of interest in functional materials. With the current developments surrounding their conductive properties, there are many remaining questions and some new ones. For example, the investigation of dynamic conductivity at domain walls is an exciting aspect [87]. This addresses important factors: a possible electric-field-induced distortion of the polarization structure at the domain wall; the dependence of conductivity on the degree of distortion; and weak-pinning scenarios of the distorted wall. The domain wall is very likely not a rigid electronic conductor, instead offering a quasi-continuous spectrum of voltage-tunable electronic states [81]. This is different from ferroelectric domains, where switching may give rise to discrete (often only two) conductance levels [17, 18]. The intrinsic dynamics of domain walls and other topological defects are expected not only to influence future theoretical and experimental interpretations of the electronic phenomena but also to pose a possibility to find unique properties of multiferroic domain walls, for example, magnetization and magnetoresistance within an insulating antiferromagnetic matrix [88] due to order parameter coupling and localized secondary order parameters [52]. Of obvious future interest is the question of what sets the limits to the current transport behavior at walls: can one "design" the topological structure of the domain wall to controllably induce electronic phase transitions within the wall arising from the correlated electron nature? Is it possible to trigger an Anderson transition by doping of domain walls or straining them? The observation of superconductivity in ferroelastic walls of WO_3 certainly points to various exciting and unexplored areas of domain-boundary physics [56].

The investigation of topological defects beyond the classic domain walls is also a new exciting area of research. Recent work by Hong *et al.* [89] shows that arrays of ferroelectric nanowires have switchable quadrupoles and thus potential as nanodevices. Exotic topological defects in nanostructures (vertices, vortices, quadrupoles, etc.) are currently a very active area of research. For example, vortex cores in $BiFeO_3$ have been demonstrated to be dynamic conductors controlled by the coupled response of polarization and electron–mobile-vacancy subsystems with external bias [90].

Several applications have been suggested to make use of domain walls in ferroelectric materials, on the basis of their additional functionalities as well as their effects on existing devices. Uses that have been mentioned are as a local strain sensor incorporated on an AFM probe or a multilevel resistance-state device that is written by an electrical current [91]. Other possibilities include nonvolatile memories, piezoelectric actuators, ultrasound transducers, surface acoustic wave devices, and optical applications [49]. For existing devices, the discovery of conducting domain walls stimulates engineers to prevent their products from having the "wrong" domain walls that could cause leakage and prevent their use in ferroelectric memories.

Experimental results and theoretical investigations in recent years have convincingly demonstrated that certain transition-metal oxides and some other materials have dominant properties driven by spatial inhomogeneity. Strongly correlated materials incorporate physical interactions (spin, charge, lattice, and/or orbital hybridization), allowing complex interactions between electric and magnetic properties, resulting in ferromagnetic and antiferromagnetic phase transitions. Of even higher interest are the heterointerfaces formed between correlated materials showing new state properties. Domain walls are only one example of "naturally" occurring interfaces in such materials. The challenge is to determine whether such complex interactions can be controlled in those materials or heterointerfaces at sufficiently high speeds and densities to enable new logic device functionality at the nanometer scale. Parameters such as interface energy, switching speed, and threshold, tunability, dynamics of the states, and size dependencies need to be quantified to determine whether domain boundary materials could be employed as building blocks for information processing systems.

In summary, we have provided an overview on nanoscale electronic conduction phenomena in complex functional oxides. The ferromagnetic properties of ferroelectric walls in paramagnetic and antiferromagnetic materials suggest that much more R&D should be done on domain walls in multiferroics as well as on the dynamics of domain walls in these materials [92–94]. Artificially engineered oxide interfaces may pave the way to novel tailored states of matter with a wide range of electronic properties. Many ferroelectric relaxor-like systems and nanoscale phase-separated materials are interesting as well, because they have intrinsic nanodomains. Domain-wall electronics, particularly with ferroelectrics and multiferroics, may become interesting for nanotechnology [95].

References

1. Imada, M., Fujimori, A., and Tokura, Y. (1998) *Rev. Mod. Phys.*, **70**, 1039–1263.
2. Ogale, S.B. (2005) *Thin Films and Heterostructures for Oxide Electronics*, Springer, New York.
3. Ohtomo, A., Muller, D.A., Grazul, J.L., and Wang, H.Y. (2002) *Nature*, **419**, 378–380.
4. Dagotto, E. (2007) When oxides meet face to face. *Science*, **318**, 1076–1077.
5. Mannhart, J. and Schlom, D.G. (2010) Oxide interfaces – an opportunity for electronics. *Science*, **327**, 1607–1611.
6. Yamada, H. *et al.* (2004) Engineered interface of magnetic oxides. *Science*, **395**, 646–648.
7. Zubko, P., Gariglio, S., Gabay, M., Ghosez, P., and Triscone, J.-M. (2011) Interface physics in complex oxide heterostructures. *Annu. Rev. Condens. Matter Phys.*, **2**, 141–165.
8. Eng, L.M. (1999) Nanoscale domain engineering and characterization of ferroelectric domains. *Nanotechnology*, **10**, 405.
9. Kalinin, S.V. *et al.* (2010) *Rep. Prog. Phys.*, **73**, 056502.
10. Gruverman, A., Rodriguez, B.J., Dehoff, C., Waldrep, J.D., Kingon, A.I., Nemanich, R.J., and Cross, J.S. (2005) Direct studies of domain switching dynamics in thin film ferroelectric capacitors. *Appl. Phys. Lett.*, **87**, 082902.
11. Rodriguez, B.J., Jesse, S., Baddorf, A.P., Zhao, T., Chu, Y.H., Ramesh, R., Eliseev,

E.A., Morozovska, A.N., and Kalinin, S.V. (2007) Spatially resolved mapping of polarization switching behavior in nanoscale ferroelectrics. *Nanotechnology*, **18**, 405701.
12. Jungk, T., Hoffmann, A., and Soergel, E. (2007) Impact of elasticity on the piezoresponse of adjacent ferroelectric domains investigated by scanning force microscopy. *J. Appl. Phys.*, **102**, 084102.
13. Choudhury, S. *et al.* (2008) The influence of 180° ferroelectric domain wall width on the threshold field for wall motion. *J. Appl. Phys.*, **104**, 084107.
14. Yang, C.-H., Seidel, J., Kim, S.Y., Rossen, P.B., Yu, P., Gajek, M., Chu, Y.-H., Martin, L.W., Holcomb, M.B., He, Q., Maksymovych, P., Balke, N., Kalinin, S.V., Baddorf, A.P., Basu, S.R., Scullin, M.L., and Ramesh, R. (2009) Electric modulation of conduction in multiferroic Ca-doped BifeO$_3$ films. *Nat. Mater.*, **8**, 485.
15. Balke, N., Jesse, S., Morozovska, A., Eliseev, E., Chung, D., Kim, Y., Adamczyk, L., and Garcia, R. (2010) Nanoscale mapping of ion diffusion in a lithium-ion battery cathode. *Nat. Nanotechnol.*, **5**, 749–754.
16. Kumar, A., Ciucci, F., Morozovska, A.N., Kalinin, S.V., and Jesse, S. (2011) *Nat. Chem.*, **3**, 707–713.
17. Garcia, V., Fusil, S., Bouzehouane, K., Enouz-Vedrenne, S., Mathur, N.D.,

Barthelemy, A., and Bibes, M. (2009) *Nature*, **460**, 81.

18. Maksymovych, P., Jesse, S., Yu, P., Ramesh, R., Baddorf, A.P., and Kalinin, S.V. (2009) *Science*, **324**, 1421.

19. Rodriguez, B.J., Chu, Y.H., Ramesh, R., and Kalinin, S.V. (2008) *Appl. Phys. Lett.*, **93**, 142901.

20. Watanabe, Y. (2007) *Ferroelectrics*, **349**, 190.

21. Seidel, J., Martin, L.W., He, Q., Zhan, Q., Chu, Y.-H., Rother, A., Hawkridge, M.E., Maksymovych, P., Yu, P., Gajek, M., Balke, N., Kalinin, S.V., Gemming, S., Wang, F., Catalan, G., Scott, J.F., Spaldin, N.A., Orenstein, J., and Ramesh, R. (2009) Conduction at domain walls in oxide multiferroics. *Nat. Mater.*, **8**, 229.

22. Seidel, J., Maksymovych, P., Katan, A.J., Batra, Y., He, Q., Baddorf, A.P., Kalinin, S.V., Yang, C.-H., Yang, J.-C., Chu, Y.-H., Salje, E.K.H., Wormeester, H., Salmeron, M., and Ramesh, R. (2010) Domain wall conductivity in La-doped BiFeO$_3$. *Phys. Rev. Lett.*, **105**, 197603.

23. Kim, S. *et al.* (2005) Optical index profile at an antiparallel ferroelectric domain wall in lithium niobate. *Mater. Sci. Eng. B*, **120**, 91–94.

24. Yang, T.J. (1999) Direct observation of pinning and bowing of a single ferroelectric domain wall. *Phys. Rev. Lett.*, **82**, 4106–4109.

25. Tselev, A. *et al.* (2010) *ACS Nano*, **4** (8), 4412–4419.

26. Fan, W., Cao, J., Seidel, J., Gu, Y., Yim, J.W., Barrett, C., Yu, K.M., Ji, J., Ramesh, R., Chen, L.Q., and Wu, J. (2011) Large kinetic asymmetry in the metal-insulator transition nucleated at localized and extended defects. *Phys. Rev. B*, **83**, 235102.

27. Maggio-Aprile, I., Rennet, C., Erb, A., Walker, E., and Fischer, O. (1997) Critical currents approaching the depairing limit at a twin boundary in YBa2Cu3O (7-δ). *Nature*, **390**, 487–490.

28. Wiessner, A., Kirschner, J., Schafer, G., and Berghaus, T. (1997) Design considerations and performance of a combined scanning tunneling and scanning electron microscope. *Rev. Sci. Instrum.*, **68**, 3790.

29. Yang, B., Park, N.J., Seo, B.I., Oh, Y.H., Kim, S.J., Hong, S.K., Lee, S.S., and Park, Y.J. (2005) *Appl. Phys. Lett.*, **87**, 062902.

30. Garcia, R.E., Huey, B.D., and Blendell, J.E. (2006) *J. Appl. Phys.*, **100**, 064105.

31. Chiu, Y.-P., Chen, Y.-T., Huang, B.-C., Shih, M.-C., Yang, J.-C., He, Q., Liang, C.-W., Seidel, J., Chen, Y.-C., Ramesh, R., and Chu, Y.-H. (2011) The evolution of local electronic structure across multiferroic domain walls. *Adv. Mat.*, **23**, 1530.

32. Choi, T. *et al.* (2010) Insulating interlocked ferroelectric and structural antiphase domain walls in multiferroic YMnO$_3$. *Nat. Mater.*, **9**, 253–258.

33. Meier, D., Seidel, J., Cano, A., Delaney, K., Kumagai, Y., Mostovoy, M., Spaldin, N.A., Ramesh, R., and Fiebig, M. (2012) Anisotropic conductance at improper ferroelectric domain walls. *Nat. Mater.*, **11**, 284.

34. Palai, R., Katiyar, R.S., Schmid, H., Tissot, P., Clark, S.J., Robertson, J., Redfern, S.A.T., Catalan, G., and Scott, J.F. (2008) β phase and γ-β metal-insulator transition in multiferroic BiFeO$_3$. *Phys. Rev. B*, **77**, 014110.

35. Lubk, A., Gemming, S., and Spaldin, N.A. (2009) First-principles study of ferroelectric domain walls in multiferroic bismuth ferrite. *Phys. Rev. B*, **80**, 104110.

36. Meyer, B. and Vanderbilt, D. (2002) Ab initio study of ferroelectric domain walls in PbTiO$_3$. *Phys. Rev. B*, **65**, 104111.

37. Hong, L., Soh, A.K., Du, Q.G., and Li, J.Y. (2008) Interaction of O vacancies and domain structures in single crystal BaTiO$_3$: two-dimensional ferroelectric model. *Phys. Rev. B*, **77**, 094104.

38. Borisevich, A.Y. *et al.* (2010) *ACS Nano*, **4**, 6071.

39. Zhao, T., Scholl, A., Zavaliche1, F., Lee, K., Barry, M., Doran, A., Cruz, M.P., Chu, Y.H., Ederer, C., Spaldin, N.A., Das, R.R., Kim, D.M., Baek, S.H., Eom, C.B., and Ramesh, R. (2006) Electrical control of antiferromagnetic domains in multiferroic BiFeO$_3$ films at room temperature. *Nat. Mater.*, **5**, 823.

40. Lebeugle, D., Colson, D., Forget, A., Viret, M., Bataille, A.M., and Gukasov, A. (2008) Electric-field-induced spin flop in

BiFeO$_3$ single crystals at room temperature. *Phys. Rev. Lett.*, **100**, 227602.

41. Chishima, Y., Noguchi, Y., Kitanaka, Y., and Miyayama, M. (2010) Defect control for polarization switching in BiFeO$_3$ single crystals. *IEEE Trans. Ultrason. Ferroelectr. Freq. Control*, **57**, 2233.

42. Eerenstein, W., Morrison, F.D., Scott, J.F., and Mathur, N.D. (2005) Growth of highly resistive BiMnO3 films. *Appl. Phys. Lett.*, **87**, 101906.

43. Idlis, B.G. and Usmanov, M.S. (1992) Effect of domain structure on the energy spectrum of narrow-gap ferroelectric semiconductors. *Pis'ma Zh. Eksp. Teor. Fiz.*, **56** (5), 268–271.

44. Xiao, Y., Shenoy, V.B., and Bhattacharya, K. (2005) Depletion layers and domain walls in semiconducting ferroelectric thin films. *Phys. Rev. Lett.*, **95**, 247603.

45. Gureev, T.M.Y., Tagantsev, A.K., and Setter, N. (2009) Structure and energy of charged domain walls in ferroelectrics. 18th IEEE ISAF Proceedings.

46. Scullin, M.L. *et al.* (2010) *Acta Mater.*, **58**, 457.

47. Yang, S.-Y., Seidel, J., Byrnes, S.J., Shafer, P., Yang, C.-H., Rossell, M.D., Yu, P., Chu, Y.-H., Scott, J.F., Ager, J.W. III., Martin, L.W., and Ramesh, R. (2010) Above bandgap voltages from ferroelectric photovoltaic devices. *Nat. Nanotech.*, **5143**.

48. Seidel, J., Fu, D., Yang, S.-Y., Alarcòn-Lladò, E., Wu, J., Ramesh, R., and Ager, J.W. (2011) Efficient photovoltaic current generation at ferroelectric domain walls. *Phys. Rev. Lett.*, **107**, 126805.

49. Seidel, J., Yang, S.-Y., Alarcòn-Lladò, E., Ager, J.W., and Ramesh, R. (2012) Nanoscale probing of high photovoltages at 109° domain walls, *Ferroelectrics*, **433**, 123.

50. Kudo, A. and Miseki, Y. (2009) Heterogeneous photocatalyst materials for water splitting. *Chem. Soc. Rev.*, **38**, 253.

51. Gopalan, V., Dierolf, V., and Scrymgeour, D.A. (2007) Defect–domainwall interactions in trigonal ferroelectrics. *Ann. Rev. Mater. Res.*, **37**, 449–489.

52. Shilo, D., Ravichandran, G., and Bhattacharya, K. (2004) Investigation of twin-wall structure at the nanometre scale using atomic force microscopy. *Nat. Mater.*, **3**, 453–457.

53. Lee, W.T., Salje, E.K.H., and Bismayer, U. (2005) Influence of point defects on the distribution of twin wall widths. *Phys. Rev. B*, **72**, 104116.

54. Salje, E.K.H. and Zhang, H. (2009) Domain boundary engineering. *Phase Transitions*, **82** (6), 452.

55. Zeng, H.R. *et al.* (2008) Domain wall thickness variations of ferroelectric BaMgF4 single crystals in the tip fields of an atomic force microscope. *Phys. Status Solidi RRL*, **2** (3), 123–125.

56. Fan, W., Cao, J., Seidel, J., Gu, Y., Yim, J.W., Barrett, C., Yu, K.M., Ji, J., Ramesh, R., Chen, L.Q., and Wu, J. (2011) Large kinetic asymmetry in the metal-insulator transition nucleated at localized and extended defects. *Phys. Rev. B*, **83**, 235102.

57. He, L. and Vanderbilt, D. (2003) First-principles study of oxygen-vacancy pinning of domain walls in PbTiO$_3$. *Phys. Rev. B*, **68**, 134103.

58. Aird, A. and Salje, E.K.H. (1998) Sheet superconductivity in twin walls: experimental evidence of WO3-x. *J. Phys. Condens. Matter*, **10**, L377.

59. Bartels, M., Hagen, V., Burianek, M., Getzlaff, M., Bismayer, U., and Wiesendanger, R. (2003) Impurity-induced resistivity of ferroelastic domain walls in doped lead phosphate. *J. Phys. Condens. Matter*, **15**, 957–962.

60. Lee, Y.-H., Wu, J.-M., and Lai, C.-H. (2006) *Appl. Phys. Lett.*, **88**, 042903.

61. Qi, X., Dho, J., Tomov, R., Blamire, M.G., and MacManus-Driscoll, J.L. (2005) *Appl. Phys. Lett.*, **86**, 062903.

62. Kim, J.K., Kim, S.S., Kim, W.-J., Bhalla, A.S., and Guo, R. (2006) *Appl. Phys. Lett.*, **88**, 132901.

63. Ko, K.T., Jung, M.H., Lee, J.H., Woo, C.S., Chu, K., Seidel, J., Chu, Y.H., Jeong, Y.H., Ramesh, R., Park, J.H., and Yang, C.-H. (2011) Concurrent transition of ferroelectric and magnetic ordering around room temperature. *Nat. Commun.*, **2**, 567.

64. Ramirez, M. *et al.* (2009) *Appl. Phys. Lett.*, **94**, 161905.

65. Ramirez, M.O. *et al.* (2008) *Appl. Phys. Lett.*, **92**, 022511.
66. Ohtomo, A., Muller, D.A., Grazul, J.L., and Wang, H.Y. (2002) *Nature*, **419**, 378–380.
67. Reyren, N., Thiel, S., Caviglia, A.D., Fitting Kourkoutis, L., Hammerl, G., Richter, C., Schneider, C.W., Kopp, T., Rüetschi, A.-S., Jaccard, D., Gaboy, M., Muller, D.A., Triscone, J.-M., and Mannhart, J. (2007) *Science*, **317**, 1196–1199.
68. Logvenov, G., Gozar, A., and Bozovic, I. (2009) *Science*, **326**, 699–702.
69. Ye, J.T., Inoue, S., Kobayashi, K., Kasahara, Y., Yuan, H.T., Shimotani, H., and Iwasa, Y. (2010) Liquid-gated interface superconductivity on an atomically flat film. *Nat. Mater.*, **9**, 125–128.
70. Takahashi, K.S., Kawasaki, M., and Tokura, Y. (2001) *Appl. Phys. Lett.*, **79**, 1324–1326.
71. Koida, T., Lippmaa, M., Fukumura, T., Itaka, K., Matsumoto, Y., Kawasaki, M., and Koinuma, H. (2002) *Phys. Rev. B*, **66**, 144418.
72. Chakhalian, J., Freeland, J.W., Habermeier, H.U., Cristiani, G., Khaliullin, G., Veenendaalvan, M., and Keimer, B. (2007) *Science*, **318**, 1114–1117.
73. Lyuksyutov, I. and Pokrovsky, V. (2005) *Adv. Phys.*, **54**, 67–136.
74. Thiel, S., Hammerl, G., Schmehl, A., Schneider, C.W., and Mannhart, J. (2006) *Science*, **313**, 1942–1945.
75. Basletic, M., Maurice, J.-L., Carretero, C., Herranz, G., Copie, O., Bibes, M., Jacquet, E., Bouzehouane, K., Fusil, S., and Barthelemy, A. (2008) *Nat. Mater.*, **7**, 621.
76. Cen, C., Thiel, S., Hammerl, G., Schneider, C., Andersen, K., Hellberg, C., Mannhart, J., and Levy, J. (2008) *Nat. Mater.*, **7**, 298–302.
77. Cen, C., Thiel, S., Mannhart, J., and Levy, J. (2009) *Science*, **323**, 1026–1030.
78. Tsymbal, E.Y. and Kohlstedt, H. (2006) *Science*, **313**, 181.
79. Zhuravlev, M.Y., Sabirianov, R., Jaswal, S.S., and Tsymbal, E.Y. (2005) *Phys. Rev. Lett.*, **94**, 246802.
80. Salamon, M.B. and Jaime, M. (2001) *Rev. Mod. Phys.*, **73**, 583.
81. Tokura, Y. (2006) *Rep. Prog. Phys.*, **69**, 797–851.
82. Zhang, L., Israel, C., Biswas, A., Greene, R.L., and de Lozanne, A. (2002) *Science*, **298**, 805.
83. Moshnyaga, V., Sudheendra, L., Lebdev, O.I., Köster, S.A., Gehrke, K., Shapoval, O., Belenchuk, A., Damaschke, B., van Tendeloo, G., and Samwer, K. (2006) *Phys. Rev. Lett.*, **97**, 107205.
84. Fäth, M. *et al.* (1999) *Science*, **285**, 1540–1542.
85. Abad, L., Laukhin, V., Valencia, S., Gaup, A., Gudat, W., Balcells, L., and Martínez, B. (2007) *Adv. Funct. Mater.*, **17**, 3918.
86. Tanaka, H., Zhang, J., and Kawai, T. (2002) *Phys. Rev. Lett.*, **88**, 027204.
87. Maksymovych, P., Seidel, J., Chu, Y.-H., Baddorf, A., Wu, P., Chen, L.-Q., Kalinin, S.V., and Ramesh, R. (2011) Dynamic conductivity of ferroelectric domain walls. *Nano Lett.*, **11**, 1906.
88. He, Q., Yeh, CH., Yang, JC., Singh-Bhalla, G., Liang, CW., Chiu, PW., Catalan, G., Martin, LW, Chu, YH., Scott, JF, and Ramesh, R., (2012) Magnetotransport at domain walls in BiFeO$_3$, *Phys. Rev. Lett.*, **108**, 067203.
89. Hong, J., Catalan, G., Fang, D.N., Artacho, E., and Scott, J.F. (2010) Topology of the polarization field in ferroelectric nanowires from first principles. *Phys. Rev. B*, **81**, 172101.
90. Balke, N., Winchester, B., Ren, W. Chu, Y.H., Morozovska, A.N., Eliseev, E.A., Huijben, M., Vasudevan, R.K., Maksymovych, P., Britson, J., Jesse, S., Kornev, I., Ramesh, R., Bellaiche, L., Chen, L.Q., and Kalinin, S.V. (2012) Enhanced electric conductivity at ferroelectric vortex cores in BiFeO$_3$, *Nat. Phys.*, **8**, 81–88.
91. Bea, H. and Paruch, P. (2009) Multiferroics: a way forward along domain walls. *Nat. Mater.*, **8**, 168–169.
92. Goltsev, A.V., Pisarev, R.V., Lottermoser, T., and Fiebig, M. (2003) *Phys. Rev. Lett.*, **90**, 177204.
93. Daraktchiev, M., Catalan, G., and Scott, J.F. (2010) Landau theory of domain wall magnetoelectricity. *Phys. Rev. B*, **81**, 224118.

94. Skumryev, V., Laukhin, V., Fina, I., Martı, X., Sanchez, F., Gospodinov, M., and Fontcuberta, J. (2011) Magnetization reversal by electric-field decoupling of magnetic and ferroelectric domain walls in multiferroic-based heterostructures. *Phys. Rev. Lett.*, **106**, 057206.

95. Catalan, G. Seidel, J., Ramesh, R., and Scott, J.F. (2012) Domain wall nanoelectronics, *Rev. Mod. Phys.*, **84**, 119.

9
Multiferroics with Magnetoelectric Coupling

Umesh V. Waghmare

9.1
Introduction: Ferroic Materials

A ferroic material exhibits (i) a symmetry-breaking phase transition from a parent high-symmetry phase to a phase with long-range order, characterized by an order parameter, and (ii) switchability of the ordered phase from one state of the order parameter to another via application of an external field. These are equal-energy states of the material that are related to each other by symmetry operations of the parent high-symmetry phase that are lacking in the low-symmetry phase. The switchability makes a ferroic material useful as a memory element in devices. In addition, the lower symmetry of the ordered ferroic phase permits additional coupling of the order parameter with fields such as mechanical stress and facilitates their use in smart materials structures such as sensors and actuators. It is evident that symmetry principles are fundamental to the science of ferroic materials. A beautiful account of results of symmetry analysis relevant to ferroics can be found in a paper by Schmid [1].

An external field that couples directly with the ordering field or the order parameter of a ferroic defines its type (Figure 9.1); such a field is the conjugate field of the ordering field. For example, a ferroelectric material is a ferroic in which electric field couples with its order parameter, namely electric polarization. Likewise, a ferromagnet is characterized by magnetization as its order parameter, while a ferroelastic material has a strain as its order parameter. A relatively rare type of ferroic is a ferrotoroidal material, with toroidal moments as its order parameter, which couples with a cross product of electric and magnetic fields. Typically, the phase at low temperatures has a lower symmetry, which permits coupling of the ordering field with another field, making it a secondary order parameter. For example, broken inversion symmetry of a ferroelectric allows a nonzero *piezoelectric* coupling of polarization with strain, which is its secondary order parameter. Thus, electric field can be used to induce mechanical strain in a ferroelectric, adding it to its functionality. Thus, ferroics form a special class of multifunctional materials.

Indeed, it is desirable to increase the functionality of a material and expand the range of its applications, and tempting to combine one type of ferroic behavior with

Functional Metal Oxides: New Science and Novel Applications, First Edition.
Edited by Satishchandra B. Ogale, Thirumalai V. Venkatesan, and Mark G. Blamire.
© 2013 Wiley-VCH Verlag GmbH & Co. KGaA. Published 2013 by Wiley-VCH Verlag GmbH & Co. KGaA.

Figure 9.1 Fields and couplings in *smart functional materials.*

another [2, 3]. This led to the idea of multiferroic materials [4]. Since 2000, there has been intense research on multiferroic materials, and several reviews as well as popular and viewpoint articles have been published [5–11] that discuss various mechanisms, materials issues, bulk versus thin films, and future prospects. The goal of this chapter is to focus on a specific class of multiferroics which involves a combination of electric and magnetic ordering (hence a magnetoelectric coupling) and introduce the reader to the basic concepts and general principles that facilitate an understanding of these biferroic materials. This chapter is *not* intended to be a review of the vast literature on the subject, but represents a simple way of thinking about the interesting physics and chemistry of magnetoelectric multiferroics, with examples of those with perovskite-based structure.

9.2
Principles of Symmetry Analysis

We consider here two order parameters: electric polarization, P, and magnetization, M. Polarization is the electric dipole moment per unit volume, which can be expressed as the first moment of the charge density of a finite and neutral material. Similarly, magnetization is the magnetic dipole moment per unit volume. To develop an intuitive picture of their symmetry properties, we note that polarization relates to the expectation value of the position operator r, while magnetic moment relates to that of angular momentum operator $r \times p$ in quantum mechanical sense, where p is the momentum. While both P and M are vector quantities, one should note that (i) P involves an odd power of r and is independent of time (i.e., involves 0) or an even power of time and (ii) M involves an even power of r and an odd power of time. This determines transformation properties of P and M under spatial and time inversion. Under proper rotations, P and M transform as a vector, while their transformation properties under inversion i and reflection (e.g., σ_z in the xy-plane) are distinct (Figure 9.2):

$$
\begin{aligned}
\hat{i}P &= -P, & \hat{i}M &= M \\
\hat{\sigma}_z P_z &= -P_z, & \hat{\sigma}_z M_z &= M_z \\
\hat{\sigma}_z P_x &= P_x, & \hat{\sigma}_z M_x &= -M_x
\end{aligned}
$$

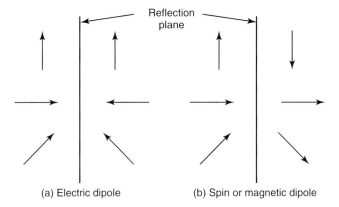

Figure 9.2 Symmetry properties of (a) electric and (b) magnetic dipoles under a plane of reflection.

Secondly, P and M transform under time-reversal τ as

$$\hat{t} P = P, \hat{t} M = -M \tag{9.1}$$

We note that the properties of transformation of position vector r or its spatial derivatives (gradient vector) are the same as those of P.

To write a continuum Hamiltonian or a Landau free energy of a ferroic, we use its parent high-symmetry phase as a reference state, in which the order parameter is zero. This corresponds to paraelectric and paramagnetic phases in the context of ferroelectric and ferromagnetic materials, respectively. Hamiltonian of a material can be written as a symmetry-invariant (under symmetry operations of the parent high-symmetry phase) Taylor expansion in the order parameter and its spatial derivatives truncated at a suitable power (say 4 or 6) in order parameter. In group-theoretical terms, this involves projection of an arbitrary term in the *Hamiltonian* of a given order with no specific symmetry onto the identity representation of the symmetry group of the parent phase. This essentially amounts to explicit symmetrization of a given term. For example, $H_4 = P_x{}^2 P_y{}^2$ is a fourth-order term in the *Hamiltonian or free energy of a ferroelectric*, and its symmetrization can be achieved as

$$H_{4s} = \frac{1}{g} \sum_i^g \hat{O}_i H_4 \tag{9.2}$$

where O_i is the ith symmetry of the g symmetry operations in the symmetry group of the parent phase. For the parent phase with cubic structure, it can be readily shown that $H_{4s} = P_x{}^2 P_y{}^2 + P_x{}^2 P_z{}^2 + P_z{}^2 P_y{}^2$. Although such an analysis for a general term involving P, M, and their spatial derivatives can be tedious, it is quite straightforward to carry out.

If symmetrization of a given term yields a vanishing expression, it means that the term of that type is not allowed in Hamiltonian. A simple application of this to a term $P.M$ for a ferroelectric shows that cubic symmetry of the parent phase cannot have a term of the type $P.M$. This is because P is odd under inversion, while M is even. Similarly, P is even under time reversal, while M is odd. As a result, $P.M$ transforms to $-P.M$ under inversion and cancels the starting term! A term $P.M$ can survive on symmetrization when the material lacks inversion and time-reversal symmetries (or other symmetries that transform these order parameters to their opposite orientation). The $P.M$ term represents a linear magnetoelectric coupling of a material. This clearly shows that it is desirable to have broken inversion and time-reversal symmetries in a material if it were to exhibit a linear magnetoelectric coupling [12]. Since inversion and time-reversal symmetries are broken in a ferroelectric and a ferromagnetic material, respectively, the symmetry of a biferroic with both electrical and magnetic ordering naturally permits a nonzero magnetoelectric coupling. However, the strength of this coupling is not determined by symmetry considerations presented here.

9.3
Magnetoelectric Couplings and a Landau-Like Theory

A clear idea about various compliances or response properties of a material can be obtained from a free-energy-based description. While absolute free energies are hard to determine experimentally or theoretically, their differences or derivatives correspond typically to measured or calculated properties. The free energy density (free energy per unit volume) can be expressed as

$$F = F_0 - H.M - P.E - \sum_{\alpha\beta}^{3} X_{\alpha\beta}^{ME} E_\alpha H_\beta - \frac{1}{2}\sum_{\alpha\beta}^{3} X_{\alpha\beta}^{e} E_\alpha E_\beta - \frac{1}{2}\sum_{\alpha\beta}^{3} X_{\alpha\beta}^{m} H_\alpha H_\beta + \cdots$$

From this, a change in polarization from its spontaneous value is obtained as a derivative of free energy with respect to electric field and takes the form

$$\Delta P_\alpha = \sum_{\beta}^{3} X_{\alpha\beta}^{ME} H_\beta + \sum_{\beta}^{3} X_{\alpha\beta}^{e} E_\beta,$$

where X^e and XME are the dielectric and magnetoelectric susceptibilities, respectively. From our earlier discussion on symmetries, it is clear now that XME is nonzero only when a material lacks both inversion and time-reversal symmetries.

A simple Landau theory with two order parameters can be expressed with

$$F = F_0 - H.M - P.E + a(T)P^2 + bP^4 + a'(T)M^2 + b'M^4 + cP^2M^2 +$$

$$d\sum_{\alpha}^{3} P_\alpha^3 P_\alpha^4 + d'\sum_{\alpha}^{3} M_\alpha^4 \cdots$$

where the terms with coefficients d represent the cubic anisotropy, while the rest are isotropic in vectors P and M. The term with coefficient c is a fourth-order coupling between polarization and magnetization. The free energy at a given temperature needs to be minimized with respect to order parameters. As a result of symmetry analysis, one can show that the following term is allowed by symmetry of a cubic material [13]:

$$H = \gamma P \times [M(\nabla.M) - (M.\nabla)M], \tag{9.3}$$

where γ is a constant and P and M are spatially dependent vector fields. Such a term, which is linear in gradient, is also known as a *Lifshitz invariant*. It is clear that the term is even in power of spatial vector r (owing to the presence of P and spatial derivative) and even in power of time (as M occurs twice); hence it is invariant under inversion and time-reversal symmetries. At temperatures where both P and M are nonzero (both inversion and time-reversal symmetries are broken), the term with coefficient c yields a linear magnetoelectric coupling.

We note that derivatives of a free energy function with respect to external or order parameter fields represent measurable quantities. A consequence of γ in the free energy expressed with Eq. (9.3) is a magnetoelectric coupling, but *not* the linear one! It gives electric field (hence polarization) that is induced by spatially inhomogeneous magnetic ordering with *quadratic* dependence on the magnitude of magnetization.

Finally, it is important to note that a material *does not* have to be a biferroic to exhibit a linear magnetoelectric coupling. If magnetism is introduced while maintaining insulating nature of a pyroelectric, such as ZnO or BeO, the resultant material does have a linear magnetoelectric coupling allowed by symmetry.

9.4
Chemical Considerations

Both ferroelectric and ferromagnetic materials have been known for over a century, and many of them are ABO_3 oxides with perovskite structure (Figure 9.3). Perovskite structure, thus, provides a common structural basis to develop a comparable understanding of the two.

Examples of ferroelectric materials include $KNbO_3$, $BaTiO_3$, and $PbTiO_3$, in which ferroelectricity arises from cooperative off-centering of B-cations (and also of A-cation in Pb-based compounds) from their high-symmetry positions.

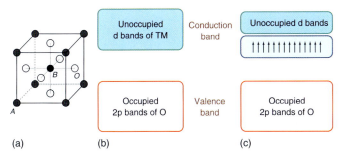

(a) (b) (c)

Figure 9.3 ABO_3 compound in the cubic perovskite structure (a), and a sketch of electronic structure of a perovskite oxide-based (b) ferroelectric with d^0 and (c) ferromagnet with d^n transition metal ion at the B-site.

The B-cation [typically a transition metal (TM)] is characterized by the oxidation state in which *none* of its d-orbitals are occupied with electrons: this is called d^0-*ness* of ferroelectrics. Of course, not all perovskite oxides with the d^0 property are ferroelectric. An example of this is $BaZrO_3$, which remains paraelectric all the way down to 0 K. It is interesting then to know which chemical mechanisms in a material control or determine its tendency to be ferroelectric.

In a seminal paper using first-principles calculations based on density functional theory, Cohen [14] showed with examples of $BaTiO_3$ and $PbTiO_3$ that the hybridization between d states of the TM and p states of oxygen is critical to the electronic origin of ferroelectricity in these compounds, in addition to an interesting role played by Pb^{2+} in the latter. Ferroelectricity in these compounds is phonon-related: collective ordering of electric dipoles in the low-symmetry phase is associated with a polar zone center (wave vector $q = (0,0,0)$) phonon. With respect to structural distortions corresponding to this normal mode, the high-symmetry cubic structure becomes unstable below the ordering temperature. Marking an important step to identify the signature of such unstable modes of the cubic phase, Zhong *et al.* [15] showed that the longitudinal optical-transverse optical (LO-TO) phonon splitting is the largest (and anomalously large) for the unstable mode of the cubic structure. Physically, LO–TO splitting reflects on the strength of long-range electric dipolar interactions in an insulator. The LO–TO splitting depends quadratically on the mode (or Born) effective charge and inversely on the electronic dielectric constant (square of the refractive index). The latter typically ranges from six to nine in the insulating d^0 oxides. Thus, a primary indicator of the tendency of a cubic structure to exhibit a ferroelectric instability through ionic displacement is the Born effective charge (BEC) Z^*.

The fact that anomalous Z^* interestingly reflects on the chemical tendency of an ion to move off-center and yield ferroelectric dipole moment is not surprising, as Z^* is the coupling between ionic (dipolar) displacement (u) and electric field: $H' = -Z^*uE$. An anomalously large Z^* essentially means that a large off-centering force acts on an ion even because of a small applied electric field. Alternatively, Z^* also gives an electric dipole moment associated with off-centering displacements of an ion: $p = Z^*u$, and is kind of a measure of *"ferroactivity"* of a given ion [16]. For example, Z^* of Ti in $BaTiO_3$ is 7.5, while that of Nb in $KNbO_3$ is about 9.3, indicating an important role of these cations in the phonon-related ferroelectricity in these compounds. In $PbTiO_3$, Z^* of Pb is about 3.9 (deviating significantly from its nominal ionic charge), indicating its relevance and contribution to ferro-electricity in $PbTiO_3$.

Chemical mechanisms responsible for the anomalously large Z^*s in $BaTiO_3$ and $PbTiO_3$ has been investigated using projection of Wannier functions on atomic orbitals, called *Wannier orbital overlap population* [17], to obtain a local picture (Figure 9.4) of adiabatic or quasi-static transfer of charge during off-centering of a TM. This analysis showed that a fraction of electronic charge is translated through an entire unit cell from a d orbital of one TM ion to that of another TM ion hopping via π-bonded p orbitals of the oxygen, in the direction opposite to that of displacement of the TM. Owing to negative charge of an electron and a large

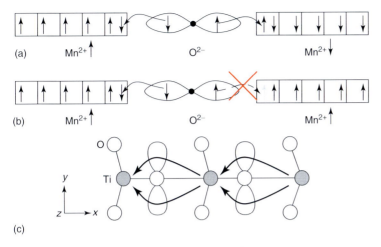

Figure 9.4 Superexchange interaction between Mn^{2+} ions favors (a) anti-parallel alignment of spins on Mn ions connected by oxygen to parallel one (b) A similar electronic transfer mechanism is involved in ferroelectric structural distortion in $ATiO_3$ (Source: Ref. [17].)

translation, it effectively gives rise to a large positive electric dipole moment, and hence a large Z^*. In fact, this is quite analogous to the double-exchange mechanism in magnetism (or electron-hopping-based superexchange [18]), but involves only the electronic charge and no spin. This mechanism is more pronounced when (i) most of the d states of the TM are unoccupied (in the conduction band) and (ii) the energy gap between the valence and conduction bands is small.

In contrast, magnetism in ABO_3 oxides is based on localized spins on TM, which exist only if d orbitals of a TM are occupied. The type of ordering between these localized spins located on the lattice of TM ions is determined by the nature of super or double-exchange interactions between nearest neighbor TM, which depend on various factors such as the d orbital energies of a TM ion as affected by crystal field, its spin state, and the TM–O–TM bond angle along which hopping electronic spin occurs. For example, the interaction between spins on the two TM ions with d^5 configuration (such as Mn^{2+}), in octahedral coordination, connected via $180°$ bond is antiferromagnetic (AFM) (Figure 9.3). Owing to scarcity of multiferroic materials with ferroelectricity and ferromagnetism, the term *multiferroic* has been liberalized to include a combination of ferroelectricity and antiferromagnetism.

From this, it is clear that the chemical ingredients essential for the emergence of ferroelectricity and ferromagnetism in ABO_3 oxides are mutually conflicting, that is, d^0 versus d^n ($n \neq 0$) electronic configuration of the TM ion at the B site. As a result, it seemed in 2000 that there would be very few materials that are simultaneously ferroelectric and ferromagnetic [19]. Indeed, this is quite true if one restricts oneself to these conventional mechanisms of ferroelectricity and ferromagnetism. One route out of this conflict is to use different cations of ABO_3 oxides for the two ferroic orders. Learning from the example of $PbTiO_3$, in which stereochemically active lone pair of electrons of Pb is responsible for its off-centering [16, 17], Pb and Bi ions at the A site can be quite effective in driving ferroelectricity, while a TM ion with nonzero occupation of its d orbital can be the source of magnetism. Thus, $BiMnO_3$ [20] and $BiFeO_3$ [21] are the classic examples of single-phase materials with independent mechanisms of ferroelectricity (Bi) and magnetism (Mn and Fe with d^4 and d^5 configurations, respectively), with fairly high transition temperatures [20, 21].

9.5
Classification of Multiferroics

One scheme of classification [10] of multiferroics is based on the nature of the fundamental mechanism through which inversion symmetry is broken and ferroelectricity arises. Fundamentally, the degrees of freedom of a solid material are associated with electrons and nuclei: electronic charge, electronic orbital, electronic spin, and phonons (associated with deviation in the lattice structure, and also referred to as the *lattice degrees of freedom*). Emergence of ferroelectricity, as discussed earlier, is associated with inversion-symmetry-breaking phase transitions in the parent high-symmetry phase of the material. In such a transition, one or more specific combinations of these degrees of freedom are instrumental (or relevant) in reducing the symmetry. If the field describing broken inversion symmetry (e.g., polarization) depends *linearly* on the degrees of freedom relevant to the transition, the material is considered to be a *proper* ferroelectric. If this dependence is *nonlinear* (typically quadratic or higher order), the material is termed as an *improper* ferroelectric. The nonlinearity, in general, may be associated with two distinct fields that order at two different temperatures.

Ferroelectricity in $BaTiO_3$ or similar d^0 compounds is an example of proper ferroelectrics in which inversion symmetry is broken directly by a polar optical phonon associated with the B cation of the cubic perovskite structure. Similarly, ferroelectricity in $PbTiO_3$ [22] or $BiFeO_3$ is associated with polar optical phonons associated with the A cation of Pb and Bi, respectively. There is greater variety of improper ferroelectrics: (i) in compounds such as $YMnO_3$, ferroelectricity arises from a combination zone-boundary phonons and is essentially improper [23, 24]; (ii) a superposition of two distinct charge-ordered states breaks inversion symmetry and yields ferroelectricity in $LuFe_2O_4$, which is an "electronic" improper ferroelectric [25]; and (iii) certain type of spin-ordering (often characterized by two components or two order parameters) breaks inversion symmetry in materials such as $TbMnO_3$ [26, 27], YMn_2O_5, or $TbMn_2O_5$ [28], which are termed *magnetic ferroelectrics* [26–28].

In another scheme of classification, suggested by Khomskii [10], the magnetic ferroelectrics defined above are termed *type II multiferroics*, and the rest as type I multiferroics. The logic in this scheme of classification is that two distinct or independent mechanisms give rise to ferroelectricity and (ferro)magnetism in type I category, and as a result the magnetoelectric coupling is not inherently strong. In contrast, the magnetoelectric coupling (albeit second order, as we will see later) in the type II multiferroics is typically much stronger, as the magnetic order itself is the driving force behind ferroelectric ordering. However, because of the smaller energy scales involved, such ordering is seen at lower temperatures.

It is interesting that these classification schemes (particularly the first one) are ferroelectric-centric. The reason for this is that an ordering of magnetic moments, either ferromagnetic or AFM, is always possible given localized spins in d^n TMs. On the other hand, the origin of local electric dipoles and their ordering is perhaps more subtle owing to the balance between short- and long-range interactions (indicated in the LO–TO splitting) and sensitivity to strain. One must wonder why analogous type II multiferroics, in which magnetic ordering results from electric ordering, are not yet seen. The symmetry arguments do help in understanding this: any *static* electric order is invariant under time-reversal symmetry. As a result, generalization of magnetoferroelectrics to electroferromagnets is not quite so simple or straightforward. However, it is expected to be quite able to give rise to dynamical (frequency-dependent magnetoelectric) properties.

9.6
Multiferroic Materials and Mechanisms

Mechanisms at many energy scales are involved in the determination of the magnetoelectric properties of a material. Zeeman splitting is typically less than a millielectron volts, while energies of phonons in oxides range up to a tenth of an electronvolt. Charge-transfer gaps and crystal field splittings are typically of the order of 1 electronvolt, and spin–orbit interaction (necessary for noncollinear ordering of spins) is of the order of a few-tenths of an electronvolt. The interplay between these makes the physics of magnetic multiferroic materials very interesting.

9.6.1
Lone-Pair Chemistry and d Magnetism

As discussed earlier in Section 9.4, Type I multiferroics with proper ferroelectricity exhibit independent sources of the two ferroic ordering. Lone pair of electrons, typically in the 6s level of the A site, is stereochemically active. While this pair of electrons does not participate in bonding of an A cation with other atoms, it gives the A site the ability to have an additional directional bond even in the absence of another atom in that direction. This leads to off-centering of the A cation to create space and a "bond" there with localized 6s electrons in it. During the process of off-centering, inversion symmetry is broken and the 6p electrons also get involved [16]. Owing to involvement of the 6s and 6p electronic states, local polarizability of an A cation becomes an important contributor to ferroelectricity [17], in contrast to that in materials such as $BaTiO_3$ or $KNbO_3$, where the local atomic polarizability is less relevant. Typically, cations of late elements in groups IV and V exhibit lone pair with such chemistry; thus Pb^{2+} and Bi^{3+} ions are highly "ferroactive."

Multiferroic materials with such mechanism of ferroelectricity include $PbVO_3$, $BiMnO_3$, and $BiFeO_3$. As magnetic and electric ordering in these materials are essentially independent in origin, they couple with each other rather weakly. To increase their coupling, it is probably desirable to modify the host material such that the temperature gap between the transitions of the two ordering becomes smaller, so that the fluctuations in electric and magnetic order are strong in the same (or overlapping) temperature range, giving rise to stronger magnetoelectric coupling. Another interesting property of these materials is that they couple rather strongly with strain, which is also related to the role of A cation and its lone pair. This is also evident in a significantly stronger-polarization–strain coupling in $PbTiO_3$ [22] than in $BaTiO_3$. As a result, ferroelectric transition in a material with A-site-driven ferroelectricity is typically of a strong first-order character and hence of a large latent heat. We note that $BiFeO_3$ exhibits G-type AFM and a weak ferromagnetism associated with the canting of spins (Section 9.7). On the other hand, $BiMnO_3$ is the only multiferroic known [20] to exhibit both ferroelectricity and FM (20), though its ferroelectric properties are weak and sensitive to stoichiometry of the compound.

9.6.2
Phonon-Driven Improper Ferroelectricity

In hexagonal manganites (such as $InMnO_3$ and $YMnO_3$), ferroelectricity arises as a secondary order, with zone-boundary (*K*-point) phonons involving tilts of MnO polyhedra as the primary structural instabilities or order parameters Q_Ks [24], and the coupling between the two is expressed with free energy as

$$F = F_0 + \gamma P.Q_k^3 + A(T)Q_k^2 + BQ_{K+...}^4$$

where γ is a fourth-order coupling factor between phonons at the zone-boundary point K and polarization (a polar phonon at the Γ point). The phase transition occurs at fairly high temperatures in these materials, and their dielectric constant (i) has a moderate magnitude and (ii) does not exhibit an anomaly at the transition, because a polar mode at Γ is not primarily relevant to ferroelectricity and does not soften. As a result, polarization switching in these materials requires high electric fields and they may not very useful for memory applications. The improper ferroelectricity in these materials was termed as *geometric ferroelectricity* by Van Aken et al. [23].

Magnetism in these hexagonal manganites is associated with noncollinear arrangements of spins on Mn site in an interesting way to get around the magnetic frustration associated with triangular lattice in the *ab*-plane. Spins exhibit components only in the *ab*-plane, indicating that the *c*-axis is a hard axis for magnetism. Magnetic transition in $YMnO_3$ also occurs at a reasonably high temperature but much below the ferroelectric transition temperature. As the structural symmetry is broken to polar group before the magnetic transition, there is a strong coupling between magnetic ordering and the same infrared (IR)-active mode that is responsible for the ferroelectric order, and there is a large discontinuity in the structure at the magnetic transition, termed as a *giant magnetoelastic effect* [29].

9.6.3
Improper Ferroelectricity from Charge Ordering

Efremov *et al.* [25] showed that a combination of bond- and charge-centered charge ordering (Figure 9.5) breaks inversion symmetry, giving rise to ferroelectricity. So far, such a scenario is believed to have been observed in $Pr_{1-x}Ca_xMnO_3$ and $LuFe_2O_4$. Electron correlations play an important role in the stabilization of such charge-ordered states.

9.6.4
Magnetically Driven Improper Ferroelectricity (Type II Multiferroics)

A great excitement in the field of multiferroics in the last several years has been in this class of multiferroics, where an inhomogeneously ordered configuration of magnetic moments (called *magnetic spirals* [11, 13]) or spins give rise to ferroelectric order in these materials. The basic idea is very simple: the inhomogeneously ordered spin configuration should be noncentrosymmetric. In a way, it is similar to the inhomogeneous structural ordering associated with the zone-boundary phonon in hexagonal manganites! In both the systems, broken inversion symmetry means the presence of an effective electric field that generates a polar response through (i) electronic charge, or (ii) electronic spins, or (iii) structure, that is, IR-active phonon. In fact, one would expect an electric dipolar field (polar response) to arise from each of these degrees of freedom, if available, as the secondary order parameter. Indeed, the contribution of the softest of these degrees of freedom (i.e., the ones with least stiffness, $E = \frac{1}{2}J_{stiff}d^2$, d being a degree of freedom) to electric polarization will be the most dominant.

Physics of the polar order associated with the spiral magnetic state (Figure 9.6) is captured by the following continuum free energy [13]:

$$H = \gamma P \times [M(\nabla.M) - (M.\nabla)M]$$

Specific configuration of a magnetic spiral (e.g., a cycloidal magnetic order in Figure 9.6d) can be viewed as a combination of a longitudinal spin wave and a transverse

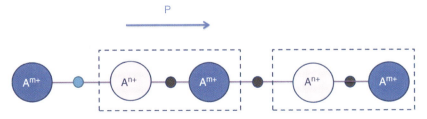

Figure 9.5 How different charges on the same elemental ionic species occurring in charge ordered systems break inversion symmetry (Source: see Ref. [10]).

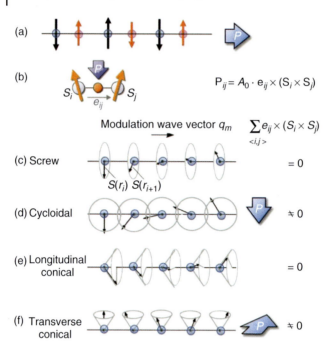

Figure 9.6 Various inhomogeneously ordered magnetic states (Source: Taken from Ref. [7]).

spin wave propagating with a wave vector $\frac{2\pi}{\lambda}$, which are separated by a phase of $\frac{\pi}{2}$. While the longitudinal spin wave contributes to both terms, the transverse wave contributes only to the second term. The key point is that the inversion symmetry is broken such that the polarity is in the same direction at all points in space, though the magnetic ordering is inhomogeneous. It is important to note that (i) the overall magnetization vanishes and that (ii) the induced polarization is proportional to product of amplitudes of transverse and longitudinal spin waves; that is, it is second order in magnetic order parameters, and a nonzero polarization requires both the amplitudes (magnetic orders) to be nonzero.

Multiferroic materials such as $TbMnO_3$ [26] exhibit magnetically driven ferroelectricity in which (i) first, a longitudinal sinusoidal spin wave freezes at a higher temperature (40 K) and (ii) ferroelectricity emerges at a lower locking temperature when a transverse spin wave freezes in. A similar mechanism is involved in multiferroic $Ni_3V_2O_8$ [27]. It is clear that energy scales of this mechanism are very small, and multiferroic behavior is observed only at very low temperatures. On a positive note, the magnetoelectric coupling is quadratic, but very strong. It is strong enough that it allows switching of polarization via application of magnetic fields of the order of a few teslas. This characteristic and the fundamentally interesting magnetic ordering have generated an interest in such materials.

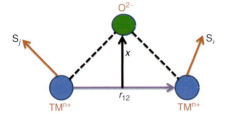

Figure 9.7 Dzyaloshinskii-Moriya Interaction between spins of two transition metals connected by an oxygen (Source: Taken from Ref. [10]).

It is indeed instructive to start with a continuum free energy model given earlier and derive the spin model it reduces to. To this end, we consider two noncollinear spins S_i and S_j in the xy plane located at $(00a)$ and (000) along the x-axis (Figure 9.7), and use a finite-difference formula to determine the continuum expression at the midpoint of the locations of two spin vectors S_i, S_j. Using

$$M = \frac{S_i + S_j}{2}, \nabla.M = \frac{S_i - S_j}{a}$$

it can be shown that

$$H = \gamma P \times M(\nabla.M) - (M.\nabla)M], \propto P.\hat{y}(S_i \times S_j)_z \propto P.\hat{x} \times (S_i \times S_j)$$

which is consistent with the microscopic model based on spin current or the Dzyaloshinskii–Moriya (DM) interaction (see later text). This is not surprising, as the formal expression of these interactions are essentially determined by symmetry considerations, and that there exists only one symmetry invariant term at the third order. Physics of these microscopic interactions will be discussed in the next section.

9.6.5
Ferroelectricity from Collinear Magnetism (Exchange Striction)

In YMn_2O_5, inversion symmetry is broken by an ordering of collinear spins [11, 26]. As the temperature is reduced, a sinusoidal incommensurate ordering of Mn spins occurs first, and is followed by a commensurate AFM ordering at a slightly lower temperature, which is accompanied by ferroelectric ordering manifested in a dielectric anomaly. Most of these materials exhibit complex magnetic phenomena and are magnetically frustrated. A simple model system that brings out this point is an Ising system on a one-dimensional lattice with ferromagnetic first-neighbor and AFM second-neighbor interactions (Figure 9.8), with a strong enough second-neighbor interaction. As is seen, there are two types of neighboring sites in the ground state: one with parallel spins and the other with antiparallel ones. Two different types of bonds naturally result in different changes in the interatomic forces (termed as *exchange striction*), leading to different bond lengths and hence an overall electric dipole moment signifying broken inversion symmetry. We note

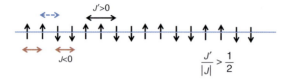

Figure 9.8 Ordering of collinear (Ising) spins that breaks inversion symmetry, also leading to exchange striction. J and J' are the strengths of ferromangetic and antiferromagnetic interactions. (Source: See Ref. [10].)

that exchange striction essentially reflects on the spin–phonon (lattice) coupling (see the next section).

9.6.6
Phonon-Mediated Linear ME Coupling

There are multiferroic materials in which symmetry allows an IR-active (polar) phonon to couple with (i) spin and (ii) electric field at the lowest order (higher order terms are more readily present). For such a phonon degree of freedom u, the free energy can be expressed as

$$F = \frac{1}{2} K u^2 - ZEu - LSu + JS^2$$

where L is the spin–phonon coupling, Z is the Born (or mode) dynamical charge, K is the spring constant of mode u, and J is the exchange interaction strength. Minimizing energy with respect to u, an effective free energy is obtained: $F_{\text{eff}} = -\frac{Z^2}{2K} E^2 + \left(J - \frac{L^2}{2K} \right) S^2 - \frac{ZL}{K} ES$, where the three terms are the phonon contribution to the dielectric constant, the phonon-renormalized exchange coupling, and the phonon-mediated linear magnetoelectric coupling, respectively. While the spin–phonon coupling at the third order is quite common even in high-symmetry systems [30], the linear spin–phonon coupling can arise in materials such as $FeAlO_3$ or $FeGaO_3$ [31], which have antisite disorder between symmetry inequivalent site of Fe and Al [31, 32]. Exchange striction mechanism [11, 26], relevant to $TbMn_2O_5$, can also be formulated in terms of spin–phonon coupling [30].

9.7
Microscopic Mechanisms of Magnetic Ferroelectricity: Type II Multiferroics

In the context of magnetic ferroelectrics, we already motivated the form of a microscopic spin model that is consistent with a continuum Landau theory [13]. The physics of such spin model derives from the classic DM interactions [33, 34]. The DM interaction arises from a relativistic correction to the superexchange interaction between transition-metal ions via bridging oxygen anions, and its strength (determined by the spin–orbit interaction) is typically weak (about 1 millielectron volts or less). Its role in magnetoferroelectrics has been analyzed by

Dagotto *et al.* [35]. The DM interaction between two spins separated by r_{12} via an anion off-centered by x (Figure 9.7) is

$$H_{\mathrm{DM}} = D(x \times r_{12}){\cdot}(S_1 \times S_2)$$

While it is weak in magnitude, it favors ordering of spins in a noncollinear manner. Commonly seen weak ferromagnetism associated with canting of AFM spins can be an effect of DM interaction in noncentrosymmetric systems (for instance, spiral ordering of spins in $BiFeO_3$ at low temperatures is likely to be a result of DM interaction).

As the DM interaction depends linearly on the displacements (x) of the oxygen ions that bridge two transition-metal ions, the inverse effect of DM interaction is a force on the oxygen atom $\left(F = -\frac{\partial E}{\partial x}\right)$ arising from noncollinear spiral ordering of spins, giving rise to oxygen displacements (exchange striction) [11] and hence a polarization. It also contributes to the stabilization of helical magnetic structures at low temperature.

Formally, similar expression for polarization coupling with noncollinear spins is obtained in terms of the spin–current model [36]. In this model, the sign of P depends on a clockwise or counterclockwise rotation of the spin (called *spin helicity*) as the wave propagates along the spiral axis. In summary, the direct effect of the DM interaction in a noncentrosymmetric bond is to cause canting of spins, while it causes exchange striction from noncollinear spins conversely, and displaces the bridging anion to generate local electric dipole (perpendicular to the TM–TM axis).

Fundamental to such complex noncollinear and collinear ordering of spins is the frustration (it could be from the competing first- and second-neighbor interactions, or from the geometric frustration in the structure or disorder, e.g., $GaFeO_3$ or $AlFeO_3$), or from DM interaction.

9.8
Domains and Metal–Insulator Transition

Our arguments of how polarization arises from certain inhomogeneously ordered magnetic states apply equally well to magnetic domains. For example, the Bloch-type domain wall between magnetic domains is similar to a sinusoidal ordering and does not give any electric dipole moments. In contrast, a Nèel-type domain wall is very similar in structure to the cycloidal magnetic order and does give rise to a local electric dipole moment. Indeed, an external electric field is expected to couple locally with the domain wall through this dipole, and have observable structural and magnetic effects. Thus, such domain walls are seats of local magnetoelectric coupling.

Second, the chemical conflict between d^0 and d^n criteria for multiferroics also translates into a competition between metallic and insulating states, both of which are affected by the structure or phonons. As a result, most multiferroics (both type I and II) are on the brink of a metal–insulator transition. Theoretical analysis of $BiFeO_3$ showed that it is metallic for any magnetic ordering in the cubic structure and a gap opens up in its electronic structure only in the phase with G-type AFM ordering and structural distortions involving rotations of FeO_6 octahedra [37]. As a magnetic domain wall involves exchange striction and subsequent change in the local structure, we also expect the electronic bandgap to vary sensitively at the domain wall, and influence its electric conductivity locally.

Another observable consequence of the vicinity of a multiferroic (particularly a material such as $BiFeO_3$) is that its Raman tensors diverge for the modes corresponding to these specific structural distortions that close or open the electronic gap. As a result, some of these phonon modes should be observable as prominent second-order peaks in their Raman spectra, while they may be silent in the first-order Raman spectra. This reflects on the inherent nonlinearity of the coupling between electrons, their spin and phonons in multiferroic materials.

Summary

Multiferroic materials exhibit a variety of interesting phenomena that involve the coupling between electronic charge, spin, orbital, and phonons. Owing to interesting science of multiferroics and their technological potential, a great deal of progress has been made in the last 10 years in (i) identifying several multiferroic materials with magnetoelectric coupling and (ii) understanding its mechanisms. While type I multiferroics exhibit high T_c of ordering, their magnetoelectric coupling is often weak. On the other hand, most of the type II multiferroics exhibit rather strong magnetoelectric coupling, but it is often at very low temperatures and with a weak electric polarization. Indeed, there is still not a single material that exhibits coupled ferromagnetism and ferroelectricity at room temperature. It would be interesting to look for a material in which the magnetic properties can be effectively controlled by an electric field. There is, of course, scope for other mechanisms of magnetoelectric behavior to be discovered, which may involve orbital magnetism and electronic topology.

References

1. Schmid, H. (2008) *J. Phys. Condens. Matter*, **20**, 434201.
2. (a) Smolenskii, G.A. *et al.* (1971(in Russian)) *Segnetoelectrics and Antisegnetoelectrics*, Nauka Publishers, Leningrad. (b) Smolenskii, G.A. and Chupis, I.E. (1982) *Sov. Phys. Uspekhi*, **25**, 475.
3. Venevtsev, Y.N. and Gagulin, V.V. (1994) *Ferroelectrics*, **162**, 23.
4. Schmid, H. (1994) *Ferroelectrics*, **162**, 317.
5. (a) Ahn, C.H., Rabe, K.M., and Triscone, J.-M. (2004) *Science*, **303**, 488. (b) Chu, Y.-H., Martin, L.W., Holcomb, M.B., and Ramesh, R. (2007) *Mater. Today*, **10**, 16.
6. Martin, L.W., Crane, S.P., Chu, Y.-H., Holcomb, M.B., Gajek, M., Huijben, M., Yang, C.-H., Balke, N., and Ramesh, R. (2008) *J. Phys. Condens. Matter*, **20**, 434220.
7. Tokura, Y. and Seki, S. (2010) *Adv. Mater.*, **22**, 1554.
8. Ramesh, R. and Spaldin, N.A. (2007) *Nat. Mater.*, **6**, 21.
9. Khomskii, D. (2009) *Physics*, **2**, 20.
10. Cheong, S.-W. and Mostovoy, M. (2007) *Nat. Mater.*, **6**, 13.

11. Mostovoy, M. (2012) *Physics*, **5**, 16.
12. Dzyaloshinskii, E. (1959) *Sov. Phys. JETP*, **10**, 628.
13. Mostovoy, M. (2006) *Phys. Rev Lett.*, **96**, 067601.
14. Cohen, R.E. (1992) *Nature*, **358**, 136.
15. Zhong, W., King-Smith, R.D., and Vanderbilt, D. (1994) *Phys. Rev. Lett.*, **72**, 3618.
16. Waghmare, U.V., Spaldin, N.A., Kandpal, H.C., and Seshadri, R. (2003) *Phys. Rev. B*, **67**, 125111.
17. Bhattacharjee, J. and Waghmare, U.V. (2010) *Phys. Chem. Chem. Phys.*, **12**(7), 1564–1570.
18. Gopal, P., Spaldin, N.A., and Waghmare, U.V. (2004) *Phys. Rev. B*, **70**, 205104.
19. Hill, N.A. (2000) *J. Phys. Chem.*, **104**, 6694.
20. dos Santos, A.M., Cheetham, A.K., Atou, T., Syono, Y., Yamaguchi, Y., Ohoyama, K., Chiba, H., and Rao, C.N.R. (2002) *Phys. Rev. B*, **66**, 064425.
21. Wang, J. *et al.* (2003) *Science*, **299**, 1719.
22. Waghmare, U.V. and Rabe, K.M. (1997) *Phys. Rev. B*, **55**, 6161.
23. Van Aken, B.B. *et al* (2004) *Nat. Mater.*, **3**, 164.
24. Fennie, C. and Rabe, K.M. (2005) *Phys. Rev. B*, **77**, 100103.
25. Efremov, D.V. *et al.* (2004) *Nat. Mater.*, **3**, 853.
26. Kimura, T. *et al.* (2003) *Nature*, **426**, 55.
27. Lawes, G. *et al.* (2005) *Phys. Rev. Lett.*, **95**, 087205.
28. Sushkov, A.B., Valdés Aguilar, R., Park, S., Cheong, S.-W., and Drew, H.D. (2007) *Phys. Rev. Lett.*, **98**, 027202.
29. Lee, S. *et al* (2007) *Nature* **451**, 805.
30. Ray, N. and Waghmare, U.V. (2008) *Phys. Rev. B*, **77**, 134112.
31. Popov, Y.F. *et al.* (1998) *JETP*, **87**, 146.
32. (a) Pradeep, K., Achintya, B., Muthu, D.V.S., Shirodkar Sharmila, N., Rana, S., Shireen, A., Sundaresan, A., Waghmare, U.V., Sood, A.K., and Rao, C.N.R. (2012) *Phys. Rev. B*, **85**, 13449–13455. (b) Saha, R., Shireen, A., Shirodkar, S.N., Waghmare, U.V., Sundaresan, A., and Rao, C.N.R. (2012) *Solid State Commun.*, **152**, 1964.
33. Dzyaloshinskii, I. (1964) *Sov. Phys. JETP*, **19**, 960.
34. Moriya, T. (1960) *Phys. Rev.*, **120**, 91.
35. Sergienko, A. and Dagotto, E. (2006) *Phys. Rev. B*, **73**, 094434.
36. Katsura, H., Nagaosa, N., and Balatsky, A.V. (2005) *Phys. Rev. Lett.*, **95**, 057205.
37. Neaton, J.B., Ederer, C., Waghmare, U.V., Spaldin, N.A., and Rabe, K.M. (2005) *Phys. Rev. B*, **71**, 014113.

Part V
Interfaces and Magnetism

Functional Metal Oxides: New Science and Novel Applications, First Edition.
Edited by Satishchandra B. Ogale, Thirumalai V. Venkatesan, and Mark G. Blamire.
© 2013 Wiley-VCH Verlag GmbH & Co. KGaA. Published 2013 by Wiley-VCH Verlag GmbH & Co. KGaA.

10
Device Aspects of the $SrTiO_3$–$LaAlO_3$ Interface; Basic Properties, Mobility, Nanostructuring, and Potential Applications

Hans Hilgenkamp

10.1
Introduction

In their 1987 Nobel Prize lecture, Bednorz and Müller [1] introduced their discovery of high-T_c superconductivity by stating: *"At IBM's Zurich Research Laboratory, there had been a tradition of more than two decades of research efforts in insulating oxides. The key materials under investigation were perovskites like $SrTiO_3$ and $LaAlO_3$, used as model crystals to study structural and ferroelectric phase transitions".* Their research on the interplay between electronic, structural, and orbital degrees of freedom when doping these wide-bandgap, highly insulating perovskite oxides ultimately led to the discovery of the most tantalizing conducting materials currently known – the high-T_c superconducting cuprates. At that time, surely, no one would have foreseen that 20 years later, the focus of attention would turn again to $SrTiO_3$ and $LaAlO_3$, even in such a prominent way as being included in the Science Magazine Top 10 Breakthroughs of the Year 2007 [2]. The reason for this was the remarkable discovery of, and subsequent research flurry on, the appearance of a high-mobility two-dimensional conducting state at the interface between $SrTiO_3$ and $LaAlO_3$ (Figure 10.1) [3]. Although it was already well appreciated that the electronic and magnetic properties of insulating or (super)-conducting oxides could be strongly modified at external or internal interfaces, Ohtomo and Hwang, supplemented in a follow-up paper by Nakagawa *et al.* [4], brought the awareness to a larger scientific community that for very basic reasons, to be discussed later, the interplay between these two insulating oxides might lead to internal charge redistributions and the formation of novel phases at their contact. Although demonstrated for the $SrTiO_3$/$LaAlO_3$ case, it was immediately clear that many more materials combinations could show related effects leading to novel behavior, and an exciting new research field on complex oxide "interphases" was opened up.

Meanwhile, this research field has lead to many hundreds of publications on the structural, electronic, magnetic, and optical properties of $LaAlO_3$/$SrTiO_3$ and related complex oxide interfaces, as well as on application-related aspects such as field-effect devices, electron mobility enhancement, and nanostructuring. Several review papers and prospectives have meanwhile been reported, describing

Functional Metal Oxides: New Science and Novel Applications, First Edition.
Edited by Satishchandra B. Ogale, Thirumalai V. Venkatesan, and Mark G. Blamire.
© 2013 Wiley-VCH Verlag GmbH & Co. KGaA. Published 2013 by Wiley-VCH Verlag GmbH & Co. KGaA.

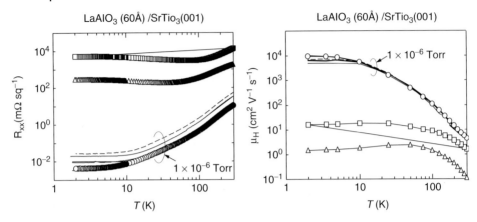

Figure 10.1 The key graphs from the 2004 Nature article of Ohtomo and Hwang pointing out the high-mobility conducting phase formed when growing LaAlO₃ on SrTiO₃. Shown are the temperature dependence of the sheet resistivity R_{xx} and the carrier mobility μ_H for a system of 6 nm of LaAlO₃ grown on a TiO₂-terminated (001)-oriented SrTiO₃ single crystal. The samples were grown and cooled down at different oxygen pressures, ranging from 10^{-4} to 10^{-6} mbar. The dashed line indicates a sample grown at 10^{-6} mbar, followed by annealing in 1 atm of O₂ at 400 °C for 2 h. (Source: From Ref. [3].) NB: The y-axis legend of the left graph has later been corrected to be R_{xx} (Ω sq^{-1}) [3].

progress in the research on these interfaces [5–11]. This chapter aims particularly to address the device aspects, without claiming completeness. As a start, however, a brief introduction to the basic characteristics of the conducting LaAlO₃/SrTiO₃ interfaces will be given, especially aimed at novices in this area of research. It is noted that this book also contains a chapter specifically on X-ray spectroscopic studies of such conducting interfaces by Wadati and Fujimori.

10.2
The LaAlO₃/SrTiO₃ Interface: Key Characteristics and Understanding

For many purposes, it is very insightful to view the perovskite oxides, with structural unit formula ABO₃, as little atomic LEGO™ blocks, from which new multicomponent, artificial, thin film crystals can be synthesized with great versatility, for example, using advanced epitaxial film growth techniques [12]. In this formula, A is typically an element from the groups of alkaline-earth metals (such as Ca, Sr, or Ba) or rare-earth metals (La, Ce, Sm, Gd, Sc, Y, etc.) and B a transition-metal element (Ti, V, Mn, Fe, Cu, Ru, etc.). Also, A and B can represent mixtures of different elements, such as La$_x$Sr$_{1-x}$ and Cu$_y$Mn$_{1-y}$. For the discussion of the interface effects, one needs to look one level deeper however, as a key aspect is to be found *within* these blocks. To elucidate this, we consider the perovskite crystals as stacks of atomic layers of the composition ... AO–BO₂–AO–BO₂ .... . For the cases of the seemingly similar wide-bandgap insulators SrTiO₃ (3.2 eV) and LaAlO₃ (5.6 eV), this layered picture immediately reveals an essential difference; when viewing the standard formal valencies, the $\{Sr^{2+}O^{-2}\}^0$ and $\{Ti^{4+}(O^{2-})_2\}^0$ layers are both charge-neutral, while the LaAlO₃ with its $\{La^{3+}O^{-2}\}^{+1}$ and $\{Al^{3+}(O^{2-})_2\}^{-1}$ layers exhibits an internal polarization.

This difference in internal polarization has profound consequences when these materials are combined. Maintaining the . . . AO–BO₂–AO–BO₂–AO . . . stacking sequence in an epitaxial heterostructure, one ends up with either a TiO₂:LaO or a SrO:AlO₂ interface. As was pointed out by Nakagawa *et al.* [4], without any reconstruction, the polarization discontinuity that arises at these interfaces would result in a build up of electrostatic potential from the interface into the polar LaAlO₃ material, analogous to the known case of polar–nonpolar heterojunctions in semiconductors, as analyzed, for example, for Ge–GaAs combinations [13]. As this potential build up increases with every LaAlO₃ unit cell, it diverges with increasing LaAlO₃ thickness, a situation referred to as the *polarization catastrophe*. However, before this internal potential can reach unphysically high values, nature has mechanisms in place to compensate for it. For the semiconductor interfaces, these typically present themselves in the form of structural surface reconstructions [13]. But interestingly, merely a charge transfer to the polar interface would already be sufficient to circumvent the potential build up. This happens, for example, in the case of K-doped C_{60}, as was described by Hesper *et al.* [14], who introduced the term *electronic reconstruction* for this.

For the SrTiO₃–LaAlO₃ case, the required charge transfer to avoid a polarization catastrophe would be half an electron charge per unit cell area, projected to the interface. As the lattice parameters of SrTiO₃ and LaAlO₃ are 3.905 and 3.793 Å, respectively, this equals $\sim 3 \times 10^{14}$ cm^{-2}. For the case of the $(TiO_2)^0/(LaO)^+$ interface, a negative (electrons) compensating interface charge would be needed, and for the $(SrO)^0/(AlO_2)^-$ case, a positive (holes), with the associated surface charge of the opposite polarity residing at the LaAlO₃ top surface.

Experimentally, it is seen that a well-conducting state appears only at the n-type TiO₂/LaO interfaces. The sheet resistivities of the SrO/AlO₂ interfaces are several orders of magnitude larger (Figure 10.2). For future nanoscopic electronic devices, it is very interesting to note that the difference between a conducting and an insulating interface is just based on a single atomic layer, being LaO or SrO, placed between the topmost substrate TiO₂ layer and the first AlO₂ layer. Likely, the compensating mechanism for the "p-type" interfaces is a structural reconstruction involving the formation of largely immobile oxygen vacancies, each oxygen vacancy representing two positive charges compared to the fully oxidized case. Most of the research, and also most of the remainder of this chapter, is focused on the n-type TiO₂/LaO interfaces.

Advances in the sample fabrication, especially facilitated by the control over the substrate surface termination [16, 17] and by the *in situ* monitoring with reflective high-energy electron diffraction (RHEED) in pulsed laser deposition (PLD) [18], have enabled studies on the properties of complex oxide thin films and heterostructures with atomic-layer precision. With this, it was discovered that the conductivity of the n-type interfaces sets in only after passing a threshold of

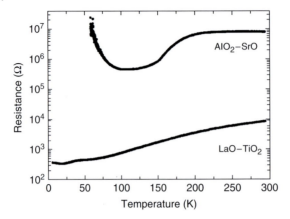

Figure 10.2 Temperature dependence of the sheet resistance for 26 unit cells thick LaAlO₃ films on SrTiO₃ substrates with a TiO₂-terminated surface (TiO₂/LaO interface) and a SrO-terminated surface (SrO/AlO₂ interface). (Source: From Refs. [5, 15].)

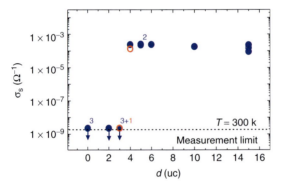

Figure 10.3 Sheet conductance of the LaAlO₃/SrTiO₃ interfaces plotted as a function of the number of their LaAlO₃ unit cells, showing the critical threshold for conductance of four unit cells of LaAlO₃. This result holds for the case that no additional capping is applied to the LaAlO₃. The samples grown at 770 and 815 °C are shown in blue and red, respectively. All samples were grown at $pO_2 = 2 \times 10^{-5}$ mbar and cooled down in 400 mbar of oxygen. The numbers next to the data points indicate the number of samples with values that are indistinguishable in this plot. (Source: From Ref. [19].)

four LaAlO₃ unit cells (Figure 10.3) [19]. This appears to be well in line with the suggested polarization catastrophe scenario, because, between three and four unit cells of LaAlO₃, the internal potential may be high enough to overcome the 3.2 eV bandgap of SrTiO₃. This is required to transfer electrons from the oxygen 2p valence band states at the LaAlO₃ surface to the Ti 3d conduction band states at the TiO₂/LaO interface. Density functional theory (DFT) calculations predict such a transition at around four unit cells of LaAlO₃ (Figure 10.4) [20, 21].

It is of interest to compare the insulator-to-conductor transition at four LaAlO₃ unit cells to the conclusions from SrTiO₃–LaAlO₃–SrTiO₃ multilayer studies,

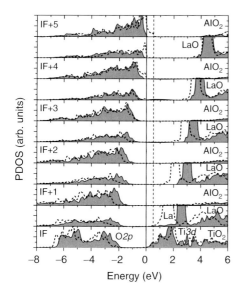

Figure 10.4 Calculated layer-resolved density of states of a five unit cells LaAlO₃ layer on SrTiO₃ showing the expected internal potential build up in the LaAlO₃. The gray-shaded areas are of interest here, which were calculated using relaxed atom positions. (Source: From Ref. [21].) NB: It is noted that DFT techniques notoriously underestimate the bandgap, which is known to be 5.6 eV for bulk LaAlO₃.

which showed that the conductivity is fully developed with an LaAlO₃ layer thickness exceeding about six unit cells [22]. Remarkably, however, in the multilayer case, conductance is observed – albeit with a reduced carrier density – down to the limit of a single LaAlO₃ unit cell sandwiched between SrTiO₃ [22, 23]. This hints at the importance of additional aspects for the above-mentioned insulator-to-conductor transition, relating to the coverage of the LaAlO₃ surface; for example, air/vacuum, specific molecular adsorbates, or capping layers such as other perovskite oxides. This will be a topic of more discussion later in this chapter.

A key unresolved issue has been the question why the expected large potential buildup inside the LaAlO₃ is not observed in spectroscopy. Such strong internal electric fields, with expected values of close to 1 V per unit cell, should, for example, give rise to clearly detectable core level shifts in X-ray photoelectron spectroscopy. However, at most about 10% of this value has been seen [24, 25]. This has cast doubt on whether the polarization catastrophe is really the driving mechanism for the interface conductance. Simultaneously, various studies indicated the important role of oxygen vacancies as a potential alternative doping source and driving mechanism for the conductance [26–28].

To shed further light on this matter, Brinkman *et al.* [29] studied the properties of LaAlO₃/SrTiO₃ interfaces grown by PLD under different oxygen pressures (Figure 10.5). From this and related studies; three different regimes could be distinguished. The first are films grown at very high vacuum conditions, typically 10^{-6} mbar. These have the lowest resistances. However, on analyzing the carrier

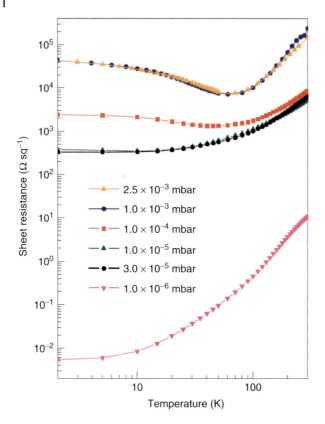

Figure 10.5 Temperature dependence of the sheet resistance of n-type SrTiO$_3$/LaAlO$_3$ (26 unit cells) interfaces grown and cooled down at the indicated oxygen partial pressures. (Source: From Ref. [29].)

densities as obtained from Hall effect measurements, it is clear that in these samples the conductivity is not confined to the interface alone, as was already noted in Ref. [3]. The carrier densities are up to 10^{17} cm^{-2}, which are orders of magnitude above the expected self-limiting value of 0.5 electrons/unit cell area from the polarization discontinuity model. Interestingly though, the mobilities of these samples are quite high, about 10^4 cm^2 V^{-1} s^{-1} at low temperatures.

The second regime is at growth pressures of around 1–6 $\times$ 10^{-5} mbar. These samples show clearly lower conductivity, primarily resulting from lower carrier densities below the 3 $\times$ 10^{14} cm^{-2} mark. Most samples reported in the literature have been grown at these oxygen pressures, in most cases with a subsequent cooling down from the deposition temperature (typically in the range of 700–800 °C) to the room temperature in a high-pressure (several tenths of a bar) oxygen environment. Superconductivity was found for such interfaces below a temperature of about 200–300 mK [30]. The (super)conductivity is clearly confined in this case to a nanoscopically thin layer near the interface, as was revealed, for example, by field-effect experiments [31] and cross-sectional conducting AFM studies [32]. A typical thickness for the electron gas of about 10 nm was reported at high and low temperatures [33]. From related studies, the authors derived a typical Fermi wavelength in the conducting $SrTiO_3$ of 16 nm at low temperatures [34], which would imply that the interfaces are on the verge of two-dimensionality.

When further increasing the oxygen pressure during growth, to typical values of about $1–5 \times 10^{-3}$ mbar, the samples no longer become superconducting but instead show magnetic activity. The resistance versus temperature characteristics shows a Kondo-like upturn in the resistance for decreasing temperature, with a minimum at the remarkably high temperature for such effects of about 70 K. Also, these samples show a strong magnetoresistance [29, 35, 36], and hysteresis effects have been observed in the resistance versus field curves at low temperatures [29]. It is noted that this hysteresis has a strong field-sweep-rate dependence and does not have the standard characteristics of ferromagnetism. Currently, studies are being carried out to further elucidate the source of this hysteresis. It is of interest to note that the magnetic effects for samples grown under similar conditions were found to be less pronounced when decreasing the $LaAlO_3$ thickness [37]. Recent scanning superconducting quantum interference device (SQUID) microscopy studies [38] indicate a critical thickness of about three unit cells for the occurrence of small, distributed ferromagnetic patches in n-type $LaAlO_3–SrTiO_3$ interfaces. Similar ferromagnetic regions were observed for p-type interfaces, indicating that conductivity is not a prerequisite for the occurrence of this ferromagnetism.

Magnetic effects have been explained at the $LaAlO_3/SrTiO_3$ interface as a possible result of $GdFeO_3$-like rotations of the TiO_6 octahedra at the interface and the close proximity of the antiferromagnetic insulating state to a ferromagnet metallic state [39]. Such an explanation based on structural deformations suggests that the magnetic properties could depend sensitively on the local strain properties at the interfaces. Another recent study has suggested the magnetism to be mediated by oxygen vacancies at or near the interface [40].

While most studies have concentrated on the above-mentioned growth regimes, with or without a postanneal at high oxygen pressures directly after growth (a comparison between both situations has been reported in Ref. [41]), a study by the NUS–Singapore group has shown that the magnetic effects become even stronger when the PLD growth pressure is increased to the 10^{-2} to 10^{-1} mbar range, with the simultaneous occurrence of different electronic/magnetic phases [42]. Shortly afterward, the coexistence of superconducting and magnetic phases in single samples, possibly at different positions, was reported by various techniques [43–45].

We can thus conclude that, under different forms of doping, whether resulting from the polarity of the $LaAlO_3$, from the growth-induced creation of defects such as oxygen vacancies in the $SrTiO_3$, from applied electric fields – discussed later – or even from photoexcitation, multiple electronic/magnetic phases can be formed at the $SrTiO_3/LaAlO_3$ interface. There appears to be a confining potential formed by band bending at the $LaAlO_3/SrTiO_3$ interface of nanoscopic (on the order of 10 nm or smaller) thickness in which the charges originating from these different sources are collected. The characteristics of these interfaces can be further influenced by aspects such as cation defects and interdiffusion, which are known to be present at these interfaces [8, 46–48], as well as by strain and local deformations. As a resulting picture, a tentative phase diagram was proposed (Figure 10.6) [5], keeping in mind that in practical samples the local doping profile and the just-mentioned additional influencing aspects need not be homogeneous. The exact position of the intrinsic doping level x_{int}, that is, the doping solely resulting from the expected polarity catastrophe in the LaAlO₃ and the resulting electronic reconstruction, depicted here in the region of magnetically active two-dimensional electron gas (2DEG), is still under debate and depends also on fabrication conditions and the LaAlO₃ layer thickness.

Besides LaAlO₃/SrTiO₃, many other materials combinations have meanwhile been studied, including (but not limited to) LaTiO₃/SrTiO₃ [49], LaAlO₃/LaVO₃/LaAlO₃ [50, 51], LaGaO₃/SrTiO₃ [52], LaCrO₃/SrTiO₃ [53], DyScO₃/SrTiO₃ [54], La₂CuO₄/La₁.₅₅Sr₀.₄₅CuO₄ [55], GdTiO₃/SrTiO₃ [56], and amorphous oxides/SrTiO₃ [57]. Many of these interfaces were found to be conducting, from which it can be concluded that the polarity of the LaAlO₃ is not necessary for the occurrence of the interface conductance. A common denominator, however, in many studies is SrTiO₃, which appears to determine most of the physics observed. In fact, also vacuum-cleaved SrTiO₃ was found to exhibit a 2DEG at its surface [58].

An aspect to consider in relation to the three to four unit cell transition is the influence of potential fluctuations at the LaAlO₃ surface on the carrier mobility in the interface underneath. It has been shown by several spectroscopy studies that even for LaAlO₃ layers of less than four unit cells, Ti³⁺ states are present at the interface [24, 25], albeit much less than those for the more than four unit cells case, signaling electron doping. These electrons, which may be due to surface redox processes and oxygen vacancies, could be localized by potential fluctuations arising,

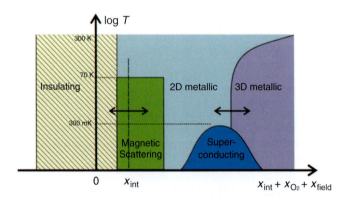

Figure 10.6 Tentative doping versus temperature phase diagram of SrTiO₃/LaAlO₃ interfaces. The doping scale consists of three possible contributions: intrinsic carrier doping from the electronic reconstruction (X_{int}); carrier doping from oxygen vacancies (X_{O_2}); and carrier doping by applied electric fields (X_{field}). (Source: From Ref. [5].)

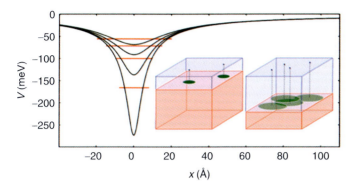

Figure 10.7 Calculated trapping potential created by a surface oxygen vacancy as seen by interface electrons versus distance for different thicknesses of the LaAlO$_3$, ranging from one (deepest) to four unit cells (shallowest). The insets show sketches of the range and density of trapped states. (Source: From Ref. [59].)

for example, from oxygen vacancies at the LaAlO$_3$ surface [59]. In Figure 10.7, it is sketched how an extended distance of the LaAlO$_3$ surface to the interface may in this scenario lead to a reduction in the trapping potential and to a delocalization of the electron charge at the LaAlO$_3$/SrTiO$_3$ interface. This would also provide a natural explanation why with an additional capping layer, which smoothens out LaAlO$_3$ surface potential fluctuations, conductivity is already observed down to one unit cell of LaAlO$_3$ [22].

10.3
Charge-Carrier Mobility

An important parameter for device applications of the conducting interface systems is the charge-carrier mobility μ. With one type of charge carrier (one-band system), the carrier mobility can straightforwardly be derived from the resistivity and Hall effect data, giving the Hall mobility μ_H. For the LaAlO$_3$–SrTiO$_3$ system, one should keep in mind, however, that the mobile carriers are expected to be derived from multiple subbands, associated with different Ti orbital states (3d$_{xy}$, 3d$_{yz}$, and 3d$_{xz}$) [58, 60] and arguably also different TiO$_2$ layers [61, 62]. The charge carriers in these subbands are expected to be characterized by different effective masses and mobilities. This manifests itself in, for example, nonlinear Hall resistivities, from which values for the carrier density and mobility can only be derived by approximate fitting. For sufficiently high mobilities, however, quantum oscillations can be discerned in the magnetoresistance of the interfaces, as a result of the Shubnikov–de Haas (SdH) effect. These SdH oscillations provide an alternative route to derive the carrier mobilities and effective masses, also in multiband systems.

Already in the 2004 Ohtomo–Hwang [3] paper, the high mobility of the conducting state at the SrTiO$_3$/LaAlO$_3$ interface was emphasized, reporting SdH oscillations and μ values exceeding $10^4 \, \text{cm}^2 \, \text{V}^{-1} \, \text{s}^{-1}$. While it is meanwhile understood that for those particular samples the $SrTiO_3$ substrates exhibit bulk conductivity owing to the growth at very low oxygen pressures (as discussed in relation to Figure 10.5, and see also Ref. [28]), it demonstrated that high-mobility conducting states are in principle attainable using these complex oxidic systems, albeit at low temperatures. In fact, comparable high mobilities were measured already long before for conducting Nb-doped $SrTiO_3$ single crystals [63]. The question now is whether such high mobilities, or even higher, can also be reached in (quasi)-2D conducting $SrTiO_3/LaAlO_3$ interface systems.

Clear SdH oscillations for $LaAlO_3/SrTiO_3$ conducting interfaces in the 2D limit were presented by Caviglia *et al.* [64]. Their samples, consisting of 5–10 unit cells of $LaAlO_3$ on $SrTiO_3$ (001), were fabricated at 1×10^{-4} mbar pO_2, with a subsequent high-pressure cooldown in $pO_2 = 200$ mbar, by PLD at a relatively low growth temperature of ~650 °C. The carrier mobilities, as derived from Hall effect measurements analyzed in a single-band approximation, reached values up to $6600 \, \text{cm}^2 \, \text{V}^{-1} \, \text{s}^{-1}$ at 1.5 K. Varying the orientation of the magnetic field from perpendicular to in-plane with the $LaAlO_3/SrTiO_3$ interface clearly showed that the oscillations and their periodicity with $1/B$ only depended on the perpendicular component of the magnetic field, indicating a two-dimensional Fermi surface.

Ben Shalom *et al.* [65] reported SdH oscillations in electric-field-gated 15 unit cells $LaAlO_3/SrTiO_3$ samples fabricated at 10^{-4} mbar pO_2, reaching values of about $6000 \, \text{cm}^2 \, \text{V}^{-1} \, \text{s}^{-1}$ at $T = 0.4$ K. They note that the carrier density as derived from the SdH-oscillation frequency was considerably lower than that from Hall effect measurements, which also show marked nonlinearities. This points to the earlier-mentioned multiband scenario, corroborating indications for this reported also in Ref. [64], in which only part of the carriers have sufficiently high mobilities to give rise to quantum oscillations.

By capping the $LaAlO_3$ layer with a stack consisting of $SrCuO_2$ and $SrTiO_3$, the Twente group showed a strong enhancement of the mobilities, with values well exceeding $10\,000 \, \text{cm}^2 \, \text{V}^{-1} \, \text{s}^{-1}$ at $T = 4.2$ K [66]. The highest value of $\mu = 50\,000 \, \text{cm}^2 \, \text{V}^{-1} \, \text{s}^{-1}$ was reached at the same temperature for samples deposited at very low oxygen pressures (10^{-6} mbar), while maintaining low carrier densities on the order of $10^{13} \, \text{cm}^{-2}$. The exact mechanism for this mobility enhancement due to the $SrCuO_2$ is not yet clear, but it appears to be related to the elimination of scattering sites that also act as charge donors with a typical activation energy of about 6 meV. This follows from Figure 10.8 a,b. The enhancement of the mobility strikingly coincides with the removal of a thermally activated charge-carrier contribution.

Manifest multiband SdH oscillations have been measured for several $SrCuO_2$-containing samples in high magnetic field measurements, as shown in Figure 10.9 [67]. These reveal four to five 2D conduction subbands with carrier effective masses between 1 and 3 m_e, quantum mobilities of the subbands for these particular samples on the order of $500–3000 \, \text{cm}^2 \, \text{V}^{-1} \, \text{s}^{-1}$ ($T = 4.2$ K), and band edges of only a few millielectron volt's below the Fermi energy.

It is conceivable that, with optimized samples, perhaps requiring MBE growth, these mobilities can be further enhanced. In this respect, it is of interest to note

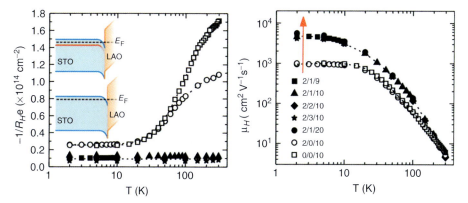

Figure 10.8 Transport properties of SrTiO$_3$-(SrCuO$_2$)-LaAlO$_3$-SrTiO$_3$(001) heterostructures. (a) Temperature dependence of $-1/R_He$, indicating the carrier density in a one-band approximation, for heterostructures with (closed symbols) and without (open symbols) SrCuO$_2$ layer. The corresponding bandstructures, in a polarization-catastrophe-driven scenario, are schematically represented in the insets, with the shaded areas indicating the bandgap regions and the red line representing the defect donor states. When the additional SrCuO$_2$ layer is introduced, the defect donor band (top inset) is eliminated (bottom inset). (b) Corresponding temperature dependence of Hall mobility μ_H. Various heterostructure configurations are given; for example, 2/1/10 represents a 2 unit cell SrTiO$_3$ layer with a 1 unit cell SrCuO$_2$ layer and a 10 unit cell LaAlO$_3$ layer on an SrTiO$_3$ substrate, and in the heterostructure indicated with 2/0/10, the SrCuO$_2$ layer is absent. Later, as yet unpublished, studies showed an enhancement up to 50 000 cm^2 V^{-1} s^{-1} for samples grown at 10^{-6} mbar while maintaining low carrier densities. (Source: From Ref. [66].)

that in La-doped SrTiO$_3$ thin films whopping values for the carrier mobilities as high as 120 000 cm^2 V^{-1} s^{-1} have been reached by the application of compressive strain [68], indicating, indeed, room for such further enhancement also in the SrTiO$_3$-based interface systems.

10.4
Micro/Nanostructuring

To create devices, patterning submicrometer structures is indispensable. Two common methods of patterning, namely lift-off and ion beam etching, are complicated by the fact that the high-temperature growth of the oxides is not compliant with the use of a standard photoresist stencil mask for lift-off, and that etching into the SrTiO$_3$ substrate by, for example, Ar ion beams can easily render the material conducting. Therefore, alternative methods have been developed. One elegant method is to make use of the insulator-to-metal transition at four unit cells of LaAlO$_3$, introduced earlier (Figure 10.3). Schneider *et al.* [69] have demonstrated this by first growing epitaxially two unit cells of LaAlO$_3$ on SrTiO$_3$ with the usual high-temperature growth techniques. Then, in a room-temperature lift-off process,

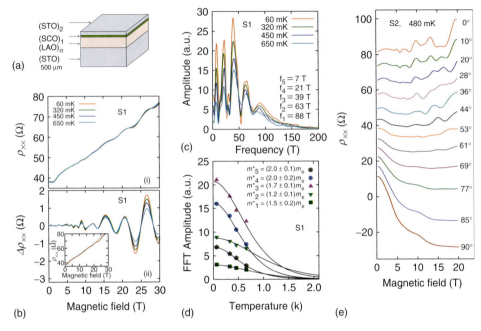

Figure 10.9 (a) Sketch of the $SrTiO_3$–$SrCuO_3$–$LaAlO_3$–$SrTiO_3$ structure. (b) (i) Temperature dependence of the longitudinal resistivity ρ_{xx}. (ii) The oscillations of (i) with background removed. The background is indicated as a dashed line in the inset. (c) Fourier transform between 4 and 30 T for the sample of (a,b). The five peaks in this graph indicate five different charge carriers (subbands) with different characteristic parameters. (d) Temperature dependence of the oscillation amplitudes, fitted with the common Lifshits–Kosevich formalisms. From these fits, the indicated effective masses m^* in units of m_e are derived. (e) Dependence of the Shubnikov–de Haas oscillations on the magnetic field orientation, revealing a 2D conductance character. The angle between the magnetic field direction and the normal to the interface is indicated beside the curves. (Source: Adapted from Ref. [67].)

an amorphous $LaAlO_3$ layer is deposited on a patterned photoresist lift-off stencil. After removal of the photoresist, an epitaxial $LaAlO_3$ layer is grown again, complementing the areas uncovered by the amorphous $LaAlO_3$ to an epitaxial $LaAlO_3$ layer with a thickness exceeding four unit cells and thus becoming conducting. This way, linewidths of 200 nm have been realized [69].

Even smaller structuring, also based on the three to four unit cell transitions, has been demonstrated by a tantalizing scanning-probe-based technique in a collaboration between the Pittsburgh and Augsburg groups (Figure 10.10) [70, 71]. Starting with an insulating sample of three unit cells of $LaAlO_3$ on $SrTiO_3$ (100), the researchers found that, on applying a sufficiently high positive potential (like 10 V) on the tip of a conducting AFM in contact mode, the interface below the tip is switched to the conducting state. Moreover, they could drive the interface insulating again by locally applying the opposite AFM tip potential. In this way, small side-gate field-effect transistors [71], dubbed sketch-field-effect transistors

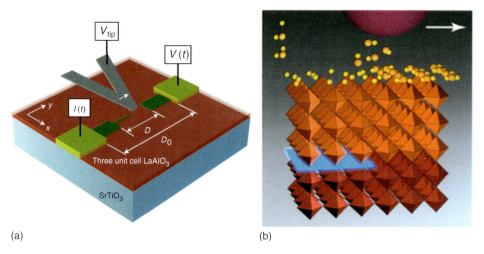

(a)   (b)

Figure 10.10 (a) Schematic diagram for writing a conducting wire in a three unit cell LaAlO$_3$ film on SrTiO$_3$. (b) Illustration of a possible mechanism for the AFM-tip-induced insulator-to-metal transition, based on the manipulation of ionic adsorbents at the LaAlO$_3$ surface. (Source: From Ref. [74].)

(FETs) after their fabrication procedure, and rectifying nanostructures [72] have been realized. Subsequent research has indicated that the mechanism for the AFM-tip-induced insulator-to-metal transition is most likely linked to the ionization of water molecules at the LaAlO$_3$ surface and the selected removal of OH$^-$ groups by a positive voltage and the neutralization of the H$^+$ ions by a negative voltage [73, 74]. The ionic charges at the LaAlO$_3$ surface then provide an electric field effect to the SrTiO$_3$/LaAlO$_3$ interface, switching its conductance. Testing a series of solvents, Xie *et al.* [75] noticed that exposure of the LaAlO$_3$ to various molecules, but only polar ones, can induce large changes in the conductivity and an insulator-to-metal transition around the critical LaAlO$_3$ thickness. For applications in the writing of (nano)electronic circuitry, the stability over time and under the influence of thermal cycling, ambient humidity, and so on is an aspect that needs further improvement.

10.5
Electric Field Gating

Modulating the transport properties by electric-field gating has been demonstrated both in a back-gate (electrode at the bottom of the SrTiO$_3$ substrate) [19, 31, 37] and in top-gate geometry [19, 76, 77]. Compared to semiconductor 2DEGs, the on–off gating of the conductance is complicated by the larger carrier densities of typically 10^{13}–10^{14} cm^{-2}, several orders of magnitude above common semiconductor 2DEGs. For a back-gate geometry, using a standard 0.5 mm SrTiO$_3$ substrate, this implies typical gate voltages of 100 V or more. Nevertheless, the Augsburg–Geneva collaboration has shown that for superconducting LaAlO$_3$/SrTiO$_3$ interfaces, the transition temperature can be spectacularly modulated by the back-gate voltage, mapping the superconducting dome and crossing a quantum critical point to the insulating regime (Figure 10.11). Such electric-field modulation of superconductivity has been a long pursued goal, allowing the realization of a superconducting field-effect transistor (SuFET).

The mechanism of the back-gating has initially been somewhat debated, partly because of the fact that prolonged prebiasing procedures seemed to be needed to reach a stable initial condition. This triggered questions on the occurrence of ion/oxygen vacancy migration at the used bias voltages, but the prebiasing may also be necessary to fill/empty traps or donor states near the interface. Also, while it is natural to discuss the field effects principally in terms of a modulation of the carrier density, it was reported by Bell *et al.* [78] that the mobility variation in their samples was in fact almost five times larger than the modulation of the sheet carrier density. The latter value ranged from about 1.4 to 2.5 × 10^{13} cm^{-2} when varying the back-gate voltage from −50 to 100 V. They discuss the mobility modulation as the result of a change of shape of the confining potential with the applied gate voltage, affecting the density distribution of the 2DEG and the distance to the more disordered interface.

More and more field-effect studies are becoming available, especially also in the top-gating configuration for which no large voltages need to be applied over the SrTiO$_3$ substrate, making clear that indeed the carrier density is strongly modulated by the electric field, and that the SrTiO$_3$–LaAlO$_3$ interface can become completely devoid of mobile charge carriers. An example is presented in Figure 10.12.

It is also good to note in this respect that the carrier density and mobility are not necessarily independent quantities in these confined, correlated electron systems,

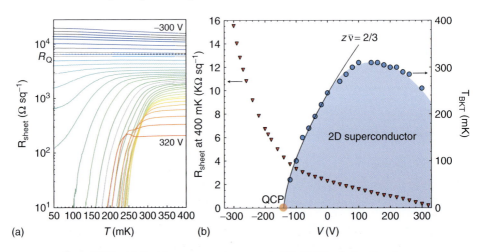

Figure 10.11 (a) Resistance versus temperature characteristics of a back-gated LaAlO$_3$/SrTiO$_3$ device. (b) Modulation of the critical temperature T_{BTK} of the superconducting phase in a back-gated LaAlO$_3$/SrTiO$_3$ device. For strong negative gating, the sample passes through a quantum critical point (QCP) into an insulating regime. (Source: From Ref. [31].)

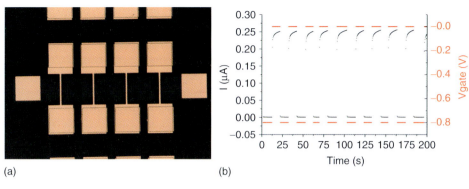

(a)

(b)

Figure 10.12 (a) Micrograph of a top-gated field-effect device. From left to right, the drain–source contacts are visible, contacting a buried $SrTiO_3/LaAlO_3$ interface patterned in a strip geometry (hardly visible in this image). Over the drain–source channel, four $20\,\mu m$ wide Au top gates are patterned. (b) The on–off switching of the current in a voltage-biased (50 mV) drain–source channel, by the application of a gate voltage over a 11 unit cell $LaAlO_3$ layer on a $SrTiO_3$ (001) substrate. The leakage current was below 1 nA in this $20 \times 100\,\mu m$ FET structure. (Source: Figures courtesy of P. Eerkes, University of Twente.)

and that the mobility modulation may arise partly as a result of the carrier density changes. Also, a selective depletion of subbands in these multiband systems can play a role, which reduces the degrees of freedom for scattering and may have an extra proportional effect on the mobility [79]. This is, however, expected to lead to a mobility enhancement with reduced carrier density, which is opposite to the findings of [78, 80].

The effects of a complete depletion of the interface system under the appropriate electric-field bias become strikingly visible in a very strong diode-like behavior for top-gated samples with three to four unit cells of $LaAlO_3$, allowing a considerable gate current by tunneling or thermal activation [81]. These diodes are characterized by very sharp nonlinear characteristics and large breakdown voltages of up to 200 V.

10.6
Applications

Having introduced now the basic properties of the $LaAlO_3/SrTiO_3$ system and techniques for fabrication, gating, and nanopatterning, which applications come in view? In terms of mobility, these interfaces are so far no match for the high-mobility semiconductor 2DEGs with μ values exceeding $30 \times 10^6\,cm^2\,V^{-1}\,s^{-1}$ at low temperatures [82] or for the MBE-grown $Mg_xZn_{1-x}O/ZnO$ oxide 2DEG systems with μ exceeding $700\,000\,cm^2\,V^{-1}\,s^{-1}$ [83], which render them less favorable for the highest-speed applications. This holds especially so, as the mobilities reduce to values around $5-10\,cm^2\,V^{-1}\,s^{-1}$ at room temperature, which leave them way behind silicon or other common semiconductors and semiconductor heterosystems, let alone graphene.

The main potential of the LaAlO$_3$/SrTiO$_3$ and related complex oxide interfaces is in aspects such as their rich phase diagram, arising partly from electronic correlations, in the opportunities for epitaxial combinations with other oxides with extraordinary properties, in the short electronic length scales facilitating the realization of electronic nanostructures, and in the sharp switching between the insulating to the conducting states. Other interesting aspects can be the optical transparency of the materials involved and the stability of the materials at elevated temperatures.

As the class of perovskite oxides also includes excellent ferroelectrics, which can be grown epitaxially on the top of the LaAlO$_3$/SrTiO$_3$ stack, an all-oxide ferroelectric random access memory is within reach. Moreover, the perovskites also include magnetic compounds and multiferroics such as BiFeO$_3$, creating the possibility to couple the interface conductivity to the magnetization of an overlying layer. Such coupling to magnetized materials also provides prospects for spintronic devices, in which the spin of the charge carriers is used as the information carrier, rather than the charge. The first successful injection of spin-polarized currents into the LaAlO$_3$/SrTiO$_3$ interface from a Co overlayer has recently been demonstrated [84]. The strong Rashba spin–orbit coupling arising from the breaking of the structural inversion symmetry at the interfaces can provide additional functionality, especially because it can be tuned by electric-field gating [80].

As described earlier, the interface transport properties can be modulated by ions or polar molecules adsorbed at the LaAlO$_3$ surface. Even placing a drop of distilled water on a 10 unit cell LaAlO$_3$/SrTiO$_3$ system markedly changes the interface transport characteristics, attributed to an electric-field effect arising from the polar water molecules [85]. This may be exploited for, for example, gas/liquid-sensing applications and has even been suggested as a basis for DNA sensors [85].

An example of a special property arising from electron–electron correlations with interesting potential for applications is a negative electronic compressibility, based on which considerable capacitance enhancements can be reached in capacitor devices [86]. The underlying principle is that for correlated electron systems, exchange and correlation energy terms within the electrodes should be included in the charging energy, in addition to the standard contributions based on the geometric capacitance [87].

There are several further aspects of potential interest for specialized applications, such as the possibility to create highly anisotropic and even 1D-like interface conductance on substrates with other than (001)-oriented surfaces [88], or by making use of step edges [89]. Also, photoconductivity effects can be of interest; when exposing the LaAlO$_3$/SrTiO$_3$ to ambient light, the conductivity has been seen to be greatly enhanced in a persistent way [15, 90]. Most likely, additional electrons that are released by photon absorption within the SrTiO$_3$ can collect in the potential well formed at the interface. By engineering the bandgap of the SrTiO$_3$, for example, by a tailored doping, this may be optimized for photovoltaics or photocatalysis applications.

In conclusion, since the seminal article by Ohtomo and Hwang [3], great efforts have been devoted to basic studies on these conducting interfaces and in their technological mastering. Progress has been made in understanding the mechanisms responsible for the conductance, which appears to be a combination of factors, but a consensus has not yet been reached. This has not hampered the progress to device development, such as all-oxide, optically transparent field-effect transistors. This chapter could only highlight some of the most important developments that have taken place in the period 2004–2012, and provide a snapshot of the situation to date. Without a doubt, the next years will bring many more insights and important steps toward applications.

Acknowledgments

The author acknowledges Alexander Brinkman, Peter Eerkes, and Sander Wenderich for their suggestions on the manuscript; Peter Eerkes also for providing Figure 10.12; and the oxide interface team in Twente for many discussions on the topic over the years. Also, the hospitality of the NanoCore Research Initiative at the National University of Singapore is gratefully acknowledged.

References

1. Bednorz, J.G. and Müller, K.A. (1988) Perovskite-type oxides – the new approach to high-T_c superconductivity. *Rev. Mod. Phys.*, **60**, 585.

2. (2007) Science magazine Top 10 breakthroughs of the year 2007. *Science*, **318**, 1844.

3. Ohtomo, A. and Hwang, H.Y. (2004) A high-mobility electron gas at the LaAlO$_3$/SrTiO$_3$ heterointerface. *Nature*, **427**, 423; **441** 7089, 120 (2006).

4. Nakagawa, N., Hwang, H.Y., and Muller, D.A. (2006) Why some interfaces cannot be sharp. *Nat. Mater.*, **5**, 204.

5. Huijben, M., Brinkman, A., Koster, G., Rijnders, G., Hilgenkamp, H., and Blank, D.H.A. (2009) Structure–property relation of SrTiO$_3$/LaAlO$_3$ interfaces. *Adv. Mater.*, **21**, 1665.

6. Pentcheva, R. and Pickett, W.E. (2010) Electronic phenomena at complex oxide interfaces: insights from first principle. *J. Phys. Condens. Matter*, **22**, 043001.

7. Zubko, P., Gariglio, S., Gabay, M., Ghosez, P., and Triscone, J.-M. (2011) Interface physics in complex oxide heterostructures. *Ann. Rev. Condens. Matter Phys.*, **2**, 141.

8. Chambers, S.A. (2011) Understanding the mechanism of conductivity at the LaAlO$_3$/SrTiO$_3$ (001) interface. *Surf. Sci.* doi: 10.1016/j.susc.2011.04.011

9. Gariglio, S. and Triscone, J.-M. (2011) Oxide interface superconductivity. *C. R. Phys.*, **12**, 591.

10. Schlom, D.G. and Mannhart, J. (2011) Interface takes charge over Si. *Nat. Mater.*, **10**, 168.

11. Hwang, H.Y., Iwasa, Y., Kawasaki, M., Keimer, B., Nagaosa, N., and Tokura, Y. (2012) Emergent phenomena at oxide interfaces. *Nat. Mater.*, **11**, 103.

12. Rijnders, G. and Blank, D.H.A. (2005) Materials science: build your own superlattice. *Nature*, **433**, 369.

13. Harrison, W.A., Kraut, E.A., Waldrop, J.R., and Grant, R.W. (1978) Polar heterojunction interfaces. *Phys. Rev. B*, **18**, 4402.

14. Hesper, R., Tjeng, L.H., Heeres, A., and Sawatzky, G.A. (2000) Photoemission evidence of electronic stabilization of polar surfaces in K$_3$C$_{60}$. *Phys. Rev. B*, **62**, 16046.

15. Huijben, M. (2006) Interface engineering for oxide electronics: tuning

electronic properties by atomically controlled growth. PhD thesis. University of Twente.

16. Kawasaki, M., Takahashi, K., Maeda, T., Tsuchiya, R., Shinohara, M., Ishiyama, O., Yonezawa, T., Yoshimoto, M., and Koinuma, H. (1994) Atomic control of the $SrTiO_3$ crystal-surface. *Science*, **266**, 1540.

17. Koster, G., Kropman, B.L., Rijnders, G.J.H.M., Blank, D.H.A., and Rogalla, H. (1998) Quasi-ideal strontium titanate crystal surfaces through formation of strontium hydroxide. *Appl. Phys. Lett.*, **73**, 2920.

18. Rijnders, G.J.H.M., Koster, G., Blank, D.H.A., and Rogalla, H. (1997) In situ monitoring during pulsed laser deposition of complex oxides using reflection high energy electron diffraction under high oxygen pressure. *Appl. Phys. Lett.*, **70**, 1888.

19. Thiel, S., Hammerl, G., Schmehl, A., Schneider, C.W., and Mannhart, J. (2006) Tunable quasi-two-dimensional electron gases in oxide heterostructures. *Science*, **313**, 1942.

20. Ishibashi, S. and Terakura, K. (2008) Analysis of screening mechanisms for polar discontinuity for $LaAlO_3/SrTiO_3$ thin films based on ab initio calculations. *J. Phys. Soc. Jpn.*, **77**, 104706.

21. Pentcheva, R. and Pickett, W.E. (2009) Avoiding the polarization catastrophe in $LaAlO_3$ overlayers on $SrTiO_3(001)$ through polar distortion. *Phys. Rev. Lett.*, **102**, 107602.

22. Huijben, M., Rijnders, G., Blank, D.H.A., Bals, S., van Aert, S., Verbeeck, J., van Tendeloo, G., Brinkman, A., and Hilgenkamp, H. (2006) Electronically coupled complementary interfaces between perovskite band insulators. *Nat. Mater.*, **5**, 556.

23. Pentcheva, R., Huijben, M., Otte, K., Pickett, W.E., Kleibeuker, J.E., Huijben, J., Boschker, H., Kockmann, D., Siemons, W., Koster, G., Zandvliet, H.J.W., Rijnders, G., Blank, D.H.A., and Hilgenkamp, H. (2010) Parallel electron–hole bilayer conductivity from electronic interface reconstruction. *Phys. Rev. Lett.*, **104**, 166804.

24. Sing, M., Berner, G., Goß, K., Müller, A., Ruff, A., Wetscherek, A., Thiel, S., Mannhart, J., Pauli, S.A., Schneider, C.W., Willmott, P.R., Gorgoi, M., Schäfers, F., and Claessen, R. (2009) Profiling the interface electron gas of $LaAlO_3/SrTiO_3$ heterostructures with hard X-ray photoelectron spectroscopies. *Phys. Rev. Lett.*, **102**, 176805.

25. Slooten, E. (2013) *et al.* Internal Electric Field in $SrTiO_3/LaAlO_3$ Heterostructures Probed with HAXPES and DFT Calculations. *Phys. Rev. B*, **87**, 085128.

26. Kalabukhov, A., Gunnarsson, R., Borjesson, J., Olsson, E., Claeson, T., and Winkler, D. (2007) Effect of oxygen vacancies in the $SrTiO_3$ substrate on the electrical properties of the $LaAlO_3/SrTiO_3$ interface. *Phys. Rev. B*, **75**, 121404(R).

27. Siemons, W., Koster, G., Yamamoto, H., Harrison, W.A., Lucovsky, G., Geballe, T.H., Blank, D.H.A., and Beasley, M.R. (2007) Origin of charge density at $LaAlO_3$ on $SrTiO_3$ heterointerfaces: possibility of intrinsic doping. *Phys. Rev. Lett.*, **98**, 196802.

28. Herranz, G., Basletić, M., Bibes, M., Carrétéro, C., Tafra, E., Jacquet, E., Bouzahouane, K., Deranlot, C., Hamzić, A., Broto, J.-M., and Barthélémy, A. (2007) High mobility in $LaAlO_3/SrTiO_3$ heterostructures: origin, dimensionality, and perspectives. *Phys. Rev. Lett.*, **98**, 216803.

29. Brinkman, A., Huijben, M., van Zalk, M., Huijben, J., Zeitler, U., Maan, J.C., van der Wiel, W.G., Rijnders, G., Blank, D.H.A., and Hilgenkamp, H. (2007) Magnetic effects at the interface between non-magnetic oxides. *Nat. Mater.*, **6**, 493.

30. Reyren, N., Thiel, S., Caviglia, A.D., Kourkoutis, L.F., Hammerl, G., Richter, C., Schneider, C.W., Kopp, T., Ruetschi, A.S., Jaccard, D., Gabay, M., Muller, D.A., Triscone, J.M., and Mannhart, J. (2007) Superconducting interfaces between insulating oxides. *Science*, **317**, 1196.

31. Caviglia, A.D., Gariglio, S., Reyren, N., Jaccard, D., Schneider, T., Gabay, M., Thiel, S., Hammerl, G., Mannhart, J., and Triscone, J.-M. (2008) Electric field

control of the LaAlO$_3$/SrTiO$_3$ interface ground state. *Nature (London)*, **456**, 624.

32. Basletic, M., Maurice, J.-L., Carrétéro, C., Herranz, G., Copie, O., Bibes, M., Jacquet, É., Bouzehouane, K., Fusil, S., and Barthélémy, A. (2008) Mapping the spatial distribution of charge carriers in LaAlO$_3$/SrTiO$_3$ heterostructures. *Nat. Mater.*, **7**, 621.

33. Copie, O., Garcia, V., Bödefeld, C., Carrétéro, C., Bibes, M., Herranz, G., Jacquet, E., Maurice, J.-L., Vinter, B., Fusil, S., Bouzehouane, K., Jaffrès, H., and Barthélémy, A. (2009) Towards two-dimensional metallic behavior at LaAlO$_3$/SrTiO$_3$ interfaces. *Phys. Rev. Lett.*, **102**, 216804.

34. Herranz, G., Basletic, M., Bibes, M., Ranchal, R., Hamzic, A., Tafra, E., Bouzehouane, K., Jacquet, E., Contour, J.P., Barthélémy, A., and Fert, A. (2006) Full oxide heterostructure combining a high-Tc diluted ferromagnet with a high-mobility conductor. *Phys. Rev. B*, **73**, 064403.

35. Ben Shalom, M., Tai, C.W., Lereah, Y., Sachs, M., Levy, E., Rakhmilevitch, D., Palevski, A., and Dagan, Y. (2009) Anisotropic magnetotransport at the SrTiO$_3$/LaAlO$_3$ interface. *Phys. Rev. B*, **80**, 140403.

36. Ben Shalom, M., Sachs, M., Rakhmilevitch, D., and Dagan, Y. (2010) Tuning spin-orbit coupling and superconductivity at the SrTiO$_3$/LaAlO$_3$ interface: a magnetotransport study. *Phys. Rev. Lett.*, **104**, 126802.

37. Bell, C., Harashima, S., Hikita, Y., and Hwang, H.Y. (2009) Thickness dependence of the mobility at the LaAlO$_3$/SrTiO$_3$ interface. *Appl. Phys. Lett.*, **94**, 222111.

38. Kalisky, B., Bert, J.A., Klopfer, B.B., Bell, C., Sato, H.K., Hosoda, M., Hikita, Y., Hwang, H.Y., and Moler, K.A. (2012) Critical thickness for ferromagnetism in LaAlO$_3$/SrTiO$_3$ heterostructures. *Nat. Commun.*, **3**, 922.

39. Zhong, Z.C. and Kelly, P.J. (2008) Electronic-structure-induced reconstruction and magnetic ordering at the LaAlO$_3$:SrTiO$_3$ interface. *Europhys. Lett.*, **84**, 27001.

40. Pavlenko, N., Kopp, T., Tsymbal, E.Y., Sawatsky, G.A., and Mannhart, J. (2012) Magnetic and superconducting phases at the LaAlO$_3$/SrTiO$_3$ interface: the role of interfacial Ti 3d electrons. *Phys. Rev. B*, **85**, 020407(R).

41. Cancellieri, C., Reyren, N., Gariglio, S., Caviglia, A.D., Fête, A., and Triscone, J.-M. (2010) Influence of the growth conditions on the LaAlO$_3$/SrTiO$_3$ interface electronic properties. *Europhys. Lett.*, **91**, 17004.

42. Ariando , Wang, X., Baskaran, G., Liu, Z.Q., Huijben, J., Yi, J.B., Annadi, A., Roy Barman, A., Rusydi, A., Dhar, S., Feng, Y.P., Ding, J., Hilgenkamp, H., and Venkatesan, T. (2011) Electronic phase separation at the LaAlO$_3$/SrTiO$_3$ interface. *Nat. Commun.* doi: 10.1038/ncomms 1992.

43. Dikin, D.A., Mehta, M., Bark, C.W., Folkman, C.M., Eom, C.B., and Chandrasekhar, V. (2011) Coexistence of superconductivity and ferromagnetism in two dimensions. *Phys. Rev. Lett.*, **107**, 056802.

44. Li, L., Richter, C., Mannhart, J., and Ashoori, R. (2011) Coexistence of magnetic order and two dimensional superconductivity at the LaAlO$_3$/SrTiO$_3$ interfaces. *Nat. Phys.*, **7**, 762.

45. Bert, J.A., Kalisky, B., Bell, C., Kim, M., Hikita, Y., Hwang, H.Y., and Moler, K.A. (2011) Direct imaging of the coexistence of ferromagnetism and superconductivity at the LaAlO3/SrTiO3 interface. *Nat. Phys.*, **7**, 767.

46. Willmott, P.R., Pauli, S.A., Herger, R., Schleputz, C.M., Martoccia, D., Patterson, B.D., Delley, B., Clarke, R., Kumah, D., Clonca, C., and Yacobi, Y. (2007) Structural basis for the conducting interface between LaAlO3 and SrTiO3. *Phys. Rev. Lett.*, **99**, 155502.

47. Vonk, V., Huijben, J., Kukuruznyak, D., Stierle, A., Hilgenkamp, H., Brinkman, A., and Harkema, S. (2012) Polar-discontinuity-retaining A-site intermixing and vacancies at SrTiO3/LaAlO3 interfaces. *Phys. Rev. B*, **85**, 045401.

48. Qiao, L., Droubay, T.C., Shutthanandan, V., Zhu, Z., Sushko, P.V., and Chambers, S.A. (2010) Thermodynamic instability at the stoichiometric

LaAlO3/SrTiO3 (001) interface. *J. Phys. Condens. Matter*, **22**, 312201.

49. Ohtomo, A., Muller, D.A., Grazul, J.L., and Hwang, H.Y. (2002) Artificial charge-modulation in atomic-scale perovskite titanate superlattices. *Nature*, **419**, 378.

50. Higuchi, T., Hotta, Y., Susaki, T., Fujimori, A., and Hwang, H.Y. (2009) Modulation doping of a Mott quantum well by a proximate polar discontinuity. *Phys. Rev. B*, **79**, 075415.

51. Takizawa, M., Hotta, Y., Susaki, T., Ishida, Y., Wadati, H., Takata, Y., Horiba, K., Matsunami, M., Shin, S., Yabashi, M., Tamasaku, K., Nishino, Y., Ishikawa, T., Fujimori, A., and Hwang, H.Y. (2009) Spectroscopic evidence for competing reconstructions in polar multilayers LaAlO3/LaVO3/LaAlO3. *Phys. Rev. Lett.*, **102**, 236401.

52. Perna, P., Maccariello, D., Radovic, M., Scotti di Uccio, U., Pallecchi, I., Codda, M., Marr, D., Cantoni, C., Gasquez, J., Varela, M., Pennycook, S.J., and Granozio, F.M. (2010) Conducting interfaces between band insulating oxides: the LaGaO3/SrTiO3 heterostructure. *Appl. Phys. Lett.*, **97**, 152111.

53. Chambers, S.A., Qiao, L., Droubay, T.C., Kaspar, T.C., Arey, B.W., and Sushko, P.V. (2011) Band alignment, built-in potential, and the absence of conductivity at the LaCrO3/SrTiO3 heterojunction. *Phys. Rev. Lett.*, **107**, 206802.

54. Li, D.F., Wang, Y., and Dai, J.Y. (2011) Tunable electronic transport properties of DyScO3/SrTiO3 polar heterointerfaces. *Appl. Phys. Lett.*, **98**, 122108.

55. Gozar, A., Logvenov, G., Kourkoutis, L.F., Bollinger, A.T., Giannuzzi, L.A., Muller, D.A., and Bozovic, I. (2008) High-temperature interface superconductivity between metallic and insulating copper oxides. *Nature*, **455**, 782.

56. Moetakef, P., Cain, T.A., Oullette, D.G., Zhang, J.Y., Klenov, D.O., Janotti, A., van de Walle, C.G., Rajan, S., Allen, J.S., and Stemmer, S. (2011) Electrostatic carrier doping of GdTiO3/SrTiO3 interfaces. *Appl. Phys. Lett.*, **99**, 232116.

57. Chen, Y.Z., Pryds, N., Kleibeuker, J.E., Koster, G., Sun, J.R., Stamate, E., Shen, B.G., Rijnders, G., and Linderoth, S. (2012) Metallic and insulating interfaces of amorphous SrTiO3-based oxide heterostructures. *Nano Lett.*, **11**, 3774.

58. Santander-Syro, A.F., Copie, O., Kondo, T., Fortuna, F., Pailhès, S., Weht, E., Qiu, X.G., Bertran, F., Nicolaou, A., Taleb-Ibrahimi, A., Le Fevre, P., Herranz, G., Bibes, M., Reyren, N., Apertet, Y., Lecoeur, P., Barthélémy, A., and Rozenberg, M.J. (2011) Two-dimensional electron gas with universal subbands at the surface of SrTiO3. *Nature*, **469**, 189.

59. Bristowe, N.C., Littlewood, P.B., and Artacho, E. (2011) Surface defects and conduction in polar heterostructures. *Phys. Rev. B*, **83**, 205405.

60. Tokura, Y. and Nagaosa, N. (2000) Orbital physics in transition-metal oxides. *Science*, **288**, 462.

61. Popović, Z.S., Satpathy, S., and Martin, R.M. (2008) Origin of the two-dimensional electron gas carrier density at the LaAlO3 on SrTiO3 interface. *Phys. Rev. Lett.*, **101**, 256801.

62. Breitschaft, M., Tinkl, V., Pavlenko, N., Paetel, S., Richter, C., Kirtley, J.R., Liao, Y.C., Hammerl, G., Eyert, V., Kopp, T., and Mannhart, J. (2010) Two-dimensional electron liquid state at LaAlO3-SrTiO3 interfaces. *Phys. Rev. B*, **81**, 153414.

63. Frederikse, H.P.R., Hosler, W.R., Thurber, W.R., Babiskin, J., and Siebenmann, P.G. (1967) Shubnikov-de Haas effect in SrTiO3. *Phys. Rev.*, **158**, 775.

64. Caviglia, A.D., Gariglio, S., Cancellieri, C., Sacépe, B., Fête, A., Reyren, N., Gabay, M., Morpurgo, A.F., and Triscone, J.-M. (2010) Two-dimensional quantum oscillations of the conductance at LaAlO3/SrTiO3 interfaces. *Phys. Rev. Lett.*, **105**, 236803.

65. Ben Shalom, M., Ron, A., Palevski, A., and Dagan, Y. (2010) Shubnikov – De Haas oscillations in SrTiO3/LaAlO3 interface. *Phys. Rev. Lett.*, **105**, 206401.

66. Huijben, M., Koster, G., Molegraaf, H.J.A., Kruize, M.K., Wenderich, S., Kleibeuker, J.E., McCollam, A., Guduru, V.K., Brinkman, A., Hilgenkamp, H., Zeitler, U., Maan, J.C., Blank, D.H.A., and Rijnders, G. (2010) High Mobility

Interface Electron Gas by Defect Scavenging in a Modulation Doped Oxide Heterostructure, *Adv. Funct. Mater.*, ArXiv 1008.1896v1 (accepted).

67. McCollam, A., Wenderich, S., Kruize, M.K., Guduru, V.K., Molegraaf, H.J.A., Huijben, M., Koster, G., Blank, D.H.A., Rijnders, G., Brinkman, A., Hilgenkamp, H., Zeitler, U., and Maan, J.C. (2012) Quantum Oscillations and Subband Properties of the Two-Dimensional Electron Gas at the LaAlO3/SrTiO3 Interface, ArXiv:1207.7003.

68. Jalan, B., Allen, S.J., Beltz, G.E., Moetakef, P., and Stemmer, S. (2011) Enhancing the electron mobility of SrTiO3 with strain. *Appl. Phys. Lett.*, **98**, 132102.

69. Schneider, C.W., Thiel, S., Hammerl, G., Richter, C., and Mannhart, J. (2006) Microlithography of electron gases formed at interfaces in oxide heterostructures. *Appl. Phys. Lett.*, **89**, 122101.

70. Cen, C., Thiel, S., Hammerl, G., Schneider, C.W., Andresen, K.E., Hellberg, C.S., Mannhart, J., and Levy, J. (2008) Nanoscale control of an interfacial metal-insulator transition at room temperature. *Nat. Mater.*, **7**, 298.

71. Cen, C., Thiel, S., Mannhart, J., and Levy, J. (2009) Oxide nanoelectronics on demand. *Science*, **323**, 5917.

72. Bogorin, D.F., Bark, C.W., Jang, H.W., Cen, C., Folkman, C.M., Eom, C.-B., and Levy, J. (2010) Nanoscale rectification at the LaAlO3/SrTiO3 interface. *Appl. Phys. Lett.*, **97**, 013102.

73. Bi, F., Bogorin, D.F., Cen, C., Bark, C.W., Park, J.-W., Eom, C.-B., and Levy, J. (2010) Water-cycle mechanism for writing and erasing nanostructures at the LaAlO3/SrTiO3 interface. *Appl. Phys. Lett.*, **97**, 173110.

74. Stephanos, C., Breitschaft, M., Jany, R., Kiessig, B., Paetel, S., Richter, C., and Mannhart, J. (2012) Writing nanowires with large conductivity ratios in LaAlO3/SrTiO3 interfaces. *J. Phys. Soc. Jpn.*, **81**, 064703.

75. Xie, Y., Hikita, Y., Bell, C., and Hwang, H.Y. (2011) Control of electronic conduction of an oxide heterointerface using polar adsorbates. *Nat. Commun.*, **2**, 494.

76. Singh-Bhalla, G., Bell, C., Ravichandran, J., Siemons, W., Hikita, Y., Salahuddin, S., Hebard, A.F., Hwang, H.Y., and Ramesh, R. (2011) Built-in and induced polarization across LaAlO3/SrTiO3 heterojunctions. *Nat. Phys.*, **7**, 80.

77. Förg, B., Richter, C., and Mannhart, J. (2012) Field-effect devices utilizing LaAlO3-SrTiO3 interfaces. *Appl. Phys. Lett.*, **100**, 053506.

78. Bell, C., Harashima, S., Kozuka, Y., Kim, M., Kim, B.G., Hikita, Y., and Hwang, H.Y. (2009) Dominant mobility modulation by the electric field effect at the LaAlO3/SrTiO3 interface. *Phys. Rev. Lett.*, **103**, 226802.

79. Störmer, H.L., Gossard, A.C., and Wiegmann, W. (1982) Observation of intersubband scattering in a 2-dimensional electron system. *Solid State Commun.*, **41**, 707.

80. Caviglia, A.D., Gabay, M., Gariglio, S., Reyren, N., Cancellieri, C., and Triscone, J.-M. (2010) Tunable Rashba spin-orbit interaction at oxide interfaces. *Phys. Rev. Lett.*, **104**, 126803.

81. Jany, R., Breitschaft, M., Hammerl, G., Horsche, A., Richter, C., Paetel, S., Mannhart, J., Stucki, N., Reyren, N., Gariglio, S., Zubko, P., Caviglia, A.D., and Triscone, J.-M. (2010) Diodes with breakdown voltages enhanced by the metal-insulator transition of LaAlO3-SrTiO3 interfaces. *Appl. Phys. Lett.*, **96**, 183504.

82. Kumar, A., Csáthy, G.A., Manfra, M.J., Pfeiffer, L.N., and West, K.W. (2010) Nonconventional odd-denominator fractional quantum Hall states in the second landau level. *Phys. Rev. Lett.*, **105**, 246808.

83. Falson, J., Maryenko, D., Kozuka, Y., Tsukazaki, A., and Kawasaki, M. (2011) Magnesium doping controlled density and mobility of two-dimensional electron gas in MgxZn1-xO/ZnO heterostructures. *Appl. Phys. Express*, **4**, 091101.

84. Reyren, N., Bibes, M., Lesne, E., George, J.-M., Deranlot, C., Collin, S., Barthélémy, A., and Jaffrès, H. (2012) Gate-controlled spin injection at LaAlO3/SrTiO3 interfaces. *Phys. Rev. Lett.*, **108**, 186802.

85. Au, K., Li, D.F., Chan, N.Y., and Dai, J.Y. (2012) Polar liquid molecule induced transport property modulation at LaAlO3/SrTiO3 heterointerface. *Adv. Mater.*, **24**, 2598.

86. Li, L., Richter, C., Paetel, S., Kopp, T., Mannhart, J., and Ashoori, R.C. (2011) Very large capacitance enhancement in a two-dimensional electron system. *Science*, **332**, 825.

87. Kopp, T. and Mannhart, J. (2009) Calculation of the capacitance of conductors: perspectives for the optimization of electronic devices. *J. Appl. Phys.*, **106**, 064504.

88. Annadi, A., Wang, X., Gopinadhan, K., Lü, W.M., Roy Barman, A., Liu, Z.Q., Srivastava, A., Saha, S., Zhao, Y.L., Zeng, S.W., Dhar, S., Tuzla, N., Olsson, E., Zhang, Q., Gu, B., Yonoki, S., Maekawa, S., Hilgenkamp, H., Venkatesan, T., and Ariando (2012) Unexpected Anisotropic Two Dimensional Electron Gas at the LaAlO3/SrTiO3 (110) Interface, ArXiv:1208.6135.

89. Bristowe, N.C., Fix, T., Blamire, M.G., Littlewood, P.B., and Artacho, E. (2012) Proposal of a one-dimensional electron gas in the steps at the LaAlO3-SrTiO3 interface. *Phys. Rev. Lett.*, **108**, 16802.

90. Tebano, A., Fabbri, E., Pergolesi, D., Balestrino, G., and Traversa, E. (2012) Room-temperature giant persistent photoconductivity in SrTiO3/LaAlO3 heterostructures. *ACS Nano*, **6**, 1278.

11
X-Ray Spectroscopic Studies of Conducting Interfaces between Two Insulating Oxides

Hiroki Wadati and Atsushi Fujimori

11.1
Introduction

Interfaces between two different transition-metal oxides often exhibit novel electronic properties that are not present in the component materials. Particularly prominent among these is electronic conduction at the interfaces between two insulators. Notable examples are the $SrTiO_3/LaAlO_3$ (STO/LAO) [1–3], $SrTiO_3/LaTiO_3$ (STO/LTO) [4], and $SrTiO_3/LaVO_3$ (STO/LVO) [5] interfaces. Here, STO and LAO are band insulators, and LTO and LVO are Mott insulators. Each side of the interface is a perovskite-type oxide with the chemical composition ABO_3 consisting of an alternating stack of AO and BO_2 planes along the (001) direction, as shown in Figure 11.1. In the case of STO, where A is Sr^{2+} and B is Ti^{4+}, both AO and BO_2 planes are charge neutral, making STO a nonpolar stack of layers. On the other hand, in the case of LAO, LTO, and LVO, where A is La^{3+} and B is Al^{3+}, Ti^{3+}, or V^{3+}, AO and BO_2 planes have $+e$ and $-e$ charges per two-dimensional unit cell, respectively, making these materials polar and necessitating some kind of reconstruction to avoid huge electric potential difference between the two sides of the film and the resulting accumulation of huge electrostatic energy (the so-called polar catastrophe) when the ABO_3 layer is thick. Figure 11.2 shows different types of reconstruction: the charges of the outermost layer are reduced to half [panel (a)] as observed at the surfaces of polar K_xC_{60} films by Hesper *et al.* [6], the structure of the outermost layer is distorted along the (001) direction [panel (b)], or charged atoms or molecules are adsorbed on, or removed from, the surface [panel (c)]. At the interface between polar and nonpolar layers, too, electronic reconstruction mechanism has been proposed as depicted in Figure 11.2a, [6, 7] and the interfacial electronic conduction has been explained. As shown in Figure 11.3, electronic reconstruction, that is, transfer of half the change, eliminates the potential divergence. On the other hand, many counterarguments have also been made against the proposal of the electronic reconstruction [8]. In this chapter, we focus on the studies of the electronic structures of three interfaces, LAO/LVO, LTO/STO, and LAO/STO, by photoemission spectroscopy and related spectroscopic techniques.

Functional Metal Oxides: New Science and Novel Applications, First Edition.
Edited by Satishchandra B. Ogale, Thirumalai V. Venkatesan, and Mark G. Blamire.

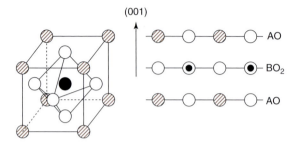

Figure 11.1 Crystal structure of the perovskite structure ABO_3.

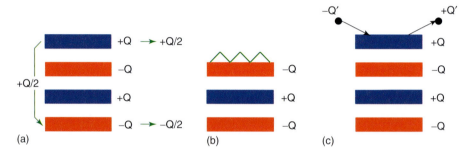

Figure 11.2 Electronic potential within a polar thin film and its elimination through electronic reconstruction. Different types of reconstruction: (a) electronic reconstruction, (b) ionic reconstruction, and (c) chemical reconstruction.

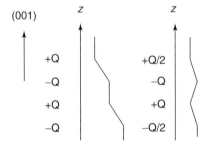

Figure 11.3 Electronic potential within a polar thin film and its elimination through electronic reconstruction.

11.2
Photoemission Measurements of Interfaces

Photoemission spectroscopy has been extensively used to study the electronic structures of interfaces. Soft X-ray photoemission spectroscopy measurements are performed using either laboratory X-ray sources of Mg $K\alpha$ line ($h\nu = 1253.6$ eV) and Al $K\alpha$ line ($h\nu = 1486.6$ eV) or soft X-ray beamlines in synchrotron radiation facilities. Since one uses a "photon-in-electron-out" process in photoemission

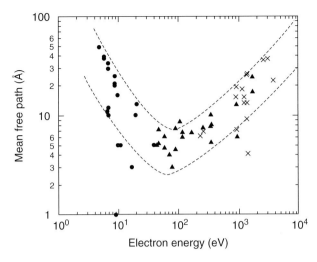

Figure 11.4 Mean free path of electrons in solids as a function of electron kinetic energy. Dashed curves indicate the approximate range of the distribution of experimental data [9].

spectroscopy, its probing depth is determined by the electron mean free path of photoelectrons, as shown in Figure 11.4 [9]. The mean free path of photoelectrons is only ∼5–10 Å in the region of 100–1000 eV (soft X-ray), which is below approximately three unit cells of the perovskite structures. Here, we would like to note that it is possible to obtain the information only from interfaces when we measure the core-level photoemission spectra of elements in the bottom of the interface. We also note that, using this method, one can effectively increase the probing depth beyond the photoelectron mean free path because the core-level signals, even though weak, are usually not buried in other emission signals. In order to obtain the information about deeply buried interfaces, hard X-ray photoemission spectroscopy (HXPES) [10–12] is a powerful experimental technique. In HXPES, the mean free path can be increased up to ∼100 Å. By combining the relatively surface-sensitive soft X-rays and the bulk-sensitive hard X-rays, one can study the electronic structures of buried interfaces quite efficiently.

The electronic structure of interfaces near the Fermi level (E_F) can be directly measured by photoemission spectroscopy if the top layer is an n-type semiconductor because the n-type semiconductor is transparent in its bandgap region. For example, STO is an n-type semiconductor and is "transparent" between the Fermi level and the O $2p$ band maximum (∼3.2 eV below E_F). Therefore, by depositing thin enough STO overlayers on top of, for example, LTO layers, one can study the photoemission spectra near E_F of STO/LTO interfaces. Measurements using this technique were first implemented in Ref. [23].

The surface sensitivity can also be changed by changing the emission angle of photoelectrons θ, as shown in Figure 11.5 [14]. The probing depth of photoelectrons can be written as $\lambda \cos\theta$, where λ is the electron mean free path.

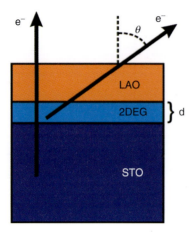

Figure 11.5 Emission angles θ of photoelectrons [14].

11.3
Interfaces between a Mott Insulator and a Band Insulator: LaAlO$_3$/LaVO$_3$ and LaTiO$_3$/SrTiO$_3$

LAO/LVO is an interface between a band insulator LAO and a Mott insulator LVO. LVO has the same polarity as that of LAO, and hence, there is no polar discontinuity at this interface. Nevertheless, interesting phenomena have been observed as functions of LAO overlayer thickness. Higuchi *et al.* [15] studied the transport properties of LVO sandwiched by LAO in trilayer structures. LVO has the same polar structure (LaO$^+$/VO$_2^-$/LaO$^+$/VO$_2^-$/....) as that of LAO (LaO$^+$/AlO$_2^-$/LaO$^+$/....), and hence, no polar discontinuity exists in the LAO/LVO/LAO trilayer. However, they found an exponential drop in resistance when the LAO overlayer became thin, and a polar AlO$_2$-terminated surface of LAO was brought in close proximity to the LVO/LAO interface. Furthermore, the measured positive thermopower voltage indicated holelike carriers, which scales with the doping dependence of bulk LVO. These results indicate that polarity can be used for the tunable doping of holes, which cannot arise from oxygen vacancies introduced during the growth process.

Hotta *et al.* [16] investigated the electronic structure of LVO layers of one to five unit cell thicknesses sandwiched by LAO layers by means of soft X-ray photoemission spectroscopy. They performed angle-dependent measurements of V 2$p_{3/2}$ core levels at emission angles θ of 30°, 55°, and 70° (measured from the surface normal), as shown in Figure 11.6. For each spectrum, the experimental curves were well fitted to two components, V^{3+} and V^{4+}. As θ increased from 30° to 70°, the spectra became increasingly sensitive to the topmost LVO layer, and they increasingly exhibited a V^{4+} contribution. This indicates that V^{4+} is not uniformly distributed in the LVO layer but is preferentially distributed in the upper part of the layer. The existence of V^{4+} is expected in the scenario of electronic reconstruction, that is, the increased valence of V in the topmost layer, which screens the potential

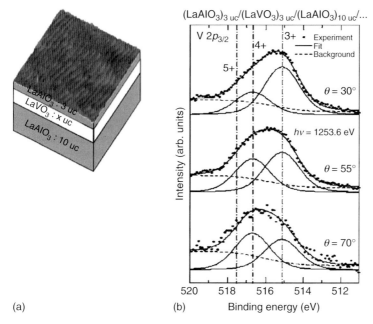

$(LaAlO_3)_{3\,uc}/(LaVO_3)_{3\,uc}/(LaAlO_3)_{10\,uc}/...$

Figure 11.6 (a) Schematic figure of the LAO /LVO heterostructures and AFM image of a typical surface morphology. (b) Angle-dependent XPS spectra of the V $2p_{3/2}$ core level of the LAO (3 unit cells)/LVO (3 unit cells)/LAO (10 unit cells) heterostructure. Dots and solid curves show the experimental data and the fitting curves consisting of V^{3+} and V^{4+} components, respectively [16].

slope in the polar layer. Although this model would predict 50% V^{4+}, the result showed 75% V^{4+} contribution probably as result of oxygen off-stoichiometry.

Wadati *et al.* [17] investigated the electronic structure of the same interfaces by means of HXPES ($hv = 7937$ eV). They found that the valence of V in LVO was partially converted from V^{3+} to V^{4+} only at the LAO/LVO interface on the top side of the LVO layer and that the amount of V^{4+} increased with LVO layer thickness. They also studied the valence-band spectra and found that the Mott-Hubbard gap of LVO remained open at the interface, as shown in Figure 11.7, indicating that the interface remained insulating.

The LAO capping layer thickness dependence of the valence redistribution of V in LAO/LVO/LAO was studied by Takizawa *et al.* [13], using HXPES. They found that the V valence was intermediate between V^{3+} and V^{4+} for thin LAO cap layers, decreases with increasing cap-layer thickness, and finally recovers the bulk value of V^{3+} at 10 unit-cell thickness as shown in Figure 11.8.

Wadati *et al.* [18] performed an X-ray absorption study of the dependence of the V oxidation state on the thickness of LVO and capping LAO layers in the multilayer structure of LVO sandwiched by LAO and found a similar trend of the decrease of the valence of V with increasing LAO cap-layer thickness, but the results are complicated by the presence of a considerable amount of V^{4+} in the bulk of the thicker LVO layers.

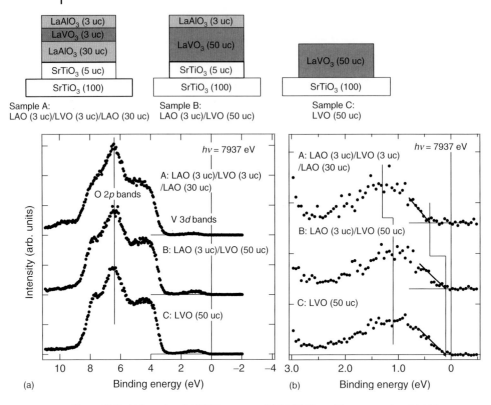

Figure 11.7 Valence-band HXPES spectra of LAO/LVO multilayer samples. (a) Valence-band spectra over a wide energy range. (b) V 3*d* band region [17].

All the earlier-mentioned experimental results can be understood if atomic reconstruction (oxygen vacancy formation) at the polar LAO surface or electronic reconstruction (V valence change) occurs at the LAO/LVO interface, depending on the LAO layer thickness so that the polar catastrophe is avoided at the cost of minimum energy. Schematic illustrations of such atomic and electronic reconstructions are shown in Figure 11.9. Panel (a) shows the case without any reconstruction where the heterostructure with all polar planes suffers from the polar catastrophe. If atomic reconstruction occurs at the top surface of LAO as shown in panel (b), extra electrons (half an electron per 2D unit cell) will be transferred from the surface layer to the bottom LAO/STO interface, resulting in the electronic reconstruction $Ti^{4+} \rightarrow Ti^{3.5+}$ at the LAO/STO interface at the bottom, thereby providing a compensating macroscopic dipole. Panel (c) is an alternative case where the valence change ($V^{3+} \rightarrow V^{3.5+}$) at the LAO/LVO interface creates extra electrons, resulting in purely electronic reconstructions on both sides of the trilayer. When the LAO overlayer is thin, electronic reconstruction (c) occurs, but in the limit of thick top LAO layer, the polar catastrophe within the LAO layer has to be avoided and therefore process (b) overcomes process (c).

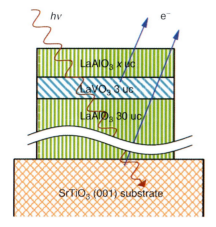

LaAlO$_3$(x uc)/LaVO$_3$(3 uc)/LaAlO$_3$ (30 uc)/SrTiO$_3$ substrate

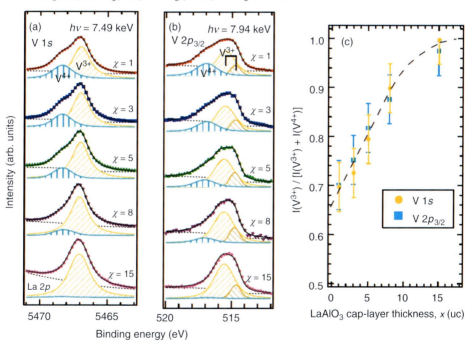

Figure 11.8 LAO cap-layer thickness dependence of the V core-level spectra of the LAO (x unit cell)/LVO (3 unit cells)/LAO (30 unit cells) trilayer samples grown on STO substrates. (a) V 1s core-level spectra and their line shape analyses. (b) V $2p_{3/2}$ core-level spectra and their line shape analyses. (c) LAO cap-layer thickness dependence of the V^{3+} intensity relative to the total V intensity (V^{3+} + V^{4+}) in the V 1s and $2p_{3/2}$ core-level spectra. The dashed curve is a guide to the eye [13].

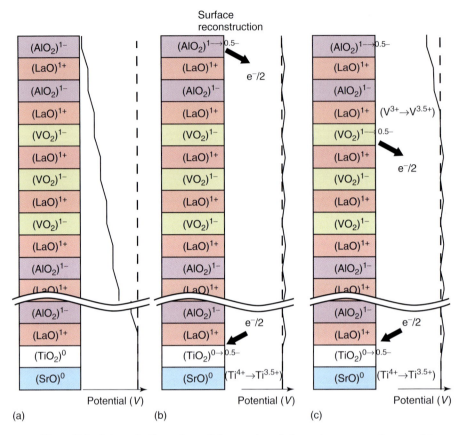

Figure 11.9 Schematic illustrations of the atomic and electronic reconstructions resolving the polar catastrophe in the all polar LAO/LVO/LAO trilayers. The details are described in the text [13].

Another band insulator/Mott insulator interface STO/LTO has been studied extensively. The interface between the band insulator STO and the Mott insulator LTO has a polar discontinuity unlike LAO/LVO. Ohtomo *et al.* [4] have made an atomic-resolution electron-energy-loss spectroscopy study of one to two layers of LTO embedded in STO and found that, in spite of the chemically abrupt interfaces, Ti $3d$ electrons are not completely confined within the LTO layer but are spread over the neighboring STO up to ~2.5 layers. Here, bulk LTO has the d^1 configuration and shows antiferromagnetism below T_N ~140 K [19], while STO has the d^0 configuration, that is, the empty d band. Okamoto and Millis [20, 21] have studied the spectral function of such systems by a model Hartree-Fock and dynamical-mean-field-theory (DMFT) calculation and explained the metallic behavior at the interface between the two insulators. STO/LTO shows metallic conductivity, and the conductivity is attributed to electronic reconstruction [4, 22]. If such an electronic reconstruction occurs and the interface becomes metallic, it

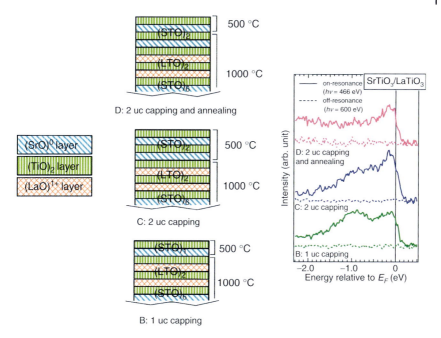

Figure 11.10 Ti $2p \rightarrow 3d$ resonance photoemission spectra taken at 466 eV of the STO/LTO superlattice samples; sample B is capped with one unit cell STO, sample C is with two unit cells STO, and sample D is with two unit cells STO and annealing. For comparison, off-resonant photoemission spectrum taken at 600 eV is also shown below each resonance photoemission spectrum [23].

would be observed as a Fermi cutoff in the photoemission spectra. Takizawa *et al.* [23] measured photoemission spectra of STO/LTO superlattices with a topmost STO layer of variable thickness. Finite coherent spectral weight with a clear Fermi cutoff was observed at chemically abrupt STO/LTO interfaces as shown in Figure 11.10, indicating that an "electronic reconstruction" indeed occurs at the interface between the Mott insulator LTO and the band insulator STO.

11.4
Interfaces between Two Band Insulators: LaAlO₃/SrTiO₃

The interface between two band insulators STO and LAO is an especially interesting system in which not only metallic conductivity [1] but also superconductivity [2] as well as ferromagnetism [24–26] have been reported. In the STO/LAO heterostructures, the LaO/TiO₂/SrO interface (the so-called n-type interface) is metallic while the SrO/AlO₂/LaO interface (the so-called p-type interface) is insulating. There has been an intense debate about the origin of the metallicity, that is, whether it is due to oxygen vacancies (extrinsic) [27, 28] or due to the polar discontinuity between the LAO and STO layers, which could result in an electronic reconstruction.

The influence of the LAO thickness on the electronic properties of STO/LAO interfaces was reported by Thiel *et al.* [29]. They found the critical LAO thickness of four unit cells, where the sheet conductance jumps from $< 10^{-9}\,\Omega^{-1}$ to $\sim 10^{-5}\,\Omega^{-1}$, and the carrier density from ~ 0 to 10^{13} (cm^{-2}). These results support the scenario of electronic reconstruction because the divergence of electrostatic potential depends on the LAO thickness while oxygen vacancies do not. In the STO/LAO interface, there is asymmetry between the two polarities. The LaO/TiO$_2$/SrO interface is metallic, while the SrO/AlO$_2$/LaO interface is insulating [1]. This observation is consistent with the electronic reconstruction mechanism in which extra electrons are accommodated in the Ti 3d conduction band of STO for the n-type interface. The existence of electrons in the Ti 3d band, that is, the existence of Ti^{3+} species, was proved in Sing *et al.*'s HXPES results [14], as shown in Figure 11.11. From the Ti 2p signal and its angle dependence, they derived that the thickness of the electron gas is of the same order as the probing depth of 4 nm. Therefore, their results support the electronic reconstruction due to the LAO overlayer as the driving mechanism for the conducting interface. However, the sheet carrier density estimated from the Ti^{3+} signal intensity $\sim 10^{14}$ cm^{-2} is an order of magnitude larger than that deduced from the transport measurements $\sim 10^{13}$ cm^{-2} [29].

The depth profile of the Ti^{3+} distribution at the STO/LAO interface was studied by using X-ray evanescent waves in HXPES measurements [30]. When the grazing angle, defined as the *angle* between the X-ray beam and the sample surface, is smaller than the critical angle 0.4°, incident X-rays are totally reflected and their intensities exponentially decay into the sample as evanescent waves. In this case, photoelectrons mainly arise from the surface region. As the grazing angle increases across 0.4°, the penetration of X-rays dramatically increases and the contribution of

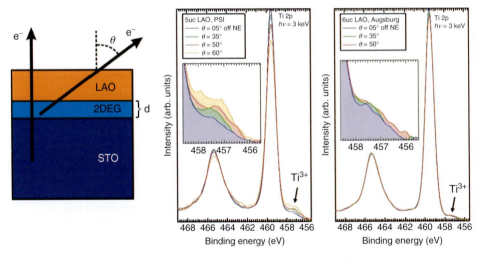

Figure 11.11 Ti 2p spectra of two different LAO/STO samples for various emission angles θ [14]. Ti^{3+} signal was observed as a lower-binding-energy feature. The thickness of the two-dimensional electron gas was estimated to be one and eight unit cells, and the carrier density was 2.0×10^{14} and 1.1×10^{14} cm^{-2}, respectively.

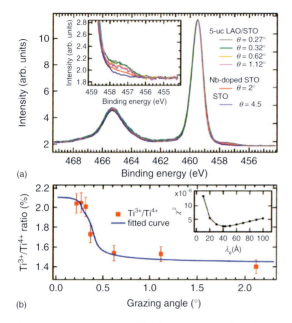

(a)

(b)

Figure 11.12 (a) Ti $2p$ core-level HXPES spectra with selected grazing angles measured from samples of the LAO/STO bilayer, bare SrTiO₃, and SrTiO₃ doped with 5% Nb by weight. (b) Measured and simulated Ti^{3+}/Ti^{4+} ratio of Ti $2p$ HXPES of LAO/STO as a function of grazing angle. The inset shows the variation of the chi-square value of the fitting as a function of λ_e [30].

the near-surface region is substantially reduced. Figure 11.12a shows Ti $2p$ spectra from samples of STO, Nb-doped STO, and LAO/STO heterostructure [30]. All these spectra show the existence of Ti^{3+}. They obtained that Ti^{3+} is distributed within 4.8 nm, consistent with Ref. [14].

Systematic core-level photoemission studies of the LAO/STO interface were performed by Takizawa *et al.* [31]. They used soft X-ray photoemission spectroscopy and studied the LAO-layer thickness dependence of the spectra as well as the difference of the spectra between the n-type and the p-type interfaces. In the Ti $2p$ core-level spectra of the conducting n-type interfaces, Ti^{3+} signals were observed, while they were absent for the insulating p-type interfaces, consistent with the electronic reconstruction scenario. The core-level shifts as functions of the LAO-layer thickness, however, were much smaller than the values predicted from the polar catastrophe model described in the introduction. Similarly, small core-level shifts have also been reported by Segal *et al* [32]. Takizawa *et al.* also found that the Ti^{3+} signals increased with the LAO thickness, but the increase started well below the critical thickness of four unit cells of the metallic transport as shown in Figure 11.13. The intensity of the Ti^{3+} signals was consistent with the HXPES studies [14, 30] and again significantly higher than that expected from the transport carrier densities, indicating that a large portion of the Ti $3d$ carriers at the interfaces are localized. The increase of the Ti^{3+} signal intensity already well below the

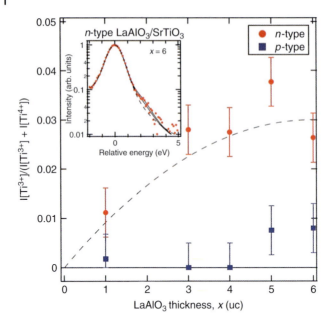

Figure 11.13 Ti^{3+} intensity ratio in the Ti 2p core-level spectra of the LAO/STO bilayer as a function of the LAO overlayer thickness x. The dashed curve is a guide to the eye. Inset: Ti 2p spectrum of the $x = 6$ sample and its fitted curves. Two solid curves show those with different parameters yielding the range of error bars. The dashed curve shows the Ti^{4+} spectrum [31].

critical thickness of four unit cells may be explained by surface defects/adsorbates, which provides charges to the interface even below the critical thickness. This mechanism was first proposed by the theoretical study of the LAO/STO interface by Bristowe *et al.* [33]. A recent transport study of LAO/STO with various surface adsorbates have shown that the interfacial conductivity sensitively depends on the polarizability of the surface adsorbates [34]. The existence of localized Ti^{3+} carriers at the interface below the critical thickness is consistent with the recent report that the conductivity transition of LAO/STO is governed by changes in the mobility rather than by the carrier number [35].

As for the valence-band structure of the LAO/STO interfaces, Yoshimatsu *et al.* [36] have studied LAO/STO samples using *in situ* soft X-ray photoemission spectroscopy. They revealed that a band bending exists at the metallic n-type LAO/TiO$_2$-STO interface but not at the insulating p-type LAO/SrO-STO interface as shown in Figure 11.14. Their results indicate that the metallic states originate from the accumulation of carriers on the length scale of band bending of ∼10 unit cells as shown in Figure 11.15, roughly consistent with the core-level photoemission studies [14, 30–32]. They also measured the valence-band spectra and found that the intensity of the Ti 3d band is much smaller than the case where 0.5$e^−$ of Ti 3d electrons from electronic reconstruction is confined within one unit cell of the

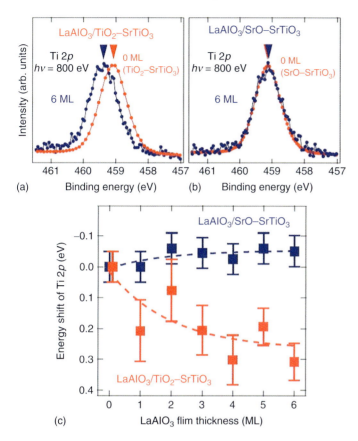

Figure 11.14 Band bending of STO layers for the metallic n-type and insulating p-type LAO/STO interfaces: (a) Ti 2p core-level spectra of a TiO₂-terminated STO and the metallic p-type interface. (b) Spectra of an SrO-terminated STO and the insulating p-type interface. (c) Plots of the energy shift of the Ti 2p core-level peaks for the metallic and insulating LAO/STO interfaces as a function of the LAO overlayer thickness [36].

surface. This result is in accordance with the above-mentioned conclusion that Ti 3d electrons are distributed within ∼10 unit cells of the surface.

Another powerful spectroscopic technique for such interfaces is resonant inelastic x-ray scattering (RIXS). Recently, RIXS studies of STO/LAO were reported in Refs [37, 38]. Berner *et al.* [37] applied RIXS to the single interface of LAO/STO. From the Ti 3d $t_{2g} \rightarrow e_g$ crystal-field excitations measured at the Ti $2p_{3/2} \rightarrow 3d$ resonance, they not only derived information about the local geometry at the interface but also obtained the evolution of the sheet carrier density with the thickness of the LAO overlayer, as shown in Figure 11.16. These results show that the charge density exceeds the one derived from Hall-effect measurements, again indicating the coexistence of itinerant and localized Ti 3d electrons at the interface. They observed a saturation of the charge-carrier concentration above an LAO thickness of six unit cells at 1×10^{14} cm^{-2}, well below the canonical value

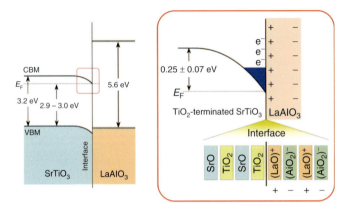

Figure 11.15 Band diagram of the metallic n-type LAO/STO interface determined by photoemission spectroscopy. As a consequence of downward band bending, a notched structure is formed in the STO layer in the interfacial region. [36]

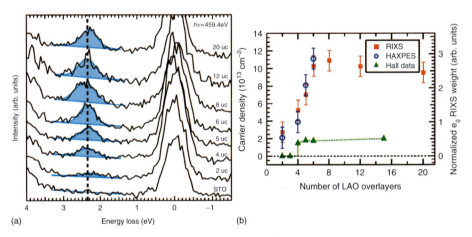

Figure 11.16 (a) RIXS spectra from several samples with varying thicknesses of LAO overlayers. For reference, the spectrum of bare STO is also shown. (b) Comparison of the normalized weights of the inelastic e_g peak and the sheet carrier density resulting from HXPES and Hall measurements [37].

for ideal electronic reconstruction. Zhou *et al.* performed further RIXS studies of LAO/STO interfaces with higher energy resolution [38]. They studied the excitations between t_{2g} energy levels and determined that the interface $Ti^{3+}O_6$ octahedra are orthorhombically distorted and quantify the crystal-field splitting energies.

11.5
Summary

In this chapter, we have made an overview of the photoemission studies of the electronic structures of oxide interfaces that exhibit conductivity transitions as a function of overlayer thickness. We have focused on three interfaces, LAO/LVO, LTO/STO, and LAO/STO. The information obtained by the spectroscopic studies includes the valence of transition-metal atoms, core-level shifts, and the existence of the Fermi cutoff. Especially we would like to note that the existence of electronic reconstruction was confirmed by photoemission spectroscopy but not in the original form. In the case of the LAO/STO interface, the electronic reconstruction takes place gradually with increasing LAO thickness, and charge transfer from the LAO surface to the interface plays an essential role. Also, a significant portion of the doped electrons at the n-type interface is localized. Further detailed information is desired for about the localized Ti^{3+} and the existence of Ti^{3+} below the critical thickness. We have also shown that photoemission spectroscopy and RIXS give complementary information, and we expect that the combination of these X-ray spectroscopic techniques will lead to further understanding and detection of novel phenomena in interface physics.

The authors thank M. Takizawa, A. Chikamatsu, K. Yoshimatsu, H. Kumigashira, M. Oshima, Y. Takata, S. Shin, Y. Hotta, T. Susaki, S. Tsuda, H. Y. Hwang, and G. A. Sawatzky for continuous collaboration and the Japan Society for the Promotion of Science (JSPS) and the Council for Science and Technology Policy (CSTP) for financial support [Grants-in-Aid for Scientific Research and ''Funding Program for World-Leading Innovative R&D on Science and Technology (FIRST Program)''].

References

1. Ohtomo, A. and Hwang, H.Y. (2004) *Nature*, **427**, 423.
2. Reyren, N., Thiel, S., Caviglia, A.D., Kourkoutis, L.F., Hammer, G., Richter, C., Schneider, C.W., Kopp, T., Ruetschi, A.-S., Jaccard, D., Gabay, M., Muller, D.A., Triscone, J.-M., and Mannhart, J. (2007) *Science*, **317**, 1196.
3. Brinkman, A., Huijben, M., Zalk, M.V., Huijben, J., Zeitler, U., Maan, J.C., Wiel, W.G.V.D., Rijnders, G., Blank, D.H.A. and Hilgenkamp, H. (2007) *Nat. Mater.*, **6**, 493.
4. Ohtomo, A., Muller, D.A., Grazul, J.L., and Hwang, H.Y. (2002) *Nature*, **419**, 378.
5. Hotta, Y., Susaki, T., and Hwang, H.Y. (2007) *Phys. Rev. Lett.*, **99**, 236805.
6. Hesper, R., Tjeng, L.H., Heeres, A., and Sawatzky, G.A. (2000) *Phys. Rev. B*, **62**, 16046.
7. Nakagawa, N., Hwang, H.Y., and Muller, D.A. (2006) *Nat. Mater.*, **5**, 204.
8. Herranz, G., Basletic, M., Bibes, M., Carretero, C., Tafra, E., Jacquet, E., Bouzehouane, K., Deranlot, C., Hamzic, A., Broto, J.-M., Barthelemy, A., and Fert, A. (2007) *Phys. Rev. Lett.*, **98**, 216803.
9. Brundle, C.R. (1974) *J. Vac. Sci. Technol.*, **11**, 212.
10. Takata, Y., Tamasaku, K., Tokushima, T., Miwa, D., Shin, S., Ishikawa, T., Yabashi, M., Kobayashi, K., Kim, J.J., Yao, T., Yamamoto, T., Arita, M., Namatame, H., and Taniguchi, M. (2004) *Appl. Phys. Lett.*, **84**, 4310.
11. Takata, Y., Tamasaku, K., Nishino, Y., Miwa, D., Yabashi, M., Ikenaga, E., Horiba, K., Arita, M., Shimada, K., Namatame, H., Nohira, H., Hattori, T., Sodergren, S., Wannberg, B., Taniguchi, M., Shin, S., Ishikawa, T., and Kobayashi, K. (2005) *J. Electron Spectrosc. Relat. Phenom.*, **144–147**, 1063.
12. Takata, Y., Yabashi, M., Tamasaku, K., Nishino, Y., Miwa, D., Ishikawa, T., Ikenaga, E., Horiba, K., Shin, S., Arita, M., Shimada, K., Namatame, H., Taniguchi, M., Nohira, H., Hattori, T., Sodergren, S., Wannberg, B., and Kobayashi, K. (2005) *Nucl. Instrum. Methods Phys. Res., Sect. A*, **547**, 50.

13. Takizawa, M., Hotta, Y., Susaki, T., Ishida, Y., Wadati, H., Takata, Y., Horiba, K., Matsunami, M., Shin, S., Yabashi, M., Tamasaku, K., Nishino, Y., Ishikawa, T., Fujimori, A., and Hwang, H.Y. (2009) *Phys. Rev. Lett.*, **102**, 236401.

14. Sing, M., Berner, G., Goss, K., Muller, A., Ruff, A., Wetscherek, A., Thiel, S., Mannhart, J., Pauli, S.A., Schneider, C.W., Willmott, P.R., Gorgoi, M., Schafers, F., and Claessen, R. (2009) *Phys. Rev. Lett.*, **102**, 176805.

15. Higuchi, T., Hotta, Y., Susaki, T., Fujimori, A., and Hwang, H.Y. (2009) *Phys. Rev. B*, **79**, 075415.

16. Hotta, Y., Mukunoki, Y., Susaki, T., Hwang, H.Y., Fitting, L., and Muller, D.A. (2006) *Appl. Phys. Lett.*, **89**, 031918.

17. Wadati, H., Hotta, Y., Fujimori, A., Susaki, T., Hwang, H.Y., Takata, Y., Horiba, K., Matsunami, M., Shin, S., Yabashi, M., Tamasaku, K., Nishino, Y., and Ishikawa, T. (2008) *Phys. Rev. B*, **77**, 045122.

18. Wadati, H., Hawthorn, D.G., Geck, J., Regier, T.Z., Blyth, R.I.R., Higuchi, T., Hotta, Y., Hikita, Y., Hwang, H.Y., and Sawatzky, G.A. (2009) *Appl. Phys. Lett.*, **95**, 023115.

19. Tokura, Y. (1992) *J. Phys. Chem. Solids*, **53**, 1619.

20. Okamoto, S. and Millis, A.J. (2004) *Nature (London)*, **428**, 630.

21. Okamoto, S. and Millis, A.J. (2004) *Phys. Rev. B*, **70**, 241104(R).

22. Shibuya, K., Ohnishi, T., Kawasaki, M., Koinuma, H., and Lippmaa, M. (2004) *Jpn. J. Appl. Phys.*, **43**, L1178.

23. Takizawa, M., Wadati, H., Tanaka, K., Hashimoto, M., Yoshida, T., Fujimori, A., Chikamatsu, A., Kumigashira, H., Oshima, M., Shibuya, K., Mihara, T., Ohnishi, T., Lippmaa, M., Kawasaki, M., Koinuma, H., Okamoto, S., and Millis, A.J. (2006) *Phys. Rev. Lett.*, **97**, 057601.

24. Li, L., Richter, C., Mannhart, J., and Ashoori, R.C. (2011) *Nat. Phys.*, **7**, 763.

25. Bert, J.A., Kalisky, B., Bell, C., Kim, M., Hikita, Y., Hwang, H.Y., and Moler, K.A. (2011) *Nat. Phys.*, **7**, 767.

26. Dikin, D.A., Mehta, M., Bark, C.W., Folkman, C.M., Eom, C.B., and Chandrasekhar, V. (2011) *Phys. Rev. Lett.*, **107**, 056802.

27. Siemons, W., Koster, G., Yamamoto, H., Harrison, W.A., Lucovsky, G., Geballe, T.H., Blank, D.H.A., and Beasley, M.R. (2007) *Phys. Rev. Lett.*, **98**, 196802.

28. Kalabukhov, A., Gunnarsson, R., Borjesson, J., Olsson, E., Claeson, T., and Winkler, D. (2007) *Phys. Rev. B*, **75**, 121404(R).

29. Thiel, S., Hammerl, G., Schmehl, A., Schneider, C.W., and Mannhart, J. (2006) *Science*, **313**, 1942.

30. Chu, Y.Y., Liao, Y.F., Tra, V.T., Yang, J.C., Liu, W.Z., Chu, Y.H., Lin, J.Y., Huang, J.H., Weinen, J., Agrestini, S., Tsuei, K.-D., and Huang, D.J. (2011) *Appl. Phys. Lett.*, **99**, 262101.

31. Takizawa, M., Tsuda, S., Susaki, T., Hwang, H.Y., and Fujimori, A. (2011) *Phys. Rev. B*, **84**, 245124.

32. Segal, Y., Ngai, J.H., Reiner, J.W., Walker, F.J., and Ahn, C.H. (2009) *Phys. Rev. B*, **80**, 241107(R).

33. Bristowe, N.C., Littlewood, P.B., and Artacho, E. (2011) *Phys. Rev. B*, **83**, 205405.

34. Xie, Y., Hikita, Y., Bell, C., and Hwang, H.Y. (2011) *Nat. Commun.*, **2**, 494.

35. Bell, C., Harashima, S., Kozuka, Y., Kim, M., Kim, B.G., Hikita, Y., and Hwang, H.Y. (2009) *Phys. Rev. Lett.*, **103**, 226802.

36. Yoshimatsu, K., Yasuhara, R., Kumigashira, H., and Oshima, M. (2008) *Phys. Rev. Lett.*, **101**, 026802.

37. Berner, G., Glawion, S., Walde, J., Pfaff, F., Hollmark, H., Duda, L.-C., Paetel, S., Richter, C., Mannhart, J., Sing, M., and Claessen, R. (2010) *Phys. Rev. B*, **82**, 241405(R).

38. Zhou, K.-J., Radovic, M., Schlappa, J., Strocov, V., Frison, R., Mesot, J., Patthey, L., and Schmitt, T. (2011) *Phys. Rev. B*, **83**, 201402(R).

12
Interfacial Coupling between Oxide Superconductors and Ferromagnets

Mark G. Blamire, Jason W.A. Robinson, and Mehmet Egilmez

12.1
Introduction

With the exception of a few materials that may show intrinsic p-wave superconductivity such as Sr_2RuO_4, all superconductors exhibit singlet pairing in which the Cooper pair is formed from electrons with antiparallel spins. The application of a magnetic field imposes a different energy on the two electrons and destabilizes the pairing; this is the ultimate origin of the critical field of a superconductor.

The exchange splitting of the conduction bands in a ferromagnet has an equivalent effect on singlet pairs so that the proximity effect between a ferromagnet and a superconductor is normally extremely short-ranged. Indeed, detailed analysis reveals a rapidly oscillating decay of the pairing order parameter at a superconductor/ferromagnet interface, which is a signature of the interference between the different spatial decays of the spin-up and -down electron wavefunctions within the singlet pair.

Work in metallic systems over the past 15 years has revealed a rich field of interactions between superconductors and ferromagnets, including the direct studies of these so-called Larkin–Ovchinnikov–Ferrell–Fulde (LOFF) [1, 2] states via critical temperature variation [3] and Josephson junctions with ferromagnetic barriers [4].

With very few exceptions, the metallic devices used for such studies are polycrystallines, the superconductors are s-wave and isotropic, and the ferromagnets are itinerant with moderate spin polarization. It is reasonable to suppose, therefore, that interfaces between oxide superconductors, with their large anisotropy and complex pairing symmetries, and oxide ferromagnets, which are frequently virtually 100% spin-polarized, should reveal even more exotic effects. Even without the current understanding of the underlying physics of superconductor/ferromagnet interactions, the very similar crystal structures of high-temperature superconductors (HTS) and double-exchange ferromagnets such as $La_{0.7}Sr_{0.3}MnO_3$ (LSMO) made the growth of oxide heterostructures immediately interesting to researchers once both materials systems could be grown individually. There have been extensive experimental work in the field and a number of interesting results, which are the subjects of this chapter. Despite those developments, while writing this chapter, it

Functional Metal Oxides: New Science and Novel Applications, First Edition.
Edited by Satishchandra B. Ogale, Thirumalai V. Venkatesan, and Mark G. Blamire.
© 2013 Wiley-VCH Verlag GmbH & Co. KGaA. Published 2013 by Wiley-VCH Verlag GmbH & Co. KGaA.

is reasonable to regard oxide superconductor/ferromagnet studies as being in their infancy, and another aspect of this chapter is to attempt to explain the material problems underlying this and make suggestions as to how they may be overcome.

12.2
Experimental Results

In this section, we provide an overview of the various experimental effects observed at oxide superconductor/ferromagnet interfaces. For simplicity, these are arranged in chronological order.

12.2.1
Spin Injection

In the absence of triplet pairing, it is obvious that an excess of spin within a superconductor will inhibit pairing and so weaken superconductivity. Spin transport and accumulation lie at the heart of giant magnetoresistance (GMR) and spintronics [5], and so the investigation of spin injection into oxide superconductors was an obvious step following the discovery of large magnetoresistive effects in the doped lanthanum manganites (e.g., $La_{0.7}Ca_{0.3}MnO_3$ (LCMO)). This work was pioneered by Vasko *et al.* [6], who demonstrated a reduction in the critical current of a $DyBa_2Cu_3O_{3-d}$ track when current was simultaneously injected from an LSMO contact (Figure 12.1) through a thin buffer layer of La_2CuO_4. This work was quickly

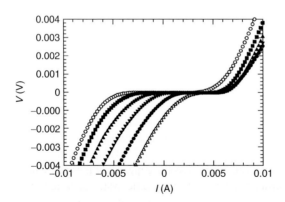

Figure 12.1 Voltage–current characteristics of a LSMO/La_2CuO^{4-y} $DyBa_2Cu_3O_7$ heterostructure at 50 K showing the critical current suppression when the parallel current in the ferromagnet equals 0 mA (open circles), 2 mA (squares), 4 mA (solid triangles), 6 mA (inverted triangles), 8 mA (solid circles), and 10 mA (open triangles). (Source: Reprinted with permission from [6]. Copyright 1997 by the American Physical Society).

followed up by other groups who demonstrated critical current suppression effects, which appeared to be considerably larger than were obtained in the absence of spin injection when currents were injected from nonmagnetic contacts [7–9]. Since spin-polarized current injection generally took place through interfacial buffer layers or tunnel barriers, direct proximity effects could be discounted and so the behavior was explained as an extension of the existing theory of nonequilibrium quasiparticle injection in BCS superconductors. Briefly, injected carriers enter quasiparticle states over a range of energies controlled by the injection voltage; these quasiparticles then relax rapidly to the gap edge where their density determines the suppression of the energy gap via the BCS gap equation. The suppression is then determined by a dynamic equilibrium between the injection rate and the pair recombination rate. Simple models suggest that this recombination rate is substantially reduced if the injected electrons are spin-polarized and so the gap-edge quasiparticle density and hence gap suppression are enhanced in the case of spin-polarized current injection.

This interpretation of the existing experimental results was challenged by Gim *et al.* [10], who argued that the dominant effects in the experiments then reported were likely to be a direct current-summation effect in which the injected charge current necessarily reduced the apparent transport critical current. Heating and a variable current transfer length that may prevent the entirety of the superconductor being affected are also significant extrinsic effects. More recent experiments, for example [9], took some account of these concerns, but nevertheless no definitive direct spin injection effect that satisfies all of the concerns of Gim *et al.* has been reported. Indeed, even in metallic, nonsuperconducting systems, tunnel injection (as opposed to diffusive flow) of spin-polarized currents has been hard to prove convincingly.

12.2.2
Proximity Effects

The proximity effect is a classic manifestation of superconductivity and can be viewed as demonstrating a passive exchange of paired and unpaired electrons between a superconductor and a normal metal. In contrast to spin injection, no transport current is, in principle, required and the effect is maximized by a clean interface between the two materials. There are four standard ways of determining the proximity coupling between materials: critical temperature measurements, superconductor/normal/superconductor (SNS) junction critical current measurements, Andreev reflection measurements and tunnel spectroscopy. All of these have been performed on oxide heterostructures but, unlike metallic systems, results have been inconclusive and somewhat unspectacular. Of these, only the first three have yielded significant results.

12.2.2.1 Critical Temperature
Provided the superconductor is thinner than its coherence length, the critical temperature (T_c) of a superconductor/normal metal bilayer should decrease

with increasing thickness of the normal metal. Correspondingly, the critical temperature of the normal metal should increase, although this is harder to determine experimentally. While such experiments are straightforward for metallic systems – see for example [11, 12] – there are significant complications for high-temperature oxide superconductors. First, the coherence lengths of these materials are very short (typically only a few nanometers), particularly in the c-axis direction, which are strongly favored for thin-film and heterostructure growth. Second, the growth of ultrathin films, which is necessitated by the short coherent length, is complicated by the sensitivity of superconducting properties to strain and stoichiometry variations. Finally, for cuprates at least, the superconductors themselves are magnetically active with an antiferromagnetic ground state; it has been demonstrated that the surface Cu ions at a cuprate/manganite interface are magnetically coupled to the Mn lattice [13].

The first study of T_c suppression in oxide superconductor/ferromagnet heterostructures was conducted by Sefrioui *et al.* [14, 15]. This study found a rapid suppression of T_c with ferromagnet thickness for five unit-cell-thick $YBa_2Cu_3O_{7-d}$ (YBCO) layers, which was qualitatively but not quantitatively consistent with theories developed for s-wave metallic systems. This has been investigated further by several groups; for example, Werner *et al.* [16] found that the proximity-induced T_c suppression in YBCO/LSMO was even smaller than had been measured earlier but that their material showed a large interfacial magnetic exchange between Cu and Mn ions, which was interpreted as imposing a barrier on carrier exchange. The issue of magnetic exchange has been investigated in parallel by magnetization measurements as a function of temperature [17–19].

Active measurements of the proximity effects have been proposed in which the superconductor is sandwiched by two magnetic layers configured so that their magnetic alignment can be changed using an external field between parallel and antiparallel [20]. These proximity-effect-driven spin-switch effects [20, 21] have been observed in metallic heterostructures [22–24], although contradictory results have also been obtained.

An additional complication has been the presence of transport currents in the usual resistive measurement of T_c. Just as for current-in-plane (CIP) measurements of GMR, the measured resistance and potentially the superconductivity itself can be affected by differential spin scattering of electrons at interfaces. Since resistive measurements of T_c are necessarily taken in a finite resistance state and at temperatures close to T_c, a significant fraction of the current in the superconductor is carried via quasiparticles. Unlike singlet pairs, quasiparticles can transport spin and so significant magnetoresistance effects can be seen in such measurements, for example, in ferromagnet/superconductor/ferromagnet trilayers [25, 26], which are distinct from the purely proximity-effect-driven spin-switch effects [20, 21] observed in metallic heterostructures [22]. Spin transport is discussed in more detail in Section 12.2.3.

12.2.2.2 SNS Junction Critical Current Measurements

In metallic systems, experiments involving diffusive Josephson junctions with ferromagnetic barriers have achieved numerous breakthroughs, including the realization of LOFF pi-junctions [4] and the creation of spin-triplet supercurrents [27–29]. The original observation of triplet currents [29] using half-metallic CrO_2 as a barrier between metallic superconducting electrodes stimulated theoretical work [30], which demonstrated that coherent tunneling under such circumstances required electron spin reorientation at superconductor/ferromagnet interfaces to form spin-aligned triplet pairs. Many oxide ferromagnets, particularly the manganites, are half metallic and so there is an obvious attraction in creating oxide heterostructure devices to explore triplet pairing.

Unfortunately, despite an enormous research effort, the reliable processing of oxide superconductor junctions even with artificial nonmagnetic barriers has proved highly problematic. Early devices, for example, $YBCO/SrRuO_3/YBCO$ [31] and YBCO/LCMO/YBCO [32], although showing supercurrents with apparently long coherence lengths, did not display the standard properties expected for a Josephson device, such as microwave and magnetic field modulation. More recent work on YBCO/manganite devices [33], albeit with a metallic superconducting counterelectrode has demonstrated more distinctly junction-like properties only if the barrier is nonmagnetic. However, this result is inconsistent with a recent scanning tunneling spectroscopy study [34] on LCMO thin films (<40 nm) that were proximity coupled to YBCO, which revealed signatures of a superconducting state in the LCMO up to 30 nm from the LCMO/YBCO interface (Figure 12.2). In this case, the length scale of the proximity coupling was interpreted as evidence for induced triplet pairing.

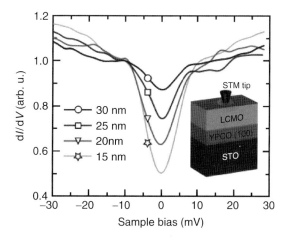

Figure 12.2 Averaged tunneling spectra at 4.2 K measured on a LCMO/YBCO bilayer with varying LCMO thickness, as indicated. (Source: From Ref. [34]).

12.2.2.3 Andreev Reflection

In an N/S proximity system, where N is a normal metal in good electrical contact with a superconductor S, superconducting correlations are induced in N over a length scale defined as the normal metal coherence length (ξ_N) while they are weakened in the S side over a scale of the superconducting coherence length (ξ_S). The mechanism underlying the proximity effect at N/S interfaces is Andreev reflection [35]: when electrons are impinged on the interface from the N side, hole-like quasiparticles are retroreflected as electron-like quasiparticles with inverse spin (maintaining their phase coherence over $\xi_N = \sqrt{\hbar D/k_B T}$, where D is the diffusion coefficient of the normal metal), while destroying Cooper pairs in the S side.

Andreev reflection also plays an important role in understanding the transport properties of a junction consisting of a ferromagnet and a superconductor (F/S). The Andreev reflection near the Fermi level preserves energy and momentum but does not preserve spin. This means the incoming electron and the reflected hole have opposite spin. This is not important for the transport across N/S interfaces because of spin-rotation symmetry in normal metals [36]. On the other hand, at F/S interfaces, because the spin-up and -down bands in ferromagnets are different, the spin flipping changes the conductance profile. In a fully spin-polarized metal, all carriers have the same spin and the Andreev refection is totally suppressed, and, consequently at zero bias voltage, the normalized conductance becomes zero. However, when the spin polarization is less than unity, a gradual suppression of the zero bias conductance peak (ZBCP) with increasing spin polarization is observed. Soulen *et al.* [36] have shown that from the reduction of the ZBCP, it is possible, with a modified Blonder–Tinkham–Klapwijk (BTK) model [37], to estimate the polarization.

Theoretical work analogous to the LOFF state in ferromagnetic superconductors [1, 2, 38] predicts a rapid and oscillatory decay of the superconducting order parameter in a ferromagnet of the form $\exp(-x/\xi_F)\cos(x/\xi_F)$ in the dirty limit (where x is the distance into the ferromagnet from the S/F interface). The corresponding coherence length in a ferromagnet is $\xi_F = \sqrt{\hbar D/2E_{ex}}$, where E_{ex} is the exchange energy of the ferromagnet. For transition-metal ferromagnets, such as Co, Fe, or Ni, ξ_F is ~1 nm at low temperatures, which is much smaller than the coherence of a Cooper pair in a normal metal ξ_N, such as Cu (~100 nm). Recently, however, much longer ranged ferromagnetic coherence lengths have been observed in YBCO/LCMO [34] (Figure 12.2) and YBCO/SrRuO$_3$ [39] systems. In these experiments, the penetration of superconductivity was 20–30 nm, which seems to defy the possibility of conventional spin singlet pairing.

The long-range effect reported in YBCO/SrRuO$_3$ bilayers was explained in terms of a crossed Andreev reflection effect (CARE) [39]. In CARE, a spin-polarized hole arriving from one magnetic domain is Andreev-reflected as an electron in an adjacent domain having the opposite spin polarization [40]. In order for CARE to occur, the width of a domain wall must be smaller than the superconducting coherence length ξ_S.

Kupferschmidt and Brouwer [41] have shown theoretically that Andreev reflection is possible between a half metal and a superconductor if the magnetism at S/F interface is spatially nonuniform with respect to the magnetism within the bulk half metal. In this situation, a wide range of novel effects are possible.

12.2.3
Spin Transport

A further manifestation of superconductor/ferromagnet coupling in oxide heterostructures is the appearance of very large magnetoresistance effects around T_c. This has become known as the *"inverse"* spin-valve or spin-switch effect to distinguish it from the proximity effect predicted by Tagirov [20], in which T_c is more strongly suppressed when the two ferromagnetic layers sandwiching a superconductor are parallel than when they are antiparallel. However, this terminology is somewhat confusing because the magnetoresistance that is measured in the vicinity of T_c has the same field dependence as that of a GMR spin valve (rather than the inverse). As discussed in Section 12.2.2.1, this is not an effect that appears to be significant in equivalent metallic structures, possibly because metallic ferromagnets are not 100% spin-polarized.

The primary effect, reported initially by Peña *et al.* [25] and Pang *et al.* [26], is of a CIP magnetoresistance in a ferromagnet/superconductor/ferromagnet trilayer, which is small or zero above the T_c of the superconductor and is large in the superconducting state close to T_c (Figure 12.3). The favored interpretation of the results [42–44] is that it is genuinely a GMR effect associated with spin-polarized quasiparticles differentially scattering depending on whether the magnetic layers are parallel or antiparallel. An alternative view [45] is that this is an unconventional proximity effect associated with the antiferromagnetic interfacial coupling discussed in Section 12.2.2.1. This work has been extended to consider the possibility of triplet current generation at the S/F interface [46].

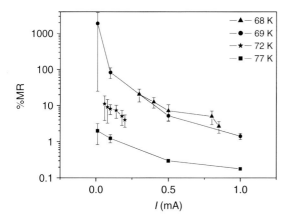

Figure 12.3 The magnetoresistance of an LSMO/YBCO/LCMO trilayer versus bias current at various temperatures, derived from magnetic-field-induced changes of the magnetic alignment of the two manganite layers. (Source: From Ref. [26]).

Cuoco *et al.* [47] discovered an unusual nonlocal voltage effect in LCMO tracks coupled to a YBCO conductor via an insulating tunnel barrier that may be related to these effects.

12.3
Materials Considerations

There are oxide superconductors, for example, $SrTiO_3$ and $Ba_{1-x}K_xBiO_3$, that have isotropic crystal structures and are believed to show BCS-like s-wave superconductivity. However, the vast majority of oxide superconductors has highly anisotropic crystal structures and properties and at least a substantial contribution of d-wave pairing symmetry. Their properties are strongly sensitive to strain and changes in electronic doping, which means that although heterostructure growth with other perovskite-based materials is relatively straightforward, maintaining bulk properties at interfaces is not. Many review articles (e.g., [48, 49]) have been written about the growth and potential of oxide materials; here we will concentrate on aspects that relate specifically to superconductor/ferromagnet interfaces. Finally, the design, growth, and understanding of oxide superconductor interfaces are clearly hampered by the continued absence of a complete theory for the superconductivity in high-temperature superconductors.

12.3.1
Growth Texture

Layered, high-temperature superconductors, such as YBCO, grow most stably along the *c*-axis direction. This means that this is also the natural growth direction for oxide heterostructures. However, the extreme anisotropy of the high-temperature materials makes proximity coupling in this direction extremely weak; not only is the *c*-axis coherence length very short (less than the lattice parameter in may cases), but the d-wave pairing symmetry means that the energy gap is effectively zero along the *c*-axis.

Although many experiments have been performed in this geometry, a full understanding of the problems associated with the *c*-axis has directed HTS device research toward structures in which coupling is possible along the ab planes of the crystal structure. In the ramp junction structure [50], a shallow ramp is etched into a *c*-axis film on which is grown a barrier and counter electrode materials. Provided full cube-on-cube epitaxy can be maintained (which in practice requires a certain range of ramp angles and careful reconstruction of the milled ramp edge), then reproducible device properties can be obtained. Most notably, the Twente group was able to make YBCO/Nb junctions that provided a demonstration of the coupling of d- and s-wave superconductivity and the fabrication of intrinsic π-squids [51]. Attempts to extend this process to all-oxide YBCO/manganite devices have so far been hampered by high interfacial resistances [52].

12.3.2
Lattice-Mismatch Strain

There is ample evidence that the properties of oxide superconductors are strongly strain-dependent (see, e.g., [53]); but this is also true for the manganites [54, 55] – the most widely applied oxide ferromagnets in interface studies. Heterointerfaces, even within well-lattice-matched structures, are sources of strain and can thus have properties that are very different to the bulk. For example, heteroepitaxial magnetic tunnel junctions provide a direct measurement of the interface polarization of manganite films and there is widespread evidence that this is significantly suppressed compared with the bulk, unless interfacial strain is carefully controlled [56, 57].

Thickness-dependent strain in heterostructures has been shown to have significant effects on the magnetic properties of manganites [58] and has been implicated in the anomalous proximity effects observed in manganite/cuprate heterostructures [59].

12.3.3
Charge Transfer

Both the lanthanum manganites and the usual cuprates such as YBCO are hole conductors. Nevertheless, there is the potential for charge transfer at the interface, which can significantly alter the magnetic and superconducting properties [60].

12.4
Conclusions

The coupling of superconducting and magnetic oxides is a highly topical field. Indeed, interest in coupled materials extends beyond the device physics, which is the subject of this review, to include magnetic pinning for HTS conductors [61, 62]. However, because of the problems that have been thrown up by the research to date, it is increasingly clear that significant material challenges have to be tackled before clear-cut results will be possible.

References

1. Larkin, A. and Ovchinnikov, A. (1965) *Soviet Phys. JETP*, **20**, 762.
2. Fulde, P. and Ferrell, R.A. (1964) Superconductivity in a strong spin-exchange field. *Phys. Rev.*, **135**, A550–A563.
3. Jiang, J.S., Davidovic, D., Reich, D.H., and Chien, C.L. (1996) Superconducting transition in Nb/Gd/Nb trilayers. *Phys. Rev. B*, **54**, 6119–6121.
4. Ryazanov, V.V., Oboznov, V.A., Rusanov, A.Y., Veretennikov, A.V., Golubov, A.A., and Aarts, J. (2001) Coupling of two superconductors through a ferromagnet: evidence for a pi junction. *Phys. Rev. Lett.*, **86**, 2427–2430.
5. Valet, T. and Fert, A. (1993) Theory of the perpendicular magnetoresistance in magnetic multilayers. *Phys. Rev. B*, **48**, 7099–7113.
6. Vasko, V.A., Larkin, V.A., Kraus, P.A., Nikolaev, K.R., Grupp, D.E., Nordman, C.A., and Goldman, A.M. (1997) Critical

current suppression in a superconductor by injection of spin-polarized carriers from a ferromagnet. *Phys. Rev. Lett.*, **78**, 1134–1137.

7. Dong, Z.W., Ramesh, R., Venkatesan, T., Johnson, M., Chen, Z.Y., Pai, S.P., Talyansky, V., Sharma, R.P., Shreekala, R., Lobb, C.J., and Greene, R.L. (1997) Spin-polarized quasiparticle injection devices using $Au/YBa_2Cu_3O_7/LaAlO_3/Nd0.7Sr0._3/MnO_3$ heterostructures. *Appl. Phys. Lett.*, **71**, 1718–1720.

8. Mikheenko, P., Colclough, M.S., Severac, C., Chakalov, R., Welhoffer, F., and Muirhead, C.M. (2001) Effect of spin-polarized injection on the mixed state of $YBa_2Cu_3O_7$-delta. *Appl. Phys. Lett.*, **78**, 356–358.

9. Fu, C.C., Huang, Z., and Yeh, N.C. (2002) Spin-polarized quasiparticle transport in cuprate superconductors. *Phys. Rev. B*, **65**, 224516.

10. Gim, Y., Kleinsasser, A.W., and Barner, J.B. (2001) Current injection into high temperature superconductors: does spin matter? *J. Appl. Phys.*, **90**, 4063–4077.

11. Deutscher, G., Lindenfeld, P., and Wolf, S. (1969) Low-temperature saturation of superconducting properties induced in silver by proximity effect. *Phys. Rev. Lett.*, **23**, 1102–1104.

12. Smith, P.H., Shapiro, S., Miles, J.L., and Nicol, J. (1961) Superconducting characteristics of superimposed metal films. *Phys. Rev. Lett.*, **6**, 686–688.

13. Chakhalian, J., Freeland, J.W., Srajer, G., Strempfer, J., Khaliullin, G., Cezar, J.C., Charlton, T., Dalgliesh, R., Bernhard, C., Cristiani, G., Habermeier, H.U., and Keimer, B. (2006) Magnetism at the interface between ferromagnetic and superconducting oxides. *Nat. Phys.*, **2**, 244–248.

14. Sefrioui, Z., Arias, D., Pena, V., Villegas, J.E., Varela, M., Prieto, P., Leon, C., Martinez, J.L., and Santamaria, J. (2003) Ferromagnetic/superconducting proximity effect in $La0.7Ca0.3MnO_3/YBa_2Cu_3O_7$-delta superlattices. *Phys. Rev. B*, **67**, 214511.

15. Pena, V., Visani, C., Garcia-Barriocanal, J., Arias, D., Sefrioui, Z., Leon, C., Santamaria, J., and Almasan, C.A. (2006) Spin diffusion versus proximity effect at ferromagnet/superconductor $La0.7Ca0.3MnO_3/YBa_2Cu_3O_7$-delta interfaces. *Phys. Rev. B*, **73**, 104513.

16. Werner, R., Raisch, C., Ruosi, A., Davidson, B.A., Nagel, P., Merz, M., Schuppler, S., Glaser, M., Fujii, J., Chasse, T., Kleiner, R., and Koelle, D. (2010) $YBa_2Cu_3O_7/La0.7Ca0.3MnO_3$ bilayers: Interface coupling and electric transport properties. *Phys. Rev. B*, **82**, 224509.

17. Senapati, K. and Budhani, R.C. (2005) Superconducting and normal-state interlayer exchange coupling in $La0.67Sr0.33MnO_3$-$YBa_2Cu_3O_7$-$La0.67Sr0.33MnO_3$ epitaxial trilayers. *Phys. Rev. B*, **71**, 224507.

18. Perez, F., Gross, K., Baca, E., Saidarriaga, W., Prieto, P., Gomez, M.E., Moran, O., Hott, R., Grube, K., Fuchs, D., and Schneider, R. (2005) Superconducting and ferromagnetic properties of epitaxial YBa_2Cu_3O7-delta/$La_2/3Ca_1/3MnO_3$ bilayers. *Phys. C: Supercond. Appl.*, **432**, 275–280.

19. Hoppler, J., Stahn, J., Niedermayer, C., Malik, V.K., Bouyanfif, H., Drew, A.J., Rossle, M., Buzdin, A., Cristiani, G., Habermeier, H.U., Keimer, B., and Bernhard, C. (2009) Giant superconductivity-induced modulation of the ferromagnetic magnetization in a cuprate-manganite superlattice. *Nat. Mater.*, **8**, 315–319.

20. Tagirov, L.R. (1999) Low-field superconducting spin switch based on a superconductor/ferromagnet multilayer. *Phys. Rev. Lett.*, **83**, 2058–2061.

21. Oh, S., Youm, D., and Beasley, M.R. (1997) A superconductive magnetoresistive memory element using controlled exchange interaction. *Appl. Phys. Lett.*, **71**, 2376–2378.

22. Gu, J.Y., You, C.Y., Jiang, J.S., Pearson, J., Bazaliy, Y.B., and Bader, S.D. (2002) Magnetization orientation dependence of the superconducting transition temperature in ferromagnet-superconductor-ferromagnet system: CuNi/Nb/CuNi. *Phys. Rev. Lett.*, **89**, 267001.

23. Moraru, I.C., Pratt, W.P., and Birge, N.O. (2006) Magnetization-dependent *T*-c shift in ferromagnet/superconductor/ferromagnet trilayers with a strong ferromagnet. *Phys. Rev. Lett.*, **96**, 037004.

24. Moraru, I.C., Pratt, W.P., and Birge, N.O. (2006) Observation of standard spin-switch effects in ferromagnet/superconductor/ferromagnet trilayers with a strong ferromagnet. *Phys. Rev. B*, **74**, 220507.

25. Peña, V., Sefrioui, Z., Arias, D., Leon, C., Santamaria, J., Martinez, J.L., te Velthuis, S.G.E., and Hoffmann, A. (2005) Giant magnetoresistance in ferromagnet/superconductor superlattices. *Phys. Rev. Lett.*, **94**, 057002.

26. Pang, B.S.H., Bell, C., Tomov, R.I., Durrell, J.H., and Blamire, M.G. (2005) Pseudo spin-valve behavior in oxide ferromagnet/superconductor/ferromagnet trilayers. *Phys. Rev. A*, **341**, 313–319.

27. Robinson, J.W.A., Witt, J.D.S., and Blamire, M.G. (2010) Controlled injection of spin-triplet supercurrents into a strong ferromagnet. *Science*, **329**, 59–61.

28. Khaire, T.S., Khasawneh, M.A., Pratt, W.P., and Birge, N.O. (2010) Observation of spin-triplet superconductivity in co-based josephson junctions. *Phys. Rev. Lett.*, **104**, 137002.

29. Keizer, R.S., Goennenwein, S.T.B., Klapwijk, T.M., Miao, G., Xiao, G., and Gupta, A. (2006) A spin triplet supercurrent through the half-metallic ferromagnet CrO_2. *Nature*, **439**, 825–827.

30. Eschrig, M. and Löfwander, T. (2008) Triplet supercurrents in clean and disordered half-metallic ferromagnets. *Nat. Phys.*, **4**, 138–143.

31. Antognazza, L., Char, K., Geballe, T.H., King, L.L.H., and Sleight, A.W. (1993) Josephson coupling of $YBa_2Cu_3O_{7-x}$ through a ferromagnetic barrier $SrRuO_3$. *Appl. Phys. Lett.*, **63**, 1005–1007.

32. Lawler, J.F., Lunney, J.G., and Coey, J.M.D. (1994) Tunneling like behavior in trilayer heterostructures of $YBa_2Cu_3O_7$-delta/La1-xCaxMnO$_3$/$YBa_2Cu_3O_7$-delta. *Physica C*, **235**, 737–738.

33. Ovsyannikov, G.A., Constantinian, K.Y., Kislinski, Y.V., Shadrin, A.V., Zaitsev, A.V., Petrzhik, A.M., Demidov, V.V.,

Borisenko, I.V., Kalabukhov, A.V., and Winkler, D. (2011) Proximity effect and electron transport in oxide hybrid heterostructures with superconducting/magnetic interfaces. *Supercond. Sci. Technol.*, **24**, 055012.

34. Kalcheim, Y., Kirzhner, T., Koren, G., and Millo, O. (2011) Long-range proximity effect in La$_2$/3Ca1/3MnO$_3$/(100)YBa$_2$Cu$_3$O$_7$- delta ferromagnet/superconductor bilayers: evidence for induced triplet superconductivity in the ferromagnet. *Phys. Rev. B*, **83**, 064510.

35. Andreev, A.F. (1964) *Soviet Phys. JETP*, **19**, 1228.

36. Soulen, R.J., Byers, J.M., Osofsky, M.S., Nadgorny, B., Ambrose, T., Cheng, S.F., Broussard, P.R., Tanaka, C.T., Nowak, J., Moodera, J.S., Barry, A., and Coey, J.M.D. (1998) Measuring the spin polarization of a metal with a superconducting point contact. *Science*, **282**, 85–88.

37. Blonder, G.E., Tinkham, M., and Klapwijk, T.M. (1982) Transition from metallic to tunneling regimes in superconducting microconstrictions: excess current, charge imbalance, and supercurrent conversion. *Phys. Rev. B: Condens. Matter*, **25**, 4515.

38. Buzdin, A.I. (2005) Proximity effects in superconductor-ferromagnet heterostructures. *Rev. Mod. Phys.*, **77**, 935–976.

39. Asulin, I., Yuli, O., Koren, G., and Millo, O. (2006) Evidence for crossed Andreev reflections in bilayers of (100) YBa$_2$Cu$_3$O$_7$ and the itinerant ferromagnet SrRuO$_3$. *Phys. Rev. B*, **74**, 092501.

40. Deutscher, G. and Feinberg, D. (2000) Coupling superconducting-ferromagnetic point contacts by Andreev reflections. *Appl. Phys. Lett.*, **76**, 487–489.

41. Kupferschmidt, J.N. and Brouwer, P.W. (2011) Andreev reflection at half-metal/superconductor interfaces with nonuniform magnetization. *Phys. Rev. B*, **83**, 014512.

42. Visani, C., Pena, V., Garcia-Barriocanal, J., Arias, D., Sefrioui, Z., Leon, C., Santamaria, J., Nemes, N.M., Garcia-Hernandez, M., Martinez, J.L., Velthuis, S., and Hoffmann, A. (2007)

Spin-dependent magnetoresistance of ferromagnet/superconductor/ferromagnet La0.7Ca0.3MnO$_3$/YBa$_2$Cu$_3$O$_7$-delta/La0.7Ca0.3MnO$_3$ trilayers. *Phys. Rev. B*, **75**, 054501.

43. Nemes, N.M., Garcia-Hernandez, M., Velthuis, S., Hoffmann, A., Visani, C., Garcia-Barriocanal, J., Pena, V., Arias, D., Sefrioui, Z., Leon, C., and Santamaria, J. (2008) Origin of the inverse spin-switch behavior in manganite/cuprate/manganite trilayers. *Phys. Rev. B*, **78**, 094515.

44. Mandal, S., Budhani, R.C., He, J.Q., and Zhu, Y. (2008) Diverging giant magnetoresistance in ferromagnet-superconductor-ferromagnet trilayers. *Phys. Rev. B*, **78**, 094502.

45. Salafranca, J. and Okamoto, S. (2010) Unconventional proximity effect and inverse spin-switch behavior in a model manganite-cuprate-manganite trilayer system. *Phys. Rev. Lett.*, **105**, 256804.

46. Hu, T., Xiao, H., Visani, C., Sefrioui, Z., Santamaria, J., and Almasan, C.C. (2009) Evidence from magnetoresistance measurements for an induced triplet superconducting state in La0.7Ca0.3MnO$_3$/YBa$_2$Cu$_3$O$_7$-delta multilayers. *Phys. Rev. B*, **80**, 060506.

47. Cuoco, M., Saldarriaga, W., Polcari, A., Guarino, A., Moran, O., Baca, E., Vecchione, A., and Romano, P. (2009) Nonlocal voltage effects in La$_2$/3Ca$_1$/3MnO$_3$/La1/3Ca$_2$/3MnO$_3$/YBa$_2$Cu$_3$O$_7$ trilayers. *Phys. Rev. B*, **79**, 014523.

48. Norton, D.P. (2004) Synthesis and properties of epitaxial electronic oxide thin-film materials. *Mater. Sci. Eng. R: Rep.*, **43**, 139–247.

49. Mannhart, J. and Schlom, D.G. (2010) Oxide interfaces—an opportunity for electronics. *Science*, **327**, 1607.

50. Verhoeven, M.A.J., Gerritsma, G.J., Rogalla, H., and Golubov, A.A. (1996) Ramp-type junction parameter control by Ga doping of PrBa$_2$Cu$_3$O$_7$-delta barriers. *Appl. Phys. Lett.*, **69**, 848–850.

51. Hilgenkamp, H., Ariando, Smilde, H.J.H., Blank, D.H.A., Rijnders, G., Rogalla, H., Kirtley, J.R., and Tsuei, C.C. (2003) Ordering and manipulation of the magnetic moments in large-scale superconducting pi-loop arrays. *Nature*, **422**, 50–53.

52. van Zalk, M., Brinkman, A., Aarts, J., and Hilgenkamp, H. (2010) Interface resistance of YBa2Cu3O7-delta/La0.67Sr0.33MnO3 ramp-type contacts. *Phys. Rev. B*, **82**, 134513.

53. Pavarini, E., Dasgupta, I., Saha-Dasgupta, T., Jepsen, O., and Andersen, O.K. (2001) Band-structure trend in hole-doped cuprates and correlation with *T*-c (max). *Phys. Rev. Lett.*, **87**, 047003.

54. Paranjape, M., Raychaudhuri, A.K., Mathur, N.D., and Blamire, M.G. (2003) Effect of strain on the electrical conduction in epitaxial films of La0.7Ca0.3MnO3. *Phys. Rev. B*, **67**, 214415.

55. Blamire, M.G., Teo, B.S., Durrell, J.H., Mathur, N.D., Barber, Z.H., Driscoll, J.L.M., Cohen, L.F., and Evetts, J.E. (1999) Strain-induced time-dependent magnetic disorder in ultra-thin La0.7Ca0.3MnO$_3$ films. *J. Magn. Magn. Mater.*, **191**, 359–367.

56. Jo, M.H., Mathur, N.D., Todd, N.K., and Blamire, M.G. (2000) Very large magnetoresistance and coherent switching in half-metallic manganite tunnel junctions. *Phys. Rev. B*, **61**, R14905–R14908.

57. Bowen, M., Bibes, M., Barthelemy, A., Contour, J.P., Anane, A., Lemaitre, Y., and Fert, A. (2003) Nearly total spin polarization in La2/3Sr1/3MnO3 from tunneling experiments. *Appl. Phys. Lett.*, **82**, 233–235.

58. Jo, M.H., Mathur, N.D., Evetts, J.E., Blamire, M.G., Bibes, M., and Fontcuberta, J. (1999) Inhomogeneous transport in heteroepitaxial La0.7Ca0.3MnO3/SrTiO3 multilayers. *Appl. Phys. Lett.*, **75**, 3689–3691.

59. Mani, A., Kumary, T.G., Hsu, D., and Lin, J.G. (2008) Strain enhanced spin polarization in Nd0.43Sr0.57MnO3/YBa2Cu3O7 bilayers. *J. Appl. Phys.*, **104**, 053910.

60. Yunoki, S., Moreo, A., Dagotto, E., Okamoto, S., Kancharla, S.S., and Fujimori, A. (2007) Electron doping of

cuprates via interfaces with manganites. *Phys. Rev. B*, **76**, 064532.

61. Wimbush, S.C., Durrell, J.H., Blamire, M.G., and MacManus-Driscoll, J.L. (2011) Strong flux pinning by magnetic interlayers compatible with YBa2Cu3O7-delta. *IEEE Trans. Appl. Supercond.*, **21**, 3159–3161.

62. Wimbush, S.C., Durrell, J.H., Tsai, C.F., Wang, H., Jia, Q.X., Blamire, M.G., and MacManus-Driscoll, J.L. (2010) Enhanced critical current in YBa2Cu3O7-delta thin films through pinning by ferromagnetic YFeO3 nanoparticles. *Supercond. Sci. Technol.*, **23**, 045019.

Part VI
Devices and Applications

Functional Metal Oxides: New Science and Novel Applications, First Edition.
Edited by Satishchandra B. Ogale, Thirumalai V. Venkatesan, and Mark G. Blamire.
© 2013 Wiley-VCH Verlag GmbH & Co. KGaA. Published 2013 by Wiley-VCH Verlag GmbH & Co. KGaA.

13
Metal-Oxide Nanoparticles for Dye-Sensitized Solar Cells

Frédéric Sauvage, Mohammad K. Nazeeruddin, and Michael Grätzel

13.1
TiO$_2$: Polymorphism, Optoelectronic Properties, and Bandgap Engineering

Titanium dioxide (TiO$_2$), also commonly called "titania," exists mainly in three known polymorphs: the rutile, the anatase, and the brookite. Moreover, there also exist three metastable synthetic forms (TiO$_2$-B; H, hollandite; and R, ramsdellite) and five other high-pressure/high-temperature structures (columbite, baddeleyite, fluorite, pyrite, and cotunnite). Discovered by Abraham Gottlob Werner in 1803, the rutile structure is thermodynamically the most stable with a difference of $\sim$5–12 kJ mol^{-1} in free enthalpy of formation ΔG_f° compared to the anatase [1]. Bulk brookite and anatase irreversibly convert into the rutile at a temperature of 750 and $\sim$700 °C under atmospheric pressure, respectively. This transition temperature can vary significantly with the particle size as a result of combined phenomena such as surface defects, easier strains relaxation, or energy surface, which predominates when getting toward the nanosize. This is what happens when the particle size ranges below a threshold of 14 nm. A kinetic model has been developed to account for the anatase-to-rutile phase transformation, and calorimetric studies underlines the existence of a crossover point in this established relative polymorphism stability [2, 3]. As a result, these analyses suggest that the anatase becomes more stable than the rutile in nanoparticles, the crossover being governed by the lower surface energy of the anatase (0.4 ± 0.1 J m^{-2}) compared to the rutile (2.2 ± 0.2 J m^{-2}). As a corollary of this lower energy surface in anatase, we have experimentally given evidence of this intersect by combining *in situ* X-ray diffraction and *in situ* Raman spectroscopy at different temperatures, where we highlighted an important increase in temperature for the anatase-to-rutile phase transformation. Indeed, whereas a typical onset transition at 700 °C is noticed for particles of 20 nm, we found this temperature to increase significantly to 1000 °C when the particles are about 4 nm in size, together with a noticeable delay in the grain coarsening process [4].

This rich structural polymorphism in TiO$_2$ provides this material with a variety of band structures and therefore of physical properties (Table 13.1).

The crystal structure of anatase and rutile adopts a tetragonal lattice cell, in which Ti^{4+} is hexacoordinated with oxygen. The difference between these two

Functional Metal Oxides: New Science and Novel Applications, First Edition.
Edited by Satishchandra B. Ogale, Thirumalai V. Venkatesan, and Mark G. Blamire.
© 2013 Wiley-VCH Verlag GmbH & Co. KGaA. Published 2013 by Wiley-VCH Verlag GmbH & Co. KGaA.

Table 13.1 Evolution of the crystal structure, lattice cell parameters, density, and optical bandgap as a function of TiO$_2$ polymorph.

	Crystal system	Space group	a (Å)	b (Å)	c (Å)	ρ (g cm^{-3})	E_g (eV)
Rutile	Tetragonal	$P4_2/mmm$	4.58	4.58	2.95	4.23	3.0
Anatase	Tetragonal	$I4_1/amd$	3.79	3.79	9.51	3.90	3.2
Brookite	Orthorhombic	$Pbca$	5.46	9.18	5.14	4.10	3.4

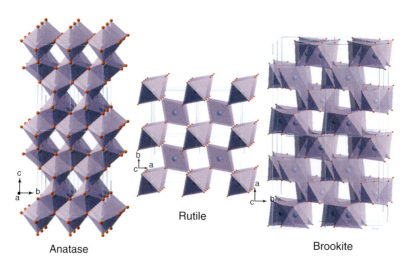

Rutile

Anatase

Brookite

Figure 13.1 Representation of the crystal structure for anatase, rutile, and brookite TiO$_2$.

structures stems from a specific arrangement of the [TiO$_6$] octahedra in the lattice with different degrees of distortion. For instance, in the anatase, chains of [TiO$_6$] octahedra are formed by sharing the oxygen apex along [100] and [010] directions while four edges are also shared, forming cubic close-packed oxygen sublattices (Figure 13.1). By contrast, the rutile is formed by isolated chains of [TiO$_6$] octahedra along [001] direction connected only by edges. This arrangement of the octahedra makes two-dimensional (2D) channels suitable for ion mobility along the [100] and [010] directions in the case of the anatase, and only 1D channel along the [001] direction for the rutile. Combined with Ti(+IV/+III) redox activity, lithium conduction has been found in these two structures, making them particularly attractive as negative electrodes for lithium-ion batteries with a redox activity at $E \approx 1.85$ V versus Li$^+$/Li [5–8].

Brookite adopts a lower lattice cell symmetry owing to its shorter order range in the lattice. The [TiO$_6$] octahedron is particularly distorted, resulting in six different Ti–O distances. This structure can be described as zigzag chains of [TiO$_6$] octahedra sharing edges along the [001] direction.

The electronic structure significantly differs depending on the polymorph. This can be illustrated by the difference in values in the bandgap energy, which are 3.0 eV (indirect), 3.2 eV (indirect), 3.4 eV (direct) for the rutile, anatase, and brookite, respectively [8–10]. The occupied states of the valence band are essentially composed of O 2s (deep in the valence band (VB) and O 2p orbitals in contrast to the conduction band, which is predominantly composed of Ti 3d orbitals. As a result of the crystal field, created with the surrounding oxygen ions, the energy of these orbitals splits into three low-lying t_{2g} and two high-lying e_g components. The bandgap modification mainly originates from a movement of the conduction band edge. These materials show n-type conductivity triggered not only by the presence of native donor impurities but also because TiO_2 is typically a slightly oxygen-deficient material leading to the formation of Ti^{3+} punctual defects. Because of this n-type conduction, both anatase and rutile can be used as a photoanode in photoelectrochemical devices, albeit photoactivity in rutile is relatively scarce compared to anatase, which is most likely due to lower carrier mobility and high effective electron mass (Table 13.2). Rutile polymorph is in turn preferred for optical applications because of its higher refractive index. It has been used therefore as a light-scattering layer for dye-sensitized solar cells (DSCs) [11].

The use of the brookite polymorph remains also relatively sparse besides recent reports relative to DSCs showing an interesting 6% power conversion efficiency (PCE) while using the N719 dye without the scattering layer [10].

With their large bandgap value, light capture by TiO_2 is effective only in the narrow UV-A portion of the solar spectrum (365–415 nm), which accounts for only about 4% of the total solar irradiation and <0.1% of indoor lightning. Under bandgap excitation, electron/hole (e^-/h^+) pairs are formed in the material. In contact with an electrolyte, these pairs are further separated within the space charge layer confined to the surface of TiO_2. In this case, the ensuing electric field resulting in band bending drives the main charge carrier (e^-) toward the inner part of the particles while the holes move in the opposite direction toward the surface, where a photoanodic reaction can occur (e.g., oxidation of water, alcohol, aromatic hydrocarbons, phenols, etc.). Photocatalytic splitting of water using TiO_2 electrodes was envisioned by Honda *et al.* [21] back in 1972, which opened up new perspectives for heterogeneous photocatalysis using inorganic or hybrid semiconducting materials.

Table 13.2 Values for electron mobility and effective mass for rutile and anatase polymorph measured on single crystals or polycristalline films.

	Electron mobility in single crystal $(cm^2\ V^{-1}{\cdot}s^{-1})$	Electron mobility in polycrystalline film $(cm^2\ V^{-1}{\cdot}s^{-1})$	Electron effective mass (m_e)	Refractive index $(\lambda = 600\ nm)$
Rutile	0.1–1 [12, 13]	0.1 [12, 14]	9–32 [12, 15, 16]	2.9 [17]
Anatase	15 [18]	0.1–4 [19]	1.0 [15]	2.5 [20]

The interesting feature of TiO_2 also concerns its high dielectric constant and, above all, the $3d^0$ electronic configuration adopted by Ti^{4+}. This renders its optoelectronic properties extremely sensitive to point defects; the latter can be created subsequent to the formation of oxygen vacancies or titanium interstitials introduced by post-annealing in a reducing atmosphere or by means of aliovalent doping using s, p, d, or even f class of elements. Although TiO_2 is well established to be a white material, it is probably the only oxide semiconductor for which the absorption edge can be tuned as high as 2.3 eV, ranging from a blue shift to 3.8 eV by electron/hole pair confinement effect below 2 nm particle size [22] to a red shift leading to a black color and a bandgap as low as 1.54 eV under H_2 post-annealing from 1 to 20 bar [23–26]. This is illustrated in Figure 13.2, which shows the color modification experienced in the case of Fe^{3+} doping in 5-nm anatase TiO_2 particles or the black TiO_2 obtained after 5 days hydrogenation at 450 °C under 15 bars pressure.

Another means to render TiO_2 optically active toward visible light is the sensitization by either a chromophore (Figure 13.3) or inorganic quantum dots covalently anchored to the acidic surface of TiO_2, an approach that forms the basis of the DSC technology [27].

The basic idea to dope nanostructured TiO_2 has been extensively investigated with the aim of improving the photoactivity to natural light. Indeed, the doping approach gets largely democratized as it allows both increasing the free-charge-carrier concentration as well as reducing the material's bandgap as a result of the creation of new energy levels – either donor levels below the conduction band (n-type doping) or acceptor levels above the valence band (p-type doping). This enables, therefore, in a more optimal way to make use of the visible and near-infrared (NIR) light for photocatalysis or photoelectrochemical reactions. Further, according to Poisson's relationship given later in the spherical coordinate system, the increase in charge-carrier concentration (N) favors band bending ($d\psi$) and therefore charge separation (Eq. (13.1)). It also reduces the width of the depletion layer (W), enabling

15bars H_2 Anatase TiO_2 Fe^{3+} doping

Figure 13.2 Color modification of anatase TiO_2 powder induced by Fe^{3+} doping or H_2 posttreatment.

Figure 13.3 Sensitized anatase TiO_2 photoanode by different organic donor–acceptor chromophores. (Source: Courtesy of Dr. Jun-ho Yum – EPFL.)

the minor reactive charge carriers to get confined into the material surface (Eq. (13.2)). For application in DSCs, irrespective of whether or not the band-bending takes place under illumination – discussion still under debate for intrinsic TiO_2 – a too high carrier concentration in the conduction band is an unwished condition, as it will promote the recombination process on the one hand and possibly the degradation of the cell with time on the other hand, resulting from an excessive bandgap excitation.

$$\frac{1}{r^2}\frac{d}{dr}\left(r^2\frac{d\psi}{dr}\right) = \frac{qN}{\varepsilon\varepsilon_0} \tag{13.1}$$

$$W = \left(\frac{2\Delta\psi\varepsilon\varepsilon_0}{qN}\right)^{\frac{1}{2}} \tag{13.2}$$

Cationic doping has been achieved successfully with a plethora of aliovalent and isovalent transition or rare-earth metals (e.g., Cu^{2+}, Zr^{4+}, Zn^{2+}, $Co^{3+/2+}$, Ni^{2+}, Cr^{3+}, Mn^{3+}, Mo^{6+}, Nb^{5+}, V^{5+}, Fe^{3+}, Y^{3+}, Ga^{3+}, Ru^{4+}, W^{5+}, Sb^{3+}, Sn^{4+}, Au^+, Ag^+, Pt^{4+}, La^{3+}, Ce^{3+}, Er^{3+}, Pr^{3+}, Gd^{3+}, Nd^{3+}, or Sm^{3+}) [9, 28–43]. However, it is worth mentioning that the introduction of punctual defects is also responsible for important changes in the thermodynamics of the material, which conducts in some cases to thermal instability, leading in turn to phase segregation and/or simply to the formation of metastable phases. These phases also separate with time or on exposing to light as in the case of niobium-doped TiO_2, which was investigated as a new transparent conduction oxide alternative to indium-doped tin oxide [9].

Similarly, doping in anionic sites has been widely investigated (e.g., N, S, C, B, P, I, and F) [40, 44–58], even though it is largely believed that the solubility of a dopant anion in the lattice is more restricted owing to the higher free enthalpy energy for the substitution reaction in the anion sites. By contrast, it is worth mentioning that, depending on the crystallographic location of the dopant (in substitution or in interstice), it has been reported that low concentration of interstitial nitrogen can drastically decrease the activation energy to form oxygen vacancies from 4.2 to 0.6 eV [59, 60], thus also contributing to the red shift of the absorption edge and facilitating the anion solubility.

13.2
Principle and Basis of Dye-Sensitized Solar Cell (DSC) Technology

In this section, we introduce the DSC technology and cover its basic principles of operation. We deliberately restrict this chapter to n-type DSCs with liquid electrolytes, which have attracted the most research activity and are suitable for commercialization. The related technologies, such as p-type DSC [61] or the solid-state version including a hole conductor electrolyte (e.g., Spiro-OMeTAD) [62], are referenced for further information. But first, it should be remembered that sensitization of a large-bandgap semiconductor, for which inspiration is originally drawn from natural photosynthesis (e.g., the leaf), can be carried out by mimicking the principle of one photosystem, in which Chlorophyll P680 or P700 plays the role of a light absorber, and the thylakoid membrane transports the photoinduced charge species. This is also the underlying basis of the monochromic photographic process discovered by Hermann Vogel in 1873, who optimized an aniline-based dye and a silver halide emulsion to provide panchromatic sensitivity to visible light.

DSC is the only PV technology that relies on electrochemical processes and, therefore, can be considered as a photoelectrochemical galvanic cell. It comprises two electrodes, namely a photoanode and a cathode, separated by a low-viscosity liquid electrolyte containing among additives a redox, basically I_3^-/I^- (triiodide/iodide), to ensure ionic conductivity and dye regeneration functionality (Figure 13.4). The photoanode is basically made of a screen-printed, thin, transparent layer of TiO_2 on a transparent glass conducting oxide. The cathode is supported by an electrochemical catalyst made of nanoparticles of Pt prepared, for instance, by thermal degradation of H_2PtCl_6 in isopropanol to assist I−I bond cleavage, thereby considerably increasing the fill factor (ff). Less expensive alternatives have been proposed such as first, carbon black or multiwalled carbon nanotubes [63–66] or, more recently, conductive polystylenesulfonate polymers (e.g., PEDOT) [67] or semimetallic compounds such as CoS [68]. A new step forward into power conversion efficiency (PCE was recently published by the Ecole Polytechnique Fédérale de Lausanne (EPFL) with a record value of 12.3% (AM 1.5G) [69], by combining two absorbers in which the respective light absorption is complementary and an alternate redox couple based on Co(+III/+II) tris(2,2′ bipyridyl) [70]. The advantage of this redox couple stems from its higher chemical potential than I_3^-/I^- ($E_{(Co+III/+II)} = 0.56$ V versus NHE − $E°(I_3^-/I^-) \approx 0.32$ V versus NHE (normal hydrogen electrode)) and lower recombination rate than the I_3^-/I^- system, thus improving the open-circuit photovoltage. DSC modules of 100 cm^2 achieved 10.0% efficiency [71], and similar efficiencies were maintained in laboratory devices under

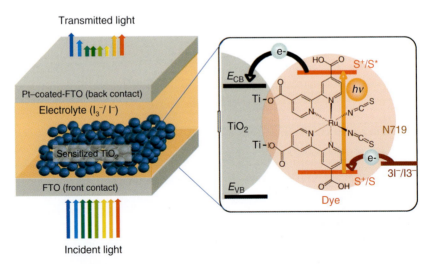

Figure 13.4 Schematic representation of a dye-sensitized solar cell. (Inset) N719 dye anchored onto TiO_2 with the energetic position.

$60\,°C/100\,\text{mW cm}^{-2}$, accelerating tests with PCE retention over 95% after 1000 h aging [72]. Such excellent results obtained after 20 years of intensive research from both academia and industry stress once again the viability of such technology as well as its maturity for the PV market.

When exposed to light, irrespective of whether an external load is connected or not, a sequence of charge transfers takes place in the cell (Figure 13.5). Let us first consider the reactions at the photoanode part. The light capture by the dye S induces a photoexcitation leading to an excited state S^*

$$S + h\nu \longrightarrow S^*\,(\text{photoexcitation})$$

This excited state, depending on its lifetime, has two possible pathways. The first is an ultrafast charge injection toward the acceptor levels contained in TiO_2 (of femtosecond (fs) range). This step is promoted when the injection is significantly faster than the lifetime of the dye-excited state. This is the case, for instance, for the ruthenium$(+II)$–polypyridyl complexes for which the triplet MLCT (metal-to-ligand charge transfer) excited states have lifetimes in the range of $\tau = 10$–$100\,\text{ns}$.

$$S^* \longrightarrow e^-{}_{CB} + S^+\,(\text{injection})$$

On the contrary, when the rate constant for injection becomes comparable to or lower than the excited-state lifetime, injection can compete with a rapid radiative or nonradiative decay to return to its fundamental energy. This will give rise to

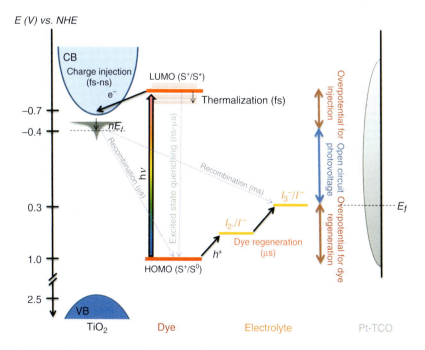

Figure 13.5 Representation of the main photoelectrochemical processes involved in dye-sensitized solar cells.

nonquantitative overall quantum efficiency.

$$S^* \longrightarrow S + h\nu \text{(radiative quenching)}$$

Note that, if the cell under illumination is in open-circuit condition, we will have the condition where $j_{inj} = -j_{rec}$ without any charge transport in TiO_2 and in the electrolyte. The oxidized dye S^+ after having injected one electron into the conduction band (CB) of TiO_2 can either be regenerated to its initial redox state by electron recapture from the conduction band of TiO_2 (recombination reaction), or react with the reductive redox couple contained within the electrolyte (e.g., I^-). This competing reaction is kinetically favored toward the regeneration, which occurs 10–100 times faster than the recombination process. This is likely the case when the iodide concentration in the electrolyte is sufficient. The kinetics of dye regeneration is particularly crucial because, along with the injection rate, these two parameters control the e^-/h^+ charge separation, and therefore the overall efficiency of the cell.

$$e^-_{CB} + S^+ \longrightarrow S \text{ (recombination)}$$

$$2S^+ + 3I^- \longrightarrow 2S + I_3^- \text{(regeneration)}$$

Finally, slower processes also take place; the first is concerned with the transport of the photon-induced electron throughout the mesoporous layer of TiO_2. The electron transport, which requires *ms* range to get collected, is ensured by both diffusion and migration control. By using nanoparticles of TiO_2, although perfectly crystallized, the existence of a continuum of intraband energy states lying below the bottom of the conduction band explicitly renders the energy and the conduction of these electrons more complicated than in a small polaron-hopping transport scheme. These intraband energy states localized at the particle's surface have their origin from surface point defects created by the unsatisfied coordination sphere of the Ti(+IV). The optimized size of TiO_2 for DSC is about 20 nm in diameter, which represents a Brunauer, Emmett, and Teller (BET) surface area of about $80 \, m^2$ g^{-1} (N_2 sorption). These characteristics represent a surface-to-volume ratio of 15% considering a spherical model, as we will see later in the chapter, whose morphology is more or less close to that of the synthesized particles. These intermediate states, which act as electron traps, have been characterized by different complementary techniques such as thermoluminescence, electron paramagnetic resonance (ESR), and electrochemical impedance spectroscopy [73–76]. The examination of the density as a function of energy suggests a single exponential distribution between 200 and 500 mV below the bottom of the conduction band. Their energy and their distribution are strongly dependent on the synthetic conditions to obtain the mesoporous TiO_2 or the electrolyte composition, for example, if it includes a Lewis base such as the *ter*-butyl pyridine or the *N*-butyl-benzimidazole, which deprotonates the surface of TiO_2 and, as a consequence, upshifts the energy of these trap states. Finally, the dye polarity and electron density in the ligand also play an important role in their distribution [77]. The photon-induced electron is therefore transported toward the current collector by a random walk trapping–detrapping mechanism for which, again, the density of state versus energy has a direct control on the transport time as well as on the recombination rate, and thus finally on the charge-collection efficiency (CCE) [78].

The slower process in DSC is the redox mass transport, most particularly tri-iodide, which takes place in the milliseconds to seconds range time domain. Indeed, iodide is oxidized to triiodide during the dye regeneration process, which is an electrochemical reaction with a second-order rate constant. Triiodide thus diffuses toward the counterelectrode before being reduced by a two-electron process from the external circuit. Iodide diffuses toward the dyes, where it is consumed. This shuttle reaction loops the overall electrochemical processes involved in DSC.

13.3
Progress in TiO₂ Engineering for Improved Charge-Collection Efficiency and Light Confinement

In this section, we describe the recent progress made at EPFL on the TiO₂ photoanode in order to improve the CCE, and also the synthesis and the integration of mesoporous TiO₂ beads; these combine both functionalities of high surface area, enabling strong light capture by the photoanode, as well as light scattering ability to improve light confinement in the photoanode. This progress has, for example, led to DSC devices achieving well above 10% efficiency while including only one type of particles.

Doping has long been considered to have a harmful effect on DSC performance because the increase in charge-carrier concentration promotes charge recombination. However, we found a trade-off level before the semiconductor turns conductive and where the doping is beneficial to increase the CCE. In particular, we have first investigated cationic aliovalent doping by means of either supervalent elements (i.e., Nb^{5+}) or trivalent elements (Ga^{3+} and Y^{3+}). Note that the term "doping", which is typically reserved in the specific case where the dopant is fully ionized into the lattice, should be herein corrected by the term "substituent" because Rietveld refinement carried out from X-ray diffractograms suggests a substitution mechanism with the Ti^{4+} and the element is not or barely ionized, leading to a limited increase in free charge-carrier concentration.

Anatase TiO₂ can be substituted for niobium by hydrothermal synthesis using titanium isopropoxide and an adequate amount of niobium pentachloride [9]. For this, we followed the reported procedure for the synthesis and preparation of the screen-printing paste [79]. The interest in niobium in the literature comes from successive reports showing almost 100% ionization of Nb^{5+} as a result of strong hybridization of the 4d orbitals of Nb^{5+} with the 3d orbitals of the Ti^{4+}. This gives rise to a significant interest in Nb:TiO₂ materials for lower cost transparent conducting oxide (TCO) materials combining high transparency, that is, 90% transmission of visible light, with a low room-temperature resistivity of $2-3 \times 10^{-4}$ Ω cm [80].

The demonstration for the effective doping stems from the evolution of X-ray diffractograms, which show the diffraction peaks shifting as a function of the doping level from 0.5 to 2 at%. This is confirmed by the Rietveld refinement, which verifies Vegard's law, that is, a linear evolution of the lattice cell parameter as a function of doping level (Figure 13.6). The refined lattice cell parameters for the anatase TiO_2 are $a = 3.7856(1)$ Å and $c = 9.5013(3)$ Å. The incorporation of Nb induces a slight cell expansion for a and c, leading to $a = 3.7888(3)$ Å and $c = 9.5052(8)$ Å for 2% Nb-doped sample, which can be ascribed to the larger ionic radius of Nb^{5+} ($r_{(Nb5+)} = 0.64$ Å versus $r_{(Ti4+)} = 0.605$ Å [81]). It also led to a textural modification of the particles with a slight decrease in their sizes, which in turn increases the BET surface area from 77 to 113 m^2 g^{-1}. Note the slight deviation between the concentration of dopant in the solution and its effective introduction into the lattice. For example, 2 at% of $NbCl_5$ in solution leads to about 1.5% dopant successfully introduced. The incorporation of niobium gives rise to a slight morphological evolution; more elongated and bulged elliptical shaped particles are obtained in comparison to the undoped sample. This can be visualized from the 3D modeling for the crystallize size/shape obtained from the X-ray diffraction (XRD) Rietveld refinement and confirmed by transmission electron microscopy (TEM). The introduction of donor states results in the Burstein–Moss effect visible as a slight blue shift of the bandgap from 3.22 to 3.28 eV [9]. The doping mechanism has been attributed as follows:

$$Nb^{5+}_{solvated} + Ti^{x}_{Ti}(TiO_2) \longrightarrow \left[Nb^{\bullet}_{Ti} + \frac{1}{4}V''''_{Ti} \right](TiO_2)$$

Niobium introduction, which causes point defects within the anatase lattice, exerts a remarkable influence on the $J-V$ characteristics. For instance, Figure 13.7 shows the curve recorded by using only a 7 μm thick transparent electrode (i.e., without the scattering layer) sensitized with the heteroleptic ruthenium(+II) complex coded C101. The dye structure is reported for information in the inset, and all optoelectronic information about this dye can be found in Ref. [82]. Increasing the concentration of Nb^{5+} results in a significant drop in the photovoltage from 721 to 656 mV. However, it is largely compensated by an increase in short-circuit photocurrent from 13.6 to 15.2 mA cm^{-2} before declining to 14.4 mA cm^{-2}. It is therefore not surprising that the ff decreases from 74.6% to 69%. With 0.5% doping level, a maximum PCE of 8.1% is obtained under the standard illumination conditions (AM 1.5G). This should be compared to the 7.4% PCE of pristine, undoped TiO_2. The increased photocurrent recorded with such a low amount of niobium is consistent with the IPCE comparison which shows superior photon-to-electron conversion efficiency between 390 and 650 nm. The strong point of low doping amount with niobium comes from the fact that not only the photocurrent can be improved but also the photovoltage, which can be maintained to 721 mV. Improvement in the PCE can be achieved subsequently by the so-called $TiCl_4$ treatment, which hampers the fast recombination kinetics between the injected electron and triiodide [83]. By this means, a PCE of 8.7% has been reached under standard illumination.

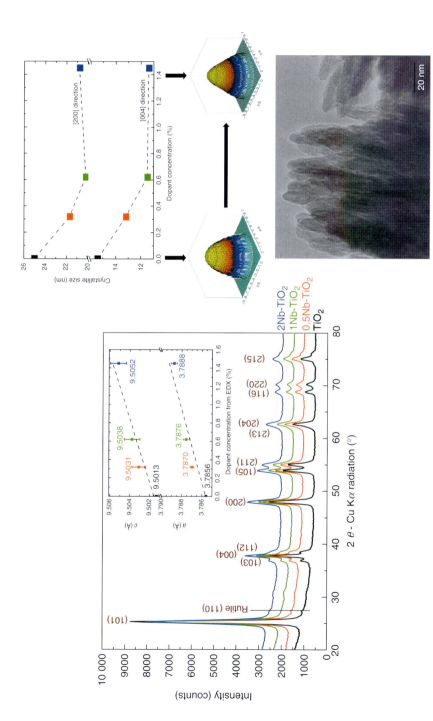

Figure 13.6 Superposition of the X-ray diffractograms for undoped TiO_2 and the different Nb-doped samples including the variation of the lattice cell parameters verifying Vegard's law. The evolution of the crystallite size and shape is shown, as well as a TEM micrograph of the Nb-doped sample (2%).

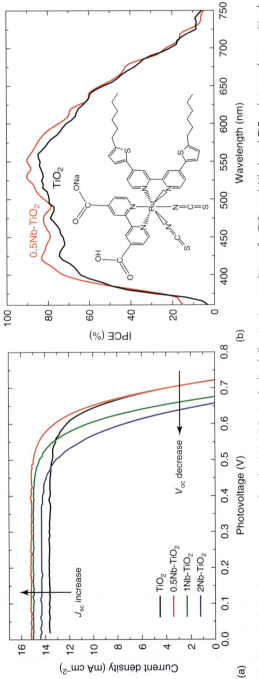

Figure 13.7 (a) (*J*–*V*) curves measured under AM 1.5G standardized illumination conditions for TiO₂ and Nb-doped TiO₂ photoanodes sensitized with the C101 dye (without TiCl₄₍aq₎ posttreatment). (b) The IPCE comparison between TiO₂ and 0.5% Nb:TiO₂ is also shown.

Charge extraction measurements, which consist in filling the intraband states with electrons by means of a white light bias under open-circuit conditions and then switching off the light while short-circuiting the cell at the same time to collect the charges, show that the introduction of niobium induces an important increase in the density of states below the conduction band (Figure 13.8). This can be understood by the creation of donor levels as a result of this n-type doping mechanism. For 2%, we can even notice a deviation from a single exponential distribution of traps, which looks more like a bi-sigmoid distribution.

As mentioned previously, this modification of the trap states distribution has a strong influence on the cell kinetics, in particular for the charge transport and the dynamics of recombination, both processes being in competition within the millisecond timescale. Analysis of the photovoltage (OCV condition) or photocurrent transient decay within the milliseconds range (SC condition) allows the measurement of the electron lifetime and electron transport, respectively. These two parameters give the CCE (η_{CCE}), determined as the ratio of the transport rate (k_t) and the sum of the transport and recombination rate constants (k_{rec}). Figure 13.9 shows the evolution of electron lifetime and transport rate as a function of charge density within the film. Note that it is also compared to 17-nm TiO$_2$ in order to rule out a possible size effect because the doped samples exhibit a size close to 17 nm. The deeper trap states increase noticeably the electron lifetime by more than one order of magnitude. This can be understood by (i) the lower overpotential between the Fermi level in TiO$_2$ and the chemical potential of I$_3^-$/I$^-$ redox couple, and

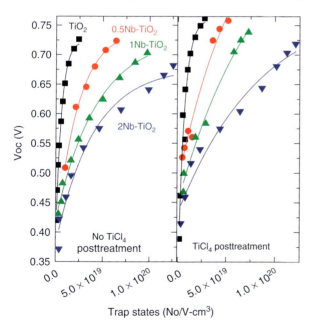

Figure 13.8 Distribution of intraband trap states for TiO$_2$ and Nb-doped TiO$_2$ with or without TiCl$_4$ posttreatment.

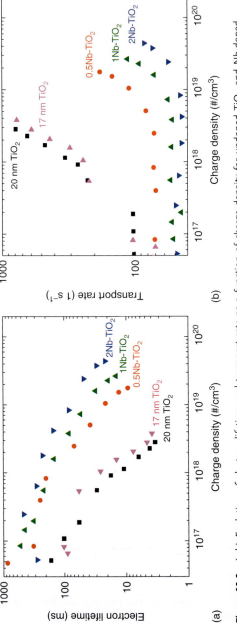

Figure 13.9 (a,b) Evolution of electron lifetime and transport rate as a function of charge density for undoped TiO$_2$ and Nb-doped TiO$_2$.

(ii) the stronger effect of electron trapping. However, as a consequence of electron trapping strengthening, it induces a slower electron transport toward charge collection. Through this example, we could experimentally illustrate the influence of these traps (i.e., their density and energy) on the critical parameter CCE.

The most interesting observation is that the increased electron lifetime is not completely compensated by a loss of transport. This leads then to a noticeable amelioration in the CCE when the mesoporous film of TiO₂ is modified by Nb^{5+} doping (Figure 13.10). This clearly shows that, contrary to the preconceived idea, doping at low level (causing more structural defects than tremendous charge-carrier density raise) is an efficient approach to increase the CCE via the electron lifetime in DSC. The best conversion performances were obtained by adding 0.5% $NbCl_5$ to the solution, which in the end corresponded to a composition within the solid of 0.35 at%.

As we have seen above, the introduction of supervalent elements entails deeper trap states, which penalizes the output photovoltage. By contrast, the incorporation of subvalent elements, such as Ga^{3+} and Y^{3+}, does not lead to this effect because their introduction induces the creation of acceptor levels. This is exactly what we have observed experimentally by charge extraction measurements where the doping in the case of these trivalent elements has no influence either on the distribution or on the energy of the trap states lying below the conduction band (Figure 13.11). Note that for conciseness reasons, the whole characterization of the materials obtained is not herein detailed as much as for Nb^{5+}; however, more information can be found in Ref. [84]. We narrowed the description to 1% Ga^{3+} and 1% Y^{3+}, compositions that offered the best performance in terms of PCE. Experimentally, the solubility of these two elements is significantly lower than Nb^{5+} as a result of the important mismatch of ionic radii for Y^{3+} and Ga^{3+} ($r_{(Ga3+)} = 0.61$ Å and

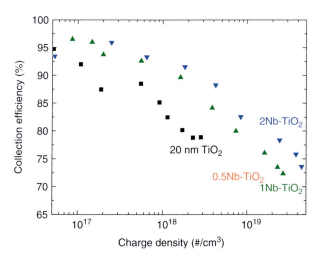

Figure 13.10 Evolution of the charge collection efficiency as a function of charge density for undoped TiO₂ and Nb-doped TiO₂.

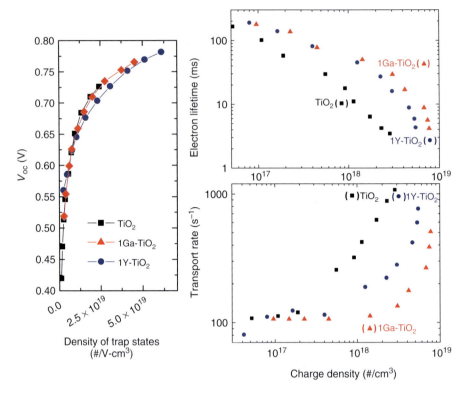

Figure 13.11 Evolution of the intraband trap states distribution, electron lifetime, and transport rate as a function of charge density for undoped TiO$_2$ and 1% Ga^{3+}- and Y^{3+}-doped TiO$_2$.

$r_{(Y3+)} = 0.90$ Å) and their doping mechanism leading to the formation of oxygen vacancies in the material. From elementary quantification and XRD analysis, the 1% sample was found to lead to a ratio of Ga/Ti $= 0.31$ and Y/Ti ≈ 0.05. The doping mechanism can be written, using the Kröger–Vink notation, as

$$Ga^{3+}_{solvated} + Ti^x_{Ti}(TiO_2) \longrightarrow \left[Ga'_{Ti} + \frac{1}{2}V_{\ddot{O}} \right] (TiO_2)$$

From the single exponential decay of photovoltage at open-circuit condition and short-circuit current density, the electron lifetime and transport rate constant were extracted as a function of the charge density. Similar observations can be made with trivalent dopants: that is to say, two- to fourfold improvement of the electron lifetime, which translates into a lower recombination rate constant, and retardation of electron transport due to the lowering of the transport rate (Figure 13.11). For example, at a given charge density of 10^{18} e$^-$ cm^{-3}, the electron lifetime moves up from 18 to 55 ms, and the transport rate decreases from 379 to 171 s^{-1} for Y^{3+} and 111 s^{-1} for Ga^{3+}.

From the combination of electron lifetime and transport rate, we demonstrate once again that the introduction of point defects within the intrinsic anatase TiO$_2$ is a beneficial approach to improve the CCE within the mesoporous anatase TiO$_2$ photoanode (Figure 13.12).

By sensitizing with the C101 dye, the incorporation of a trivalent element also demonstrates its ability to reduce the dynamics of electron recombination, however, this time without modifying the energy and distribution of the trap states. The photovoltage is significantly improved from 721 to 739 mV for Y^{3+} and even 755 mV for Ga^{3+}, and the ff from 75 to 77% for Y^{3+} and 79% for Ga^{3+} while maintaining the photocurrent from 13.6 to 13.4 mA cm^{-2} for Ga^{3+} and even increasing to 15.9 mA cm^{-2} for Y^{3+}. This improvement in short-circuit current density can also be clearly observed in the IPCE spectrum as the Y^{3+}-doped TiO$_2$ can attain as high as 90% conversion efficiency at 550 nm where the dye absorbs the most in energy (Figure 13.13). It is extremely rare, and therefore worth highlighting, that the successful incorporation of an ε of Y^{3+} into the anatase lattice can exert an immediate improvement of the three-cell characteristics (J_{sc}, V_{oc}, and ff). It then leads, not surprisingly, to a noticeable improvement of the PCE from 7.4 to 8.1% and to 9.0% for Ga^{3+} and Y^{3+}, respectively. Another interesting feature from a fundamental and an application point of view is that applying this posttreatment to the 1% Ga/Y-TiO has absolutely no effect on the PCE [85]. Finally, also worth pointing out in this section is that noticeable improvements were also realized by introducing these new materials for TiO$_2$/PbS/Au solid-state heterojunction solar cells. Indeed, the introduction of point defects by trivalent doping shows significant advantages to increase the short-circuit current density and ff of such kinds of cells from 5.2 to 11.1 mA cm^{-2} and from 25 to 35% for TiO$_2$ and 1% Ga-TiO$_2$, respectively. This leads to a PCE enhancement from 0.7 to 1.9% under standard

Figure 13.12 Evolution of the charge collection efficiency as a function of charge density for undoped TiO$_2$ and Ga^{3+}/Y^{3+}-doped TiO$_2$. A SEM cross-sectional image is shown in the inset to illustrate the mesoporous film screen-printed on FTO.

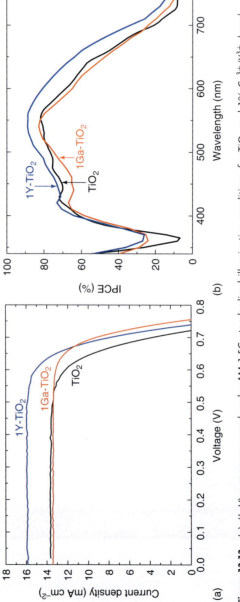

Figure 13.13 (a) (*J–V*) curves measured under AM 1.5G standardized illumination conditions for TiO$_2$ and 1% Ga^{3+}/Y^{3+}-doped TiO$_2$ photoanodes sensitized with the C101 dye (without TiCl$_{4(aq)}$ posttreatment). (b) The IPCE comparison between TiO$_2$ and 1% Ga^{3+}/Y^{3+}-doped TiO$_2$ is also shown.

illumination conditions. Note that 1% Y-TiO$_2$ exhibits improved performance too, compared to regular TiO$_2$, with 1.0% PCE. More details can be found in Ref. [85].

In general, the photoanode structure is composed of two layers. The first, typically about 7 μm thick, is composed of 20-nm-based nanoparticles, whose high roughness factor enables high dye loading and therefore more effective light absorption. However, where the dye absorbs less, generally in the range of 600–750 nm for ruthenium dyes, it is required to cover this first layer with a scattering layer composed of larger (about 400 nm) particles. The thickness of this layer is around 5 μm (sometimes reported as the 7 + 5 configuration). From the results described earlier, we have deliberately excluded the use of a scattering layer in order to more clearly highlight the influence of doping. The difference in terms of PCE between the two configurations differs significantly according to the dye used. For ruthenium dyes, which have a lower molar extinction coefficient with respect to organic sensitizers $(7000 < \varepsilon_{MLCT} < 20\,000\,\mathrm{l\,M^{-1}\,cm^{-1}})$ and a broader energy absorption range, the PCE difference is 1–1.5% increase due to the presence of the scattering layer. However, for roll-to-roll or line processes in DSC manufacturing, this two-layer configuration is something difficult to bring into play from a technical point of view. This difficulty has been circumvented by the introduction of a single-layer configuration containing an adequate mixture of 20-nm-based nanoparticles and scattering particles, for instance, the commercially available DSL18NR-AO (Active Opaque) paste sold by Dyesol (Figure 13.14).

Another approach to combine strong scattering ability and well-packed inter-connected anatase TiO$_2$ nanoparticles is the synthesis of mesoporous spheres of TiO$_2$ exhibiting a high surface area similar to the 20-nm-based nanoparticles (i.e., ~80 m^2 g^{-1}) with a diameter in the range of 400 nm to diffuse light, according to Mie scattering theory. This can be achieved by a two-step sol–gel process using hexadecylamine as a surfactant to direct the mesoporosity and a mixture of water/ethanol for the solvothermal treatment to crystallize the anatase crystal structure. More information about the effect of the synthetic parameters on the beads' morphology is available in Refs. [86, 87]. Figure 13.15 illustrates an example of the obtained morphology after optimization. XRD analysis confirms that the nanocrystals are made up of the anatase polymorph. The particles are extremely well monodispersed with a size centered at about 800 nm in diameter, which

Figure 13.14 Scanning electron micrographs of a single layer of anatase TiO$_2$ containing a mixture of 20-nm-based particles and scattering particles screen printed on FTO glass.

Figure 13.15 Scanning electron micrograph and transmission electron micrographs of porous TiO$_2$ beads.

is large enough to effectively scatter the whole portion of visible light [88]. As shown in the ultramicrotome micrograph (second picture from left), the beads are totally filled inside forming mesoporosity with densely packed, well-interconnected anatase nanoparticles. The high-resolution image presents evidence of an inter-growth mechanism between the anatase crystallites. The BET surface area of the particles reaches 95 m^2 g^{-1} with type IV isotherm typical of mesoporous materials. The pore distribution using the Barrett–Joyner–Halenda (BJH) method depicts a narrow distribution centered at 18 nm corresponding to intrabead porosity.

The PV performances using the mesoporous beads were evaluated still using the C101 dye. In contrast to the doped TiO$_2$ samples, the TiCl$_4$ posttreatment has a clear impact on the PV characteristic of the cell, with a total enhancement of about 1.5% PCE obtained through an increase in the three-cell characteristics (Figure 13.16). After this treatment, a single layer of beads with a thickness of about 12 μm exhibit a remarkable 10.6% PCE under AM 1.5G standard illumination conditions ($J_{sc} = 18.4$ mA cm^{-2}, $V_{oc} = 745$ mV, and $ff = 76.7\%$). For comparison, an analogous configuration using the commercial P25 particles was giving "only" 8.5% PCE ($J_{sc} = 15.0$ mA cm^{-2}, $V_{oc} = 761$ mV, and $ff = 74.7\%$), a value still below those of the beads even without TiCl$_4$ posttreatment. This notable increase in photocurrent stems from the drastic enhancement of the incident photon-to-electron IPCE (Figure 13.16). When constituted by the beads, the IPCE is superior over the whole range of light wavelength, in which the C101 dye absorbs effectively. It even reaches a maximum of 92% at 570 nm, stressing a close to 100% conversion of incident photons to electricity if we consider the light losses ascribed to the TCO glass absorption. Note that P25 at this wavelength was giving 77%. The superiority of the beads compared to P25 can be seen even in the NIR region: that is, 54% IPCE at 700 nm (32% for P25) due to the high scattering ability of the beads [86]. This duality, namely high scattering properties and strong light capture/conversion, is once more demonstrated when comparing the PV characteristic of this layer with the optimized 8 + 5 μm double-layer configuration, for which the best result obtained was 10.5% PCE (Figure 13.17). Regardless of the dye used (i.e., N719, C101, or C106), the very strong point of the beads, as we can notice below, concerns the high ff typically obtained. It lies between 76% to even more than 80% in some cases, which is an advantage worth highlighting for the application in modules where this characteristic becomes more critical. This underlines the potential and advantages of using the beads to obtain both performance and process efficiencies.

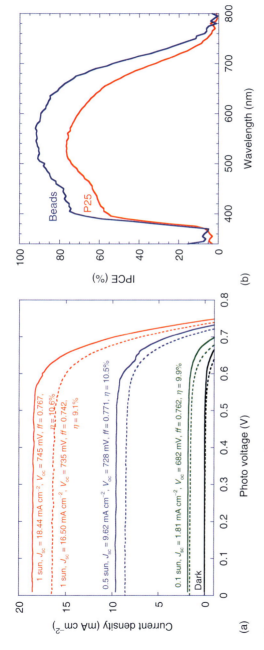

Figure 13.16 (a) (*J*–*V*) curves measured under AM 1.5G standardized illumination conditions for 12 μm thick beads electrode sensitized with C101 without (dashed line) and with (full line) TiCl$_4$ posttreatment. (b) The IPCE comparison between beads and P25 sensitized by C101 is also reported.

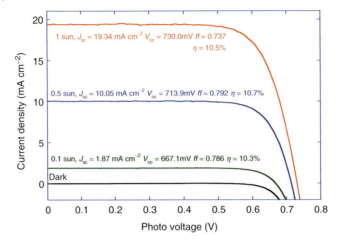

Figure 13.17 (*J–V*) curves measured under AM 1.5G standardized illumination conditions for $8+5\,\mu m$ thick double-layer configuration sensitized by C101.

Electron lifetime and electron diffusion coefficient were parameters extracted and compared to P25 as a function of the charge density within the mesoporous film (Figure 13.18). For the whole range of charge densities, the electron lifetime with the beads is superior to that with P25 by a factor of about 2, confirming the lower reactivity of the photoinduced electrons of TiO_2 with respect to either I_3^- or the oxidized state of the dye. Similarly, the electron diffusion coefficient is superior as well. This benefit in terms of collection time can be attributed to the well-packed TiO_2 nanocrystals which are favorable in providing an efficient electron conduction pathway toward collection. However, this good characteristic is marred when the charge density is greater than 2×10^{18} e$^-$ cm^{-3}, where there is a crossover point with P25. This intersect is attributed to the lower interconnection between each

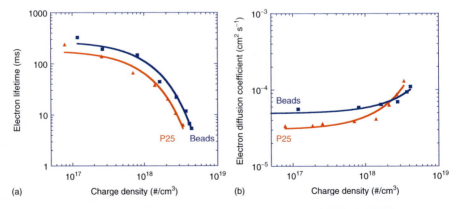

Figure 13.18 (a,b) Evolution of electron lifetime and electron diffusion coefficient as a function of charge density for P25 (in red) and beads (in blue) with $TiCl_4$ posttreatment.

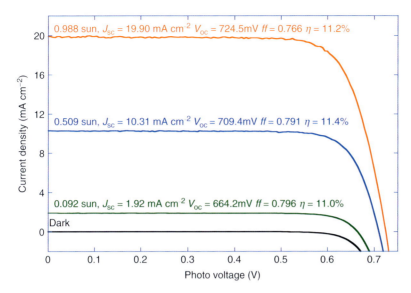

Figure 13.19 (J–V) characteristics of C106 dye for different light intensities (1 sun, 0.5 sun, 0.1 sun, and dark) of a triple-layer photoanode configuration consisting of a first layer of 20-nm-based transparent electrode + mesoporous beads + CCIC scattering layer.

bead. One way to strengthen the beads' interconnection is the application of the $TiCl_4$ posttreatment, which enables the consolidation of their interconnection by depositing small nanocrystals of anatase at the boundary between two beads.

Finally, by tailoring an optimized photoanode geometry perfectly suited to the optoelectronic properties of the closely related C106 dye, we have achieved as high as 11.2% PCE under $100 \, mW \, cm^{-2}$ illumination and even 11.4% efficiency under half illumination ($50 \, mW \, cm^{-2}$) (Figure 13.19). This dye has the peculiarity to bear one thio-thiophene unit instead of a simple thiophene ring in the ancillary ligand [85], and to date it has achieved the best performances in stability [72].

Such achievements emphasize the recent progress realized in the photoanode architecture for obtaining high-PCE DSC PV technology. In particular, we have herein stressed that inorganic chemistry throughout materials science engineering can greatly impact the performances of DSCs and more particularly by simple chemical approaches, and that after a stepwise optimization, PCE well beyond 11% can be reached (beads). This laboratory effort now requires implementation into the DSC modules to close the gap between academic research in DSC and its practical application.

13.4
Development of Molecular Sensitizers Suited to TiO₂ Optoelectronic Properties

Sensitization of large bandgap semiconductors such as TiO_2 to the visible and the NIR solar spectrum can be achieved by anchoring molecular absorbers. In DSCs, the sensitizer not only should absorb visible and near-IR light but also exhibit thermal and photochemical stability. The anchoring groups of the sensitizer should have strong binding properties, and the lowest unoccupied molecular orbital (LUMO) of the sensitizer and the conduction band of conductor oxide should overlap efficiently, facilitating electron transfer from the dye's excited state to the conduction band of the inorganic mesoporous semiconductor layer. The sensitizer in the excited state should have directionality so that the excited electron migrates toward anchoring group and the positive charge on the opposite side. Also, the sensitizer should have the LUMO just above the conduction band and the highest occupied molecular orbital (HOMO) below the working function of the hole transporting material or the chemical potential of the redox couple. Moreover, the sensitizer must have electrochemical robustness to undergo millions of cycles. To incorporate all these requirements, engineering of sensitizers at the molecular level is required, which is the purpose of this section.

The photophysical and photochemical properties of group VIII metal complexes using terpyridine and bipyridine ligands have been thoroughly investigated during the past three decades [89]. The main thrust behind these studies is to understand the energy- and electron-transfer processes in the excited state and to apply this knowledge to potential practical applications such as DSCs and light-driven information processing [90]. Ruthenium(II) complexes have been extensively used as charge-transfer sensitizers on nanocrystalline TiO_2 films [91]. The choice of ruthenium metal is of special interest for a number of reasons: (i) because of its octahedral geometry, one can introduce specific ligands in a controlled manner; (ii) the photophysical, photochemical, and the electrochemical properties of these complexes can be tuned in a predictable way; and (iii) the ruthenium metal possess stable and accessible oxidation states from I to IV [92]. Although iron is an inexpensive and abundant metal, the photophysical and electrochemical properties of its coordination complexes are difficult to tune in a predictable manner [93]. The other notable disadvantage of this metal is the weaker ligand-field splitting compared to ruthenium and osmium. On the other hand, osmium, being in the third row of the transition-metal ions, has a stronger ligand-field splitting compared to ruthenium. Moreover, the spin–orbit coupling in osmium complexes leads to enhanced response in the red region [94]. But, the low abundance of this metal restricts its use for large-scale applications.

13.4.1
Ruthenium Sensitizers

While several transition-metal complexes as well as metal-free organic dyes have been tested [95–105], the best photovoltaic performances in terms of both conversion yield and long-term stability have been achieved so far with polypyridyl complexes of ruthenium. The ruthenium complex *cis*-$RuL_2(NCS)_2$ (**1**) (where L = 2,2′-bipyridyl-4,4′-dicarboxylic acid), known as the N3 dye, has become the paradigm of heterogeneous charge-transfer sensitizers for DSCs [106]. The role of the carboxylate groups is to immobilize the sensitizer to the TiO_2 film surface via the formation of bidendate coordination, while the NCS groups enhance visible light absorption by destabilizing the HOMO of the metal t_{2g} orbitals. The N3 dye exhibits metal-to-ligand charge-transfer transitions (MLCTs) at 400–535 nm with molar extinction coefficients of 1.45×10^4 and $1.41 \times 10^4\ M^{-1}\ cm^{-1}$, respectively. The MLCT excitation of the dye involves the transfer of one electron from the metal t_{2g} orbital to the π^* orbital of the ligand, which is anchored onto the inorganic surface through carboxylic acid groups. The N3 dye, on anchoring, releases its protons, which afterward is specifically adsorbed onto the TiO₂ surface. Since the position of the Fermi level in TiO₂ is known to have a Nernstian dependence on proton activity (pH) [107, 108], it is expected that the dye, which contains initially four protons (depending on the number of groups anchored), influences the energy level in TiO₂ and hence the efficiency of the device. The fully protonated N3 sensitizer **1** charges the TiO₂ surface positively by transferring its protons on adsorption. The electric field associated with the surface dipole generated in this manner enhances the adsorption of the anionic ruthenium complex and assists electron injection from the excited state of the sensitizer into the titania conduction band, favoring high photocurrents (18–19 mA cm^{-2}). However, the open-circuit potential (0.65 V) is lowered owing to positive shift of the conduction band edge induced by the surface protonation. The V_{oc} is determined by the energy difference between the Fermi level of the solid under illumination and the Nernst potential of the redox couple in the electrolyte. Since η of the DSC is the product of the photocurrent density at short circuit (J_{sc}), V_{oc}, and ff of the cell divided by the incident light intensity (I_s), to obtain high light-to-electric efficiencies (η), J_{sc} and V_{oc} of the solar cell have to be optimized. V_{oc} can be tuned by either controlling the pH of electrolyte or, by extension, introducing a potential-determining cation in electrolyte such as the alkali family or the number of protons carried by the sensitizer. The other way is by tuning the redox couples' oxidation potential more positively, which will be discussed in the electrolyte section.

In order to visualize the influence of protons, the performances of the three sensitizers **1**, **2**, and **3**, which contain different degrees of protonation, were studied on nanocrystalline TiO₂ electrodes [109]. Figure 13.20 shows the photocurrent action spectra obtained with a monolayer of these complexes coated on TiO₂ films. The sensitizer **3** that carries no proton exhibited lower J_{sc} than **1**. By contrast, it has a significantly higher V_{oc} because of the relatively negative shift of the conduction band edge induced by the lower proton activity in the medium. Between those two extremes, there should be an optimal degree of protonation of the sensitizer for which the product of J_{sc} and V_{oc} is maximal.

The incident monochromatic IPCE is plotted as a function of excitation wavelength. The IPCE value in the plateau region is 80% for complex **1**, while it is only about 66% for complex **3**. This difference is even more pronounced in the red region. Thus, at 700 nm, the IPCE value is twice as high for the fully protonated complex **1** as compared to the deprotonated complex **3**. Consequently, the J_{sc} drops from 18–19 mA cm^{-2} for complex **1** to only 12–13 mA cm^{-2} for complex **3**. However, there is a trade-off in photovoltage, which is 0.9 V for complex **3** when compared to 0.65 V for complex **1**. Nevertheless, this is insufficient to compensate for the current loss.

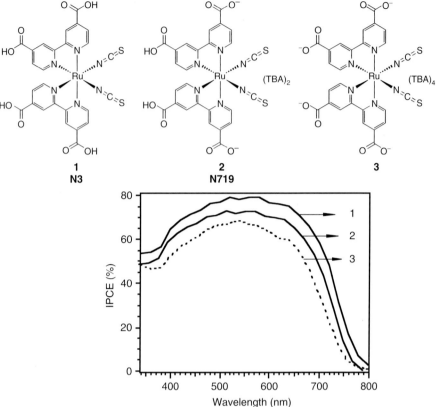

Figure 13.20 Photocurrent action spectra of nanocrystalline TiO$_2$ films sensitized by complexes **1**, **2**, and **3**. The incident photon to current conversion efficiency is plotted as a function of wavelength.

Photovoltaic performance data obtained with a sandwich cell under illumination by simulated AM 1.5G solar light using complex **2** with additives are shown in Figure 13.21. At 1 sun, the sensitized solar cell exhibited 17.73 ± 0.5 mA current, 846 mV potential, and 0.75 *ff*, yielding an overall conversion efficiency of 11.18%. Hence, the photovoltaic performance of complex **2** carrying two protons is superior to that of compounds **1** and **3** that contain four or no protons, respectively. The doubly protonated form of the complex is therefore preferred over the other two sensitizers for sensitization of nanocrystalline TiO$_2$ films.

13.4.2
Hydrophobic Ruthenium Sensitizers

Owing to the chemical nature of the anchoring groups, the water-induced desorption of the sensitizer from the TiO$_2$ surface is a crucial aspect in DSCs, as it impacts the long-term stability of the device. To overcome this problem, alkyl

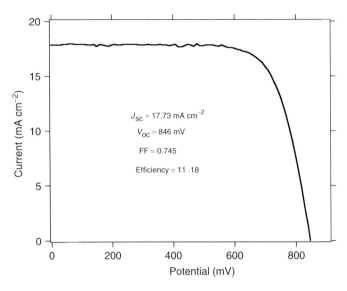

Figure 13.21 Photocurrent–voltage curve of a solar cell based on complex **2**. The cell was equipped with an antireflective coating. The conversion efficiency in full AM 1.5 sunlight illumination (100 mW cm^{-2}) is 11.18%. The cell is masked with black plastic to avoid the diffusive light leaving an active cell area of 0.158 cm^2.

chains are added onto the 2,2′-bipyridine unit to confer a hydrophobic character to the complexes (**4–8**). The absorption spectra of these complexes show broad features in the visible region and display maxima around 530 nm. The performance of these hydrophobic complexes as charge-transfer photosensitizers in nanocrystalline TiO₂-based solar cell shows excellent stability toward water-induced desorption [110].

4: R = CH$_3$
5: R = C$_6$H$_{13}$
6: R = C$_9$H$_{19}$
7: R = C$_{13}$H$_{27}$
8: R = C$_{18}$H$_{37}$

In addition, these sensitizers that have C6–C12 alkyl chains hamper considerably the recombination reactions with triiodide. Several groups have tried to reduce the recombination reaction by using sophisticated device architectures such as composite metal oxides as the semiconductor with different bandgaps [111, 112]. Gregg *et al.* [113] have examined surface passivation by deposition of insulating polymers. We have studied the influence of spacer units between the dye and the TiO₂ surface but with little success [114]. Nevertheless, by using TiO₂ films containing hydrophobic sensitizers that contain long aliphatic chains (**5–7**), the recombination reaction was suppressed considerably [115, 116]. The most likely explanation for the reduced dark current is that the long chains of the sensitizer interact laterally to form an aliphatic network, like a shield, thereby preventing triiodide from reaching the TiO_2 surface.

13.4.2.1 Sensitizers with Extended π System Showing High Molar Extinction Coefficient

Owing to its ease of synthesis and the availability of chemicals, coupled with excellent performances, complex **1** has become a paradigm in the area of dye-sensitized nanocrystalline TiO_2 films [117]. Therefore, the vast majority of sensitizers are based on its design. In spite of this, the main drawback of this sensitizer is the lack of absorption in the red region of the visible spectrum and also the relatively low molar extinction coefficient, which is $14\,500\,M^{-1}\,cm^{-1}$. Therefore, sensitizers with high molar extinction coefficients have been particularly sought after. A new series of sensitizers with high molar extinction coefficients (**9–11**) featuring alkyloxy groups have been synthesized and utilized. The purpose of the 4,4′-di-(2-(3,6-dimethoxyphenyl)ethenyl)-2,2′-bipyridine ligand that contains extended π-conjugation with substituted methoxy groups is to enhance the molar extinction coefficient of the sensitizers, and to provide directionality in the excited state by fine-tuning the LUMO level of the ligand with the electron-donating alkoxy groups.

The absorption spectra of complexes **9–11** are dominated by the MLCT in the visible region, with the lowest allowed MLCT bands appearing at 400–545 nm. The molar extinction coefficients of these bands are close to $35\,000–19\,000\,M^{-1}\,cm^{-1}$, respectively, which are significantly higher than that of N3 (Figure 13.22).

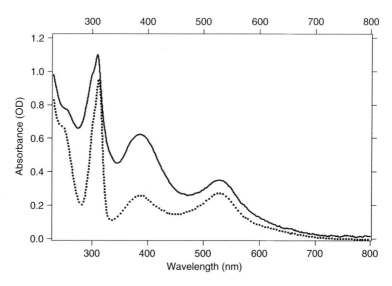

Figure 13.22 Comparison of absorption spectra of complexes **1** and **11** in ethanol. (Source: Reproduced from Ref. [118].)

The photovoltaic data of these sensitizers using an electrolyte containing 0.60 M butylmethylimidazolium iodide (BMII), 0.03 M I$_2$, 0.10 M guanidinium thiocyanate, and 0.50 M *tert*-butylpyridine in a mixture of acetonitrile and valeronitrile (volume ratio: 85 : 15) exhibited a short-circuit photocurrent density of 16.50 ± 0.2 mA cm^{-2}, with an open-circuit voltage of 790 ± 30 mV and an *ff* of 0.72 ± 0.03, resulting in an overall conversion efficiency of 9.6% under standard AM 1.5G sunlight and demonstrated stable performance under light- and heat-soaking at 80 °C [119].

Recently, a new design consisting of a ligand incorporating thiophene moieties to shift the spectral response into the near-IR region, and to enhance the molar extinction coefficient has been developed [85].

CYCB1   CYCB11

As the extinction coefficients of those sensitizers are much higher than that of N3, it is possible to decrease the thickness of the semiconductor film and thereby decrease the charge-collection time. This results in an enhanced open-circuit voltage as well as ff, translating into high efficiencies of 11.4% [85]. To enhance further the molar extinction coefficient of ruthenium sensitizers, two thiophene units were incorporated (CYC B1 and CYC B11), resulting in a slight red shift in the absorption maxima (about 552 nm) but displaying a high molar extinction coefficient values of over 24 000 M^{-1} cm^{-1} [120].

The absorption spectral properties of ruthenium sensitizers can be tuned toward the red part of the visible spectrum by introducing a ligand having a low-lying π^* molecular orbital (LUMO) and/or shifting the HOMO level by introducing strong donor ligands. The LUMO of a ligand depends on the π-conjugation, and therefore 4,4′,4″-tricarboxylic acid-2,2′;6,2″-terpyridine has a lower LUMO compared to the 4,4′-dicarboxylic acid-2,2′-bipyridine. The metal t_{2g} orbitals can be destabilized through the introduction of donor thiocyanate type of ligands. The former lowers the energy of the LUMO, while the latter destabilizes the HOMO of the sensitizer, ultimately reducing the HOMO–LUMO gap. However, the extension of the spectral response into the near-IR region by lowering the LUMO energy is limited to energy levels below which charge injection into the TiO$_2$ conduction band can no longer occur [121–123]. On the other hand, near-IR response by destabilization of Ru t_{2g} (HOMO) levels close to the redox potential of the redox mediator also proves to be not useful because of problems associated with regeneration of the oxidized dye following the electron injection into the TiO$_2$. Therefore, the optimum ruthenium sensitizers should exhibit excited-state oxidation potential of at least −0.9 V versus SCE, in order to inject electrons efficiently into the TiO$_2$ conduction band [114], while the ground-state oxidation potential should be about 0.5 V versus SCE, in order to be regenerated rapidly via electron donation from the electrolyte (iodide/triiodide redox system or a hole conductor). The panchromatic ruthenium complex N749 (the so-called black dye), in which the ruthenium center is coordinated to

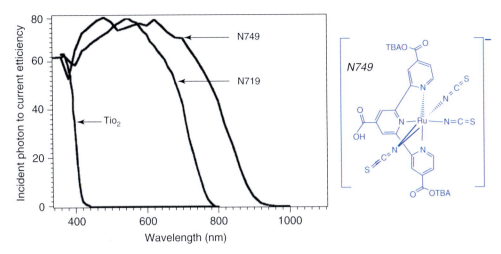

Figure 13.23 IPCE obtained with the N749 attached to nanocrystalline TiO₂ films. The incident photon to current conversion efficiency is plotted as a function of the wavelength of the exciting light. IPCE for bare TiO₂ and TiO₂ sensitized with N719 has been included for comparison.

tricarboxylic acid terpyridine ligand and three thiocyanate ligands, has been synthesized [124, 125].

Figure 13.23 shows the photocurrent action spectrum of a cell containing N719 and N749 sensitizers, where the IPCE is plotted as a function of wavelength. It is evident that the response of the N749 extends 100 nm further into the near-IR region than that of N719. The photocurrent onset is close to 920 nm, that is, near the optimal threshold for single-junction converters. The IPCE rises gradually from 340 until, at 700 nm, it reaches a plateau of over 80%. From the overlap integral of the curves in IPCE with the AM 1.5 solar emission, one predicts the short-circuit photocurrents (J_{sc}) of N719- and N749-sensitized cells to be 16.5–20.5 mA cm^{-2} [126]. Routinely, experimental photocurrents obtained with N749 are in the range of 18–21 mA cm^{-2} [125]. The V_{oc} is 720 mV, and ff was 0.7, yielding an overall solar (global AM 1.5 solar irradiance 1000 W m^{-2}) to electricity η value of 10.4% [125]. With the N749 dye, η of 11.4% has been achieved using high-haze TiO₂ electrodes by Han and colleagues [127, 128].

Ruthenium metal complexes other than 2,2′-bipyridine and 2,2′;6,2″-terpyridine have only rarely been investigated [89]. 2,2′:6′,2″:6″,2‴-Quaterpyridine (qpy) ligands as DSC sensitizers are largely unexplored, in contrast to the extensive studies on 4,4′-dicarboxylic acid 2,2′-bipyridine-based sensitizers. This is likely due to the synthetic challenges associated with the qpy ligands, and very few reports have been published on DSCs with Ru(II) sensitizers based on such tetradentate ligands [129–131]. Although their overall solar to electric power efficiencies were not among the highest-ranked, these studies have pointed out the possible panchromatic response of the corresponding complexes extending from the NIR to the UV region, rendering them as alternative promising sensitizers with enhanced solar harvesting capability over the conventional bpy-based sensitizers.

To address the common issues related to the rather weak absorption of Ru(II) dyes in the red and NIR region, a number of heteroleptic Ru(II) sensitizers with extended bpy ancillary ligands π-conjugated with electron-rich benzenoid cores carrying donor primary organic functionalities such as alkoxy have been introduced [85, 120]. Ru(II) complexes based on such ligands are generally endowed with enhanced optical properties with respect to the prototypical N3 or N719 dyes, which also translates into higher photocurrents when employed in DSCs. In particular, encouraging performances have been achieved recently by using thiophene-based derivatives as electron-rich donor end groups [120]. In order to address some of the major issues so far reported in the literature, one approach would be to synergistically combine the superior optical properties of π-donor-conjugated bpy with the panchromatic response of qpy ligands. However, π-donor conjugated 2,2′:6′,2″ : 6″,2″-qpys have so far been only rarely investigated and even more rarely used as DSC sensitizers. Recently, a first example of heteroarylvinylene π-conjugated 4,4‴-bis[(E)-2-(3,4-ethylenedioxythien-2-yl)vinyl]-4′,4″-bis(carboxy)-2,2′ was published: 6′,2″:6″,2″-qpy ligand and its *trans*-dithiocyanato Ru(II) sensitizer (coded N1044, Figure 13.24). This complex exhibits panchromatic spectral response extending from the UV throughout the entire visible spectral region [132].

The photovoltaic performance of N1044 has been explored by using a double-layer (20 nm particle layer + diffusive/reflective layer) photoanode configuration of anatase TiO_2. Chenodeoxycholic acid was added to a DMF solution of the dye as a coadsorbing agent to prevent excessive formation of dye aggregates both in solution and in the nanostructured film. Figure 13.25a presents the IPCE spectrum, and Figure 13.25b shows the photovoltaic characteristics of current–voltage curves of N1044 at different illumination intensities. The N1044 complex delivers a maximum IPCE of 65% at 646 nm. Despite this value being lower than what have been reported for other Ru(II) sensitizers exceeding 90% [117], at 1 equiv sun illumination (100 mW cm^{-2}, AM 1.5G), N1044 delivers an excellent short-circuit current density as high as $J_{sc} = 19.15$ mA cm^{-2} with an $ff = 0.67$. Nevertheless, the low V_{oc}, likely due to the presence of Li^+ ions in the electrolyte and to the dye

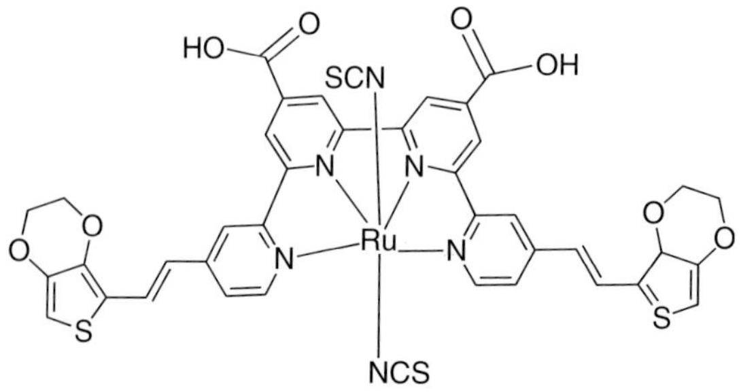

Figure 13.24 Molecular structure of N1044 sensitizer.

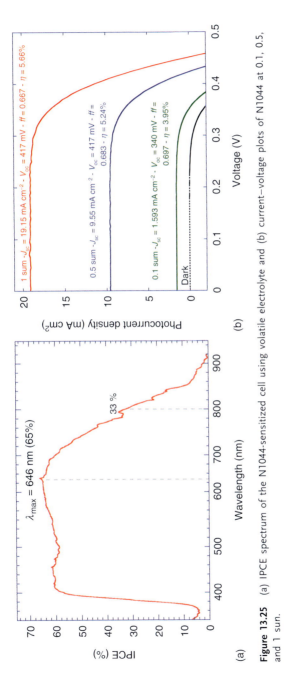

Figure 13.25 (a) IPCE spectrum of the N1044-sensitized cell using volatile electrolyte and (b) current–voltage plots of N1044 at 0.1, 0.5, and 1 sun.

adsorption characteristics discussed above, penalizes the overall cell performance leading to a PCE of 5.7%. These values can be compared with $J_{sc} = 17.6\,\mathrm{mA\,cm^{-2}}$, $V_{oc} = 849\,\mathrm{mV}$, and $ff = 0.73$, which deliver a record efficiency at 1 sun of 11.2% for the prototype dye N719 [117]. In agreement with the measured high current density and broad absorption in the visible range, the IPCE is 65% at 646 nm and still 33% at 800 nm, where the vast majority of dye sensitizers have zero or negligible response. It should be noted that the absorption tail is recorded up to 910 nm, which is deeply inside the NIR region. A bathochromic shift of about 30 nm is found as a consequence of the dye absorption onto the TiO_2 surface, which entails the formation of carboxylate group as well as the difference of solvation shell strength when using acetonitrile/valeronitrile-based electrolytes.

As mentioned earlier, the poor V_{oc} can be possibly related to the dye adsorption onto TiO_2, whereby the loose packing between dye aggregates of different beltswould lead to empty spaces for the oxidized species in the electrolyte (e.g., I_2) to access the TiO_2 surface, thus leading to increased recombination between the injected electrons and the electrolyte. We can also speculate that this class of dyes might generally suffer from sizable recombination between the injected electrons and the oxidized dye cation, because the dye HOMO, localized across the metal-NCS groups, lies very close to the surface with little screening from the dye aromatic ligands.

13.4.3
Porphyrin Sensitizers

Given the low molar extinction coefficient of ruthenium sensitizers, it was paramount to develop donor–π bridge–acceptor (D-π-A) Zn porphyrin sensitizers that absorb light over the whole visible range with high molar extinction coefficient. D-π-A sensitizers, armed with long-chain alkyloxy groups, give high open-circuit potential because of impaired interfacial back-electron-transfer reaction (Figure 13.26). Recently, we have reported a D-π-A zinc porphyrin dye as sensitizer (designated as YD2 and YD2-*o*-C8). Both porphyrins show a maximum absorbance in the 400–500 and 550–750 nm ranges. The rational design of YD2-*o*-C8 sensitizer has not only the merit to harvest sunlight across the visible spectrum but can also slow down the interfacial back electron transfer from the conduction band of the nanocrystalline TiO_2 film to the oxidized form of the redox couple. As a consequence, it can generate large photocurrents and also higher open-circuit potential when associated with a suitable cobalt-based redox couple. By cosensitizing YD2-*o*-C8 with another organic dye whose absorption is strong into the 550 nm region, an enhancement of the device performances and record PCE could be achieved. Note that the increased short-circuit current density and ff were also resulting from the cosensitized film rather than from YD2-*o*-C8 alone. This increase in J_{sc} and ff results in an unprecedented efficiency of 12.3% at AM 1.5G full sun by using the Co(II/III)tris(bipyridyl)-based redox shuttle, with even a higher value of 13.1% at 50.9% sun [69].

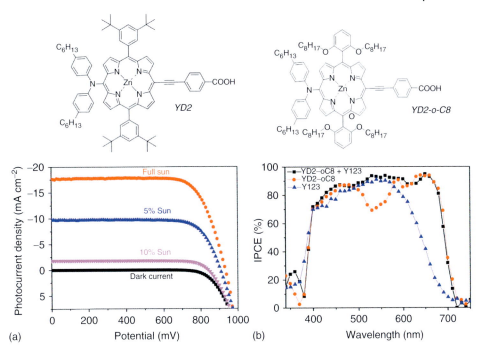

Figure 13.26 Chemical structure of YD2 and YD2-o-C8 porphyrin dyes. (a) Photocurrent–voltage curves of YD2-o-C8 + Y123 under various light intensities of AM 1.5 sunlight. (b) IPCE spectrum of Y123 (blue line), YD2-o-C8 (red line), and YD2-o-C8 + Y123 (black line).

13.5
Development of Redox Mediators

In DSCs, light is absorbed by a dye monolayer located at the junction between a nanostructured electron transporting material (n-type) and a hole-acceptor species such as I^-. The main drawbacks of the commonly used triiodide/iodide redox (I_3^-/I^-) system are its coloration and its large energy mismatch between its oxidation potential $(E^0 (I_3^-/I^-) = 0.32$ V vs the NHE) and the oxidation potential of the sensitizer $(E^0 (S^+/S) = \sim 1.0$ V vs NHE), which limits the open-circuit voltage (V_{oc}) to 0.7–0.8 V [133]. The large energy cost for efficient dye regeneration is due to the complex regeneration kinetics with the I_3^-/I^- redox couple involving the formation of intermediates, that is, the I_2^- radical as shown in Figure 13.5 [134]. Additionally, these redox couples are very corrosive toward metals such as Ag, Au, and Cu, thereby excluding the use of such materials as current collectors in DSC modules for long-term stability. Therefore, the development of noncorrosive redox mediators, with reduced mismatch between the oxidation potential of the dye and the redox couple, to increase the open-circuit potential.

Previous studies to replace the I_3^-/I^- redox system by cobalt–polypyridine complexes have attracted attention because of their low visible light absorption, higher redox potential, and reduced corrosiveness toward metallic conductors [135, 136]. However, the overall efficiency obtained with these cobalt electrolytes were inferior compared to that with the I_3^-/I^- couple, especially under full sunlight, owing to mass transport control [136], faster back reaction of the photo-injected electrons with the oxidized redox species [137], and slow regeneration kinetics of the Co(II) species at the counterelectrode. Recently, cobalt redox shuttles have attracted renewed attention after Hagfeldt *et al.* reported increased PCE at 6.7% by employing a newly designed D-π-A organic sensitizer, coded D35, in conjunction with the cobalt (III/II) tris-bipyridyl complex, $[Co(III)(bpy)_3](PF_6)_3/[Co(II)(bpy)_3](PF_6)_2$ couple [70]. The authors were able to achieve devices of more than 1 V in association with cobalt phenanthroline redox shuttles. However, only a modest PCE was obtained (below 3.6%) [118]. It is a well-known fact that I_3^-/I^- redox shuttle is a limiting factor in the achievable efficiencies of DSCs, and recently mesoscopic solar cells with Co(II/III)tris(bipyridyl)-based redox shuttle using donor-π-bridge-acceptor zinc porphyrin dye as sensitizer (designated YD2-*o*-C8) have yielded a record PCE of 12.3% under simulated AM 1.5 global sunlight. The rational design of YD2-*o*-C8 slows down the rate of interfacial back electron transfer from the conduction band of the mesoscopic TiO_2 film to the oxidized cobalt mediator, resulting in high photovoltages, close to 1000 mV.

Since cobalt belongs to the first-row transition-metal series, dissociation and exchange of ligands may occur rapidly in the case of Co(II). Tridentate ligands are likely to improve significantly the stability of the cobalt complex compared to bidentate ligands. Another advantage of tridentate ligands over bidentate ligands is the absence of isomers. Thus, in the case of dissymmetric bidentate ligands, useful for fine-tuning of the redox properties, facial and meridional isomers coexist. One particular attractive feature of cobalt complexes is the facile tuning of their redox potential, which can be adjusted to match the oxidation potential of the sensitizer, thereby minimizing energy loss in the dye regeneration step. A series of cobalt complexes with various polypyridyl ligands were prepared, whose oxidation potential could be tuned from 0.17 to 0.34 V versus ferrocene [118]. Addition of electron-acceptor groups on pyridyl or by replacing pyridyl by pyrazole was used to stabilize the HOMO of the cobalt complexes. The higher oxidation potential of pyrazole-containing cobalt complexes was attributed to the presence of pyrazole, which stabilizes the HOMO of the complex more than a pyridine group. Weakly donating substituents (methyl groups) on the backbone of the ligand in a complex result in a slight increase in the standard potential to 0.34 V versus ferrocenium/ferrocene. Most recently, Yum *et al.* have reported a redox relay system $[Co(III/II)(bpy-pz)_2](PF_6)_{3/2}$ [bpy-pz = 6-(1H-pyrazol-1-yl)-2,2'-bipyridine]), and used it as a redox mediator in DSCs. An unprecedented output voltage >1000 mV due to its high oxidation potential of 0.86 V versus NHE of the cobalt redox mediator was achieved [138]. The designed complex $[Co(bpy-pz)_2]^{3+/2+}$ (Figure 13.27), whose redox potential is offset by only 230 mV from that of the 3-{6-{4-[bis(2',4'-dihexyloxybiphenyl-4-yl)amino-]phenyl}-4,4-dihexyl-cyclopenta-[2,1-b:3,4-b]dithiophene-2-yl}-2-cyanoacrylic acid (coded Y123) dye, gave a PCE of >10% under different solar intensities from 10 to 100 mW cm^{-2}. The cobalt(III) and

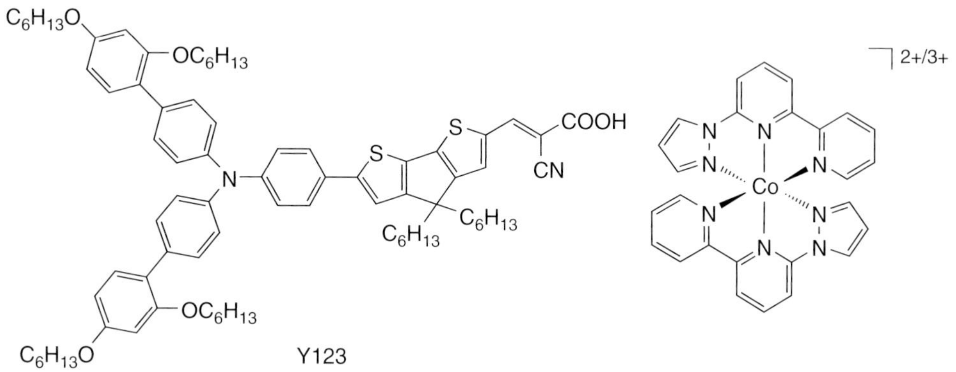

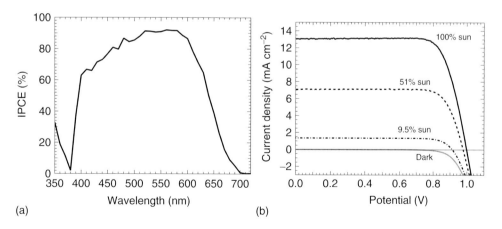

Figure 13.27 Chemical structures of Y123 dye and tridentate cobalt complex.

(II) complexes based on 6-(1*H*-pyrazol-1-yl)-2,2'-bipyridine tridentate ligand, used as redox mediators, in combination with the high molar extinction coefficient sensitizer Y123 in mesoscopic DSCs gave a very high open-circuit voltage of ~1000 mV along with a short-circuit current of 13.06 mA cm^{-2} and *ff* of 0.77, resulting in a PCE of above 10% (Figure 13.28) under 1 sun condition. The high open-circuit voltage was obtained by stabilizing the HOMO level of the cobalt (II) complex in order to minimize the mismatch between the redox couple and the sensitizer HOMO level to just 230 mV. This resulted in higher V_{oc} and PCE compared to the triiodide/iodide redox couple. These promising results confirm that the molecularly engineered cobalt redox shuttles are an alternative to the commonly used I_3^-/I^- redox shuttle.

Figure 13.28 (a) IPCE and (b) *J*–*V* characterization of the DSC employing the double layered TiO$_2$ (5.6 + 5 μm) with Pt counterelectrode based on [Co(bpy-pz)$_2$]$^{3+/2+}$. (Source: Reproduced from Ref. [138].)

13.6
Conclusions

DSCs have become a credible alternative to solid-state p–n junction devices. Conversion efficiencies in excess of 12% have already been achieved at laboratory scale with single-junction cells, but there is ample room for further improvement in the PCE. Future research will focus on combining ruthenium sensitizers with one-electron redox mediators to obtain $J_{sc} > 22\,mA\,cm^{-2}$ and an open-circuit potential of 1 V. With a possible ff of 0.75, these numbers will lead to PCE close to 16.5%. DSCs with 16.5% efficiency armed with hue of colors and flexibility are well suited for a whole realm of applications ranging from the low-power market to large-scale installations. Their excellent performance in diffuse light gives them a competitive edge over silicon in providing electric power for stand-alone electronic equipment both for indoor and outdoor applications (reaching ~26% PCE). Integration of DSCs into building architecture has already started and will become a fertile field of future commercial development.

Acknowledgments

FS would like to acknowledge the contributors to the TiO_2 work, Aravind Kumar Chandiran, Pascal Comte, Leo-Philip Heiniger, Dr Patra Snehangshu (LRCS), Dr Christian Andria (LRCS), Dr Rachel Caruso, and Pr. Yi-Bing Cheng (Monash University). FS also acknowledge Dr Claudia Barolo and Dr Nadia Barbero for their contribution to the synthesis of N1044 dye. The authors are also indebted to the European FP7 project "Robust DSC" with Grant Agreement No. 212792 and RS2E hub for energy storage through the ANR "Store-ex."

References

1. Fahmi, A., Minot, C., Silvi, B., and Causa, M. (1993) *Phys. Rev. B*, **47** (18), 11717–11724.
2. Gribb, A.A. and Banfield, J.F. (1997) *Am. Mineral.*, **82**, 717–728.
3. Navrotsky, A. (2003) *Geochem. Trans.*, 4 (6), 34–37.
4. Snehansghu, P., Davoisne, C., Sauvage, F. submitted.
5. Ohzuku, T., Kodama, T., and Hirai, T. (1985) *J. Power Sources*, **14**, 153–166.
6. Kavan, L., Fattakhova, D., and Krtil, P. (1999) *J. Electrochem. Soc.*, **146** (4), 1375–1379.
7. Baudrin, E., Cassaignon, S., Koelsch, M., Jolivet, J.-P., Dupont, L., and Tarascon, J.-M. (2007) *Electrochem. Commun.*, **9** (2), 337–342.
8. Kavan, L., Graetzel, M., Gilbert, S.E., Klemenz, C., and Scheel, H.J. (1996) *J. Am. Chem. Soc.*, **118**, 6716.
9. Chandiran, A.K., Sauvage, F., Casas-Cabanas, M., Comte, P., and Graetzel, M. (2010) *J. Phys. Chem. C*, **114** (37), 15849–15856.
10. Magne, C., Cassaignon, S., Lancel, G., and Pauporté, T. (2011) *ChemPhysChem*, **12** (13), 2461–2467.
11. Koo, H.J., Park, J., Yoo, B., Yoo, K., Kim, K., and Park, N.G. (2008) *Inorg. Chim. Acta*, **361**, 677–683.
12. Breckenridge, R.G. and Hosler, W.R. (1953) *Phys. Rev.*, **91** (4), 793–802.

13. Chen, X. and Mao, S.S. (2007) *Chem. Rev.*, **107** (7), 2891–2959.

14. Tang, H., Prasad, K., Sanjinés, R., Schmid, P.E., and Levy, F. (1994) *J. Appl. Phys.*, **75** (4), 2042–2047.

15. Frederikse, H.P.R. (1961) *J. Appl. Phys.*, **32** (10), 2211–2215.

16. Acket, G.A. and Volger, J. (1966) *Physica*, **32** (10), 1680–1692.

17. Jellison, G.A., Modine, F.A., and Boatner, L.A. (1997) *Opt. Lett.*, **22** (23), 1808–1810.

18. Forro, L., Chauvet, O., Emin, D., Zuppiroli, L., Berger, H., and Levy, F. (1994) *J. Appl. Phys.*, **75** (1), 633–635.

19. Tang, H. (1994) Electronic properties of anatase TiO_2 investigated by electrical and optical measurements on single crystals and thin films. Département de Physique, Ecole Polytechnique Fédérale de Lausanne EPFL. Lausanne.

20. Hosaka, N., Sekiya, T., Satoko, C., and Kurita, S. (1997) *J. Phys. Soc. Jpn.*, **66**, 877–880.

21. Fujushima, A. and Honda, K. (1972) *Nature*, **238** (5358), 37–38.

22. Satoh, N., Nakashima, T., Kamikura, K., and Yamamoto, K. (2008) *Nat. Nanotechnol.*, **3**, 106–111.

23. Chen, X., Liu, L., Yu, P.Y., and Mao, S.S. (2011) *Science*, **331**, 746–750.

24. Tao, J., Luttrell, T., and Batzill, M. (2011) *Nat. Chem.*, **3**, 296–300.

25. Diebold, U. (2011) *Nat. Chem.*, **3**, 271–272.

26. Naldoni, A., Allieta, M., Santangelo, S., Marelli, M., Fabbri, F., Cappelli, S., Bianchi, C.L., Psaro, R., and Dal Santo, V. (2012) *J. Am. Chem. Soc.*, **134** (18), 7600–7603.

27. O'Regan, B. and Graetzel, M. (1991) *Nature*, **353**, 737–740.

28. Wang, Y., Hao, Y., Cheng, H., Ma, J., Xu, B., Li, W., and Cai, S. (1999) *J. Mater. Sci.*, **34** (12), 2773–2779.

29. Dvoranova, D., Brezova, V., Mazur, M., and Malati, M.A. (2002) *Appl. Catal., B*, **37** (2), 91–105.

30. Umebayashi, T., Yamaki, T., Itoh, H., and Asak, K. (2002) *J. Phys. Chem. Solids*, **63** (10), 1909–1920.

31. Davydov, L., Reddy, E.P., France, P., and Smirniotis, P.G. (2001) *J. Catal.*, **203** (1), 157–167.

32. Anpo, M. (2000) *Pure Appl. Chem.*, **72** (9), 1787–1792.

33. Fuerte, A., Hernandez-Alonso, M.D., Maira, A.J., Martinez-Arias, A., Fernandez-Garcias, M., Conesa, J.C., and Soria, J. (2001) *Chem. Commun.*, **24**, 2718–2719.

34. Yamashita, H., Harada, M., Misaka, J., Takeuchi, M., Ichihashi, Y., Goto, F., Ishida, M., Sasaki, T., and Anpo, M. (2001) *J. Synchrotron Radiat.*, **8** (2), 569–571.

35. Wilke, K. and Breuer, H.D. (1999) *J. Photochem. Photobiol., A*, **121** (1), 49–53.

36. Wilke, K. and Breuer, H.D. (1999) *J. Phys. Chem.*, **213**, 135–140.

37. Hong, N.H., Sakai, J., Prellier, W., and Ruyter, A. (2005) *J. Appl. Phys.*, **38** (6), 816–821.

38. Narayanan, B.N., Koodathil, R., Gangadharan, T., Yaakob, Z., Saidu, F.K., and Chandralayam, S. (2010) *Mater. Sci. Eng., B*, **168** (1–3), 242–244.

39. Xu, A.W., Gao, Y., and Liu, H.Q. (2002) *J. Catal.*, **207** (2), 151–157.

40. Moon, J., Takagi, H., Fujishiro, Y., and Awano, M. (2001) *J. Mater. Sci.*, **36** (4), 949–955.

41. Li, J. and Zeng, H.C. (2007) *J. Am. Chem. Soc.*, **129** (51), 15839–15847.

42. Shen, Y., Xiong, T., Li, T., and Yang, K. (2008) *Appl. Catal., B*, **83** (3–4), 177–185.

43. Chandiran, A.K., Sauvage, F., Lioz, E., and Graetzel, M. (2011) *J. Phys. Chem. C*, **115** (18), 9232–9240.

44. Burda, C., Lou, Y.B., Chen, X.B., Samia, A.C.S., Stout, J., and Gole, J.L. (2003) *Nano Lett.*, **3** (8), 1049–1051.

45. Asahi, R., Morikawa, T., Ohwaki, T., Aoki, K., and Taga, Y. (2001) *Science*, **293** (5528), 269–271.

46. Ihara, T., Miyoshi, M., Iriyama, Y., Matsumoto, O., and Sugihara, S. (2003) *Appl. Catal., B*, **42** (4), 403–409.

47. Sakthivel, S. and Kisch, H. (2003) *ChemPhysChem*, **4** (5), 487–490.

48. Irie, H., Watanabe, Y., and Hashimoto, K. (2003) *Chem. Lett.*, **32** (8), 772–773.

49. Park, J.H., Kim, S., and Bard, A.J. (2006) *Nano Lett.*, **6** (1), 24–28.

50. Yamaki, T., Umebayashi, T., Sumita, T., Yamamoto, S., Maekawa, M., Kawasuso, A., and Itoh, H. (2003)

Nucl. Instrum. Methods Phys. Res., Sect. B, **206**, 254–258.

51. Ohno, T., Mitsui, T., and Matsumura, M. (2003) *Chem. Lett.*, **32** (4), 364–365.

52. Liu, Y., Chen, X., Li, J., and Burda, C. (2005) *Chemosphere*, **61** (1), 11–18.

53. Yu, J.C., Zhang, L., Zhang, Z., and Zhao, J. (2003) *Chem. Mater.*, **15** (11), 2280–2286.

54. Lei, L., Su, Y., Zhou, M., Zhang, X., and Chen, X. (2007) *Mater. Res. Bull.*, **42** (12), 2230–2236.

55. Kang, I.C., Zhang, Q., Yin, S., Stao, T., and Saito, F. (2008) *Appl. Catal., B*, **80** (1–2), 81–87.

56. Bettinelli, M., Dallacasa, V., Falcomer, D., Fornasiero, P., Gombac, V., Montini, T., Romano, L., and Speghini, A. (2007) *J. Hazard. Mater.*, **146** (3), 529–534.

57. Liu, C., Tang, X., Mo, C., and Qiang, Z. (2008) *J. Solid State Chem.*, **181** (4), 913–919.

58. Kim, S.W., Khan, R., Kim, T.J., and Kim, W.J. (2008) *Bull. Korean Chem. Soc.*, **29** (6), 1217–1223.

59. Di Valentin, C., Pacchioni, G., Selloni, A., Livraghi, S., and Giamello, E. (2005) *J. Phys. Chem. B*, **109**, 11414.

60. Zhao, L., Jiang, Q., and Lian, J. (2008) *Appl. Surf. Sci.*, **254**, 4620.

61. He, J., Lindström, H., Hagfeldt, A., and Lindquist, S.E. (1999) *J. Phys. Chem. B*, **103**, 8940–8943.

62. Bach, U., Lupo, D., Comte, P., Moser, J.-E., Weissortel, F., Salbeck, J., Spreitzer, H., and Graetzel, M. (1998) *Nature*, **395** (6702), 583–585.

63. Kay, A. and Graetzel, M. (1996) *Sol. Energy Mater. Sol. Cells*, **44**, 99.

64. Murakami, T.N. and Ito, S. (2006) *J. Electrochem. Soc.*, **153** (12), A2255–A2261.

65. Imoto, K., Takahashi, K., Yamaguchi, T., Komura, T., Nakamura, J., and Murata, K. (2003) *Sol. Energy Mater. Sol. Cells*, **79**, 459.

66. Lee, W.J., Ramasamy, E., Lee, D.Y., and Song, J.S. (2009) *ACS Appl. Mater. Interfaces*, **1**, 1146.

67. Saito, Y., Kitamura, T., Wada, Y., and Yanagida, S. (2002) *Chem. Lett.*, **31**, 1060.

68. Wang, M., Anghel, A.M., Marsan, B., Cevey, N.L., Pootrakulchote, N., Zakeeruddin, S.M., and Graetzel, M. (2009) *J. Am. Chem. Soc.*, **131**, 15976.

69. Yella, A., Lee, H.W., Tsao, H.N., Yi, C., Chandiran, A.K., Nazeeruddin, M.K., Diau, E.W.G., Yeh, C.Y., Zakeeruddin, S.M., and Graetzel, M. (2011) *Science*, **334** (6056), 629–634.

70. Feldt, S.M., Gibson, E.A., Gabrielsson, E., Sun, L., Boschloo, G., and Hagfeldt, A. (2010) *J. Am. Chem. Soc.*, **132**, 16714–16724.

71. Noda, K. (2010) Presented at DSC-International conference, 2010, Colorado Spring, USA.

72. Sauvage, F., Chhor, S., Marchioro, A., Moser, J.-E., and Graetzel, M. (2011) *J. Am. Chem. Soc.*, **133**, 13103–13109.

73. Howe, R.F. and Graetzel, M. (1985) *J. Phys. Chem.*, **89**, 4495–4499.

74. Graetzel, M. and Howe, R.F. (1990) *J. Phys. Chem.*, **94**, 2566–2572.

75. Siripala, W. and Tomkievicz, M. (1982) *J. Electrochem. Soc.*, **129**, 1240–1245.

76. Bisquert, J. (2003) *Phys. Chem. Chem. Phys.*, **5**, 5360–5364.

77. De Angelis, F., Fantacci, S., Selloni, A., Graetzel, M., and Nazeeruddin, M.K. (2007) *Nano Lett.*, **7** (10), 3189–3195.

78. Peter, L. (2009) *Acc. Chem. Res.*, **42** (11), 1839–1847.

79. Ito, S., Murakami, T.N., Comte, P., Liska, P., Graetzel, C., Nazeeruddin, M.K., and Graetzel, M. (2008) *Thin Solid Films*, **516**, 4613.

80. Furubayashi, Y., Hitosugi, T., Yamamoto, Y., Inaba, K., Kinoda, G., Hirose, Y., Shimada, T., and Hasegawa, T. (2005) *Appl. Phys. Lett.*, **86**, 252101.

81. Shannon, R.D. (1976) *Acta Crystallogr.*, **A32**, 751.

82. Gao, F., Wang, Y., Shi, D., Zhang, J., Wang, M., Jing, X., Humphry-Baker, R., Wang, P., Zakeeruddin, S.M., and Graetzel, M. (2008) *J. Am. Chem. Soc.*, **130**, 10720.

83. O'Regan, B., Durrant, J.R., Sommeling, P.M., and Bakker, N.J. (2007) *J. Phys. Chem. C*, **111**, 14001.

84. Chandiran, A.K., Sauvage, F., Etgar, L., and Graetzel, M. (2011) *J. Phys. Chem. C*, **115**, 9232–9240.

85. Cao, Y.M., Bai, Y., Yu, Q.J., Cheng, Y.M., Liu, S., Shi, D., Gao, F., and Wang, P. (2009) *J. Phys. Chem. C*, **113**, 6290–6297.

86. Chen, D.H., Huang, F.Z., Cheng, Y.B., and Caruso, R.A. (2009) *Adv. Mater.*, **21**, 2206–2210.

87. Chen, D.H., Cao, L., Huang, F.Z., Imperia, P., Cheng, Y.B., and Caruso, R.A. (2010) *J. Am. Chem. Soc.*, **132**, 4438–4444.

88. Van de Hulst, H.C. (1957) *Light Scattering by Small Particles*, John Wiley & Sons, Inc., New York.

89. Yuris, A., Balzani, V., Barigelletti, F., Campagna, S., Belser, P., and Von Zelewsky, A. (1988) *Coord. Chem. Rev.*, **84**, 85.

90. Nazeeruddin, M.K. and Graetzel, M. (2007) *Photofunctional Transition Metals Complexes*, Structure and Bonding, Vol. 123, Springer-Verlag, Berlin, pp. 113–175.

91. Moser, J.E., Noukakis, D., Bach, U., Tacchibana, Y., Klug, D.R., Durrant, J.R., Humphry-Baker, R., and Graetzel, M. (1998) *J. Phys. Chem. B*, **102**, 3649–3650.

92. Meyer, T.J. (1986) *Pure Appl. Chem.*, **50**, 1293.

93. Ferrere, S. (2000) *Chem. Mater.*, **12**, 1083.

94. Fleming, C.N., Maxwell, K.A., De Simone, J.M., Meyer, T.J., and Papanikolas, J.M. (2001) *J. Am. Chem. Soc.*, **123**, 10336–10347.

95. Bessho, T., Constable, E.C., Graetzel, M., Redondo, A.H., Housecroft, C.E., Kylberg, W., Nazeeruddin, M.K., Neuburger, M., and Schaffner, S. (2008) *Chem. Commun.*, 3717.

96. Hagberg, D.P., Yum, J.H., Lee, H., De Angelis, F., Marinado, T., Karlsson, K.M., Humphry-Baker, R., Sun, L.C., Hagfeldt, A., and Graetzel, M. (2008) *J. Am. Chem. Soc.*, **130**, 6259.

97. Kim, C., Choi, H., Kim, S., Baik, C., Song, K., Kang, M.S., Kang, S.O., and Ko, J. (2008) *J. Org. Chem.*, **73**, 7072.

98. Hara, K., Sato, T., Katoh, R., Furube, A., Ohga, Y., Shinpo, A., Suga, S., Sayama, K., Sugihara, H., and Arakawa, H. (2003) *J. Phys. Chem. B*, **107**, 597.

99. Islam, A., Sugihara, H., Hara, K., Singh, L.P., Katoh, R., Yanagida, M., Takahashi, Y., Murata, S., and Arakawa, H. (2001) *Inorg. Chem.*, **40**, 5371.

100. Geary, A.M., Yellowless, L.J., Jack, L.A., Oswald, I.D.H., Parsons, S., Hirata, N., Durrant, J.R., and Robertson, N. (2005) *Inorg. Chem.*, **44**, 242.

101. Wong, W.Y., Wang, X.Z., He, Z., Chan, K.K., Djurisic, A.B., Cheung, K.Y., Yip, C.T., Alan, M.C.N., Yan, Y.X., Mak, C.S.K., and Chan, W.K. (2007) *J. Am. Chem. Soc.*, **129**, 14372.

102. Ferrere, S. and Gregg, B.A. (1998) *J. Am. Chem. Soc.*, **120**, 843.

103. Sauve, G., Cass, M.E., Doig, S.J., Lauermann, I., Pomykal, K., and Lewis, N.S. (2000) *J. Phys. Chem.*, **104**, 3488.

104. Asbury, J.B., Hao, E., Wang, Y., and Lian, T. (2000) *J. Phys. Chem. B*, **104**, 11957.

105. Baranoff, E., Yum, J.H., Jung, I., Vulcano, R., Graetzel, M., and Nazeeruddin, M.K. (2010) *Chem. Asian J.*, **5** (3), 496–499.

106. Nazeeruddin, M.K., Kay, A., Rodicio, I., Humphry-Baker, R., Muller, E., Liska, P., Vlachopoulos, N., and Graetzel, M. (1993) *J. Am. Chem. Soc.*, **115**, 6382.

107. Yan, S.G. and Hupp, J.T. (1996) *J. Phys. Chem.*, **100**, 6867.

108. Tachibana, Y., Haque, S.A., Mercer, I.P., Moser, J.E., Klug, D.R., and Durrant, J.R. (2001) *J. Phys. Chem. B*, **105**, 7424–7431.

109. Nazeeruddin, M.K., Zakeeruddin, S.M., Humphry-Baker, R., Jirousek, M., Liska, P., Vlachopoulos, N., Shklover, V., Fischer, C.H., and Graetzel, M. (1999) *Inorg. Chem.*, **38**, 6298.

110. Nazeeruddin, M.K., Pechy, P., Renouard, T., Zakeeruddin, S.M., Humphry-Baker, R., Comte, P., Liska, P., Le, C., Costa, E., Shklover, V., Spiccia, L., Deacon, G.B., Bignozzi, C.A., and Graetzel, M. (2001) *J. Am. Chem. Soc.*, **123**, 1613.

111. Tennakone, K., Kumara, G.R.R.A., Kottegoda, I.R.M., and Perera, V.P.S. (1999) *Chem. Commun.*, 15.

112. Chandrasekharan, N. and Kamat, P.V. (2000) *J. Phys. Chem. B*, **104**, 10851.

113. Gregg, B.A., Pichot, F., Ferrere, S., and Fields, C.L. (2001) *J. Phys. Chem. B*, **105**, 1422–1429.

114. Nazeeruddin, M.K. and Graetzel, M. (2002) *Encyclopedia of Electrochemistry: Semi-Conductor Electrodes and Photoelectrochemistry*, Vol. 6, Wiley-VCH Verlag GmbH, Weinheim, pp. 407–431.

115. Schmidt-Mende, L., Kroeze, K.J., Durrant, J.R., Nazeeruddin, M.K., and Graetzel, M. (2005) *Nano Lett.*, **5**, 1315–1320.

116. Nazeeruddin, M.K., Klein, C., Liska, P., and Graetzel, M. (2005) *Coord. Chem. Rev.*, **248**, 1317.

117. Nazeeruddin, M.K., De Angelis, F., Fantacci, S., Selloni, A., Viscardi, G., Liska, P., Ito, S., Bassho, T., and Graetzel, M. (2005) *J. Am. Chem. Soc.*, **127**, 16835–16847.

118. Feldt, S.M., Wang, G., Boschloo, G., and Hagfeldt, A. (2011) *J. Phys. Chem. C*, **115**, 21500–21507.

119. Wang, P., Klein, C., Humphry-Baker, R., Zakeeruddin, S.M., and Graetzel, M. (2005) *Appl. Phys. Lett.*, **86**, 123503–123508.

120. Chen, C.Y., Wang, M.K., Li, J.Y., Pootrakulchote, N., Alibabaei, L., Ngoc-Le, C.H., Decoppet, J.D., Tsai, J.H., Graetzel, C., Wu, C.G., Zakeeruddin, S.M., and Graetzel, M. (2009) *ACS Nano*, **3**, 3103–3109.

121. Hara, K., Sugihara, H., Tachibana, Y., Islam, A., Yanagida, M., Sayama, K., Arakawa, H., Fujihashi, G., Horiguchi, T., and Kinoshita, T. (2001) *Langmuir*, **17**, 5992–5999.

122. Yanagida, M., Yamaguchi, T., Kurashige, M., Hara, K., Katoh, R., Sugihara, H., and Harakawa, H. (2003) *J. Photochem. Photobiol., A*, **158**, 131–138.

123. Islam, A., Sugihara, H., and Arakawa, H. (2003) *J. Photochem. Photobiol., A Chem.*, **158**, 131–138.

124. Nazeeruddin, M.K., Pechy, P., and Graetzel, M. (1997) *Chem. Commun.*, 1705–1706.

125. Nazeeruddin, M.K., Pechy, P., Renouard, T., Zakeeruddin, S.M., Humphry-Baker, R., Comte, P., Liska, P., Cevey, C., Costa, E., Shklover, V., Spiccia, L., Deacon, G.B., Bignozzi, C.A., and Graetzel, M. (2001) *J. Am. Chem. Soc.*, **123**, 1613–1624.

126. Graetzel, M. (2004) *J. Photochem. Photobiol., A*, **168**, 235.

127. Chiba, Y., Islam, A., Watanabe, Y., Komiya, R., Koide, N., and Han, L.Y. (2006) *Jpn. J. Appl. Phys.*, **45**, L638–L640.

128. Islam, A., Chen, H., Malapaka, C., Chiranjeevi, B., Zhang, S., Yang, X., and Yanagida, M. (2012) *Energy Environ. Sci.*, **5**, 6057.

129. Barolo, C., Nazeeruddin, M.K., Fantacci, D., Di Censo, D., Ito, S., Comte, P., Liska, P., Viscardi, G., Quagliotto, P., De Angelis, P., and Graetzel, M. (2006) *Inorg. Chem.*, **45**, 4642–4653.

130. Renouard, T., Fallahpour, R.A., Nazeeruddin, M.K., Humphry-Baker, R., Gorelsky, S.I., Lever, A.B.P., and Graetzel, M. (2002) *Inorg. Chem.*, **41**, 367.

131. Dupau, P., Renouard, T., and Bozec, H.L. (1996) *Tetrahedron Lett.*, **37**, 7503.

132. Abbotto, A., Sauvage, F., Barolo, C., De Angelis, F., Fantacci, S., Graetzel, M., Manfredi, N., Marinzi, C., and Nazeeruddin, M.K. (2011) *Dalton Trans.*, **40**, 234–242.

133. Boschloo, G. and Hagfeldt, A. (2009) *Acc. Chem. Res.*, **42**, 1819–1826.

134. Clifford, J.N., Palomares, E., Nazeeruddin, M.K., Graetzel, M., and Durrant, J.R. (2007) *J. Phys. Chem. C*, **111**, 6561–6567.

135. Sapp, S., Elliott, C.M., Contado, C., Caramori, S., and Bignozzi, C.A. (2002) *J. Am. Chem. Soc.*, **124**, 11215–11222.

136. Nusbaumer, H., Moser, J.E., Zakeeruddin, S.M., Nazeeruddin, M.K., and Gratzel, M. (2002) *J. Phys. Chem. B*, **105**, 10461.

137. Wang, H.X., Nicholson, P.G., Peter, L., Zakeeruddin, S.M., Nazeeruddin, M.K., and Graetzel, M. (2010) *J. Phys. Chem. C*, **114**, 14300–14306.

138. Yum, J.H., Baranoff, E., Kessler, F., Moehl, T., Ahmad, S., Bessho, T., Marchioro, A., Ghadiri, E., Moser, J.-E., Yi, C., Nazeeruddin, M.K., and Graetzel, M. (2012) *Nat. Commun.*, **3**. doi: 10.1038/ncomms1655

139. Bisquert, J., Fabregat-Santiago, F., Mora-Sero, I., Garcia-Belmonte, G., Barea, E.M., and Palomares, E. (2008) *Inorg. Chim. Acta*, **361**, 684–698.

14
Hybrid Solar Cells from Ordered Nanostructures

Jonas Weickert and Lukas Schmidt-Mende

14.1
Introduction

One of the main challenges of our society is the growing demand for energy. Not only is the world's population growing rapidly but also the energy consumption per capita is increasing as a result of more and more nations becoming industrialized. Superlinear growth of the world energy consumption since the nineteenth century has led to today's vast energy need of approximately 469×10^{18} J year^{-1}, or, in terms of power, 15 TW, which is approximately 10 times as much as 100-years ago [1]. Most probably, additional 30 TW of new power will be needed by 2050 [2]. Figure 14.1 shows the world energy consumption during the last 45 years, stratified by geographic regions. The energy consumption has tripled since 1965. Although already highly industrialized by 1965, North America and Europe have almost doubled their energy consumption. Even stronger growth is apparent for the Middle East, South and Central America, Africa, and especially Asia. By pushing the industrialization in these regions, the power need increases accordingly. Further growth in energy consumption can be expected when keeping in mind that today Africa and large parts of Asia are industrialized only to a very small fraction. Approximately 30% of the world's population still remains with insufficient electricity, which will probably change during the next decades [2].

To date, by far the biggest fraction of the world's power supply relies on burning fossil carbon sources, leading to the emission of the greenhouse gas CO_2. Accordingly, the concentration of CO_2 in the atmosphere has increased dramatically during the past 100 years. At the same time, the world's mean temperature has also increased, suggesting that the greenhouse effect can already be observed in our days. Figure 14.2 depicts the mean temperature of the Earth's surface and the atmospheric CO_2 concentration during the past 130 years. The parallelism of the two is striking. Considering the obvious effect of atmospheric CO_2 on global warming and the rapidly growing need for additional power, there appears no alternative than switching to clean CO_2-neutral energy sources in the near future. Best-suited concepts rely on converting sunlight incident on the earth into usable energy.

Functional Metal Oxides: New Science and Novel Applications, First Edition.
Edited by Satishchandra B. Ogale, Thirumalai V. Venkatesan, and Mark G. Blamire.
© 2013 Wiley-VCH Verlag GmbH & Co. KGaA. Published 2013 by Wiley-VCH Verlag GmbH & Co. KGaA.

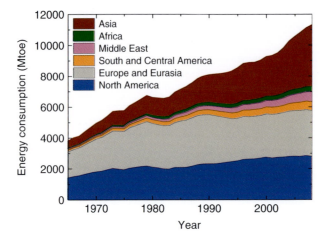

Figure 14.1 World energy consumption since 1965, stratified by geographic regions. (Source: Based on data of Ref. [1]).

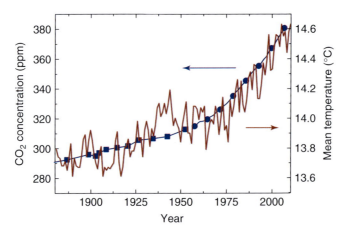

Figure 14.2 Atmospheric CO_2 concentration and mean temperature of the earth's surface since 1880. (Source: Based on data from Refs [4–6]).

The sun's mean power input on our planet's surface is approximately $165 \, \mathrm{W \, m^{-2}}$ in total average over the whole surface area and over one year. With two-thirds of the Earth's surface being oceans, where the installation of sunlight-harvesting power plants is complicated or impossible, there still remains over 1100 times more power than humankind will be consuming by 2050 [2, 3]. Finding a way to make use of this enormous power resource will be the greatest challenge for the next decades.

One of the most promising approaches for solar light harvesting is the use of photovoltaics, which directly converts incident solar light into electrical energy. Today, the photovoltaic industry is almost completely dominated by solar cells based on Si. Si photovoltaic cells exhibit high-power conversion efficiencies (PCEs) on the order of 30% for research solar cells and can be operated for several years or even decades. However, although Si is the second most abundant element on earth, it almost never occurs as free element but rather as oxide SiO_2. Extraction of elementary Si is highly energy consuming, and only high-purity and highly crystalline Si is suited for high-performance photovoltaics. Accordingly, fabrication of Si solar cells is expensive and the energy payback time – the period for which the cell must be operated to convert as much energy as its fabrication consumed – is on the order of one or two years, which appears relatively long [7]. Considering the lifetime of Si photovoltaic devices and the energy consumption during fabrication, the CO_2 equivalent of operating a Si solar cell is on the order of $100\,g\,kWh^{-1}$. This value is significantly better than for gas or coal power plants with around 400 or $900\,g\,kWh^{-1}$, respectively, but still on the same order of magnitude [8, 9]. New photovoltaics requiring less energy-intensive and less costly fabrication processes are desirable.

Since the middle of last century, alternative materials for photovoltaic devices have been extensively studied. Impressive high efficiencies have been realized with inorganic III–V semiconductor solar cells [10]. Although these photovoltaics are used for specialized applications, such as powering of orbiting satellites as in the case of GaAs solar cells, their fabrication costs as well as energy consumption are high as in the case of Si.

Today, promising candidates for low-cost alternatives to Si photovoltaics are rather based on organic compounds and metal oxides. Organic materials can be potentially synthesized at low costs, and technical-grade purities suffice for mass production. The same holds for metal oxides, which usually can be purified more easily and at lower energy consumption than elemental materials. During the past 25 years, solar cells based on both organic materials and metal oxides have reached remarkable PCEs. Fully organic photovoltaics (OPVs) offer the additional advantage that they can be realized on flexible substrates using roll-to-roll processing and low-cost techniques such as inkjet printing. The most common metal-oxide-based solar cells utilize monolayers of organic dye molecules adsorbed to large-bandgap n-type metal oxide nanoparticles, most commonly TiO_2. Materials for these so-called dye-sensitized solar cells (DSSCs) are highly abundant and can also be processed using cheap and upscalable fabrication techniques such as screen printing or doctor-blading. Currently, DSSCs exhibit higher efficiencies than OPVs and are more stable when operated at ambient conditions, but they rely on a hole-conducting liquid electrolyte, which requires rigid substrates and makes solar cells damageable by leaks. To overcome these limitations, there are also approaches to replace the electrolyte with a solid-state hole conductor in the so-called solid-state dye-sensitized solar cells (SS-DSSCs). To date, SS-DSSCs reach only about half the PCE of DSSCs. During the past few years, new concepts were introduced by combining materials from DSSCs and OPVs in the so-called hybrid solar cells (HSCs). HSCs rely on nanostructured metal oxides and absorbing hole conductors. Thus, they are supposed to benefit from the advantages of both OPVs and DSSCs. Especially, ordered nanostructures hold great promise for this novel type of photovoltaics, as will be discussed in the following sections.

Another advantage of HSCs over OPVs lies in the inverted geometry of these solar cells. The direction of current in conventional OPVs is the opposite of the current in HSCs. Therefore, noble metal top contacts can be used in HSCs compared to easily oxidizing contacts such as Al or Mg in OPVs. This inverted geometry improves the stability of solar cells if operated in ambient air and allows the fabrication of long-term-stable photovoltaics [11, 12].

In the following section, we will introduce the most common geometries and the working principles of OPVs, DSSCs, and HSCs. Besides the choice of materials, also the morphology of the photoactive layer plays a crucial role in device performance and determines limiting processes such as exciton separation and charge-carrier recombination.

In Section 14.3, we discuss different geometries of nanostructures that have been applied in HSCs. Our focus lies on TiO_2, the most commonly used metal oxide for solar cell applications. However, we will also discuss ZnO nanostructures because this material has drawn increasing attention during the past few years.

The last section gives an overview of the approaches to modify the electronic properties of metal oxides: coating with self-assembled monolayers of sensitizers, chemical treatments, and the introduction of dopants to the crystal lattice. A current approache to infiltrate organic materials into metal oxide nanostructures.

14.2
Working Mechanisms of Hybrid Solar Cells

This section describes the fundamental working mechanisms of OPVs, DSSCs, and HSCs. Typical geometries of DSSCs, SS-DSSCs, and OPVs are discussed, and the concept of HSCs is introduced. Light absorption, charge separation, and charge transport in HSCs are explained, and the most important loss mechanisms and possible workarounds are outlined.

14.2.1
Solar Cell Geometries

Since the invention of the first DSSC and the first SS-DSSC in 1991 and 1998, respectively, the concept of dye-sensitized metal oxides has been well established in the field of excitonic solar cells [13, 14]. All dye-based solar cells utilize a metal oxide, commonly with a wide bandgap as in the case of TiO_2 or ZnO, decorated with a self-assembled monolayer of a high-extinction dye. On light absorption, the dye injects an electron into the metal oxide. Subsequently, the dye cation is regenerated; that is, the hole, which resides in the dye, is transferred to the hole transporting medium (HTM) [15]. As HTM, typically a liquid electrolyte of KI is used in DSSCs, or a layer of (2,2′,7,7′-tetrakis-(N,N-di-p-methoxyphenylamine)9,9′-spirobifluorene) (Spiro-OMeTAD) in SS-DSSCs [15, 16].

Figure 14.3 shows the schematics of a typical DSSC and a SS-DSSC. Since the absorption of a dye monolayer is intrinsically low, mesoporous metal oxide layers with thicknesses of 10–20 μm are sintered from nanoparticles. Decoration of these structures with a dye monolayer results in an overall high absorption owing to the high-surface area of the mesoporous structure. In DSSCs, this mesoporous layer is usually assembled on a compact layer of TiO_2 on a transparent conductive oxide (TCO) such as fluorine-doped tin oxide (FTO) on a glass substrate. Charge separation takes place at the dye–metal oxide interface as described earlier. Most commonly, the structure is infiltrated with the KI redox electrolyte, which ensures hole transport to the semitransparent Pt-coated top electrode. Hole transport in

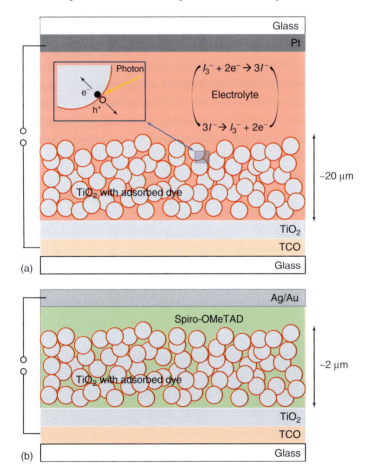

Figure 14.3 Solar cell geometries of (a) a typical DSSC and (b) a typical SS-DSSCs. In both cases, the absorbing layer consists of a few micrometers of sintered TiO_2 nanoparticles decorated with a self-assembled monolayer of dye molecules. On dye excitation, electrons are injected into TiO_2 and collected at the bottom electrode. In DSSCs, the dye is regenerated by a liquid electrolyte, typically KI. In SS-DSSCs, the electrolyte is replaced with a transparent hole-transporting medium such as Spiro-OMeTAD.

DSSCs is a cyclic reaction, as shown in Figure 14.3a, where holes are loaded to the electrolyte during the oxidation of I^- to I_3^-, and I_3^- is reduced again at the catalytic hole-collecting Pt electrode. Recently, Co-based alternative electrolytes have been introduced, which help us achieve PCEs beyond 12% [17, 18]. DSSCs based on liquid electrolytes are dealt with in Chapter 13 and not further discussed in this chapter.

SS-DSSCs are based on the same concept, with the photoactive material consisting of a dye-sensitized mesoporous metal oxide layer. However, the dye is regenerated by a transparent solid-state HTM instead of the liquid electrolyte in DSSCs, most commonly Spiro-OMeTAD [16]. Infiltration of Spiro-OMeTAD into mesoporous films is more demanding than infiltration with a liquid electrolyte, and hole transport over larger distances becomes challenging. Therefore, commonly only thinner absorbing layers (2–3 µm) are used in SS-DSSCs [19]. Spiro-OMeTAD is usually spin-coated, yielding a thin overlayer on top of the mesoporous metal oxide. Solar cells are completed by the deposition of noble metal contacts, such as Au, on top of the infiltrated structure, with the overlayer preventing direct contact of the top contact and the metal oxide, thus avoiding short circuits.

Another important class of excitonic solar cells is the OPVs. The first organic solar cells following a metal–insulator–metal principle were based on a Schottky contact between organic semiconductors and metals or TCOs such as Indium-tin oxide (ITO) and exhibited only poor performances [20, 21]. Interest in this field of research rapidly grew after the introduction of a charge-separating organic–organic junction by Tang in 1986 [22]. By combining p- and n-type organic semiconductors, which serve as the electron donor and acceptor, respectively, excitons could be separated at the donor–acceptor interface more efficiently. Further progress was made by blending the donor and acceptor together in a mixed phase film, the so-called bulk heterojunction (BHJ) [23]. Thus, more intimate contact and larger interfacial area between the donor and the acceptor than in a layered structure are achieved. Because of very limited diffusion lengths of excitons in organic materials, which are usually on the order of 10 nm, this further enhances the exciton separation yield.

Most commonly, highly absorbing p-type polymers are used in combination with n-type fullerenes, and PCEs beyond 5% have been frequently reported since the mid-2000s [24, 25]. During the past few years, near-infrared-absorbing low-bandgap polymers have emerged with efficiencies exceeding 8% [26, 27]. Although these are impressive values, further improvements require better control over the morphology of the donor–acceptor mixed phase. Good intermixing leads to efficient charge separation, but high efficiencies also demand good extraction of charge carriers from the active layer. For blended structures of polymers and fullerenes, internal quantum efficiencies close to 100% have been shown; that is, almost each absorbed photon is converted into collectable charge carriers. However, external quantum efficiencies, that is, the ratio of extracted charges and the number of incident photons, are still farther from these values [28].

Figure 14.4a shows a schematic of a BHJ with a typical unordered morphology. Ideally, the donor and acceptor domains exhibit pathways for electron and hole

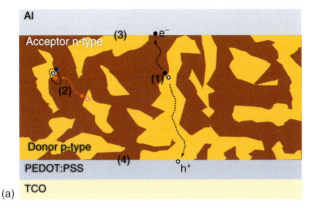

(a)

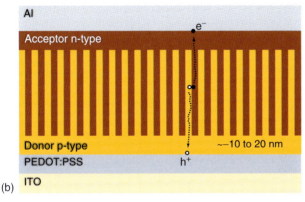

(b)

Figure 14.4 (a) Geometry of a typical fully organic bulk heterojunction solar cell. The photoactive layer comprises donor and acceptor domains in a mixed phase with relatively uncontrolled morphology. On light absorption in one of the materials, an exciton is generated. Excitons have a certain diffusion length, which is on the order of 10 nm. If the exciton encounters a donor–acceptor interface during its diffusion, it is separated into a free hole in the donor and a free electron in the acceptor. (1) If both donor and acceptor exhibit percolated pathways toward the electrodes, charge carriers can exit the device. (2) In the case of islands of either donor or acceptor, free charges cannot leave the solar cell and recombine with their electrical counterparts. (3) and (4) As a result of the processing from a blended solution, there are spots where the donor material is in contact with the electron-collecting electrode, and the acceptor material touches the hole-collecting electrode. This is supposed to hamper charge collection. (b) Ideally, the active layer is a comblike structure of interpenetrating donor and acceptor domains. Domain sizes are on the order of the exciton diffusion length, but the structure exhibits perfect percolation pathways for charge carriers.

transport toward the external contacts as in case (1). However, in real intermixed BHJs, there are always islands of donor or acceptor formation. Subsequent to exciton separation, one type of charge carriers is trapped within the island, where it stays until it recombines with a free electrical counterpart, as depicted in (2). Besides, because of the processing of the BHJ, which is commonly spin-coated from a blended solution, there are donor and acceptor domains in contact with the "wrong" electrode; that is, the donor is in direct contact with the electron-collecting electrode and the acceptor contacts the hole-collecting electrode as in (3) and (4), respectively. Exciton separation and charge transport obviously demand for different morphologies, with either a large interfacial area and complete intermixing of donor and acceptor, or consistent charge transport pathways and rather separated donor and acceptor domains. As shown in experiments and simulations, there is a trade-off between exciton separation and charge transfer, with a certain amount of phase separation between the donor and acceptor being necessary [29, 30].

According to these considerations, the active layer thickness in the BHJ concept is very limited. Since only partial control over the BHJ morphology is possible via a choice of solvents and film deposition parameters, formation of well-intermixed films with still consistent pathways throughout the whole BHJ becomes less likely in thick films. As a consequence, thicker active layers do not necessarily lead to increased device performances even though more light is absorbed. Because of these limitations, the BHJ thickness is limited to approximately 100 nm, and full photon harvesting is not possible even with strongly absorbing materials.

A more idealized morphology for a BHJ is schematically shown in Figure 14.4b. Such a comblike structure of interpenetrating donor and acceptor domains has several advantages. With the lateral domain size being on the order of the exciton diffusion length, charge separation efficiencies close to unity can be achieved. Additionally, charge carriers can be easily extracted because of perfect percolation pathways expanding in the z-direction throughout the active layer. Besides, thin over- and underlayers of donor and acceptor ensure that neither of the materials is in direct contact with the wrong external electrode.

There have been several attempts to realize interpenetrating structures as same as that in OPVs using nanoimprinting techniques or utilizing self-assembly of small molecules and polymers [31–35]. However, processing of fully organic materials into ordered structures is very challenging. Even if structuring of one component of the cell is established, infiltration of the counterpart material is problematic. Most OPV materials are soluble in the same solvents. This fact requires complicated transfer techniques or thermal evaporation in order to maintain the structuring of the donor–acceptor interface.

Over the past decade, HSCs have emerged beside OPVs and DSSCs. These devices combine large-surface-area metal oxides as used in DSSCs with highly absorbing materials known from OPV applications: that is, the transparent HTM in SS-DSSCs is substituted with absorbing materials that contribute to photo-current generation [36]. Thus, the metal oxide, which serves as electron acceptor, can be processed as a consistent and island-free network. So far, mainly strongly absorbing conjugated polymers such as poly(3-hexylthiophene) (P3HT) or polyphenylene vinylenes (PPVs) are used as HTMs because these materials are already well understood in OPVs [37, 38]. The metal oxide can also be nanostructured and thus be used as a structured acceptor as in Figure 14.4b. Typical n-type metal oxides such as TiO_2 or ZnO do not dissolve in organic solvents, making infiltration of these structures much easier than infiltration of organic nanostructures. Besides, higher aspect ratios and a broader variety of nanostructures can be realized using self-assembly processes of metal oxides, as will be further discussed in Section 14.3.

14.2.2
Light Absorption and Exciton Separation

The most fundamental prerequisite for an efficient solar cell is strong light absorption. High extinction coefficients are especially important in spectral regions where the sun's energy input on the earth surface is large, that is, mainly in the visible and the near-IR. In DSSCs and SS-DSSCs, a dye monolayer is the only absorber. Complete photon harvesting over the active spectral region demands for a high-surface area of the decorated metal oxide and active layer thicknesses of $10-20\,\mu m$, even though dyes with extremely high-extinction coefficients have been developed in the past 20 years. HSCs offer the advantage that the HTM itself contributes to the device absorption. Besides, absorption spectra of the dye and the HTM can be chosen to be complementary, yielding better overlap with the solar spectrum. Thus, PCEs beyond 3.5% have been realized in HSCs with active layer thicknesses in the submicrometer range [39].

For HSCs, currently dyes are used, which have already proved efficient in DSSCs and SS-DSSCs. Most commonly, these are Ru complex dyes such as the famous N3 dye, which gave the first efficiencies beyond 10% in DSSCs [40]. N3 absorbs strongly in the visible ($13\,900\,l\,mol^{-1}\,cm^{-1}$ at 541 nm) and up to 800 nm, allowing theoretical photocurrents beyond $25\,mA\,cm^{-2}$ under one sun illumination (AM 1.5 G) [41]. To date, there are many N3-based derivatives, such as the most widely used N719 dye; the so-called black dye N749 with a red-shifted absorption expanding up to 900 nm; and Z907, which has shown impressive efficiencies also in solid-state devices [42–44]. The latter appears to be the most promising candidate of these conventional dyes for applications in HSCs owing to its alkyl side chains, which help reduce charge-carrier recombination as discussed later. During the past years, alternatives to Ru dyes have emerged: most importantly, the fully organic metal-free indoline dyes, which exhibit higher extinction coefficients than Ru dyes and show similar performance in DSSCs and even superior performance in SS-DSSCs [41, 45, 46]. Because of the working mechanisms relatively similar to SS-DSSCs, these dyes are also interesting for application in future HSCs.

Besides the demand for strong absorption over a broad spectral range, dyes for solar cell applications have to be optimized for efficient electron injection and dye regeneration. The fundamental mechanism of charge separation in dye-based excitonic solar cells is depicted in Figure 14.5a. On light absorption, an exciton is generated in the dye and an electron–hole pair is formed at the dye–metal oxide interface, provided that highest occupied molecular orbital (HOMO) and lowest unoccupied molecular orbital (LUMO) of the dye match with valence and conduction band of the metal oxide. Generally, the injection of the electron into the metal oxide is supposed to occur on the timescale of femtoseconds, which is

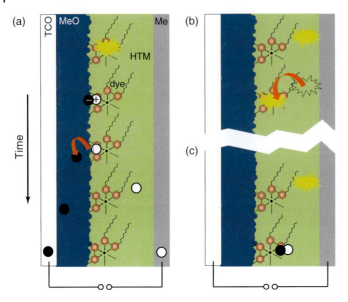

Figure 14.5 Schematic of the process of charge separation at a dye-decorated metal oxide interface. (a) The situation in DSSCs and one possible mechanism in HSCs. Incident light is absorbed by the dye and an exciton is formed. The resulting electron–hole pair is separated when the electron is quickly injected into the metal oxide. Subsequently, the dye is regenerated by the hole-transporting medium (HTM), which can be either a liquid electrolyte or a solid-state material. For absorbing HTM, it is also possible that the excited state is transferred from the HTM to the dye via an energy transfer, as shown in (b). Besides, charge separation might occur directly between excited HTM and the dye, as depicted in (c).

extremely fast compared to the lifetime of the excited state in the dye molecule of 20–60 ns [47–51]. The electron is transported away from the interface through the n-type metal oxide, which preferentially provides high-electron mobility. The dye cation is regenerated, that is, the remaining hole is transferred to the HTM. Charge separation upon dye excitation follows the same principles in HSCs, DSSCs, and SS-DSSCs, and is mainly driven by the rapid electron injection into the metal oxide. However, owing to differences in the hole transportation, dye regeneration occurs on different time scales in solid-state and liquid electrolyte devices. As mentioned earlier, the oxidation $3I^- \rightarrow I_3^- + 2e^-$ is a two-electron process and is rather slow, resulting in the back reaction, that is, charge-carrier recombination, being also slow. In contrast, dye regeneration is fast in solid-state devices because there is no ion diffusion involved and transfer of single charge carriers is possible. As a consequence, charge-carrier recombination is also fast in solid-state devices, demanding sophisticated interfacial engineering and control of the active layer's geometry.

The mechanism of charge separation between the dye and metal oxide depends not only on the energy levels of the HOMO and LUMO but also on their spatial location. Ideally, the LUMO should be located close to the metal oxide–dye interface, whereas the HOMO should point toward the HTM. Thus, electron injection becomes more likely upon dye excitation, but recombination is suppressed because electrons and holes reside at different locations within the dye. Besides, spatial separation of HOMO and LUMO enables quicker dye regeneration. This separation of the energy levels is realized in the so-called push-pull dyes. These dyes have proved efficient both in DSSCs and SS-DSSCs and constitute a sophisticated approach to enhance injection and reduce recombination [52, 53].

Three numbers, taken from a paper of O'Regan and Durrant [54], are interesting to get an impression of the situation in a dye-based solar cell under normal working conditions (i.e., illumination with one sun). In a cell with an active layer assembled of TiO_2 nanoparticles of 18 nm diameter, there are approximately 600 dye molecules located at each nanoparticle (this means a ratio of 150 TiO_2 unit cells per dye molecule). At $100\,mW\,cm^{-2}$ incident light intensity, each dye absorbs only about once every second, resulting in an electron flux of approximately $600\,s^{-1}$. Accordingly, most of the time dye molecules are in their ground state, and it is the ground state configuration that determines recombination kinetics.

In HSCs, the dye is not the sole absorber and different mechanisms of charge separation are possible. If light is absorbed in the HTM, energy transfer from the HTM to the dye is conceivable, as depicted in Figure 14.5b. Apparently, this mechanism demands for matching energy levels; that is, the difference in LUMO levels of the HTM and the dye should be smaller than the typical exciton binding energy, and the total energy of an excited state in the dye should be smaller than the energy of an exciton in the HTM. This could be possible, for instance, for a visible-light-absorbing HTM and a near-IR dye. Subsequent to energy transfer, charge separation occurs between the dye and metal oxide, as discussed in Figure 14.5a.

As shown in Figure 14.5c, it is also possible that excitons in the HTM are directly separated between the HTM and the dye. Furthermore, if the HTM can intercalate between dye molecules and be in direct contact with the metal oxide, charge separation is also possible at the HTM–metal oxide interface.

Generally, it is possible to yield at most one free electron and one free hole per incident photon. The maximum possible photocurrent is therefore directly dependent on the number of absorbed photons (in the case of loss-free photon-to-charge conversion). Under illumination, a solar cell can produce not only a current, the so-called short circuit current I_{SC}, but also a potential, the open circuit voltage V_{OC}. The exact origin of V_{OC} of excitonic solar cells is not fully understood yet, but there is evidence that it depends on the energy levels of the involved materials. In general, the potential is determined by the final energetic positions of electron and hole, that is, the conduction band of the metal oxide and the HOMO of the HTM. However, charge-carrier recombination plays an important role in lowering the open-circuit potential, and different potentials can be observed for different dyes even if the same metal oxide and HTM are used [55, 56].

14.2.3
Charge Transport

Once separated, electrons and holes are transported away from the interface through the metal oxide and the HTM, respectively. In the case of typical metal oxides such as ZnO and TiO_2, the mechanism of electron transport is still not fully understood although it is a hot topic of investigation. Charge transport is orders of magnitude faster in the case of single-crystalline materials, whereas grain boundaries between different crystallites seem to limit the mobility.

Electron mobilities up to $1\,cm^2\,V^{-1}\,s^{-1}$ have been found for the most prominent HSC material TiO_2 in its mineral form rutile. However, the presence of trap states significantly reduces the mobility. In the case of TiO_2, electron transport appears to be determined by trapping and de-trapping events between subbandgap states in the tail of the density of states and the conduction band [57, 58]. This mechanism is covered by the multiple-trapping model [59, 60]. Although it is difficult to determine their exact location, more trap states seem to be present at grain boundaries and at the surface of structures [61–63]. Accordingly, higher electron mobilities are present for larger crystal domains or even single crystals. A direct dependence of the photocurrent on the crystallite size has been shown for SS-DSSCs based on mesoporous TiO_2 [64]. The importance of crystallinity for high-charge carrier mobility has also been pointed out by Hendry *et al.* Comparison of commercially available rutile single crystals and several micrometer thick mesoporous layers assembled from TiO_2 nanoparticles has shown a 1000-fold higher electron mobility for the single crystal [65]. In general, high charge carrier mobilities are supposed to be beneficial in terms of reduced charge carrier recombination, especially in solid-state devices. Control of the metal oxide's crystallinity is therefore an important route toward efficient HSCs.

To date, hole transport in organic HTMs is well understood. In the case of small-molecule HTMs such as Spiro-OMeTAD, charge transport usually occurs as polar pair-hopping from one molecule to the other. By using additives such as lithium salts, the hole mobility of Spiro-OMeTAD can be enhanced by a factor of 10 and can reach $10^{-3}\,cm^2\,V^{-1}\,s^{-1}$ [66, 67]. However, this mechanism is not fully understood yet. In conjugated polymers such as P3HT, there is a strong overlap of π-orbitals, and electrons are delocalized along the polymer backbone or in the direction of π-stacking. On removal of an electron from the system, for example, via dye regeneration, charged units called *bipolarons* are formed, which can move through the material. Conjugated polymers typically show higher mobilities than small molecules owing to stronger delocalization of electrons and are therefore attractive candidates for efficient HTM in HSCs.

However, charge carrier mobilities do not only have to be high but should also be balanced, that is, of similar orders of magnitude for electrons and holes as pointed out by Li *et al.* [68]. Typical organic HTMs such as Spiro-OMeTAD or P3HT exhibit hole mobilities between 10^{-4} and $10^{-3}\,cm^2\,V^{-1}\,s^{-1}$, which is similar to mobilities on mesoporous TiO_2, but significantly smaller than the mobilities of single-crystalline TiO_2. However, if ordered metal oxide nanostructures are used instead of unordered mesoporous networks, enhanced hole mobilities are expected. Firstly, ordered metal oxide geometries also lead to a high order of hole conductivity, which then provides consistent charge percolation pathways. Thus, holes can move directly toward the hole-collecting electrode and exhibit a net higher mobility in the z-direction.

Additionally, confinement inside nanostructures can result in preferential ordering of crystalline materials, which has been shown for both polymers and small molecules [69, 70]. For crystalline polymers, which enable three-dimensional (3D) charge transport, molecular alignment due to confinement and the orientation of crystallites play an important role for the resulting hole mobility, as exemplarily discussed for the case of P3HT. P3HT exhibits alternating alkyl side chains attached to the thiophene units, which results in the formation of polymer sheets. These sheets stack, and the resulting crystallites can orient in three directions with respect to the substrate, as depicted in Figure 14.6. (i) Hole mobilities differ by more than an order of magnitude for different transport directions and are highest along the polymer backbone, (ii) moderate along the π-stacking direction from one sheet to the other and (iii) relatively low across the alkyl side chains [71]. Since charge transport is necessarily perpendicular to the substrate, vertical orientation of P3HT is desirable. However, P3HT orients in edge-on configuration on most substrates, which means lowest mobility in the z-direction. If P3HT is confined in nanocavities that are oriented perpendicular to the substrate and the polymer orients edge-on at the side walls of the cavities, crystallites are effectively turned by $90°$ and polymer backbones are aligned along the z-direction. Thus, enhanced hole mobilities in the direction of charge transport can be established, as shown by Coakley *et al.* [69] for the case of P3HT confinement in oriented Al_2O_3 nanopores.

14.2.4
Loss Mechanisms

As discussed in the previous sections, care is taken to optimize (i) absorption, (ii) heterojunction geometry, and (iii) charge carrier mobility. This is necessary in order to minimize each of the three main loss channels in excitonic solar cells.

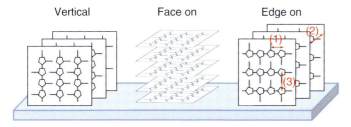

Figure 14.6 Possible orientations of P3HT on a substrate. Highest charge-carrier mobility is found for direction (1), that is, along the polymer backbone. Moderate mobilities are present in case (2), from one sheet of P3HT to the other. Transport across the alkyl side chains (3) occurs with the lowest mobility. (Source: Reproduced from Ref. [71]).

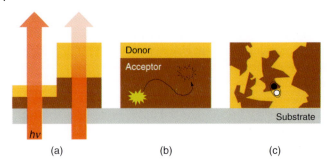

Figure 14.7 Sketch of the three most prominent loss mechanisms in excitonic solar cells. (a) Insufficient light absorption, (b) exciton recombination, and (c) charge-carrier recombination.

As depicted in Figure 14.7, these are (Figure 14.7a) insufficient light absorption, (Figure 14.7b) exciton recombination, and (Figure 14.7c) recombination of free charge carriers.

The most obvious loss mechanism in solar cells is lack of absorption. If the active layer of the solar cell does not absorb a high fraction of incident photons, high PCEs cannot be established. Besides, if only one absorbing material with a distinct band gap is used, conversion of high-energy photons is less efficient (in terms of energy conversion) than conversion of photons with the energy of the bandgap, because higher excited states are supposed to relax down to the edge of the bandgap before charge separation occurs. Thus, for one-absorber systems, there is an upper limit of PCE of around 34%, the so-called Shockley–Queisser limit [72]. In more sophisticated multijunction cells, a sequence of absorbing materials can be used instead of only one absorber. Light is coupled in from the side of the large-bandgap materials first, which finally hits the material with the smallest bandgap. Thus, less energy is lost owing to relaxation of excited states down to the band edge and the Shockley–Queisser limit can be broken [73]. In HSCs, light is typically incident through the monolayer of the dye before it hits the absorbing HTM. Absorption spectra of the dye and the HTM can be tuned independently, enhancing the overall absorption of the device.

Even in the case of strong absorption over a broad spectral range, not necessarily every photon is converted to free charge carriers owing to recombination of excitons. Excitons recombine after a certain lifetime if they do not encounter a charge-separating interface, which provides sufficient energy to overcome the exciton binding energy. Accordingly, an important prerequisite for efficient excitonic solar cells is phase separation between the donor and acceptor with domain sizes in the range of exciton diffusion lengths (typically about 10 nm). Secondly, energy levels of donor and acceptor have to provide sufficient HOMO–LUMO differences to overcome the binding energy of excitons, which is relatively high in organic materials owing to small dielectric constants and resulting strong Coulomb interactions within electron–hole pairs [74]. The latter issue can be dealt with relatively easily because HOMO and LUMO levels can be measured (e.g., using cyclic voltammetry) and suitable donor–acceptor couples identified [75]. However, control of the morphology of active layers is demanding [76]. Phase separation between the donor and acceptor cannot be directly controlled in the BHJ approach, and full control is possible only if one of the materials is prestructured and does not lose its structure when infiltrated with the other active material.

As mentioned earlier, in the BHJ approach, complete exciton separation can be achieved for very small donor and acceptor domains, but the probability of islands forming, as depicted in Figure 14.4a, is high. On the other hand, if phase separation is strong and domains are large enough to provide consistent pathways for charge transport, exciton recombination becomes a severe loss channel. If the domain size exceeds the exciton diffusion length, not all excitons can reach a separating interface, and the absorbed photon is effectively lost. In DSSCs, exciton separation and charge transport are decoupled, which is supposed to be the main reason why DSSCs are the most efficient excitonic solar cells to date. The absorber is only a monolayer of dye, resulting in complete separation of excitons. Electrons are then transported through a consistent connected network of TiO_2 and holes through the electrolyte, which also reaches every site in the network. However, because only a monolayer is absorbing light, thick active layers are needed and charges have to travel long distances before reaching the external electrodes. In HSCs, the morphology of the donor–acceptor interface can be directly controlled via structuring of the metal oxide, and the use of highly absorbing HTM allows reducing the active layer thickness and thus the distance for charge transport. This reduces the probability of recombination.

In general, recombination of free charge carriers appears to be the most important loss mechanism in excitonic solar cells. Intense research in the past decades has yielded highly absorbing material combinations with outstanding exciton-separating properties, meaning that loss mechanisms (1) and (2) can be effectively suppressed. However, recombination of already separated free charge carriers is the least understood loss mechanism, and reduction of charge recombination is difficult. Charge separation lowers the photocurrent and the open-circuit potential of solar cells and can also cause net shunts in the active layer, reducing the rectifying quality of the photodiode and the fill factor (FF) of the solar cell (FF is the ratio of $I_{SC} \times V_{OC}$ divided by the product of the current and the voltage at the point of maximum power output of the solar cell) [20].

In dye-based solar cells, the recombination mechanism depends strongly on the choice of the dye, which determines the properties of the charge-separating interface. As mentioned before, electron injection from the dye into the metal oxide is typically fast. However, if the hole remains for too long in the dye, that is, close to the interface, the probability of recombination with free electrons inside the metal oxide increases. Dyes that can be quickly regenerated are therefore desirable. In SS-DSSCs and HSCs, dye regeneration is typically much quicker than in DSSCs, but recombination of holes from the HTM with electrons in the metal oxide is in turn more likely than for the I^-/I_3^- system of DSSCs since only one electron is involved. Therefore, recombination is also relatively quick and occurs on the microsecond time scale [77]. Physical spacers such as alkyl chains coadsorbed to the metal oxide or directly attached to the dye as in the case of Z907 have been shown to enhance V_{OC} and reduce recombination [44, 78]. Another more sophisticated approach is the use of donor–acceptor dyads [79]. These molecules utilize an electron-donating domain close to the dye chromophore, and thus introduce an additional energy step at the interface. Subsequent to electron injection into the metal oxide, the chromophore is instantly regenerated by the donor domain of the dyad. Accordingly, the hole is located further away from the interface until the dyad is regenerated by the HTM. Owing to this increased spacing between electrons and holes, recombination can be slowed down significantly. Using dyads, recombination times as slow as seconds have been realized for DSSCs [80].

Besides the donor–acceptor interface, the charge transporting materials also play a crucial role for recombination. High mobilities for both electrons and holes are considered to reduce charge-carrier recombination by quickly removing charges from the region close to the interface where recombination occurs. In order to avoid the buildup of "charge jams" at the interface, these mobilities should also be balanced. If one type of carriers is transported away from the interface only slowly, space charge regions build up and recombination is promoted [81].

HSCs offer the potential to control charge-carrier recombination, especially in the case of ordered metal oxide nanostructures. As discussed earlier, high-charge carrier mobilities for both electrons and holes can be realized in ordered structures. Through the use of dyes, absorption and charge transport are partly decoupled, and the dye can be utilized as an additional recombination barrier between the metal oxide and HTM. Besides, dye and HTM can be chosen independently, which allows fine-tuning of the interfacial properties.

14.3
Nanostructures for Hybrid Solar Cells

This section deals with different types of structures for HSCs. First, we give a short overview of SS-DSSCs and HSCs based on unordered metal oxides, commonly processed from nanoparticles. We then turn to ordered metal oxides, namely nanowire and nanotubular structures, and give a short overview of core–shell structures for HSCs.

14.3.1
Nanoparticle–Polymer Blends and Unordered Metal Oxide Structures

As discussed in Section 14.2, the first HSCs were based on the same concept as that of SS-DSSCs, with Spiro-OMeTAD being replaced by an absorbing HTM. The obvious aim of this attempt was enhancement of absorption, which is relatively low in SS-DSSCs when compared to their liquid electrolyte counterparts because of thinner active layers [82]. This reduced absorption is the main reason for the currently significantly lower PCEs of SS-DSSCs compared to DSSCs, which are on the order of 5% [83–86]. In HSCs, the mesoporous metal oxide layer is typically synthesized by sintering TiO_2 nanoparticles of 10–30-nm diameter processed using sol–gels. The resulting structures are similar to those used in DSSCs, but the active layers are typically 10 times thinner and on the order of 1–3 μm. These mesoporous films exhibit porosities of about 60% and a much larger surface area than flat films.

Remarkable efficiencies of 1.5% have been realized in HSCs based on mesoporous TiO_2 and P3HT as HTM by Coakley and McGehee [87] already in 2003. In this case, no dye sensitization of TiO_2 is used, but the charge-separating interface consisted only of P3HT and TiO_2. Additives such as 4-*tert*-butylpyridine and Li salts, which are known to improve I_{SC}, V_{OC}, and, accordingly PCE, have allowed further improvements also in HSCs [77]. In 2009, efficiencies of 2.6% were realized with the same concept of mesoporous TiO_2 and P3HT as HTM by using highly absorbing metal-free dyes as sensitizers, which improve the absorption and the properties of the charge-separating interface [88]. To date, HSCs based on this concept deliver around 3% PCE [89, 90].

Another approach to realize active layers for HSCs with large but unordered interfaces of donor and acceptor is to directly blend metal oxide nanoparticles with organic HTMs. Thus, an inorganic BHJ is realized. Although percolated pathways for charge transport cannot be ensured because the metal oxide is not sintered to one consistent network, nanoparticle–polymer blends inherit certain advantages. The active layer can be solution-processed, which is potentially cost effective. Besides, complete filling of the metal oxide with the polymer, and improved contact between donor and acceptor compared to infiltrated mesoporous structures can be ensured [91].

So far, best efficiencies for hybrid blends have been shown using PbS and CdSe as inorganic acceptors. Huynh *et al.* [92] reported 1.7% PCE for blends of CdSe nanorods and P3HT. Both materials contain toxic compounds and are therefore not desirable for large-scale production in contrast to nontoxic ZnO and TiO_2. Using ZnO nanoparticles with PPV or P3HT, Beek *et al.* [37, 38] were able to show efficiencies of 1.4 and 0.9%, respectively. There are also reports on surface-treated TiO_2 particles with P3HT, but the PCEs are even lower [93]. The main reason for the relatively low efficiencies in this approach is the nonideal charge transport through the nanoparticles, which are not continuously electronically connected. Mixing ratios of nanoparticles and polymer up to 90% particles are necessary in order to achieve reasonable electron transport, and the unordered morphology of the hybrid BHJ always contains islands of donor or acceptor, which trap charges and promote charge-carrier recombination.

14.3.2
Metal Oxide Nanowires

One promising approach to realize ordered interpenetrating donor–acceptor interfaces, as in Figure 14.4b, is the use of metal oxide nanowires. These structures can be grown utilizing self-organization processes and exhibit desirable crystallinity properties, high regularity, and excellent parallel orientation. A detailed description of the application of metal oxide nanowires for energy conversion can be found in the reviews by Hochbaum and Yang [94] or by Gonzalez-Valls and Lira-Cantu [95].

By far, most work has been conducted on TiO_2 and ZnO nanowires, but other metal oxides have not been investigated that extensively to date. One of the most appealing synthesis approaches for TiO_2 wires is the so-called hydrothermal growth [96]. For this method, substrates are immersed in a vessel containing a highly concentrated precursor solution, and the vessel is closed and placed in an oven. Under conditions of high pressure and temperatures up to 200 °C, single-crystal nanowires grow as a result of different crystal growth velocities in different crystal directions: the fastest growing direction determines the z-direction of the nanowire, provided the wire's crystal lattice matches with the crystal lattice of the substrate [97].

TiO_2 typically grows as rutile single crystals. Rutile nanowires can be realized directly on FTO glass because of the FTO's preferential crystal structure [98]. Impressive PCEs of 6.9% have been realized using 33 μm long-rutile nanowires in DSSCs, and much smaller structures of only 2–3 μm already allow PCEs of 5%, pointing out the high potential of these structures for solar cell applications [97, 98]. The surface area of these wires and thus the dye loading is relatively low for these nanowires compared to mesoporous films, but high efficiencies can be reached owing to the high-electron mobilities in the single-crystalline metal oxide and to the ordered geometry.

Hydrothermal growth is an equilibrated process, which also allows homogenous doping of the resulting structures by adding dopants to the hydrothermal bath. By incorporation of Ta into the TiO_2 lattice, Feng *et al.* [99] were able to increase the V_{OC} of DSSCs by 15%, reaching 0.87 V, which is close to the theoretical maximum. Doping is also discussed in Section 14.4.2.

Only limited work has been done on HSCs based on TiO_2 rutile single-crystal nanowires, probably because application of these structures in solar cells is a relatively new approach. Besides, the electronic properties of rutile TiO_2 are supposed to be less ideal in combination with organic HTMs than anatase. Self-assembled monolayers of surface modifiers such as dyes could help overcome this challenge.

Another important issue in HSCs is the backfilling of the nanostructures with organic HTM. Complete coverage of the nanostructure with the HTM is necessary in order to take advantage of the whole charge-separating interface, whereas incomplete filling and formation of nonfilled voids promotes exciton recombination. Figure 14.8 shows cross-sectional scanning electron microscopy (SEM) images of P3HT infiltrated into rutile single-crystal nanowires. Figure 14.8a displays incomplete filling as apparent from air inclusions between the wires. In this case, the structure was wetted with a solution of P3HT in chlorobenzene and immediately spin-coated. Between wetting the sample and waiting for a few minutes before spin coating, there is sufficient time for the material to diffuse into the structure, and the filling is improved as shown in Figure 14.8b.

Besides hydrothermal growth, TiO_2 nanowires have been synthesized using template-assisted approaches. Most prominent is the use of anodized aluminum

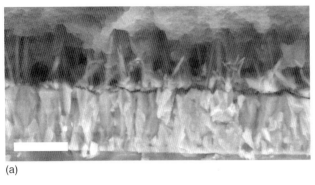

(a)

(b)

Figure 14.8 SEM cross-sectional views of TiO₂ rutile single-crystal nanowires on FTO substrates infiltrated with P3HT. Scale bars correspond to 1 μm. The images illustrate (a) incomplete filling of the nanowires and (b) almost ideal infiltration of polymer.

oxide (AAO) membranes. AAOs can be grown by anodizing Al foils or thin films of Al on supporting substrates via anodization, that is, application of an oxidizing potential in an Al_2O_3-dissolving electrolyte such as sulfuric or oxalic acid [100–102]. These structures exhibit pores of almost arbitrary length (mainly depending on the thickness of the Al feed substrate) and pore diameters between 20 and 300 nm. Infiltration of TiO_2 into these pores and subsequent removal of the Al_2O_3 yields free-standing nanowires, which can be utilized in HSCs. By filling 200°nm thick AAO membranes via a sol–gel method, Kuo et al. [103] were able to realize TiO_2 nanowires directly on ITO substrates. The resulting HSCs showed superior photocurrent over flat junction devices, but V_{OC} was dramatically reduced. This is attributed to regions between the nanowires where P3HT is in direct contact with ITO. In this case, the lack of a hole-blocking layer leads to enhanced charge-carrier recombination.

However, sputtering Al on TiO_2-coated ITO and subsequent anodization yields AAO membranes also directly on thin hole-blocking TiO_2. The AAO can also be filled via electrodeposition of TiO_2 from a $TiCl_3$ precursor, as described by Musselman et al. [104]. In their report, AAOs were grown on thin Ti and W adhesion layers. However, it is also possible to directly grow AAOs on thin TiO_2 blocking layers on ITO as shown in Figure 14.9a and fill these structures with TiO_2 via electrodeposition. The resulting free-standing TiO_2 wires on a dense TiO_2 film after removal of the AAO in NaOH solution are shown in Figure 14.9b,c.

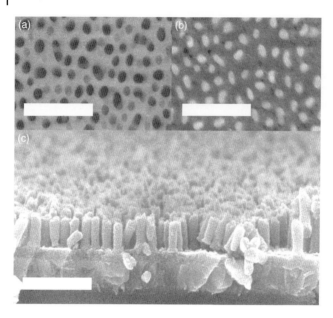

Figure 14.9 Growth of TiO$_2$ nanowires using electrodeposition of TiO$_2$ into anodized aluminum oxide (AAO) membranes. (a) SEM top view of a typical thin film AAO membrane on a TCO substrate. (b) SEM top view of TiO$_2$ nanowires grown in AAO pores after heat-curing at 450 °C and removal of the AAO in sodium hydroxide solution. (c) SEM cross-sectional view of free-standing TiO$_2$ nanowires on a thin TiO$_2$ compact layer on a TCO substrate. Scale bars are 500 nm.

The additional blocking layer underneath the nanowires ensures that the HTM is not in direct contact with the ITO when fabricating HSCs. Thus, the metal oxide electrode is electron-selective and charge-carrier recombination can be reduced.

As for TiO$_2$, hydrothermal growth is also the most prominent synthesis method for ZnO nanowires. However, in contrast to TiO$_2$, ZnO wire growth most commonly is a two-step process. First, a seed layer is deposited on the substrate, typically by spin-coating a monolayer of ZnO nanoparticles. During the subsequent hydrothermal growth, these particles serve as nucleation sites for the wire growth [105]. In contrast to TiO$_2$ wire synthesis, temperatures below 100 °C would suffice for ZnO wire growth. This fact and the possibility to control quality and dimensions of the wires via a choice of synthesis conditions make these structures extremely attractive for potential low-cost mass production [106, 107].

To date, excitonic solar cells based on ZnO nanowires exhibit only relatively low performances. A PCE of 1.5% was reported for liquid electrolyte DSSCs based on ZnO nanowires by Law *et al.* [108]. Although these efficiencies are much lower than for mesoporous TiO$_2$, the high potential of the more ordered nanowires has been convincingly pointed out. Comparison of solar cells based on active layers with different thicknesses showed that increasing photocurrents are found for increasing ZnO wire lengths, whereas saturation of the photocurrent is observed if the mesoporous TiO$_2$ layer exceeds a certain thickness. This suggests almost loss-free electron transport in the case of the nanowires, whereas charge transport seems to limit the efficiency for mesoporous films. Besides, dye loading is much smaller for nanowires than for mesoporous films, resulting in less efficient light harvesting. This renders these structures especially interesting for applications in HSCs, where the HTM contributes to the device absorption.

Nevertheless, highly efficient HSCs based on ZnO nanowires still have to be shown. So far, for P3HT, PCEs between 0.2 and 0.5% have been demonstrated. [109–111]. In order to achieve higher efficiencies up to 3%, it was necessary to add organic acceptors such as Phenyl-C_{61}-butyric acid methyl ester (PCBM) [112, 113]. However, this is attributed to nonideal wire dimensions rather than to nonideal properties as n-type material. ZnO nanowires allow the realization of an ordered interpenetrating donor–acceptor heterojunction and exhibit outstandingly high-electron mobilities, but the geometry of the wires themselves is not optimized yet. ZnO nanowires exhibit relatively low surface areas compared to mesoporous films, and wire diameters and spacings are in the order of 100 nm. Since this is about 10 times higher than typical exciton diffusion lengths, this results in prominent losses as a result of exciton recombination, which is supposed to be the main reason why additional organic acceptors can help in significantly improving the PCE of ZnO nanowire-based HSCs.

14.3.3
Metal Oxide Nanotubes

Besides nanowires, the most widely investigated ordered metal oxide nanostructures for solar cell applications are nanotubes. Arrays of nanotubes grown perpendicular to the substrate exhibit the same advantage of directed charge transport as with nanowires. Besides, nanotubes exhibit an additional surface at the inside of the tube, resulting in a significantly increased surface area depending on the tube wall thickness. For DSSCs, SS-DSSCs, and HSCs, large-area metal oxide structures are desirable, making nanotube arrays highly attractive as electron acceptors.

Most work in this field has focused on TiO_2. Details about synthesis and application of these structures can also be found in the reviews of Mor *et al.* [114], Ghicov and Schmuki [115], and Roy *et al.* [116]. TiO_2 nanotubes can be directly grown by anodization of metallic Ti in fluoride-ion-containing solutions, most prominently aqueous HF or NH_4F in ethylene glycol. Using HF, the structure length is limited because of pronounced corrosion of the structures, that is, structures grow at the bottom while being dissolved at the top [117, 118]. However, using more viscous ethylene glycol and less aggressive NH_4F, aspect ratios above 2000 have been achieved, and structure lengths up to millimeters can be reached by completely anodizing the substrate, which typically is a high-purity Ti foil [119, 120]. Besides anodization of foils, it is also possible to sputter-deposit metallic Ti on arbitrary substrates for subsequent anodization. Thus, it is possible to realize nanotubular structures on TCO substrates, such as FTO or ITO with good control over the tube length, diameter, spacing, and tube wall thickness [121, 122]. An SEM cross-sectional image of a typical TiO_2 nanotube array on ITO synthesized by anodization of sputtered Ti in NH_4F containing ethylene glycol is shown in Figure 14.10. This geometry offers the advantage to fabricate solar cells that can be illuminated from the front side, that is, through the TCO. When using anodized Ti foils, instead, solar cells have to be illuminated through the top electrode, which commonly is only semitransparent. As reported by Paulose *et al.* [123] for DSSCs based on anodized Ti foils, this results in reduced PCE owing to reduced light intensity in the active material.

The high potential of TiO_2 nanotubes for application in excitonic solar cells was demonstrated by Mor *et al.* in 2006 [124]. In their study, only 360 nm long-TiO_2 nanotube arrays were grown on FTO and used in DSSCs. Despite the relatively low absorption because of the thin active layers and the generally smaller surface area of tubes compared to mesoporous structures, they were able to realize photocurrents approaching $8\,mA\,cm^{-2}$ and PCEs of 2.9%. These impressive values could be directly attributed to the superior electron conducting properties of nanotubes compared to conventional mesoporous TiO_2.

Three years later, Mor *et al.* [39] were able to show even higher performances with HSCs based on TiO_2 nanotubes. Using a near-IR-absorbing metal-free squaraine dye and P3HT as HTM in 700 nm long nanotubes of about 35 nm diameter, they achieved PCEs up to 3.8%, which is still the record efficiency for metal oxide–dye–polymer HSCs. Nevertheless, slightly higher efficiencies have been shown by the same group when infiltrating blends of P3HT and PCBM into TiO_2 nanotubes, that is, a mixture of organic donor and acceptor, instead of only P3HT [125]. This suggests that charge separation between P3HT as HTM and the TiO_2 is still not optimized and that further fine-tuning of the interface is necessary in order

Figure 14.10 SEM cross-sectional view of TiO_2 nanotubes grown on a TCO. Ti is sputter-deposited on a TCO coated with a thin compact layer of TiO_2 and anodized in NH_4F-containing ethylene glycol. Scale bar is 500 nm.

to optimize HSCs. Additionally, reduced exciton recombination might account for the better efficiencies in the case of an additional organic acceptor.

Besides direct growth via anodization of metallic Ti, there are also template-assisted approaches to fabricate TiO_2 nanotubes. An interesting synthesis route was introduced by Foong *et al.* in 2010 [126]. The method is sketched in Figure 14.11a. AAO membranes on thin TiO_2 blocking layers on ITO are grown via anodization of sputtered Al films (1 and 2). Using atomic layer deposition (ALD), homogenous coatings of a few nanometers thick TiO_2 are grown on these structures (3), and the resulting overlayer of TiO_2 is removed via reactive ion etching or bombardment with Ar plasma (4). Finally, the AAO membrane is dissolved in NaOH solution,

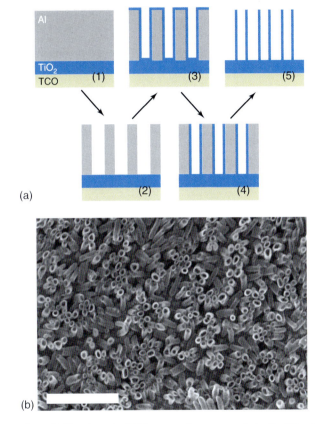

(a)

(b)

Figure 14.11 Growth of TiO_2 nanotubes on a TCO via atomic layer deposition (ALD) onto anodized aluminum oxide (AAO) membranes. (a) Schematic of the synthesis route. (1) A layer of aluminum is sputter-deposited onto a TiO_2-coated TCO and (2) anodized to yield a nanoporous structure. (3) The structure is coated with TiO_2 via ALD and (4) the overlayer of TiO_2 on top of the AAO is removed via Ar sputtering. (5) Finally, the AAO is removed in sodium hydroxide solution to yield free-standing nanotubes. (b) SEM top view of TiO_2 nanotubes grown via ALD into AAO (scale bar corresponds to 500 nm).

yielding free-standing TiO$_2$ nanotubes directly on the dense TiO$_2$ blocking layer (5). An SEM top view image of TiO$_2$ nanotubes grown via this synthesis route is shown in Figure 14.11b.

HSCs based on thus grown nanotubes and P3HT as HTM have also been presented in the study by Foong *et al.*, but comparison with flat junction cells and more detailed investigation of this system are necessary in order to reveal possible advantages of these nanotubes over flat films or mesoporous layers owing to increased interfacial area and directed charge transport.

Another template-assisted approach is the coating of nanowires (instead of nanopores) with TiO$_2$ and subsequent removal of the core of the structure. This has been demonstrated for ZnO nanowires [127]. As discussed before, these structures can be easily grown using hydrothermal methods or electrodeposition. Coating with TiO$_2$ can then be achieved via ALD, using sol–gels, or with a simple treatment in TiCl$_4$ solution, as shown by Muduli *et al.* [128].

In contrast to TiO$_2$, there are only a few studies about ZnO nanotubes. Application of nanotubular ZnO structures in DSSCs has been reported by Martinson *et al.* [129]. In this case, structures were synthesized via ALD onto AAO membranes, similar to the TiO$_2$ nanotubes discussed before. Besides, it has been shown that highly regular hexagonal ZnO nanotubes can be grown directly on Zn substrates via low-temperature liquid-phase processes [130]. Growth of extremely thin ZnO tubes has also been demonstrated directly on ZnO compact layers on Si substrates [131].

14.4
Metal Oxide Modifications

In our last section, we give short insights into the possibilities to adjust the properties of metal oxide nanostructures for HSCs. Because of the stepwise fabrication, where the electron-accepting metal oxide is synthesized first and then infiltrated with the organic HTM, fine-tuning of the metal oxide surface is possible via various chemical and physical treatments. Besides, doping of metal oxide nanostructures can influence electronic properties such as energy levels and charge-carrier mobilities.

14.4.1
Surface Treatments of Metal Oxide Nanostructures

After more than two decades of intense research on excitonic solar cells, it is a well-established fact that the donor–acceptor interface plays a crucial role in the working mechanisms of the device. In fully organic BHJ solar cells, where the active layer is synthesized from blended solutions of donor and acceptor molecules, this interface can be influenced only indirectly by changing the properties of the involved materials and via choice of the solvent. For HSCs, however, the metal oxide nanostructure is synthesized first and can be modified before the HTM is added to the system. Therefore, not only the geometry but also the electronic properties of the charge-separating interface can be adjusted to optimize I_{SC} and V_{OC} and reduce exciton- and charge-carrier recombination.

One of the most common modifications of all kinds of TiO_2 nanostructures is a treatment in an aqueous $TiCl_4$ solution. Ti^{4+} (or $Ti(OH)_2^{2+}$ in aqueous solutions) is supposed to fill vacancies in the TiO_2 lattice and repair small cracks and defects [132]. After immersion in a $TiCl_4$ bath, structures are heat-treated to form fully oxidized TiO_2 species. Typically, the dye loading, that is, the amount of dye molecules that can attach to the nanostructure, also increases upon $TiCl_4$ treatment. For ZnO, no such treatment is established yet. However, as pointed out by Musselman *et al.* [133], addition of Zn salts to synthesis baths can improve the stability of ZnO in low or high-pH solutions, thus preserving the nanostructure.

Surface treatments do not only change the properties of the metal oxide itself but also affect the inorganic–organic interface and accordingly solar cell properties such as charge separation and recombination. Suitable choice of sensitizers leads to significantly reduced charge carrier recombination and helps to increase the V_{OC}. Such effects have been shown for mesoporous TiO_2, TiO_2 nanocrystal BHJ cells, flat-junction HSCs, and HSCs based on TiO_2 nanotubes [39, 56, 93, 134]. Additionally, it is possible to enhance the efficiency of exciton separation and accordingly the photocurrent. However, to a certain extent, there seems to be a trade-off between V_{OC} and I_{SC}: if the driving force for charge separation is increased, this typically comes at the price of a lower potential of the final energetic state. Liu *et al.* [135] were able to modify the energy levels of planar TiO_2 by the introduction of self-assembled monolayers of carboxylated polythiophenes. Although this led to improved exciton separation yield, interfacial dipoles reduced the V_{OC} because of a resulting shift in band offset.

For HSCs based on P3HT, a very interesting sensitizer has been introduced by Vaynzof *et al.* [136]. They were able to show positive effects on I_{SC} and V_{OC} when using carboxylated C_{60} fullerenes as surface modifiers in flat-junction ZnO–P3HT solar cells. Similar results have also been found for TiO_2–P3HT [56]. Fullerenes are known to efficiently accept electrons from various p-type polymers and seem to introduce a beneficial energy step for charge separation between metal oxides and P3HT.

Besides modification with self-assembled monolayers, that is, sensitization, there are also well-established TiO_2 treatments with Li salts and 4-*tert*-butylpyridine (TBP). Insertion of Li atoms into the TiO_2 lattice is supposed to shift the conduction band downwards, thus increasing the driving force for charge separation at the cost of the open-circuit potential. In contrast, TBP increases the V_{OC}. Combination of these two has been conclusively shown to increase both V_{OC} and I_{SC}, and has been used in flat-junction HSCs as well as in HSCs based on mesoporous TiO_2 and TiO_2 nanotubes [39, 55, 137].

14.4.2
Doping of Metal Oxide Nanostructures

Another way of adjusting the electronic properties of metal oxide nanostructures is doping with foreign atoms. Doping can shift the conduction and valence bands of the metal oxide and lead to increased charge-carrier mobilities. This can enable more efficient charge separation as well as reduced recombination owing to the beneficially quick charge transport.

Almost doubled V_{OC}, on doping with Mg, was shown by Olson *et al.* [110] for ZnO–P3HT HSCs. Addition of Mg and formation of a $ZnMgO_2$ alloy leads to a prominently reduced band offset and allows a higher potential in HSCs. On addition of 0–25% Mg, the effective work function of the metal oxide changes from -4.2 to -3.9 eV, which results in up to 0.5 V higher V_{OC}. Similar effects have been reported for TiO_2 when using Ta or N as dopants [99, 138]. N-doping is also known to enhance the photoactivity of TiO_2 in the visible, that is, TiO_2 absorption can be extended to longer wavelengths, as shown in the case of nanotubular TiO_2 by Vitiello *et al.* [139].

14.5
Conclusion and Outlook

Hybrid photovoltaics consisting of nanostructured metal oxides and highly absorbing organic donor materials are a promising technology to compete with other excitonic solar cells. They offer the potential to combine the advantages of both fully organic BHJ photovoltaics and Grätzel cells. Using highly absorbing organic hole conductors and decoupling electron transport and charge separation, it should be possible to realize highly efficient thin-film solid-state devices, possibly even on flexible substrates. By utilizing ordered metal oxide nanostructures to determine the heterojunction geometry, complete exciton separation and almost loss-free charge transport can be reached, which has already been outlined by impressive studies on ordered structures of ZnO and TiO_2.

List of Abbreviations

ALD	Atomic layer deposition
AAO	Anodized aluminum oxide
BHJ	Bulk heterojunction
DSSC	Dye-sensitized solar cell
FTO	Fluorine-doped tin oxide
HOMO	Highest occupied molecular orbital
HSC	Hybrid solar cell
HTM	Hole transporting medium
ITO	Indium–tin oxide

LUMO   Lowest unoccupied molecular orbital
OPV   Organic photovoltaic
PCE   Power conversion efficiency
SS-DSSC Solid-state dye-sensitized solar cell
TCO   Transparent conductive oxide

References

1. BP (2010) Statistical Review of World Energy, *www.bp.com/liveassets/ bp_internet/globalbp/globalbp_uk_english/ reports_and_publications/statistical_ energy_review_2008/STAGING/local_ assets/2010_downloads/Statistical_Review_ of_World_Energy_2010.xls.*

2. Holdren, J.P. (2008) Science and technology for sustainable well-being. *Science*, **319** (5862), 424.

3. DESERTEC (2009) *Clean Power from Deserts*, 4th edn, Protext Verlag, Bonn.

4. R.F. Keeling, *et al.* 2008, Atmospheric CO_2-Curve Values (ppmv) Derived From Flask Air Samples Collected at the South Pole, *http://cdiac.ornl.gov/ftp/trends/co2/ sposio.co2* (accessed 3 January 2013).

5. Neftel, A. *et al.* (1994) Historical CO_2 Record from the Siple Station Ice Core, *http://cdiac.esd.ornl.gov/ftp/trends/co2/ siple2.013,* (accessed 3 January 2013).

6. NASA (2011) Combined Land-Surface Air and Sea-Surface Water Temperature Anomalies, *http://data.giss.nasa.gov/gistemp/ tabledata/GLB.Ts+dSST.txt* (accessed 3 January 2013).

7. Fthenakis, V. and Alsema, E. (2006) Photovoltaics energy payback times, greenhouse gas emissions and external costs: 2004−early 2005 status. *Prog. Photovoltaics Res. Appl.*, **14** (3), 275−280.

8. European Commission (2003) External Costs. Research Results on Socio-Environmental Damages Due to Electricity and Transport.

9. ACIAR Program (2004) Coal in a Sustainable Society.

10. Yamaguchi, M. (2003) III-V compound multi-junction solar cells: present and future. *Sol. Energy Mater. Sol. Cells*, **75** (1−2), 261−269.

11. Krebs, F.C. (2008) Air stable polymer photovoltaics based on a process free from vacuum steps and fullerenes. *Sol. Energy Mater. Sol. Cells*, **92** (7), 715−726.

12. Weickert, J. *et al.* (2010) Spray-deposited PEDOT:PSS for inverted organic solar cells. *Sol. Energy Mater. Sol. Cells*, **94** (12), 2371−2374.

13. Bach, U. *et al.* (1998) Solid-state dye-sensitized mesoporous TiO_2 solar cells with high photon-to-electron conversion efficiencies. *Nature*, **395** (6702), 583−585.

14. O'Regan, B. and Gratzel, M. (1991) A low-cost, high-efficiency solar-cell based on dye-sensitized colloidal TiO_2 films. *Nature*, **353** (6346), 737−740.

15. Hagfeldt, A. *et al.* (2010) Dye-sensitized solar cells. *Chem. Rev.*, **110** (11), 6595−6663.

16. Snaith, H.J. and Schmidt-Mende, L. (2007) Advances in liquid-electrolyte and solid-state dye-sensitized solar cells. *Adv. Mater.*, **19** (20), 3187−3200.

17. Yella, A. *et al.* (2011) Porphyrin-sensitized solar cells with cobalt (II/III)−based redox electrolyte exceed 12 percent efficiency. *Science*, **334** (6056), 629−634.

18. Yum, J.H. *et al.* (2012) A cobalt complex redox shuttle for dye-sensitized solar cells with high open-circuit potentials. *Nat. Commun.*, **3**, 631.

19. Schmidt-Mende, L. and Grätzel, M. (2006) TiO_2 pore-filling and its effect on the efficiency of solid-state dye-sensitized solar cells. *Thin Solid Films*, **500** (1−2), 296−301.

20. Hoppe, H. and Sariciftci, N.S. (2004) Organic solar cells: an overview. *J. Mater. Res.*, **19** (7), 1924−1945.

21. Wöhrle, D. and Meissner, D. (1991) Organic solar cells. *Adv. Mater.*, **3** (3), 129−138.

22. Tang, C.W. (1986) 2-layer organic pho-tovoltaic cell. *Appl. Phys. Lett.*, **48** (2), 183–185.

23. Shaheen, S.E. *et al.* (2001) 2.5% efficient organic plastic solar cells. *Appl. Phys. Lett.*, **78**, 841.

24. Dennler, G., Scharber, M.C., and Brabec, C.J. (2009) Polymer-fullerene bulk-heterojunction solar cells. *Adv. Mater.*, **21** (13), 1323–1338.

25. Hoppe, H. and Sariciftci, N.S. (2008) *Photoresponsive Polymers II*, Vol. 214, Springer, Berlin, Heidelberg, pp. 1–86.

26. Liang, Y. and Yu, L. (2010) A new class of semiconducting polymers for bulk heterojunction solar cells with excep-tionally high performance. *Acc. Chem. Res.*, **43**, 1127–1236.

27. Peet, J. *et al.* (2007) Efficiency enhance-ment in low-bandgap polymer solar cells by processing with alkane dithiols. *Nat. Mater.*, **6** (7), 497–500.

28. Park, S. *et al.* (2009) Bulk heterojunc-tion solar cells with internal quantum efficiency approaching 100%. *Nat. Photonics*, **3** (5), 297–302.

29. Peumans, P., Uchida, S., and Forrest, S.R. (2003) Efficient bulk hetero-junction photovoltaic cells using small-molecular-weight organic thin films. *Nature*, **425** (6954), 158–162.

30. Yang, X. and Loos, J. (2007) Toward high-performance polymer solar cells: the importance of morphology control. *Macromolecules*, **40** (5), 1353–1362.

31. Pisula, W. *et al.* (2006) Pronounced supramolecular order in discotic donor–acceptor mixtures. *Angew. Chem. Int. Ed.*, **45** (5), 819–823.

32. Chou, S.Y., Krauss, P.R., and Renstrom, P.J. (1995) Imprint of sub 25 nm vias and trenches in polymers. *Appl. Phys. Lett.*, **67**, 3114.

33. Chou, S.Y., Krauss, P.R., and Renstrom, P.J. (1996) Imprint lithography with 25-nanometer resolution. *Science*, **272** (5258), 85.

34. Hesse, H.C. *et al.* (2010) Discotic ma-terials for organic solar cells: effects of chemical structure on assembly and performance. *Sol. Energy Mater. Sol. Cells*, **94** (3), 560–567.

35. Schmidt-Mende, L. *et al.* (2001) Self-organized discotic liquid crystals for high-efficiency organic photovoltaics. *Science*, **293** (5532), 1119–1122.

36. Weickert, J. *et al.* (2011) Nanostructured organic and hybrid solar cells. *Adv. Mater.*, **23** (16), 1810–1828.

37. Beek, W.J.E., Wienk, M.M., and Janssen, R.A.J. (2004) Efficient hybrid solar cells from zinc oxide nanoparticles and a conjugated polymer. *Adv. Mater.*, **16** (12), 1009–1013.

38. Beek, W.J.E., Wienk, M.M., and Janssen, R.A.J. (2006) Hybrid solar cells from regioregular polythiophene and ZnO nanoparticles. *Adv. Funct. Mater.*, **16** (8), 1112–1116.

39. Mor, G. *et al.* (2009) Visible to near-infrared light harvesting in TiO_2 nanotube array- P3HT based hetero-junction solar cells. *Nano Lett.*, **9** (12), 4250–4257.

40. Nazeeruddin, M.K. *et al.* (1993) Conver-sion of light to electricity by *cis*-X2*bis* (2, 2′-bipyridyl-4, 4′-dicarboxylate) ruthe-nium (II) charge-transfer sensitizers (X = Cl-, Br-, I-, CN-, and SCN-) on nanocrystalline titanium dioxide elec-trodes. *J. Am. Chem. Soc.*, **115** (14), 6382–6390.

41. Horiuchi, T., Miura, H., and Uchida, K. (2003) Highly-efficient metal-free organic dyes for dye-sensitized solar cells. *Chem. Commun.*, **24**, 3036–3037.

42. Grätzel, M. (2009) Recent advances in sensitized mesoscopic solar cells. *Acc. Chem. Res.*, **42** (11), 1788–1798.

43. Nazeeruddin, M.K. *et al.* (2001) En-gineering of efficient panchromatic sensitizers for nanocrystalline TiO_2-based solar cells. *J. Am. Chem. Soc.*, **123** (8), 1613–1624.

44. Schmidt-Mende, L., Zakeeruddin, S.M., and Gratzel, M. (2005) Efficiency im-provement in solid-state-dye-sensitized photovoltaics with an amphiphilic Ruthenium-dye. *Appl. Phys. Lett.*, **86** (1), 013504.

45. Horiuchi, T. *et al.* (2004) High effi-ciency of dye-sensitized solar cells based on metal-free indoline dyes. *J. Am. Chem. Soc.*, **126** (39), 12218–12219.

46. Schmidt-Mende, L. *et al.* (2005) Organic dye for highly efficient solid state dye sensitized solar cells. *Adv. Mater.*, **17** (7), 813–815.

47. Asbury, J.B. *et al.* (1999) Femtosecond IR study of excited-state relaxation and electron-injection dynamics of Ru (dcbpy) 2 (NCS) 2 in solution and on nanocrystalline TiO$_2$ and Al2O$_3$ thin films. *J. Phys. Chem. B*, **103** (16), 3110–3119.

48. Ramakrishna, G. *et al.* (2005) Strongly coupled ruthenium–polypyridyl complexes for efficient electron injection in dye-sensitized semiconductor nanoparticles. *J. Phys. Chem. B*, **109** (32), 15445–15453.

49. Kuang, D. *et al.* (2006) High molar extinction coefficient heteroleptic ruthenium complexes for thin film dye-sensitized solar cells. *J. Am. Chem. Soc.*, **128** (12), 4146–4154.

50. He, J. *et al.* (2002) Modified phthalocyanines for efficient near-IR sensitization of nanostructured TiO$_2$ electrode. *J. Am. Chem. Soc.*, **124** (17), 4922–4932.

51. Hagfeldt, A. and Grätzel, M. (2000) Molecular photovoltaics. *Acc. Chem. Res.*, **33** (5), 269–277.

52. Clifford, J.N. *et al.* (2004) Molecular control of recombination dynamics in dye-sensitized nanocrystalline TiO$_2$ films: free energy vs distance dependence. *J. Am. Chem. Soc.*, **126** (16), 5225–5233.

53. Snaith, H.J. *et al.* (2009) Charge generation and photovoltaic operation of solid state dye sensitized solar cells incorporating a high extinction coefficient indolene based sensitizer. *Adv. Funct. Mater.*, **19** (11), 1810–1818.

54. O'Regan, B.C. and Durrant, J.R. (2009) Kinetic and energetic paradigms for dye-sensitized solar cells: moving from the ideal to the real. *Acc. Chem. Res.*, **42** (11), 1799–1808.

55. Goh, C., Scully, S.R., and McGehee, M.D. (2007) Effects of molecular interface modification in hybrid organic–inorganic photovoltaic cells. *J. Appl. Phys.*, **101** (11), 114503.

56. Weickert, J. *et al.* (2011) Characterization of interfacial modifiers for hybrid solar cells. *J. Phys. Chem. C*, **115** (30), 15081–15088.

57. Dloczik, L. *et al.* (1997) Dynamic response of dye-sensitized nanocrystalline solar cells: characterization by intensity-modulated photocurrent spectroscopy. *J. Phys. Chem. B*, **101** (49), 10281–10289.

58. Bisquert, J. and Vikhrenko, V.S. (2004) Interpretation of the time constants measured by kinetic techniques in nanostructured semiconductor electrodes and dye-sensitized solar cells. *J. Phys. Chem. B*, **108** (7), 2313–2322.

59. Tiedje, T. and Rose, A. (1981) A physical interpretation of dispersive transport in disordered semiconductors. *Solid State Commun.*, **37** (1), 49–52.

60. Orenstein, J. and Kastner, M. (1981) Photocurrent transient spectroscopy: measurement of the density of localized states in a-As_ {2} Se_ {3}. *Phys. Rev. Lett.*, **46** (21), 1421–1424.

61. Adachi, M. *et al.* (2004) Highly efficient dye-sensitized solar cells with a titania thin-film electrode composed of a network structure of single-crystal-like TiO$_2$ nanowires made by the "oriented attachment" mechanism. *J. Am. Chem. Soc.*, **126** (45), 14943–14949.

62. Schlichthörl, G., Park, N., and Frank, A. (1999) Evaluation of the charge-collection efficiency of dye-sensitized nanocrystalline TiO$_2$ solar cells. *J. Phys. Chem. B*, **103** (5), 782–791.

63. Franco, G. *et al.* (1999) Frequency-resolved optical detection of photoinjected electrons in dye-sensitized nanocrystalline photovoltaic cells. *J. Phys. Chem. B*, **103** (4), 692–698.

64. Guldin, S. *et al.* (2010) Improved conductivity in dye-sensitised solar cells through block-copolymer confined TiO$_2$ crystallisation. *Energy Environ. Sci.*, **4**, 225–233.

65. Hendry, E. *et al.* (2006) Local field effects on electron transport in nanostructured TiO$_2$ revealed by terahertz spectroscopy. *Nano Lett.*, **6** (4), 755–759.

66. Poplavskyy, D., Nelson, J., and Bradley, D. (2003) Ohmic hole injection in poly(9, 9-dioctylfluorene) polymer light-emitting diodes. *Appl. Phys. Lett.*, **83**, 707.

67. Snaith, H.J. and Grätzel, M. (2006) Enhanced charge mobility in a molecular hole transporter via addition of redox inactive ionic dopant: implication to

dye-sensitized solar cells. *Appl. Phys. Lett.*, **89**, 262114.

68. Li, G. *et al.* (2005) High-efficiency solution processable polymer photovoltaic cells by self-organization of polymer blends. *Nat. Mater.*, **4** (11), 864–868.

69. Coakley, K.M. *et al.* (2005) Enhanced hole mobility in regioregular polythiophene infiltrated in straight nanopores. *Adv. Funct. Mater.*, **15** (12), 1927–1932.

70. Kastler, M. *et al.* (2007) Nanostructuring with a crosslinkable discotic material. *Small*, **3** (8), 1438–1444.

71. Aryal, M., Trivedi, K., and Hu, W.C. (2009) Nano-confinement induced chain alignment in ordered P3HT nanostructures defined by nanoimprint lithography. *ACS Nano*, **3** (10), 3085–3090.

72. Shockley, W. and Queisser, H.J. (1961) Detailed balance limit of efficiency of p n junction solar cells. *J. Appl. Phys.*, **32** (3), 510–519.

73. Vos, A.D. (1980) Detailed balance limit of the efficiency of tandem solar cells. *J. Phys. D: Appl. Phys.*, **13**, 839.

74. Clarke, T.M. and Durrant, J.R. (2010) Charge photogeneration in organic solar cells. *Chem. Rev.*, **110** (11), 6736–6767.

75. Scharber, M.C. *et al.* (2006) Design rules for donors in bulk-heterojunction solar cells—towards 10% energy-conversion efficiency. *Adv. Mater.*, **18** (6), 789–794.

76. Moule, A.J. and Meerholz, K. (2008) Controlling morphology in polymer-fullerene mixtures. *Adv. Mater.*, **20** (2), 240–245.

77. Krüger, J. *et al.* (2001) High efficiency solid-state photovoltaic device due to inhibition of interface charge recombination. *Appl. Phys. Lett.*, **79**, 2085.

78. Wang, P. *et al.* (2003) Molecular scale interface engineering of TiO$_2$ nanocrystals: improve the efficiency and stability of dye sensitized solar cells. *Adv. Mater.*, **15** (24), 2101–2104.

79. Hirata, N. *et al.* (2004) Supramolecular control of charge-transfer dynamics on dye-sensitized nanocrystalline TiO$_2$ films. *Chemistry*, **10** (3), 595–602.

80. Haque, S.A. *et al.* (2005) Supermolecular control of charge transfer in dye-sensitized nanocrystalline TiO$_2$

films: towards a quantitative structure-function relationship. *Angew. Chem. Inter. Ed.*, **44** (35), 5740–5744.

81. Mihailetchi, V.D., Wildeman, J., and Blom, P.W.M. (2005) Space-charge limited photocurrent. *Phys. Rev. Lett.*, **94** (12), 126602.

82. Skompska, M. (2010) Hybrid conjugated polymer/semiconductor photovoltaic cells. *Synth. Met.*, **160** (1–2), 1–15.

83. Wang, M.K. *et al.* (2010) Efficient and stable solid-state dye-sensitized solar cells based on a high-motar-extinction-coefficient sensitizer. *Small*, **6** (2), 319–324.

84. Wang, M.K. *et al.* (2010) Enhanced-light-harvesting amphiphilic ruthenium dye for efficient solid-state dye-sensitized solar cells. *Adv. Funct. Mater.*, **20** (11), 1821–1826.

85. Chou, P.T., Chi, Y., and Wu, C.I. (2012) Application of F4TCNQ doped spiro-MeOTAD in high performance solid state dye sensitized solar cell. *Phys. Chem. Chem. Phys.*, **14**, 11689–11694.

86. Cai, N. *et al.* (2011) An organic D-π-A dye for record efficiency solid-state sensitized heterojunction solar cells. *Nano Lett.*, **11** (4), 1452–1456.

87. Coakley, K.M. and McGehee, M.D. (2003) Photovoltaic cells made from conjugated polymers infiltrated into mesoporous titania. *Appl. Phys. Lett.*, **83** (16), 3380–3382.

88. Zhu, R. *et al.* (2009) Highly efficient nanoporous TiO$_2$-polythiophene hybrid solar cells based on interfacial modification using a metal-free organic dye. *Adv. Mater.*, **21** (9), 994–1000.

89. Jiang, K.J. *et al.* (2009) Photovoltaics based on hybridization of effective dye-sensitized titanium oxide and hole-conductive polymer P3HT. *Adv. Funct. Mater.*, **19** (15), 2481–2485.

90. Abrusci, A. *et al.* (2011) Facile infiltration of semiconducting polymer into mesoporous electrodes for hybrid solar cells. *Energy Environ. Sci.*, **4** (8), 3051–3058.

91. Boucle, J., Ravirajan, P., and Nelson, J. (2007) Hybrid polymer-metal oxide thin films for photovoltaic applications. *J. Mater. Chem.*, **17** (30), 3141–3153.

92. Huynh, W.U., Dittmer, J.J., and Alivisatos, A.P. (2002) Hybrid nanorod-polymer solar cells. *Science*, **295** (5564), 2425–2427.

93. Boucle, J. *et al.* (2008) Hybrid solar cells from a blend of poly(3-hexylthiophene) and ligand-capped TiO$_2$ nanorods. *Adv. Funct. Mater.*, **18** (4), 622–633.

94. Hochbaum, A.I. and Yang, P.D. (2010) Semiconductor nanowires for energy conversion. *Chem. Rev.*, **110** (1), 527–546.

95. Gonzalez-Valls, I. and Lira-Cantu, M. (2009) Vertically-aligned nanostructures of ZnO for excitonic solar cells: a review. *Energy Environ. Sci.*, **2** (1), 19–34.

96. Li, Y.X. *et al.* (2009) Hydrothermal synthesis and characterization of TiO$_2$ nanorod arrays on glass substrates. *Mater. Res. Bull.*, **44** (6), 1232–1237.

97. Kumar, A., Madaria, A.R., and Zhou, C.W. (2010) Growth of Aligned Single-crystalline rutile TiO$_2$ nanowires on arbitrary substrates and their application in dye-sensitized solar cells. *J. Phys. Chem. C*, **114** (17), 7787–7792.

98. Feng, X.J. *et al.* (2008) Vertically aligned single crystal TiO$_2$ nanowire arrays grown directly on transparent conducting oxide coated glass: synthesis details and applications. *Nano Lett.*, **8** (11), 3781–3786.

99. Feng, X.J. *et al.* (2009) Tantalum-doped titanium dioxide nanowire arrays for dye-sensitized solar cells with high open-circuit voltage. *Angew. Chem. Int. Ed.*, **48** (43), 8095–8098.

100. Chu, S.Z. *et al.* (2002) Fabrication and characteristics of ordered Ni nanostructures on glass by anodization and direct current electrodeposition. *Chem. Mater.*, **14** (11), 4595–4602.

101. Chu, S. *et al.* (2003) Fabrication and characteristics of nanostructures on glass by Al anodization and electrodeposition. *Electrochim. Acta*, **48** (20–22), 3147–3153.

102. Keller, F., Hunter, M., and Robinson, D. (1953) Structural features of anodic oxide films on aluminum. *J. Electrochem. Soc.*, **100** (9), 411–419.

103. Kuo, C.Y. *et al.* (2008) Ordered bulk heterojunction solar cells with vertically aligned TiO$_2$ nanorods embedded in a conjugated polymer. *Appl. Phys. Lett.*, **93** (3), 033307.

104. Musselman, K.P. *et al.* (2008) Low-temperature synthesis of large-area, free-standing nanorod arrays on ITO/glass and other conducting substrates. *Adv. Mater.*, **20** (23), 4470–4475.

105. Greene, L.E. *et al.* (2005) General route to vertical ZnO nanowire arrays using textured ZnO seeds. *Nano Lett.*, **5** (7), 1231–1236.

106. Musselman, K., *et al.* (2010) Strong efficiency improvements in ultra-low-cost inorganic nanowire solar cells. *Adv. Mater.*, **22** (35): E254–E258.

107. Joo, J. *et al.* (2011) Face-selective electrostatic control of hydrothermal zinc oxide nanowire synthesis. *Nat. Mater.*, **10**, 596–601.

108. Law, M. *et al.* (2005) Nanowire dye-sensitized solar cells. *Nat. Mater.*, **4** (6), 455–459.

109. Olson, D.C. *et al.* (2006) Hybrid photovoltaic devices of polymer and ZnO nanofiber composites. *Thin Solid Films*, **496** (1), 26–29.

110. Olson, D.C. *et al.* (2007) Band-offset engineering for enhanced open-circuit voltage in polymer-oxide hybrid solar cells. *Adv. Funct. Mater.*, **17** (2), 264–269.

111. Ravirajan, P. *et al.* (2006) Hybrid polymer/zinc oxide photovoltaic devices with vertically oriented ZnO nanorods and an amphiphilic molecular interface layer. *J. Phys. Chem. B*, **110** (15), 7635–7639.

112. Takanezawa, K. *et al.* (2007) Efficient charge collection with ZnO nanorod array in hybrid photovoltaic devices. *J. Phys. Chem. C*, **111** (19), 7218–7223.

113. White, M.S. *et al.* (2006) Inverted bulk-heterojunction organic photovoltaic device using a solution-derived ZnO underlayer. *Appl. Phys. Lett.*, **89**, 143517.

114. Mor, G.K. *et al.* (2006) A review on highly ordered, vertically oriented TiO$_2$ nanotube arrays: fabrication, material properties, and solar energy applications. *Sol. Energy Mater. Sol. Cells*, **90** (14), 2011–2075.

115. Ghicov, A. and Schmuki, P. (2009) Self-ordering electrochemistry: a review on growth and functionality of TiO₂ nanotubes and other self-aligned MOx structures. *Chem. Commun.*, (20), 2791–2808.

116. Roy, P., Berger, S., and Schmuki, P. (2011) TiO₂ nanotubes: synthesis and applications. *Angew. Chem. Int. Ed.*, **50** (13), 2904–2939.

117. Zwilling, V., Aucouturier, M., and Darque-Ceretti, E. (1999) Anodic oxidation of titanium and TA6V alloy in chromic media. An electrochemical approach. *Electrochim. Acta*, **45** (6), 921–929.

118. Gong, D. *et al.* (2001) Titanium oxide nanotube arrays prepared by anodic oxidation. *J. Mater. Res.*, **16** (12), 3331–3334.

119. Shankar, K. *et al.* (2007) Highly-ordered TiO₂ nanotube arrays up to 220 mu m in length: use in water photoelectrolysis and dye-sensitized solar cells. *Nanotechnology*, **18** (6), 3953–3957.

120. Paulose, M. *et al.* (2007) TiO₂ nanotube arrays of 1000 m length by anodization of titanium foil: phenol red diffusion. *J. Phys. Chem. C*, **111** (41), 14992–14997.

121. Mor, G.K. *et al.* (2005) Transparent highly ordered TiO₂ nanotube arrays via anodization of titanium thin films. *Adv. Funct. Mater.*, **15** (8), 1291–1296.

122. Weickert, J. *et al.* (2011) Controlled growth of TiO2 nanotubes on conducting glass. *Chem. Mater.*, **23** (2), 155–162.

123. Paulose, M. *et al.* (2006) Backside illuminated dye-sensitized solar cells based on titania nanotube array electrodes. *Nanotechnology*, **17** (5), 1446–1448.

124. Mor, G.K. *et al.* (2006) Use of highly-ordered TiO₂ nanotube arrays in dye-sensitized solar cells. *Nano Lett.*, **6** (2), 215–218.

125. Mor, G.K. *et al.* (2007) High efficiency double heterojunction polymer photovoltaic cells using highly ordered TiO₂ nanotube arrays. *Appl. Phys. Lett.*, **91**, 152111.

126. Foong, T.R.B. *et al.* (2010) Template-directed liquid ALD growth of TiO₂ nanotube arrays: properties and potential in photovoltaic devices. *Adv. Funct. Mater.*, **20** (9), 1390–1396.

127. Na, S.I. *et al.* (2008) Fabrication of TiO₂ nanotubes by using electrodeposited ZnO nanorod template and their application to hybrid solar cells. *Electrochim. Acta*, **53** (5), 2560–2566.

128. Muduli, S. *et al.* (2011) Shape preserving chemical transformation of ZnO mesostructures into anatase TiO₂ mesostructures for optoelectronic applications. *Energy Environ. Sci.*, **4**, 2835–2839.

129. Martinson, A.B.F. *et al.* (2007) ZnO nanotube based dye-sensitized solar cells. *Nano Lett.*, **7** (8), 2183–2187.

130. Yu, H.D. *et al.* (2005) A general low-temperature route for large-scale fabrication of highly oriented ZnO nanorod/nanotube arrays. *J. Am. Chem. Soc.*, **127** (8), 2378–2379.

131. Sun, Y. *et al.* (2005) Synthesis of aligned arrays of ultrathin ZnO nanotubes on a Si wafer coated with a thin ZnO film. *Adv. Mater.*, **17** (20), 2477–2481.

132. Sommeling, P.M. *et al.* (2006) Influence of a TiCl₄ post-treatment on nanocrystalline TiO₂ films in dye-sensitized solar cells. *J. Phys. Chem. B*, **110** (39), 19191–19197.

133. Musselman, K.P. *et al.* (2011) A novel buffering technique for aqueous processing of zinc oxide nanostructures and interfaces, and corresponding improvement of electrodeposited ZnO Cu₂O photovoltaics. *Adv. Funct. Mater.*, **21** (3), 573–582.

134. Schmidt-Mende, L. *et al.* (2005) Effect of hydrocarbon chain length of amphiphilic ruthenium dyes on solid-state dye-sensitized photovoltaics. *Nano Lett.*, **5** (7), 1315–1320.

135. Liu, Y.X. *et al.* (2006) Dependence of band offset and open-circuit voltage on the interfacial interaction between TiO₂ and carboxylated polythiophenes. *J. Phys. Chem. B*, **110** (7), 3257–3261.

136. Vaynzof, Y. *et al.* (2010) Improved photoinduced charge carriers separation in organic–inorganic hybrid photo-voltaic devices. *Appl. Phys. Lett.*, **97** (3), 033309.

137. Abrusci, A. *et al.* (2011) Influence of ion induced local coulomb field and polarity on charge generation and efficiency in poly(3 hexylthiophene) based solid state dye sensitized solar cells. *Adv. Funct. Mater.*, **21** (13), 2571–2579.

138. Ma, T.L. *et al.* (2005) High-efficiency dye-sensitized solar cell based on a nitrogen-doped nanostructured ti-tania electrode. *Nano Lett.*, **5** (12), 2543–2547.

139. Vitiello, R.P. *et al.* (2006) N-doping of anodic TiO_2 nanotubes using heat treatment in ammonia. *Electrochem. Commun.*, **8** (4), 544–548.

15
Electric Field Effects in Functional Metal Oxides

Lily Mandal, Weinan Lin, James Lourembam, Satishchandra Ogale, and Tom Wu

15.1
Introduction

The field-effect transistor (FET) is the most notable microelectronic device shaping the semiconductor technology as we see today. In recent years, researchers around the world have discovered that this prototypical electronic device can be used to understand and manipulate the inherent properties of a variety of complex oxides, thereby opening up new avenues of research [1]. The fascinating physical properties of correlated electron systems, including the high-T_c superconductors and colossal magnetoresistance manganites, are a testimony to the delicate balance of the interactions between charge, spin, orbital, and lattice [2, 3]. In these complex materials, carrier density plays a key role in determining the physical properties, and often a change on the order of 0.1 electron per unit cell can lead to some exotic phase transitions [4]. For example, in high-T_c cuprates, superconductivity can be induced from the antiferromagnetic insulator phase by electrostatic gating [5]. Moreover, field-effect devices based on dilute magnetic semiconductors (DMSs) and colossal magnetoresistive materials are promising to reveal the complex correlations between charge and spin in these materials. This electrostatic approach to property modulation offers many advantages over the traditional route of chemical doping by enabling controlled and reversible tuning of carrier density without introducing a permanent chemical or structural disorder. In some cases, more carriers can be electrostatically achieved beyond the maximum limit of chemical doping.

In the conventional FET, a conducting channel connects the source and the drain, and the modulation of transport is achieved by applying a voltage on the gate. Capacitive charging and discharging across the gate dielectrics can be understood within the metal–insulator–semiconductor (MIS) model. Generally, three cases of channel modulation occur when such a transistor is biased – accumulation, depletion, or inversion. Depending on the channel material and the device configuration, different scenarios of carrier modulation and band bending at the channel–dielectric interface can be achieved, and a detailed profile of carrier distribution can be numerically solved. In oxides with high carrier densities, inversion has been rarely

Functional Metal Oxides: New Science and Novel Applications, First Edition.
Edited by Satishchandra B. Ogale, Thirumalai V. Venkatesan, and Mark G. Blamire.
© 2013 Wiley-VCH Verlag GmbH & Co. KGaA. Published 2013 by Wiley-VCH Verlag GmbH & Co. KGaA.

reported. Furthermore, notable modulations of charges often happen within a few nanometers from the interface as a result of strong screening; hence, it is important to keep the channel layer as thin as possible. In general, electrostatic doping in complex oxides is challenging as a result of high carrier concentration, but recent years have witnessed significant developments in dielectric gate materials, which help increase the modulation via realization of higher electric fields. In the following, we first review some notable achievements in developing new dielectrics, and then discuss the milestones achieved in a few key research areas.

15.2
Developments of Gate Dielectrics: from High-k Oxides to Ionic Electrolytes

The design of field-effect devices can be grouped into two major categories: top-gate and back-gate devices. In the top-gate configuration, the insulating dielectric is deposited on top of the channel, whereas in the back-gate device, the single-crystal substrate is used as the dielectric. Other device configurations, such as side gate and surrounding gate, often involve more complex fabrication procedures. In general, it is straightforward to incorporate the high-κ substrates such as $SrTiO_3$ in the bottom-gate devices, while the top-gate configuration gives more freedom to choose the gate dielectrics and has seen significant advances in recent years.

SiO_2 is the most common dielectric in FETs, as it forms excellent interface with Si. But in order to reduce the gate leakage and produce larger carrier modulation, there have been constant efforts to exploit alternative high-κ materials [7, 11]. Table 15.1 shows some examples of dielectrics and their performances in ZnO thin-film FETs. It is important to note that a high dielectric constant is not the sole criterion to select the gate dielectric; one has to consider other factors such as the breakdown voltage, band offset, interfacial defects, and chemical stability [11]. Generally, band offsets of over 1 eV are required to suppress excessive carrier

Table 15.1 Examples of oxide dielectrics used in ZnO thin-film transistors and the associated key performance parameters.

Dielectrics	μ $(cm^{-2}\,V^{-1})$	On/off ratio	Threshold voltage (V)	References
$Bi_{1.5}Zn_{1.0}Nb_{1.5}O_7$ (BZN)	0.024	2×10^4	2	[13]
HfO_2	12.2	$\sim 10^7$	2.5	[14]
$HfSiO_x$	4.5	—	1.7	[14]
Al_2O_3	17.6	$> 10^4$	6	[14]
$(Ba,Sr)TiO_3$	2.3	1.5×10^8	1.2	[15]
SiO_xN_y	70	5×10^5	1.8	[6]
$ScAlMgO_4$	70	—	—	[7]
$CaHfO_x$ w/silicon nitride	7	$> 10^6$	—	[8]

injection into the dielectric, and it is desirable to develop materials with large bandgaps and dielectric constants higher than 10 [12].

Various high-κ dielectrics such as Al_2O_3 [16, 17], Y_2O_3 [18, 19], Ta_2O_5 [20], $Zn_{0.7}Mg_{0.3}O$ [21], and HfO_2 [22] have been explored as dielectric layers in amorphous metal−oxide semiconductor (AMOS)-based thin film transistors (TFTs), a field that has rapidly developed in recent years. Amorphous $Ba_{0.5}Sr_{0.5}TiO_3$ (BST) deposited by radio frequency (RF) sputtering has also been used as the gate dielectric [23]. Many high-κ dielectrics, such as HfO_2, have problems of charge-trapping at the electrode/dielectric and dielectric/channel interfaces [24]. To avoid this problem, a buffer layer of Al_2O_3 has been employed between HfO_2 and the ZnO channel. Similarly, high-κ $Bi_{1.5}Zn_{1.0}Nb_{1.5}O_7$ (BZN) with $k > 50$ when used as single dielectric with the ZnO channel shows a leakage current and low mobility (1.5 cm^2 V^{-1} s^{-1}) [25], but the use of MgO/BZN bilayer dielectric has been shown to improve the performance [26]. To study the effect of high-κ dielectric on the mobility of indium−gallium−zinc oxide (IGZO)-based semiconductor TFTs, dielectric TiO_x was inserted between the IGZO channel and SiN_x dielectric in a bilayer dielectric configuration. The TiO_x layer was shown to decrease the mobility of the IGZO channel. In fact, it was found that the thickness of the TiO_x layer has an inverse relationship with mobility because of Coulomb and phonon scattering [27].

Ohta *et al.* have used water-infiltrated nanoporous $12CaO·7Al_2O_3$ glass (calcium aluminates with nanopores, CAN) grown by PLD (pulse laser deposition) as a gate insulator [28, 29]. They made an FET using $SrTiO_3$ as the channel. The CAN gate could generate 10^{15} cm^{-2} sheet charge density at the interface of the dielectric and the semiconductor, which is far more than the sheet charge density of 10^{13} cm^{-2} generated in conventional dielectrics. Such a large charge density is generated because of the accumulation of electrons, water electrolysis, and electrochemical reduction at the interface of $SrTiO_3$. Since $SrTiO_3$ is a thermoelectric material, the electric-field-induced two-dimensional electron gas (2DEG) is shown to produce a large enhancement in the thermopower generation. The CAN-gated $SrTiO_3$ FET generates a very thin (2 nm) and high charge density, which enhances the Seebeck coefficient |S| five times when compared to bulk.

Ferroelectric FETs (FeFETs) have been studied intensively as a result of their nonvolatile characteristics, and lead−zirconium titanate, $PbZr_xTi_{1-x}O_3$ (PZT), is the most popular ferroelectric gate insulator for oxide-based FETs. PZT has been extensively used as the gate insulator to modulate the transport of channel materials, including manganites, high-T_c superconductors, $SrRuO_3$, $SrTiO_3$, and so on [30−33]. A large polarization (15−45 μC cm^{-2}) can be routinely achieved. Using an atomic force microscope (AFM) tip as a local electrode, it has been demonstrated that charge densities in FeFETs can be locally modulated at the nanoscale [32]. Other ferroelectrics such as $Pb_{0.95}La_{0.05}Zr_{0.2}Ti_{0.8}O_3$ have also been explored [34]. FeFETs using $(Fe,Zn)_3O_4$ as the semiconductor channel with a high curie temperature and ferroelectric PZT as a dielectric have been demonstrated [35]. High-mobility zinc oxide FeFET has been made by using spin-coated PVdF−TrFE as the dielectric and sputter-deposited ZnO as the channel [36].

Table 15.2 Examples of EDLTs using various electrolytes and oxide channels.

Channel	Electrolyte	V_g (V) maximum	Carrier type	n (cm^{-2})	Modulated n (cm^{-2})	References
InO$_x$	EMI-BETI	±1.03	n	2.2×10^{13}	3.7×10^{13}	[41]
ZnO	KClO$_4$/PEO	+3	n	3.0×10^{12}	3.7×10^{13}	[42]
NiO	KClO$_4$/PEO	−0.35	p	—	5.1×10^{14}	[43]
SrTiO$_3$	KClO$_4$/PEO	5	n	—	$\sim 10^{15}$	[44]
ZrNCl	DEME-TFSI	4.5	n	0.3×10^{14}	2.5×10^{14}	[45]
SrTiO$_3$	DEME-TFSI	5	n	—	$\sim 10^{15}$	[46]
YBa$_2$Cu$_3$O$_{7-x}$	LiClO$_4$/PEO	±3	p	—	—	[47]
ZnO	DEME-TFSI	5.5	n	2.0×10^{10}	8×10^{14}	[40]
Ti$_{0.90}$Co$_{0.10}$O$_2$	DEME-TFSI	3.8	n	4.0×10^{13}	2.7×10^{14}	[48]
KTaO$_3$	DEME-BF$_4$	6	n	$\sim 10^{12}$	3.7×10^{14}	[9]
SrTiO$_3$	KClO$_4$/PEO	3.5	n	—	$>10^{14}$	[49]
NdNiO$_3$	EMI-TFSI	−4	p	$>10^{15}$	3×10^{15}	[50]
La$_{2-x}$Sr$_x$CuO$_4$	DEME-TFSI	−4.5	p	—	—	[51]

Organic polymers present interesting options as gate insulators because they promise better interfaces with less density of traps at interfaces. Nakamura *et al.* [33] used the organic polymer parylene (dielectric constant ~ 3.15) as a gate insulator to realize the highest mobility ever attained in a SrTiO$_3$-based FET till then. In another example, poly(3,4-ethylenedioxythiophene):poly(styrenesulfonate) (PEDOT:PSS) was used in polymer-gated oxide semiconductor heterostructures to construct high-performance FETs [37].

Recently, conducting ionic electrolytes have been experimented as gate dielectrics; these FETs, popularly known as electric-double-layer transistors (EDLTs), have been able to achieve unprecedented levels of carrier doping concentrations, and in some cases, even new phases have been discovered [38]. Ionic liquids have cationic and anionic components. For example, in *N,N*-diethyl-*N*-methyl-*N*-(2-methoxyethyl) ammonium bis(trifluoromethylsulfonyl) imide (DEME-TFSI), a popular ionic electrolyte, DEME is the cationic part and TFSI is the anionic species. Although this powerful method of using ionic liquids or polymer electrolytes in EDLTs is new, these electrolytes have been extensively used for a long time in various electrochemical cells such as lithium batteries, fuel cells, and energy storage elements [39]. In EDLT, electrolytes transition to a rubber phase during cooling where they can sustain higher gate voltages without breaking down [40]. Electrolytes can enable super-high carrier density modulation because they are able to sustain much higher electric fields (~ 10 MV cm^{-1}). In Table 15.2, we highlight some examples of EDLTs with oxide channels.

To fabricate a fully solution-processable FET, there is a focus on developing solution-processable dielectrics. Sol–gel-synthesized zirconium oxide has been used as high-κ dielectric for solution-processed zinc tin oxide (ZTO) channel [52]. Recently, Salleo and coworkers have reported room-temperature synthesis of zirconium oxide by sol–gel chemistry for low-voltage organic FETs [53]. The FET fabricated with PBTTT-C 14 as the channel showed an on/off ratio of 10^5 and mobility of $0.18\,cm^2\,V^{-1}\,s^{-1}$. Sol–gel-synthesized lanthanide series oxide (Gd_2O_3), Y_2O_3, and ($Gd_{1-x},Y_x)_2O_3$ films have also been fabricated [54]. These films show a high dielectric constant (11–15) and a high breakdown field of $3.5\,MV\,cm^{-1}$. Adamopoulos *et al.* [55] have synthesized high-κ Y_2O_3 and Al_2O_3 by spray pyrolysis and shown low-operating-voltage FETs with zinc oxide as channel. Recently, WO_3 has also been used as a high-κ high-temperature-stable dielectric for zinc oxide TFTs. Insulating WO_3 was grown by varying the oxygen partial pressure during PLD growth [56]. Pal *et al.* [57] have synthesized solution-processable high-κ sodium β-alumina as dielectric for fully solution-processable metal–oxide FETs. Sodium β-alumina has alternate insulating AlO_x layers, and mobile sodium ions in between the planes. These Na ions give β-alumina one-dimensional ionic conductivity between two AlO_x planes in the presence of a gate bias. The FET fabricated with sol–gel-synthesized β-alumina and solution-processable ZTO has shown field-effect mobility of $28\,cm^2\,V^{-1}\,s^{-1}$ and an on/off ratio of 10^4.

High-capacitance mesoporous SiO_2 with electrical double-layer effect has been demonstrated as gate dielectric for low-voltage (1.5 V) TFTs [58]. Spin-coated films of oleic acid-capped barium titanate and strontium titanate nanoparticles have also been shown to be effective as high-κ dielectric layers for organic–inorganic hybrid transistors [59]. Layer-by-layer assembly of self-assembled monolayers (SAMs) of AlO_x and TiO_x by cyclic molecular layer deposition (MLD) have also been used as an organic–inorganic hybrid dielectric in ZnO TFTs [60]. Similarly, self-assembled nanodielectrics (SANDs) shown in Figure 15.1 have been used as gate dielectrics in many FETs [61–64]. The SAND layers can be deposited by wet techniques at room temperature and shown to yield high capacitance. When SAND is used as dielectric, the operating voltage of FETs decreases and the performance of TFTs gets enhanced.

15.3
Electric-Field-Induced Modulation of Ferromagnetism

In modulating the physical properties of channels, FETs are more effective when dealing with semiconductors than metals because the carrier concentration is usually higher in the latter. DMSs are an important class of emerging materials that have been exploited intensively for potential applications in spintronics. Customizing the FET structures to allow gate-induced modifications in magnetic properties is an appealing approach. FETs based on oxide DMS was first realized by Zhao *et al.* [65] who used PZT to gate Co-doped TiO_2. In the same work, Zhao *et al.* detailed how a reversible modulation of saturation magnetization and coercivity can be achieved after PZT is poled with different gate voltages. In another work, Yamada *et al.* [48] demonstrated magnetization switching of a Co-doped TiO_2 channel in an EDLT. By inducing high carrier density accumulation ($>10^{14}\,cm^{-2}$) on the

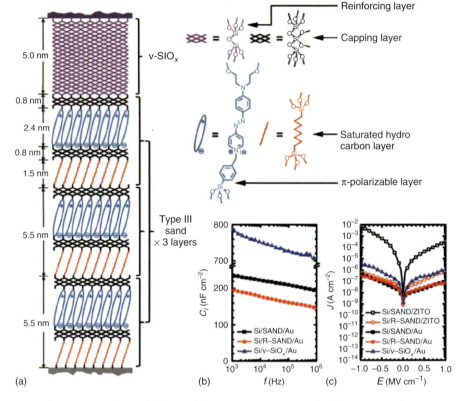

Figure 15.1 Self-assembled nanodielectric (SAND). (Reprinted with permission from [63] copyright 2011 Adv. Mat.)

DMS, a paramagnetic state was transformed into a ferromagnetic state at room temperature, which was confirmed by anomalous Hall effect (AHE) measurements.

Manganites, which are known prominently for colossal magnetoresistance, display insulator-to-metal transitions, which are accompanied by ferromagnetic transitions. Electric field effects on manganites have attracted a lot of interest, presenting fascinating opportunities to explore the rich phase diagrams of manganites. Most of these devices use a ferroelectric, notably PZT, as the gate oxide to modulate the manganite channel [30, 66–73]. Although this configuration has the advantage of effectively avoiding the gate leakage, only two discrete states of polarization are achieved. Moreover, PZT has a strong piezoelectric response; hence, strain effects are convoluted with the electrostatic effects. In a ferroelectric transistor with $La_{0.7}Ca_{0.3}MnO_3$ as the channel and $PbZr_{0.2}Ti_{0.8}O_3$ as the ferroelectric gate, Matthews *et al.* [66] found 300% modulation of conductance with several hours of retention. By improving the device design, Wu *et al.* [30] found a large electroresistance of 76% in an LCMO channel, and electric and magnetic field effects as complementary. It was proposed that the electric field amplifies the ferromagnetic phases, thereby promoting percolation in transport. In another work, Hong *et al.*

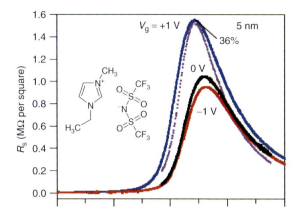

Figure 15.2 Resistivity modulation of an LCMO channel gated by the electrolyte EMIM:TFSI. (Reprinted with permission from [76] copyright 2009 by the American Physical Society.)

[69] found a 35 K reversible shift of the transition temperature in an LSMO film gated by PZT. FETs based on the all-perovskite heterostructure of PZT and LCMO fabricated on Si have also been reported [68]. Other manganite-based FETs include devices using $SrTiO_3$ as the gate dielectric, planar structures patterned by an AFM, and so on [30, 74, 75]. More recently, EDLTs with manganite (LCMO) channel and EMIM:TFSI electrolyte have been reported [76]. As shown in Figure 15.2, the resistivity modulation is about 36% under a gate voltage of 1 V.

It is more challenging to modulate the magnetism in transition metals because of their high carrier concentrations. Chiba *et al.* [77] demonstrated electrostatic tuning of the Curie temperature in Co, which may lead to a "field-effect magnet," where the magnetism can be switched electrically. In their work, they used HfO_2 as the gate insulator, and a 12 K change of Curie temperature was achieved. In another work, the coercivity and Kerr rotation of the thin films (2–4 nm) of ferromagnetic intermetallic compounds FePt and FePd were reversibly modified electrostatically by a gated electrolyte [78]. The carrier modulation amounts to 0.015 electron per unit cell, which appears to correlate well with the observed change of a small percentage in the magnetocrystalline anisotropy energy.

15.4
Electrostatic Modulation of Superconductivity

Electrostatic doping is a promising approach to probe the superconducting phase because adding carriers can induce the conversion from the antiferromagnetic phase to the superconducting phase [5]. In fact, this concept was introduced earlier for conventional superconductors such as tin and indium films in 1960s [79]. However, because of the intrinsic large carrier densities of conventional superconductors and the short screening length, the modulation of the superconducting transition temperature was quite small (on the order 10^{-4} K). On the other hand, in HTS materials, superconductivity occurs at lower carrier densities in the range of 10^{19}–10^{22} carriers per cm^3. The Thomas–Fermi screening length l_{TF} can be estimated as

$$l_{TF} = \left(\frac{\varepsilon_S E_{F0}}{4\pi^2 e^2 n_0} \right)^{\frac{1}{2}}$$

where ε_s is the dielectric constant of the superconductor, E_{F0} is the Fermi energy, and n_0 is the carrier concentration [80]. Taking YBCO as an example, its carrier concentration is $\sim 5 \times 10^{21}$ cm^{-3}, $\varepsilon_s = 26$, and thus l_{TF} is around 0.5 nm [81]. l_{TF} gives a good measure of the thickness of the channel layer that should be used in superconductor FETs.

There remain open questions regarding the possible mechanisms of the electrostatic modulation of superconductivity. It was proposed that oxygen rearrangement induced by an electrostatic field may play a role [82], while other theories considered the interactions between the electric field and charge carriers via Coulomb forces [81, 83, 84]. Recently, Salluzzo *et al.* [85] performed an *in situ* investigation of the mechanism of electrostatic modulation on the superconductor $NdBa_2Cu_3O_7$. The X-ray absorption spectroscopy data suggest that the field-injected holes mainly enter the CuO chains and are partially transferred to the CuO_2 planes, similarly to the chemical doping. However, more effort is needed to unambiguously explain the mechanism.

In the original theoretical calculations on the influence of surface charges on La_2CuO_4, it was predicted that superconductivity may be induced on the surface of La_2CuO_4 in an FET or a Schottky junction [86]. Although the concept had not come through then, a trivial modulation on the surface of $YBa_2Cu_3O_7$ film was obtained [87]. The field-effect modulation on the superconductor channel was first studied by Mannhart and his collaborators [88, 89]. In their devices, a $SrTiO_3$ thin film was deposited on the Nb-doped $SrTiO_3$, which played the role of the gate dielectric materials as well as the platform for the growth of $YBa_2Cu_3O_{7-\delta}$ film, which was adopted widely in the later field-effect experiments. On the basis of this structure, the modulation of critical current, normal state resistance, and even the transition temperature were obtained. And later it was found that, when the film contains features of weak link, the modulation of the transition temperature can even be up to 10 K [90].

In a recent flurry of activities, $SrTiO_3$, a prototypical complex oxide with high dielectric constant, found new roles in oxide electronics. In the $LaAlO_3/SrTiO_3$ heterostructures, superconductivity was observed at the interface. It is quite convenient to construct back-gate FETs based on such interface heterostructures. Researchers have found that the electric field modulates not only the carrier density but also the mobility [91]. Further the electrostatic approach may help shed light on the quantum transition between the two-dimensional superconducting and insulating states [92–94]. Caviglia *et al.* pointed out the possible existence of a quantum critical point separating the superconducting and insulating regions in the complex carrier-dictated phase diagram of LAO–STO.

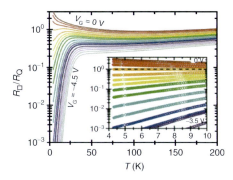

Figure 15.3 Temperature dependence of the normalized sheet resistance, showing the gate-induced superconductivity in initially underdoped and insulating $La_{2-x}Sr_xCuO_4$. (Reprinted with permission from [51] copyright 2006 Appl. Phys. lett (AIP).)

Similar to the field-effect experiments on manganites, there have been efforts using ferroelectrics as gate insulators in superconductor FETs [95]. A superconducting to insulating phase transition was demonstrated in $GdBa_2Cu_3O_{7-x}$ using PZT as dielectric material on top of the channel [96]. Furthermore, with the help of a conductive AFM, a local switching of ferroelectric polarization and therefore the superconductivity was obtained on heterostructures made of PZT and Nb-doped $SrTiO_3$ [2]. Although the transition temperature and the modulation are on the order of millikelvin, this approach is promising in the context of applications in nanoscale superconductor devices.

Because of the powerful electrostatic doping capability of electrolytes, it is not a surprise that some breakthroughs have been achieved recently in EDLTs with superconductor channels. Using this technique, superconducting transitions and modulations have been demonstrated in several systems such as $SrTiO_3$, $KTaO_3$, YBCO, and LSCO [9, 49, 51, 97]. In the case of $KTaO_3$, wherein superconductivity had not been discovered until recently, EDLT helped to obtain a carrier density that had never been reached before, above which superconductivity occurs [9]. The strong modulation capability with EDLT helps to get more insight behind the transport property. Bollinger *et al.* performed the insulator–metal transition by using EDLTs with a one-unit-cell thin $La_{2-x}Sr_xCuO_4$ channel and found that the transport conforms to the prediction of two-dimensional superconductor–insulator transition, with the critical resistance being the quantum resistance for electron pairs (Figure 15.3). They further proposed that the Bosonic model is more credible in understanding the two-dimensional quantum phase transition [51].

15.5
Transparent Amorphous Metal-Oxide Field-Effect Transistors

In the past decade, there has been a rapidly increasing interest toward the development of a transparent metal-oxide field-effect transistor. These materials

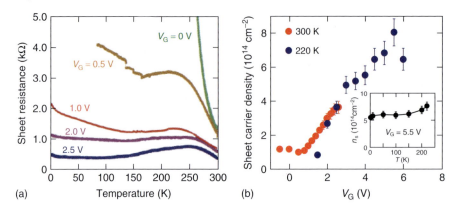

Figure 15.4 (a) Temperature dependence of sheet resistance measured in ZnO EDLT under the various gate voltages. (b) Sheet carrier densities at different gate voltages measured at 300 and 200 K. The inset shows the carrier densities at gate voltage of 5.5 V at different temperatures. (Reprinted with permission from [63]]copyright 2011 Adv. Mat.)

have inherent properties of high mobility in spite of being in the amorphous state. They have a wide bandgap, a broad transparency window, and good mechanical stress tolerance. Amorphous oxides are compatible with large-area deposition protocols and show uniformity of deposition. They are currently being used as the back-panel TFTs in flat panel displays applications. These metal oxides have flexibility of formation over a wide range of processing temperatures [98]. Metal oxides are deposited by physical vapor deposition (PVD) processes such as sputtering, PLD, chemical vapor deposition (CVD), and ion-assisted deposition (IAD).

Hoffman *et al.* [99], Carcia *et al.* [100], and Fortunato *et al.* [101] have prepared polycrystalline ZnO TFTs by sputtering. Owing to its polycrystalline nature and high density of charge carriers, the ZnO TFT has issues related to control of conductance and threshold voltage. Shimotani *et al.* [42] used $KClO_4$/PEO to induce a sheet carrier density of 4.2×10^{13} cm^{-2} in polycrystalline ZnO, which was several times larger than that achieved in the conventional ZnO FETs with oxide dielectrics. Later, the same group used a different electrolyte, DEME-TFSI, in ZnO-based EDLT, and the carrier density was increased by one order, reaching 8.0×10^{14} cm^{-2} (Figure 15.4) [40]. Misra *et al.* [41] used ionic liquids to induce a resistance modulation on the order of 10^4 in InO_x films, and the corresponding mobility was far greater than what was achieved in other FETs using oxide dielectrics. NiO is a prototypical strongly correlated p-type oxide, and its EDLT shows an estimated maximum carrier density of 5.1×10^{14} cm^{-2} [43].

Hosono *et al.* [18] have shown for the first time the use of amorphous indium gallium zinc oxide (a-InGaZnO) grown by PLD on flexible substrates as active channels in TFTs. The a-InGaZnO-based transistors have fairly high motilities of ~10 cm^2 V^{-1} s^{-1}, low subthreshold swing, low operating voltage, and, most importantly, they can be fabricated on flexible substrates [18, 98]. In the case of polycrystalline transparent semiconductors, the bulk mobility is higher than field-effect mobility as there is scattering at the channel interfaces owing to grain boundaries and defects. When a metal oxide (such as ZnO, In_2O_3, or SnO_2) film is made, it has a tendency to become crystalline even at room temperature. However, ternary oxides (such as In-Zn-O, Zn-Sn-O) can form either amorphous or polycrystalline thin films depending on the metal ions ratio in the film. With the appropriate ratio of metal ions, the ternary oxide can present an amorphous phase (which is more desirable as it does not have grain-boundary scattering). Different metal oxides have different ionic radii and charges, which help in the formation of amorphous phase in these metal oxides. Most of the amorphous metal oxides are made of the transition-metal ions Zn, Sn, and In and the non-transition-metal ion Ga. Hosono *et al.* have studied the role of these ions by comparing InZnO with InZnGaO. They showed that, while InZnO has higher electron mobility, it is hard to get low concentration of charge carriers in InZnO FET. This is due to Ga−O bond being stronger than In−O and Zn−O bonds. Thus, incorporating Ga can suppress the formation of oxygen vacancies and decrease the mobile carrier density. But at the same time, a higher concentration of gallium can adversely affect the mobility. Therefore, a precise control on the stoichiometry is important.

These amorphous transparent oxides exhibit very high electron mobility even in the amorphous phase contrary to the conventional electronics of silicon where the mobility is dependent on the crystallinity. The mobility of crystalline silicon is $1500\,cm^2\,V^{-1}\,s^{-1}$, which dramatically decreases to $1\,cm^2\,V^{-1}\,s^{-1}$ in case of hydrogenated silicon (a-Si:H). This difference is due to the difference in electronic structure of the two. While silicon has covalent bonding, in the transparent metal oxide, there is an actual charge transfer between the metal and the oxygen. In oxides, the structure is stabilized by the Madelung potential formed by the ions, raising the electronic levels in cations and lowering the levels in anions. The conduction band of these amorphous oxides is made up of the s orbital of these metal cations, and the valence band is made of the p orbital of oxygen. The charge transport occurs in the overlapped metal orbitals, while in the case of amorphous silicon, it is through hopping [102]. The low operating voltage in amorphous oxides is attributed to the low density of trap states. This is in contrast to the case of amorphous silicon, which has dangling bonds forming deep trap states when used in TFTs. Moreover, these trap states can trap both electrons and holes, which degrades the properties of amorphous silicon.

To date, the use of transparent metal oxides is restricted to driving the back panel of active matrix displays, as the circuits are mainly composed of n-channel TFTs. To widen the applications, p-type metal oxides are highly desirable. The p-type metal oxides are required to make low-power-compensation complementary metal-oxide circuits and logics. However, the performance of p-type metal oxides is not comparable to their n-type counterparts. There are only a few p-type metal oxides and it is hard to grow them into device-quality films. As the conduction band minima (CBM) of the metal oxides are formed of the transition-metal s orbital, which has a spherical spatial distribution, the electron mobility is high, whereas the valence band minima (VBM) are made up of localized oxygen p orbitals and so

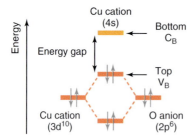

Figure 15.5 Band diagram of p-type Cu_2O. (Reprinted with permission from [63] copyright 2008 Appl. Phys. Lett. (AIP))

the hole mobility is low. The Hosono group has proposed that the hybridization of the oxygen 2p orbital and the copper 3d orbital could lead to the p-type oxide, and some transparent metal oxides of the form $CuAO_2$—where A is Al, Ga, In, La—have been proposed and examined. Recently, a p-type amorphous metal oxide Ln-M-O (Ln: lanthanide element except Ce; M: Ru or Ir) was reported [103].

Cuprous oxide as a semiconductor is known for decades. Generally, it has two oxidation states of +1 and +2, leading to two oxide forms, namely cupric oxide or tenorite (CuO), and cuprous oxide or cuprite (Cu_2O). Both CuO and Cu_2O are p-type semiconductors with bandgaps of 1.9–2.1 and 2.1–2.6 eV, respectively, but the mobility of holes in Cu_2O is far better than that in CuO. Cu_2O is a hole conductor because of the copper vacancy, which induces an acceptor level above the valence band. Contrary to other oxides whose valence band is made of oxygen 2p orbital, in Cu_2O, the valence band is a result of hybridization of the copper 3d orbital with the oxygen 2p orbital (Figure 15.5). Matsuzaki *et al.* [104] deposited single-phase epitaxial Cu_2O films by PLD, and the corresponding TFT showed field-effect mobility of $0.26\, cm^2\, V^{-1}\, s^{-1}$ and an on/off ratio of 6. The poor performance was attributed to the presence of sub-bandgap states and verified by optical measurements. Fortunato *et al.* [105] made p-type Cu_2O TFTs where the Cu_2O film was deposited by RF sputtering; the TFTs showed a mobility of $3.9\, cm^2\, V^{-1}\, s^{-1}$ and an on/off ratio of 200. Sung *et al.* fabricated a Cu_2O film by sputtering, and then annealed it in air at 200 °C to obtain a CuO film. The TFT using CuO as channel showed a mobility of $0.4\, cm^2\, V^{-1}\, s^{-1}$ and an on/off ratio of 10^4 [106]. Zou *et al.* [107] have also fabricated a Cu_2O thin film on Si/SiO_2 substrates by PLD and formed top-gate FETs using HfON as the gate dielectric. Recently, Yao *et al.* [108] have reported the synthesis of Cu_2O at room temperature by magnetron sputtering on flexible PET substrates and achieved a mobility of $20\, cm^2\, V^{-1}\, s^{-1}$ and carrier density of $10^{16}\, cm^{-3}$; in the corresponding TFT, the film showed a field-effect mobility $2.40\, cm^2\, V^{-1}\, s^{-1}$ and a current on/off ratio 3.96×10^4 (Figure 15.6).

Tin oxide is another well-known high-mobility wide-bandgap semiconductor. Tin also has two oxidation states of +2 and +4, leading to two oxide forms, namely, tin monoxide (SnO) and tin dioxide (SnO_2), respectively. SnO_2 is a well-known n-type semiconductor, whereas SnO is a p-type semiconductor. In SnO, the $Sn(5s^2)$

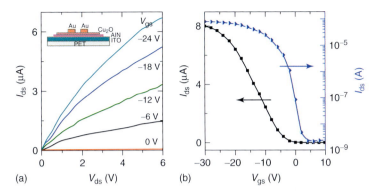

Figure 15.6 (a) $I_{ds}-V_{ds}$ and (b) $I_{ds}-V_{gs}$ curves of Cu$_2$O FET. Reprinted with permission from [63] copyright 2012 Appl. Phys. Lett. (AIP).

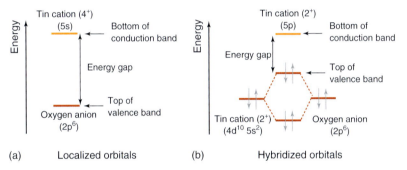

Figure 15.7 Comparison between band diagrams of n-type SnO$_2$ and p-type SnO. Reprinted with permission from [109] copyright 2012 Adv. Mat.

orbital hybridizes with oxygen 2p orbital forming the valence band minimum (VBM) (Figure 15.7) [109]. Ogo *et al.* [110] have grown SnO on yttrium-stabilized zirconium substrates at 575 °C by PLD and made TFTs which showed mobility of 1.3 cm^2 V^{-1} s^{-1} and on/off ratio of 10^2. Yabuta *et al.* [111] have fabricated a p-type SnO TFT by RF sputtering, showing a mobility of 0.24 cm^2 V^{-1} s^{-1} and an on/off ratio of 10^2. Other studies have also reported FETs based on tin oxide as the channel layer [112–114].

15.6
Solution-Processable Transparent Amorphous Metal Oxides for Thin-Film Transistors

In recent years, there has been increasing focus on making the solution-processable transparent metal oxides by wet techniques such as spin-coating or roll-to-roll printing on flexible substrates. These solution-processed metal oxides have better mobility and stability than their organic counterparts. The solution is convenient,

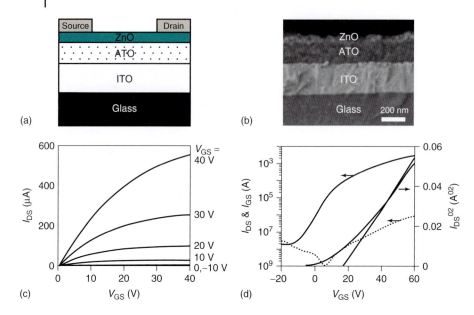

Figure 15.8 Solution-processed ZnO TFT. (Reprinted with permission from [117] copyright 2007 American Chemical society.)

inexpensive, and easy to prepare by simply dissolving the metal precursor and stabilizing agent in a suitable solvent. Generally, to make solution-processable thin films of high quality, the required annealing temperature is fairly high so as to eliminate the precursor salt completely [115]. The most recent focus of research is on lowering the processing temperature to be compatible with polymer substrates for flexible electronics where mostly organic semiconductors dominate.

In the colloid-based process, nanoparticles (~10 nm) are first synthesized and then a film is made by spin coating of the corresponding solution. The resulting film generally shows low mobility because of inefficient charge transport among the nanoparticles. The charge transport can be improved by removing the capping agent used during nanoparticle synthesis to improve the interparticle connectivity [116–118] (Figure 15.8). Ligand-free nanoparticles and nanoneedles of ZnO have been synthesized and the corresponding FETs have been tested [119]. Sol–gel process is also common, where the metal oxide precursor is made by dissolving a metal salt in an appropriate solvent, stabilizing agent, and water. Water molecules surrounding the metal ions first undergo hydrolysis, losing their proton and forming a hydroxyl or oxy-ligand, and then they undergo condensation, resulting in the M–O–M skeleton. High-temperature annealing is required to completely remove the anions of the metal salts; otherwise, residual organic impurities can hinder the formation of the metal oxide as well as the charge transport in the metal oxide film.

Solution-processed high-performance zinc oxide-based TFTs have been fabricated by direct dissolution of zinc hydroxide in aqueous ammonia solution. This solution allows for rapid metal amine dissociation, and the condensation and dehydration reaction leads to the formation of a ZnO nanocrystalline film at 140 °C by microwave-assisted annealing [120, 121]. The TFTs fabricated by using PECVD-grown SiO$_x$ dielectric on flexible polyimide substrates have been shown to withstand bendability, rollability, wearability, and foldability tests. ZTO TFTs have also been fabricated by dissolving zinc acetate and tin acetate in methoxy ethanol (0.75 M) and adding equimolar ethanolamine as a stabilizing agent. The TFTs were fabricated by spin-coating the solution on SiO$_2$-coated Si. Annealing the film at 500 °C led to a field-effect mobility of 1.1 cm^2 V^{-1} s^{-1} with an on/off ratio of 10^6. The electrical characteristics of the ZTO FET have been shown to be dependent on the Sn concentration [122]. ZTO TFTs have also been fabricated by ink-jet printing using the same solution [123].

In AMOSs, oxygen vacancy is the main source of mobile charge carriers. To obtain a high-mobility, high-performance transistor (low subthreshold swing), it is important to optimize the oxygen content. The effects of doping different metals such as Ga [124], Hf, Zr [125], and Yr [126, 127] in metal oxides have also been examined. These metal ions act as oxygen vacancy suppressors and decrease the off current of the TFTs. The role of gallium doping on the processing temperature of InGaZnO has been studied, and it has been found that the inclusion of Ga decreases the processing temperature. On the other hand, indium increases the mobility but limits the lowest processing temperature for the film formation [124]. Similarly, the effect of adding Al to indium tin oxide semiconductor has been shown to enhance the transistor performance. Al-doped indium tin oxide TFT shows a high mobility of 13.3 cm^2 V^{-1} s^{-1}, a high on/off ratio of 10^7, and a low subthreshold swing of 1.01 V dec^{-1} [128]. The effect of variation of Zn concentration on InZnGaO TFT has also been studied [129]. Similarly, in case of IZO (indium zinc oxide), the effect of the indium to zinc ratio on the TFT performance has also been examined [130].

Kim *et al.* have fabricated solution-processable indium tin oxide (ITO), by varying the In/Sn ratio and the annealing temperature such that the solution-processed ITO is compatible with flexible (polymeric) substrates. The ITO FET on SiO$_2$-coated silicon substrate shows mobility of 2 cm^2 V^{-1} s^{-1} with an on/off ratio of 10^4, whereas the same ITO FET on SAND shows mobility of 10–20 cm^2 V^{-1} s^{-1} [131]. The same group synthesized solution-processed Zn-In-Sn-O thin film transistors [132]. They have also fabricated solution-processable In$_2$O$_3$ TFTs using SiO$_2$ with high-κ organic SAND as the dielectric [61]. The In$_2$O$_3$ films were obtained by spin-coating indium chloride solution in methoxy ethanol and ethanolamine, which is followed by heating in a tube furnace at 400 °C in air for 10 min. The TFT is shown to give a saturation mobility of 43.7 cm^2 V^{-1} s^{-1} and an on/off ratio of 10^6.

Chemically assisted annealing (i.e., combustion synthesis) to obtain low-temperature-processed metal oxide films has been devised. Kim *et al.* have used urea or acetylacetonate as fuel and metal nitrate as oxidizing agent to synthesize the metal oxide film at low temperatures. The self-generating heat due to the local exothermic reaction taking place during the metal oxide formation eliminates the need of high external heat supply. This self-generating heat converts the metal precursor into the corresponding metal oxide at low processing temperatures [133]. Complete solution-processed FET with mobility of 13 cm^2 V^{-1} s^{-1} has been fabricated combining low-temperature combustion-synthesized In_2O_3 with low-temperature-synthesized Al_2O_3 as dielectric. Similarly, yttrium-doped indium oxide was synthesized by chemical-assisted annealing. Yttrium doping enhances the performance of TFTs as the Yr–O bond is stronger than the In–O bond, thus the resulting FETs having high-mobility and low carrier concentration [125].

Hydrolysis of metal alkoxide precursor-based "sol–gel on chip" was also developed to obtain low-temperature-processed metal oxides for high-mobility TFTs [134]. Whenever metal alkoxide is exposed to a humid atmosphere, hydrolysis and condensation occur, and there is transfer of proton from water to the bound alkoxide ligand, resulting in the formation of an M–O–M framework at low temperatures. The hydrolysis can be achieved by adding an aqueous catalyst (acid or base) to the metal alkoxide before film formation. Undesirable and uncontrolled precipitation can occur as a result of the high reactivity of the metal alkoxide. *In situ* hydrolysis annealing overcomes this drawback, as the hydrolysis process is directly applied on the surface of the film to make the M–O–M framework at low temperature. This approach was used to synthesize IZO TFT at 210–280 °C, showing a mobility of $4–12\,\text{cm}^2\,\text{V}^{-1}\,\text{s}^{-1}$.

Low-temperature solution-processed indium oxide thin film has also been made from indium chloride salt. The solution was made by dissolving indium chloride in acetonitrile, and then ethylene glycol was added as stabilizing agent. Finally, the spin-coated In_2O_3 film was annealed at 500 °C. The as-formed FET showed a mobility of 55 cm^2 V^{-1} s^{-1} with an on/off ratio of 10^7. When the spin-coated film was annealed in an O_2/O_3-rich atmosphere, high-quality In_2O_3 was achieved at 200–300 °C, and the FET showed a mobility of 22 cm^2 V^{-1} s^{-1} with an I_{on}/I_{off} of 10^6 [135].

More recently, Kim *et al.* have reported the synthesis of metal oxides such as IGZO, ZnO, and In_2O_3 at room temperature by photochemical oxidation of sol–gel films by deep UV irradiation. As the synthesis does not require annealing, the transistors can be made on flexible polymer substrates and are shown to be applicable for high-performance metal-oxide thin film electronic devices [136].

For the application of AMOS TFTs in display devices, there is a stringent requirement of long-term stability. The focus of recent studies has been to improve the long-term stability of these TFTs; that is, the threshold voltage and mobility should not change with operation time [137]. Even a 0.1-V change in threshold voltage of the driving TFT of the OLED can bring a 20% change in the brightness of LEDs. Various studies have been undertaken to study the effects of bias stress and passivation layer on the long-term stability of these AMOS TFTs.

The trap density in a-InGaZnO has been shown to be flat and deep when compared to that of a-Si. Further reduction in trap density is observed after annealing, which leads to a reduction in hysteresis [138]. The effects on TFT performance of the variation of metal ion ratio [125], presence of different solvents [129], and environmental conditions [139] have been studied in detail. The role of passivation layer on the bias stress stability of FETs has also been studied [135]. The effect of bias stress on the adsorption and thus the instability in the performance of spin-coated ZTO TFT with Si_3N_4 dielectric have also been investigated [140].

It was observed that the adsorbed oxygen rapidly increases the threshold voltage. High-pressure oxygen annealing is shown to improve the performance of tin-doped indium oxide transistor, as oxygen annealing leads to a reduction in oxygen vacancy concentration, which is the main cause of mobile ions in transistors [141]. The oxygen vacancies in both the channel and the dielectric were found to act as hole traps and play a role in negative bias illumination stress instability in a-InGaZnO TFTs [142].Various methods, such as photo-excited trap-charge-collection spectroscopy, have been used to measure the trap density at the semiconductor/dielectric interface [143]. ZTO transistors deposited by RF sputtering have been annealed at $300\,^\circ$C to improve their performance, but, when Al was added, annealing was required to obtain stable performance [144]. The AZTO TFT exhibited a field-effect mobility of $10.1\,\mathrm{cm^2\,V^{-1}\,s^{-1}}$, a turn-on voltage ($V_{on}$) of 0.4 V, a subthreshold swing (SS) of 0.6 V $\mathrm{dec^{-1}}$, and an on/off ratio of 10^9. The performance of the InZnO transistor made without the passivation layer has been studied in air and vacuum. The transistor shows instability under gate bias stress; that is, there is a shift in threshold voltage with gate bias. In air, the instability is found to be more significant than in vacuum because of the interaction of the film exposed to oxygen and moisture [145].

15.7
Challenges and Opportunities

Although the concept of realizing the FET by electrostatically tuning the material properties is attractive and exciting, there are scientific issues that remain to be addressed. The biggest problems in field-effect modulation in transition-metal oxides are the high charge densities in channel and therefore the short screening length. High-quality ultrathin films have been developed in recent times by using molecular beam epitaxy (MBE), but in doing so the strain and size effect introduce complications, sometimes masking the intrinsic physics. Nevertheless, this research field offers new opportunities in thin-film devices, especially in the field of transparent, low-temperature-processable transistors. One important research direction is to find new routes toward transistor devices that are not limited by the Thomas–Fermi screening length. In such devices, the carrier modulation at the dielectric/channel interface is expected to trigger a "domino" effect in the bulk of the oxide channel, leading to a collective change of ground state. Recently, this concept was realized by Nakano *et al.* [146] in the electrostatic macroscopic phase control in a VO_2 transistor. The inherent collective interaction between the electrons and the crystal lattice was proposed to be the origin of the nonlocal switching of the electronic state. Actually, complex oxides are often featured by strong interactions between charge, spin, orbital, and lattice; thus we expect more oxides to exhibit such "domino"-type field effects.

EDLT is a major progress toward unprecedented high modulation of carrier density, and more exciting results are expected as more powerful electrolytes are employed. For example, we will see works on transforming insulators to superconductors or drastically increasing the transition temperatures. On the other hand, there is an urgent need to gain more understanding on the physical mechanisms during the gating. Furthermore, dielectrics with higher breakdown strength and a wider electrochemical potential window are expected to be discovered. Besides the well-recognized stability issue, EDLTs are limited by low switching speed and sensitivity to water vapor [147, 148].

The major challenge in AMOS is to achieve high electron mobility with long-term stability under bias stress. In order for AMOS-based FETs to be viable for technological utilization, it is important to develop new and innovative solution-processing routes for the deposition of uniform and good quality films of these metal oxides on large areas, which could allow the use of techniques such as spin coating or printing. It is equally important that the processing temperatures are kept low enough to allow device fabrication on flexible polymer substrates.

Thus far, there has been very limited theoretical effort on elucidating the electric field effects on transition-metal oxides. Theoretical frameworks based on electronic band structures, first-principles methods, and device simulations are missing. Besides the commonly expected modulation of resistance and metal–insulator transition, future FET experiments probably can offer more information regarding the quantum phase transition, charge–spin–orbit interaction, magnetic ordering, and so on. This research also provides an opportunity to understand the quantum critical phenomena driven by quantum fluctuations, as electrostatic doping induces minimal disorder, and it has been acknowledged that disorder masks such quantum phase transitions [51, 149]. Last but not the least, down the road we will see more works on the critical control of phase boundaries and transitions at the nanoscale, and emerging materials will be incorporated into state-of-the-art devices.

References

1. Ahn, C.H., Triscone, J.-M., and Mannhart, J. (2003) Electric field effect in correlated oxide system. *Nature*, **424**, 1015.

2. Takahashi, K.S. *et al.* (2006) Local switching of two-dimensional superconductivity using the ferroelectric field effect. *Nature*, **441**, 195.

3. Dagotto, E. (2005) Complexity in strongly correlated electronic systems. *Science*, **309**, 257.

4. Imada, M., Fujimori, A., and Tokura, Y. (1998) Metal-insulator transitions. *Rev. Mod. Phys.*, **70**, 1039.

5. Ahn, C.H. *et al.* (2006) Electrostatic modification of novel materials. *Rev. Mod. Phys.*, **78**, 1185.

6. Fortunato, E. (2004) High field-effect mobility zinc oxide thin film transistors produced at room temperature. *J. Non-Cryst. Solids*, **338**, 806.

7. Suzuki, T.I. *et al.* (2004) Hall and field-effect mobilities of electrons accumulated at a lattice-matched ZnO/ScAlMgO$_4$ heterointerface. *Adv. Mater.*, **16**, 1887.

8. Nishii, J. *et al.* (2003) High mobility thin film transistors with transparent ZnO channels. *Jpn. J. Appl. Phys.*, **42**, L347.

9. Ueno, K. *et al.* (2011) Discovery of superconductivity in KTaO$_3$ by electrostatic carrier doping. *Nat. Nanotechnol.*, **6**, 408.

10. Inoue, I.H. (2005) Electrostatic carrier doping to perovskite transition-metal oxides. *Semicond. Sci. Technol.*, **20**, S112.

11. Robertson, J. (2004) High dielectric constant oxides. *Eur. Phys. J. Appl. Phys.*, **28**, 265.

12. Kingon, A.I., Maria, J.-P., and Streiffer, S.K. (2000) Alternative dielectrics to

silicon dioxide for memory and logic devices. *Nature*, **406**, 1032.

13. Kim, I.-D., Choi, Y., and Tuller, H.L. (2005) Low-voltage ZnO thin-film transistors with high-k $Bi_{1.5}Zn_{1.0}Nb_{1.5}O_7$ gate insulator for transparent and flexible electronics. *Appl. Phys. Lett.*, **87**, 043509.

14. Carcia, P.F., McLean, R.S., and Reilly, M.H. (2006) High-performance ZnO thin-film transistors on gate dielectrics grown by atomic layer deposition. *Appl. Phys. Lett.*, **88**, 123509.

15. Siddiqui, J. *et al.* (2006) ZnO thin-film transistors with polycrystalline (Ba,Sr)TiO$_3$ gate insulators. *Appl. Phys. Lett.*, **88**, 212903.

16. Levy, D.H., Freeman, D., Nelson, S.F., Cowdery-Corvan, P.J., and Irving, L.M. (2008) Stable ZnO thin film transistors by fast open air atomic layer deposition. *Appl. Phys. Lett.*, **92**, 192101.

17. Park, S.-H.K., Hwang, C.-S., Jeong, H.Y., Chu, H.Y., and Cho, K.I. (2008) Transparent ZnO-TFT arrays fabricated by atomic layer deposition. *Electrochem. Solid-State Lett.*, **11**, H10.

18. Nomura, K., Ohta, H., Takagi, A., Kamiya, T., Hirano, M., and Hosono, H. (2004) Room-temperature fabrication of transparent flexible thin-film transistors using amorphous oxide semiconductors. *Nature*, **432**, 488.

19. Yabuta, H., Sano, M., Abe, K., Aiba, T., Den, T., Kumomi, H., Nomura, K., Kamiya, T., and Hosono, H. (2006) High-mobility thin-film transistor with amorphous InGaZnO$_4$ channel fabricated by room temperature rf-magnetron sputtering. *Appl. Phys. Lett.*, **89**, 112123.

20. Zhang, L., Li, J., Zhang, X.W., Jiang, X.Y., and Zhang, Z.L. (2009) High performance ZnO-thin-film transistor with Ta$_2$O$_5$ dielectrics fabricated at room temperature. *Appl. Phys. Lett.*, **95**, 072112.

21. Dhananjay and Krupanidhi, S.B. (2007) Low threshold voltage ZnO thin film transistor with a $Zn_{0.7}Mg_{0.3}O$ gate dielectric for transparent electronics. *J. Appl. Phys.*, **101**, 123717.

22. Nomura, K., Ohta, H., Ueda, K., Kamiya, T., Hirano, M., and Hosono, H. (2003) Thin-film transistor fabricated in single-crystalline transparent oxide semiconductor. *Science*, **300**, 1269.

23. Kim, J.B., Fuentes-Hernandez, C., and Kippelen, B. (2008) High-performance InGaZnO thin-film transistors with high-k amorphous $Ba_{0.5}Sr_{0.5}TiO_3$ gate insulator. *Appl. Phys. Lett.*, **93**, 242111.

24. Chang, S., Song, Y.-W., Lee, S., Lee, S.Y., and Ju, B.-K. (2008) Efficient suppression of charge trapping in ZnO-based transparent thin film transistors with novel $Al_2O_3/HfO_2/Al_2O_3$ structure. *Appl. Phys. Lett.*, **92**, 192104.

25. Kim, I.-D., Lim, M.-H., Kang, K.T., Kim, H.-G., and Choi, S.-Y. (2006) Room temperature fabricated ZnO thin film transistor using high-k $Bi_{1.5}Zn_{1.0}Nb_{1.5}O_7$ gate insulator prepared by sputtering. *Appl. Phys. Lett.*, **89**, 022905.

26. Lim, M.-H., Kang, K.T., Kim, H.-G., Kim, I.-D., Choi, Y.W., and Tuller, H.L. (2006) Low leakage current—stacked $MgO/Bi_{1.5}Zn_{1.0}Nb_{1.5}O_7$ gate insulator for low voltage ZnO thin film transistors. *Appl. Phys. Lett.*, **89**, 202908.

27. Park, J.-S., Jeong, J.K., Mo, Y.-G., and Kim, S. (2009) Impact of high-k TiOx dielectric on device performance of indium-gallium zinc oxide transistors. *Appl. Phys. Lett.*, **94**, 042105.

28. Ohta, H., Sato, Y., Kato, T., Kim, S.W., Nomura, K., Ikuhara, Y., and Hosono, H. (2010) Field-induced water electrolysis switches an oxide semiconductor from an insulator to a metal. *Nat. Commun.*, **1**, 118.

29. Ohta, H., Mizuno, T., Zheng, S., Kato, T., Ikuhara, Y., Abe, K., Kumomi, H., Nomura, K., and Hosono, H. (2012) Unusually large enhancement of thermopower in an electric field induced two-dimensional electron gas. *Adv. Mater.*, **24**, 740.

30. Wu, T. *et al.* (2001) Electroresistance and electronic phase separation in mixed-valent manganites. *Phys. Rev. Lett.*, **86**, 5998.

31. Gariglio, S. *et al.* (2002) Electrostatic tuning of the hole density in $NdBa_2Cu_3O_{7-\delta}$ films and its effect on the hall response. *Phys. Rev. Lett.*, **88**, 067002.

32. Ahn, C.H. (1997) Local, nonvolatile electronic writing of epitaxial $Pb(Zr_{0.52}Ti_{0.48})O3/SrRuO_3$ Heterostructure. *Science*, **276**, 1100.

33. Nakamura, H. *et al.* (2006) Low temperature metallic state induced by electrostatic carrier doping of $SrTiO_3$. *Appl. Phys. Lett.*, **89**, 133504.

34. Watanabe, Y. (1995) Epitaxial all-perovskite ferroelectric field effect transistor with a memory retention. *Appl. Phys. Lett.*, **66**, 1770.

35. Takaobushi, J., Kanki, T., Kawai, T., and Tanaka, H. (2011) Preparation of ferroelectric field effect transistor based on sustainable strongly correlated Fe,Zn_3O_4 oxide semiconductor and their electrical transport properties. *Appl. Phys. Lett.*, **98**, 102506.

36. Lee, K.H., Lee, G., Lee, K., Oh, M.S., Im, S., and Yoon, S.-M. (2009) High-mobility nonvolatile memory thin-film transistors with a ferroelectric polymer interfacing ZnO and pentacene channels. *Adv. Mater.*, **21**, 4287.

37. Nakano, M. *et al.* (2010) Electric-field control of two-dimensional electrons in polymer-gated-oxide semiconductor heterostructures. *Adv. Mater.*, **22**, 876.

38. Prassides, K. (2011) Condensed matter physics: superconductivity at the double. *Nat. Nanotechnol.*, **6**, 400.

39. Yuan, H. *et al.* (2010) Electrostatic and electrochemical nature of liquid-gated electric-double-layer transistors based on oxide semiconductors. *J. Am. Chem. Soc.*, **132**, 18402.

40. Yuan, H. *et al.* (2009) High-density carrier accumulation in ZnO field-effect transistors gated by electric double layers of ionic liquids. *Adv. Funct. Mater.*, **19**, 1046.

41. Misra, R., McCarthy, M., and Hebard, A.F. (2007) Electric field gating with ionic liquids. *Appl. Phys. Lett.*, **90**, 052905.

42. Shimotani, H. *et al.* (2007) Insulator-to-metal transition in ZnO by electric double layer gating. *Appl. Phys. Lett.*, **91**, 082106.

43. Shimotani, H. *et al.* (2008) p-type field-effect transistor of NiO with electric double-layer gating. *Appl. Phys. Lett.*, **92**, 242107.

44. Ueno, K. *et al.* (2010) Electrostatic charge accumulation versus electrochemical doping in $SrTiO_3$ electric double layer transistors. *Appl. Phys. Lett.*, **96**, 252107.

45. Ye, J.T. *et al.* (2009) Liquid-gated interface superconductivity on an atomically flat film. *Nat. Mater.*, **9**, 125.

46. Lee, Y. *et al.* (2011) Phase diagram of electrostatically doped $SrTiO_3$. *Phys. Rev. Lett.*, **106**, 136809.

47. Dhoot, A.S. *et al.* (2010) Increased T(c) in electrolyte-gated cuprates. *Adv. Mater.*, **22** (23), 2529.

48. Yamada, Y. *et al.* (2011) Electrically induced ferromagnetism at room temperature in cobalt-doped titanium dioxide. *Science*, **332**, 1065.

49. Ueno, K. *et al.* (2008) Electric-field-induced superconductivity in an insulator. *Nat. Mater.*, **7**, 855.

50. Scherwitzl, R. *et al.* (2010) Electric-field control of the metal-insulator transition in ultrathin NdNiO films. *Adv. Mater.*, **22**, 5517.

51. Bollinger, A.T. *et al.* (2011) Superconductor-insulator transition in $La_{2-x}Sr_xCuO_4$ at the pair quantum resistance. *Nature*, **472**, 458.

52. Lee, C.-G. and Dodabalapur, A. (2010) Solution-processed zinc–tin oxide thin-film transistors with low interfacial trap density and improved performance. *Appl. Phys. Lett.*, **96**, 243501.

53. Park, Y.M., Daniel, J., Heeney, M., and Salleo, A. (2011) Room-temperature fabrication of ultrathin oxide gate dielectrics for Low-voltage operation of organic field-effect transistors. *Adv. Mater.*, **23**, 971.

54. Choi, S., Park, B.-Y., Jeong, S., and Jung, H.-K. (2011) Promising solution processed lanthanide films as high-*k* gate insulators for Low voltage-driven oxide thin film transistor. *Electrochem. Solid-State Lett.*, **14**, H426.

55. Adamopoulos, G., Thomas, S., Bradley, D.D.C., McLachlan, M.A., and Anthopoulos, T.D. (2011) Low-voltage ZnO thin-film transistors based on Y_2O_3 and Al_2O_3 high-*k* dielectrics deposited by spray pyrolysis in air. *Appl. Phys. Lett.*, **98**, 123503.

56. Lorenz, M., von Wenckstern, H., and Grundmann, M. (2011) Tungsten oxide as a gate dielectric for highly transparent and temperature-stable zinc-oxide-based thin-film transistors. *Adv. Mater.*, **23**, 5383.

57. Pal, B.N., Dhar, B.M., See, K.C., and Katz, H.E. (2009) Solution-deposited sodium beta-alumina gate dielectrics for low-voltage and transparent field-effect transistors. *Nat. Mater.*, **8**, 898.

58. Lu, A., Sun, J., Jiang, J., and Wan, Q. (2010) Low-voltage transparent electric-double-layer ZnO-based thin-film transistors for portable transparent electronics. *Appl. Phys. Lett.*, **96**, 043114.

59. Cai, Q.J., Gan, Y., Chan-Park, M.B., Yang, H.B., Lu, Z.S., Li, C.M., Guo, J., and Dong, Z.L. (2009) Solution-processable barium titanate and strontium titanate nanoparticle dielectrics for Low-voltage organic thin-film transistors. *Chem. Mater.*, **21**, 3153.

60. Cha, S.H., Oh, M.S., Lee, K.H., Im, S., Lee, B.H., and Sung, M.M. (2008) Electrically stable low voltage ZnO transistors with organic/inorganicnanohybrid dielectrics. *Appl. Phys. Lett.*, **92**, 023506.

61. Kim, H.S., Byrne, P.D., Facchetti, A., and Marks, T.J. (2008) High performance solution-processed indium oxide thin-film transistors. *J. Am. Chem. Soc.*, **130**, 12580.

62. DiBenedetto, S.A., Frattarelli, D., Ratner, M.A., Facchetti, A., and Marks, T.J. (2008) Vapor phase self-assembly of molecular gate dielectrics for thin film transistors. *J. Am. Chem. Soc.*, **130**, 7528.

63. Liu, J., Hennek, J.W., Bruce Buchholz, D., Ha, Y.-G., Xie, S., Dravid, V.P., Chang, R.P.H., Facchetti, A., and Marks, T.J. (2011) Reinforced self-assembled nanodielectrics for high-performance transparent thin film transistors. *Adv. Mater.*, **23**, 992.

64. Liu, J., Buchholz, D.B., Chang, R.P.H., Facchetti, A., and Marks, T.J. (2010) High-performance flexible transparent thin-film transistors using a hybrid gate dielectric and an amorphous zinc indium tin oxide channel. *Adv. Mater.*, **22**, 2333.

65. Zhao, T. *et al.* (2005) Electric field effect in diluted magnetic insulator anatase Co: TiO_2. *Phys. Rev. Lett.*, **94**, 126601.

66. Mathews, S. (1997) Ferroelectric field effect transistor based on epitaxial perovskite heterostructures. *Science*, **276**, 238.

67. Ogale, S.B. *et al.* (1996) Unusual electric field effects in $Nd_{0.7}Sr_{0.3}MnO_3$. *Phys. Rev. Lett.*, **77**, 1159.

68. Zhao, T. *et al.* (2004) Colossal magnetoresistive manganite-based ferroelectric field-effect transistor on Si. *Appl. Phys. Lett.*, **84** (5), 750.

69. Hong, X. *et al.* (2003) Ferroelectric-field-induced tuning of magnetism in the colossal magnetoresistive oxide La1-xSrxMnO3. *Phys. Rev. B*, **68**, 134415.

70. Hong, X., Posadas, A., and Ahn, C.H. (2005) Examining the screening limit of field effect devices via the metal-insulator transition. *Appl. Phys. Lett.*, **86** (14), 142501.

71. Hu, F.X. and Gao, J. (2006) Investigations on electroresistance effect in epitaxial manganite films using field effect configurations. *Appl. Phys. Lett.*, **88** (13), 132502.

72. Kanki, T. *et al.* (2003) Electrical-field control of metal–insulator transition at room temperature in $Pb(Zr_{0.2}Ti_{0.8})O_3/La_{1-x}Ba_xMnO_3$ field-effect transistor. *Appl. Phys. Lett.*, **83** (23), 4860.

73. Pallecchi, I. *et al.* (2008) Field effect in manganite ultrathin films: magnetotransport and localization mechanisms. *Phys. Rev. B*, **78**, 024411.

74. Sun, Y. *et al.* (2008) Electric and magnetic modulation of fully strained dead layers in La0.67Sr0.33MnO3 films. *Phys. Rev. B*, **78**, 024412.

75. Pallecchi, I. *et al.* (2003) Reversible shift of the transition temperature of manganites in planar field-effect devices patterned by atomic force microscope. *Appl. Phys. Lett.*, **83** (21), 4435.

76. Dhoot, A. *et al.* (2009) Large electric field effect in electrolyte-gated manganites. *Phys. Rev. Lett.*, **102** (13), 136402.

77. Chiba, D. *et al.* (2011) Electrical control of the ferromagnetic phase transition in

cobalt at room temperature. *Nat. Mater.*, **10**, 853.

78. Weisheit, M. *et al.* (2007) Electric field-induced modification of magnetism in thin-film ferromagnets. *Science*, **315** (5810), 349.

79. Glover, R.E. III, and Sherrill, M.D. (1960) Changes in superconducting critical temperature produced by electrostatic charging. *Phys. Rev. Lett.*, **5** (6), 248.

80. Konsin, P. and Sorkin, B. (1998) Electric field effects in high-T_c cuprates. *Phys. Rev. B*, **58** (9), 5795.

81. Xi, X.X. *et al.* (1992) Effects of field-induced hole-density modulation on normal-state and superconducting transport in $YBa_2Cu_3O_{7-x}$. *Phys. Rev. Lett.*, **68**, 1240.

82. Chandrasekhar, N., Valls, O.T., and Goldman, A.M. (1993) Mechanism for electric field effects observed in $YBa_2Cu_3O_{7-x}$ films. *Phys. Rev. Lett.*, **71** (7), 1079.

83. Frey, T. *et al.* (1995) Mechanism of the electric-field effect in the high-T_c cuprates. *Phys. Rev. B*, **51** (5), 3257.

84. Talyansky, V. *et al.* (1996) Experimental proof of the electronic charge-transfer mechanism in a $YBa_2Cu_3O_{7-x}$-based field-effect transistor. *Phys. Rev. B*, **53** (21), 14575.

85. Salluzzo, M. *et al.* (2008) Indirect electric field doping of the CuO_2 planes of the cuprate $NdBa_2Cu_3O_7$ superconductor. *Phys. Rev. Lett.*, **100** (5), 056810.

86. Brazovskii, S.A. and Yakovenko, V.M. (1988) Possible superconductivity on the junction surface of dielectric La2CuO4. *Phys. Lett. A*, **132** (5), 290.

87. Fiory, A.T. *et al.* (1990) Metallic and superconducting surfaces of $YBa2Cu_3O_7$ probed by electrostatic charge modulation of epitaxial films. *Phys. Rev. Lett.*, **65** (27), 3441.

88. Mannhart, J. *et al.* (1991) Electric field effect on superconducting $YBa_2Cu_3O_{7-\delta}$ films. *Z. Phys. B: Condens. Matter*, **83** (3), 307.

89. Mannhart, J. *et al.* (1991) Influence of electric fields on pinning in $YBa_2Cu_3O_{7-\delta}$ films. *Phys. Rev. Lett.*, **67** (15), 2099.

90. Logvenov, G.Y. *et al.* (2005) Studies of superconducting field effect transistors with sheet resistances close to the quantum resistance. *Appl. Phys. Lett.*, **86** (20), 202505.

91. Bell, C. *et al.* (2009) Dominant mobility modulation by the electric field effect at the $LaAlO_3/SrTiO_3$ interface. *Phys. Rev. Lett.*, **103** (22), 226802.

92. Caviglia, A.D. *et al.* (2008) Electric field control of the LaAlO3/SrTiO3 interface ground state. *Nature*, **456** (7222), 624.

93. Dikin, D.A. *et al.* (2011) Coexistence of superconductivity and ferromagnetism in two dimensions. *Phys. Rev. Lett.*, **107** (5), 056802.

94. Caviglia, A.D. *et al.* (2010) Tunable rashba spin-orbit interaction at oxide interfaces. *Phys. Rev. Lett.*, **104** (12), 126803.

95. Ahn, C.H. *et al.* (1995) Ferroelectric field effect in epitaxial thin film oxide $SrCuO_2/Pb(Zr_{0.52}Ti_{0.48})O_3$ heterostructures. *Science*, **269**, 373.

96. Ahn, C.H. *et al.* (1999) Electrostatic modulation of superconductivity in ultrathin $GdBa_2Cu_3O_{7-x}$ films. *Science*, **284**, 1152.

97. Leng, X. *et al.* (2011) Electrostatic control of the evolution from a superconducting phase to an insulating phase in ultrathin $YBa_2Cu_3O_7$ films. *Phys. Rev. Lett.*, **107** (2), 027001.

98. Kamiya, T. and Hosono, H. (2010) Material characteristics and applications of transparent amorphous oxide semiconductors. *NPG Asia Mater.*, **2** (1), 15.

99. Hoffman, R.L., Norris, B.J., and Wager, J.F. (2003) ZnO-based transparent thin-film transistors. *Appl. Phys. Lett.*, **82**, 733.

100. Carcia, P.F., McLean, Z.S., Reilly, M.H., and Nunes, G. (2003) Transparent ZnO thin-film transistor fabricated by rf magnetron sputtering. *Appl. Phys. Lett.*, **82**, 1117.

101. Fortunato, E.M.C., Barquinha, P.M.C., Pimentel, A., Goncalves, A.M.F., Marques, A.J.S., Pereira, L.M.N., and Martins, R.F.P. (2005) Fully transparent ZnO thin-film transistor produced at room temperature. *Adv. Mater.*, **17**, 590.

102. Hosono, H. (2006) Ionic amorphous oxide semiconductors: material design, carrier transport, and device application. *J. Non-Cryst. Solids*, **352**, 851.

103. Li, J., Kaneda, T. *et al.* (2012) P-type conductive amorphous oxides of transition metals from solution processing. *Appl. Phys. Lett.*, **101**, 052102.

104. Matsuzaki, K. *et al.* (2008) Epitaxial growth of high mobility Cu_2O thin films and application top-channel thin film transistor. *Appl. Phys. Lett.*, **93**, 202107.

105. Fortunato, E. *et al.* (2010) Thin-film transistors based on p-type Cu2O thin films produced at room temperature. *Appl. Phys. Lett.*, **96**, 192102.

106. Sung, S.-Y. *et al.* (2010) Fabrication of p-channel thin-film transistors using CuO active layers deposited at low temperature. *Appl. Phys. Lett.*, **97**, 222109.

107. Zou, X.A. *et al.* (2010) Top-gate low-threshold voltage thin-film transistor grown on substrate using a high-*k* HfON gate dielectric. *IEEE Electron Dev. Lett.*, **31**, 827.

108. Yao, Z.Q. *et al.* (2012) Room temperature fabrication of p-channel Cu_2O thin-film transistors on flexible polyethylene terephthalate substrates. *Appl. Phys. Lett.*, **101**, 042114.

109. Fortunato, E., Barquinha, P., and Martins, R. (2012) Oxide semiconductor thin-film transistors: a review of recent advances. *Adv. Mater.*, **24**, 2945.

110. Ogo, Y. *et al.* (2008) p-channel thin-film transistor using p-type oxide semiconductor, SnO. *Appl. Phys. Lett.*, **93**, 032113.

111. Yabuta, H. *et al.* (2010) Sputtering formation of p-type SnO thin-film transistors on glass toward oxide complimentary circuits. *Appl. Phys. Lett.*, **97**, 072111.

112. Liang, L.Y. *et al.* (2010) Sun phase and optical characterizations of annealed SnO thin films and their p-type TFT application. *J. Electrochem. Soc.*, **157**, H598.

113. Fortunato, E. *et al.* (2010) Transparent p-type SnOx thin film transistors produced by reactive rf magnetron sputtering followed by low temperature annealing. *Appl. Phys. Lett.*, **97**, 052105.

114. Okamura, K. *et al.* (2012) Solution-processed oxide semiconductor SnO in p-channel thin-film transistors. *J. Mater. Chem.*, **22**, 4607.

115. Jeong, S. and Moon, J. (2012) Low-temperature, solution-processed metal oxide thin film transistors. *J. Mater. Chem.*, **22**, 1243.

116. Okamura, K. *et al.* (2010) Polymer stabilized ZnO nanoparticles for low-temperature and solution processed field-effect transistors. *J. Mater. Chem.*, **20**, 5651.

117. Ong, B.S. *et al.* (2007) Stable, solution-processed, high-mobility ZnO thin-film transistors. *J. Am. Chem. Soc.*, **129**, 2750.

118. Sun, B. and Sirringhaus, H. (2005) Solution-processed zinc oxide field-effect transistors based on self-assembly of colloidal nanorods. *Nano Lett.*, **5** (12), 2408.

119. Jun, T. *et al.* (2011) Influences of pH and ligand type on the performance of inorganic aqueous precursor-derived ZnO thin film transistors. *ACS Appl. Mater. Interfaces*, **3**, 774.

120. Jun, T. *et al.* (2011) High-performance low-temperature solution-processable ZnO thin film transistors by microwave-assisted annealing. *J. Mater. Chem.*, **21**, 1102.

121. Song, K. *et al.* (2010) Fully flexible solution-deposited ZnO thin-film transistors. *Adv. Mater.*, **22**, 4308.

122. Jeong, S., Jeong, Y., and Moon, J. (2008) Solution-processed zinc tin oxide semiconductor for thin-film transistors. *J. Phys. Chem. C*, **112** (30), 11082.

123. Kim, D. *et al.* (2009) Inkjet-printed zinc tin oxide thin-film transistor. *Langmuir*, **25** (18), 11149.

124. Jeong, S. *et al.* (2010) Role of gallium doping in dramatically lowering amorphous-oxide processing temperatures for solution-derived indium zinc oxide thin-film transistors. *Adv. Mater.*, **22**, 1346.

125. Rim, Y.S. *et al.* (2010) Effect of Zr addition on ZnSnO thin-film transistors using a solution process. *Appl. Phys. Lett.*, **97**, 233502.

126. Hennek, J.W. *et al.* (2012) Exploratory combustion synthesis: amorphous indium yttrium oxide for thin-film transistors. *J. Am. Chem. Soc.*, **134**, 9593.

127. Jun, T. *et al.* (2011) Bias stress stable aqueous solution derived Y-doped ZnO thin film transistors. *J. Mater. Chem.*, **21**, 13524.

128. Jeon, J.H. *et al.* (2010) Addition of aluminum to solution processed conductive indium tin oxide thin film for an oxide thin film transistor. *Appl. Phys. Lett.*, **96**, 212109.

129. Nayak, P.K. *et al.* (2010) Zinc concentration dependence study of solution processed amorphous indium gallium zinc oxide thin film transistors using high-*k* dielectric. *Appl. Phys. Lett.*, **97**, 183504.

130. Kim, D. *et al.* (2009) Compositional influence on sol-gel-derived amorphous oxide semiconductor thin film transistors. *Appl. Phys. Lett.*, **95**, 103501.

131. Kim, H.S. *et al.* (2009) Low-temperature solution-processed amorphous indium tin oxide field-effect transistors. *J. Am. Chem. Soc.*, **131**, 10826.

132. Kim, M.-G. *et al.* (2010) High-performance solution-processed amorphous zinc-indium-Tin oxide thin-film transistors. *J. Am. Chem. Soc.*, **132**, 10352.

133. Kim, M.-G. *et al.* (2011) Low-temperature fabrication of high-performance metal oxide thin-film electronics via combustion processing. *Nat. Mater.*, **10**, 382.

134. Banger, K.K. *et al.* (2011) Low-temperature, high-performance solution-processed metal oxide thin-film transistors formed by a 'sol-gel on chip' process. *Nat. Mater.*, **10**, 45.

135. Han, S.-Y., Herman, G.S., and Chang, C.-H. (2011) Low-temperature, high-performance, solution-processed indium oxide thin-film transistors. *J. Am. Chem. Soc.*, **133**, 5166.

136. Kim, Y.-H. *et al.* (2012) Flexible metal-oxide devices made by room-temperature photochemical activation of sol-gel films. *Nature*, **489**, 11434.

137. Park, J.S. *et al.* (2012) Review of recent developments in amorphous oxide semiconductor thin-film transistor devices. *Thin Solid Films*, **520**, 1679.

138. Kimura, M. *et al.* (2008) Trap densities in amorphous-InGaZnO4 thin-film transistors. *Appl. Phys. Lett.*, **92**, 133512.

139. Kim, Y.-H. *et al.* (2010) Solvent-mediated threshold voltage shift in solution-processed transparent oxide thin-film transistors. *Appl. Phys. Lett.*, **97**, 092105.

140. Chen, Y.-C. *et al.* (2010) Bias-induced oxygen adsorption in zinc tin oxide thin film transistors under dynamic stress. *Appl. Phys. Lett.*, **96**, 262104.

141. Park, S.Y. *et al.* (2012) Improvement in the device performance of tin-doped indium oxide transistor by oxygen high pressure annealing at 150 °C. *Appl. Phys. Lett.*, **100**, 162108.

142. Ryu, B. *et al.* (2010) O-vacancy as the origin of negative bias illumination stress instability in amorphous In-Ga-Zn-O thin film transistors. *Appl. Phys. Lett.*, **97**, 022108.

143. Lee, K. *et al.* (2010) Density of trap states measured by photon probe into ZnO based thin-film transistors. *Appl. Phys. Lett.*, **97**, 082110.

144. Cho, D.-H. *et al.* (2008) Transparent Al–Zn–Sn–O thin film transistors prepared at low temperature. *Appl. Phys. Lett.*, **93**, 142111.

145. Liu, P.-T., Chou, Y.-T., and Teng, L.-F. (2009) Environment-dependent metastability of passivation-free indium zinc oxide thin film transistor after gate bias stress. *Appl. Phys. Lett.*, **95**, 233504.

146. Nakano, M. *et al.* (2012) Collective bulk carrier delocalization driven by electrostatic surface charge accumulation. *Nature*, **487**, 459.

147. Rivera-Rubero, S. and Baldelli, S. (2004) Influence of water on the surface of hydrophilic and hydrophobic room-temperature ionic liquids. *J. Am. Chem. Soc.*, **126**, 11788.

148. Welton, T. (1999) Room-temperature ionic liquids. Solvents for synthesis and catalysis. *Chem. Rev.*, **99**, 2071.

149. Coleman, P. and Schofield, A.J. (2005) Quantum criticality. *Nature*, **433**, 226.

16
Resistive Switchings in Transition-Metal Oxides

Isao H. Inoue and Akihito Sawa

16.1
Introduction

Promising candidates for the next-generation memory devices have emerged one after another for the past decade. Ferroelectric random access memories (FeRAM), magnetoresistive random access memories (MRAM), and phase-change memories (PCM) are indeed at the dawn of the international development races. Along with these three fascinating memories, we focus here on probably the most seminal candidate of the future device—resistive random access memory (RRAM® or ReRAM). ReRAM consists of a simple metal/oxide/metal sandwich structure, as shown in Figure 16.1 with myriad combinations of the metals and oxides. The sandwich shows reversible and nonvolatile changes of the electric resistance by applications of ordinary electric pulses (Figure 16.1b). This phenomenon is called "resistance change" or "resistive switching."

Because of the nondescript structure and simple switching operation, ReRAM is believed to hold potentially much better cell-size scalability than those of the established memory technologies, even including the brand-new FeRAM, MRAM, and PCM.

It is interesting to mention that the phrase of resistive switching for ReRAM was taken, in literature, in a very restricted meaning, representing a sort of phenomena *without* a phase-change, magnetoresistance, or ferroelectricity. Furthermore, this classic resistive switching has been praised as a *precocious talent* for half a century since it was discovered. The typical examples of the classic resistive switching were Al_2O_3-based and SiO_2-based sandwiches on "electroforming" [2, 3]. Then, what is the electroforming? In short, it is an application of a voltage above a certain critical value to the sandwich to draw out the latent resistive switching properties [4–8]. This electroforming or simply forming was believed to be an inevitable process for the resistive switching; we will classify several types of electroformings in the following section and suggest the necessity of the process.

Extensive reviews of the classic resistive switching had been already given around 1970s and 1980s by Dearnaley *et al.* [9], Biederman [10], Oxley [2], and Pagnia and Sotnik [11]. Those reviews successfully identified the key players: voids,

Functional Metal Oxides: New Science and Novel Applications, First Edition.
Edited by Satishchandra B. Ogale, Thirumalai V. Venkatesan, and Mark G. Blamire.
© 2013 Wiley-VCH Verlag GmbH & Co. KGaA. Published 2013 by Wiley-VCH Verlag GmbH & Co. KGaA.

Pulse voltage

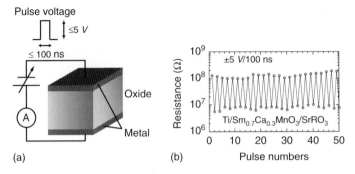

(a)   (b)

Figure 16.1 Schematic diagram of a ReRAM memory cell with a metal/oxide/metal sandwich structure and resistive switching characteristics in a Ti/$Sm_{0.7}Ca_{0.7}MnO_3$/$SrRuO_3$ cell at room temperature. By applying pulsed voltages of ± 5 V, the resistance of the cell changes reversibly between high- and low-resistance states. (Source: Adapted with permission from Ref. [1] © 2008 Elsevier.)

dislocations, and defects, which are, in a word, nonstoichiometry unavoidable in every oxide thin film. Nevertheless, around the mid-1980s, the research on this topic was rapidly left on the backburner. This was probably because of the legendary discovery of the *high-temperature superconductivity in Cu oxides* in 1980s [12], toward which most of the physicists and chemists in the field of metal-oxides steered the course of their researches. Moreover, the epoch-making invention of the *flash memory* was also done in 1980s. This charge-storage-type nonvolatile memory gave a great impact in the field of electronic engineering. The two incidents have balked further application research of the classic resistive switching phenomena.

However, since around the year of 2000, it has become the highest concern that the stronghold of the charge-based memory is being imperiled because of the technical and physical limitations of miniaturization. Then, a renewed interest of the resistive switching phenomena resurfaced [13, 21]. Among the new works, notably pioneering ones were done by Beck *et al.* [22] and Liu *et al.* [23], demonstrating resistive switchings in new types of perovskite transition-metal oxides. Because the former was reported by the research institute of IBM and because the latter researchers were in collaboration with SHARP [24, 25], the two reports were widely appreciated and the prospectivity of the metal/oxide/metal sandwiches for the future nonvolatile memory was renowned. Furthermore, Seo *et al.* in SAMSUNG [26, 27] reported another type of resistive switching using a binary oxide NiO, which has ignited new-generation experimental works [28–35] and theoretical works [36–38]. Eventually, these new works have rekindled the long-running controversy on the microscopic mechanisms behind the resistive switching phenomena.

With regard to the latest research results, the classic and rather exclusive definition of the resistive switching is considered to be antiquated or rather pointless. This is because the different phenomena encompassing the broad range of physics were only naively classified so far by the *differences of oxides*. Therefore, several recent works [1, 39–41] reviewed again the smorgasbord of resistance switching phenomena, including the latest results, and have labeled not only the materials difference but also the key features from renewed points of views.

The common key words are *uniform or local*, as well as *drift or diffusion*. If the resistive switching depends on the polarity of the external bias voltage, drift movement of charges (ions or vacancies) is the central character, causing the valence change at the interface. Meanwhile, if it is independent of the polarity, thermal diffusion is more essential for driving the resistive switching. Of course, we think it may be too early to take up a unified stand. Considerable time will have been spent before the microscopic mechanism of each resistive switching phenomenon is fully understood and incontrovertible.

Here, our aim is to add another unique point of view to those recent reviews in order to encourage more comprehensive researches on the resistive switching phenomena. We present a one-look overview of the resistive switching by schematically classifying their typical current–voltage (I–V) hystereses. For a space constraint, we especially focus on one of the classified— bipolar continuous switching and its possible device applications.

16.2
Classification of Current–Voltage Hystereses

The I–V hysteresis is, in general, a good measure of the electronic properties of materials and interfaces. The resistive switching phenomenon can be attributed to one of the four typical hystereses shown later. However, it should be noted that many samples of resistive switching are rather complex; that is, they are built-in serial and/or parallel connections of the resistive switching with different I–V hystereses. Moreover, the I–V hystereses may depend on the sweep rates of bias voltage/current, which are not specified in many experimental papers. We therefore have to preface that the classification shown in the following sections is never monolithic, and its boundaries are essentially blurred. Nevertheless, we believe that this kind of terse classification will be helpful for organizing and reconstructing a volume of research in this field.

16.2.1
Continuous and Fuse–Antifuse Switchings

At first view, I–V curves of the resistive switching phenomena are classified into two different types, as shown in Figure 16.2. The top row corresponds to the so-called fuse–antifuse switching, while the bottom one is called the *continuous switching*. In Figure 16.2a, the dashed lines correspond to the low-resistance state (LRS), while the solid lines correspond to the high-resistance state (HRS). The resistive switching from HRS to LRS is called *set*, while the opposite is called *reset*.

In general, the resistance change can be ascribed to spatiotemporal pattern formations under sufficiently high electric field or current [42–44]. When an inhomogeneity of voltage is formed, N-type negative differential resistance (NDR)

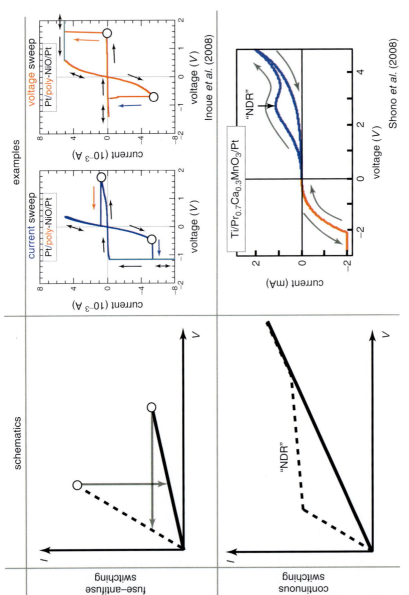

Figure 16.2 (a) Schematic *I–V* curves for the fuse–antifuse switching (top) and continuous switching (bottom). (b) Corresponding examples [49, 50] are shown. (Source: Adapted with permissions from Ref. [49] © 2008 American Physical Society and from Ref. [50] © 2008 Japan Society of Applied Physics, respectively.) For the fuse–antifuse switching, opaque circles denote the threshold points at which sudden resistance changes occur. The sudden changes are equally seen in both the *V*- and *I*-sweep measurements. The continuous switching exhibits a negative differential resistance (NDR), which reads as "NDR" because it is nonvolatile, and, in many cases, the derivative is not indeed negative.

is observed. Meanwhile, when a current filamentation is done, S-type NDR is observed. The example of the continuous switching shown in Figure 16.2 exhibits a kind of N-type NDR manifesting the formation of the inhomogeneous electric field inside the device. In general, NDR does not guarantee the nonvolatility. We denote NDR in Figure 16.2 as "NDR" to distinguish it from the classic and volatile NDR. For "NDR", the derivative is not indeed negative in several cases. It is widely believed that this "NDR" is due to a nonvolatile inhomogeneity formation, which is a Schottky barrier alternating its height/width continuously at the metal/oxide interface because of the drift of charged ions and defects [45–47]. Trap–detrap of the drifting electrons may also cause the continuous switching with the "NDR" [40, 48].

A sudden change of resistance at a "threshold" is observed in either the V- or the I-sweep measurement [51]. However, as shown in the example of the fuse–antifuse switching, the abrupt change of the resistance at the threshold (open circles in Figure 16.2) can be observed both in the V- and I-sweep measurement. This is distinctly different from the conventional threshold switching and cannot be understood on an equal footing with the continuous switching. Recently, it has been demonstrated that the current flows inhomogeneously through the oxide in the low-resistance state, while, in contrast, it remains rather homogeneous in the high-resistance state [49, 52, 53]. This suggests the existence of a *current constricting* structure, most probably at the interface. Thanks to the constriction, the current density becomes extremely large especially when the resistance state changes from low to high. A shared understanding in this research field is that the sudden change of resistance is related to the current constriction structure, which is popularly called *filament* or, in a more specific understanding, an electric *faucet* at the interface. The local current path is cut and connected by a local phase transition, and the switching is thus called a fuse–antifuse type. The local phase transition, especially a metal–insulator (MI) transition, can be either structural or electronic transition; however, the most important point is the *locality* [6, 9, 11, 28, 30, 33, 49] of the current path and the transition, which makes the switching abrupt when compared with the timescale of V or I sweep. Here, it should be noted that, even when a clear filament is formed through the oxide, the tip of the filament (faucet) at the interface with the metal electrode does not always show the fuse–antifuse action (phase transition) but sometimes shows the continuous switching. This is due to the delicate balance of the drift/diffusion (strength of the current density) of vacancies and ions and the sweep rate of the applied voltage and current; this important issue is reviewed more in the following discussion.

16.2.2
Polarity Dependence and the Origin Symmetry

In this review, we focus on the self-crossing $I–V$ curves passing through the origin of the $I–V$ plane.

The resistive switching with the self-crossing $I–V$ curves show (topologically) origin-symmetric high- and low-resistance states as schematically plotted in Figure 16.3 (typical examples are seen in Figure 16.4). If the switchings between

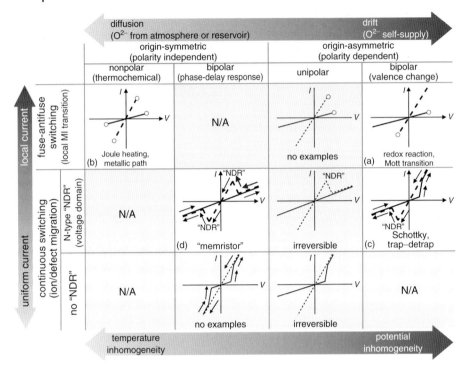

Figure 16.3 Classification of *I–V* hystereses. Dashed and solid lines correspond to low- and high-resistance states, respectively. Arrows indicate the direction of the hystereses. Origin symmetry means that the topology of the hysteresis are symmetric with respect to the origin of the *I–V* plane. The patterns that cannot be drawn even schematically, those without any obvious examples, or those irreversible are shaded. Four patters remain: (a) bipolar fuse–antifuse type due to, for example, local redox reactions and Mott transition, (b) nonpolar fuse–antifuse type related to the thermochemical reaction, (c) bipolar continuous type due to the potential inhomogeneity by the ion/defects migration, and (d) "memristor" type, where thick black arrows indicate the characteristic behavior only seen in the memristor type. The (b)-type is conventionally called *unipolar* but to be exact, it should be called *nonpolar*. See text for details.

the two states are dependent on, or independent of, the polarity, the topology of the *I–V* curves becomes origin-asymmetric or origin-symmetric, respectively.

An explanation of the origin-symmetry (polarity-independence) is given by assuming that the Joule heating and thermal diffusion of ions and defects could be dominant. This is reasonable because the Joule heating is independent of the polarity of applied voltage or current. (Here, it is interesting to note that other heatings, such as those due to tunneling-electron energy deposition, is polarity dependent.) Similarly, the origin-asymmetry (polarity-dependence) can be naturally associated with the electric-field drift of ions and defects along the direction of applied voltage or current. If the diffusion of, for example, oxygen ions is dominant, the system requires the supply of oxygens from outside, that is, from either the atmosphere or some kind of metal electrodes, which work as oxygen reservoirs [54].

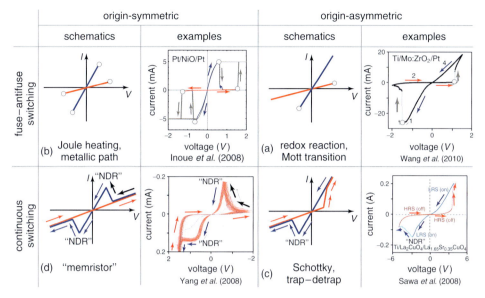

Figure 16.4 (a–d) Typical examples [1, 49, 56, 57] of the *I*–*V* hystereses shown in Figure 16.3 (Source: Adapted with permissions from Ref. [49] © 2008 American Physical Society, from Ref. [56] © 2010 Institute of Physics, from Ref. [57] © 2008 Nature Publishing Group, and from Ref. [1] © 2008 Elsevier, respectively.) In practical devices, *I*–*V* curves are given as that of serial and/or parallel connections of those *I*–*V* curves, and each may depend on the sweep rate. Thus, the examples are not disparate and sometimes they change their types over the blurred boundaries.

On the other hand, the electric-field drift of oxygen does not necessarily require the oxygen from outside; that is, the oxygens are self-supplied or recycled by the alternating electric field [54]. In fact, by controlling the strength of electroforming, by modifying the metal/oxide interface, and by changing the atmosphere, the size/amount of the local current paths is considered to be tuned [34, 55]. Those will change the diffusion/drift balance, which alters the origin-symmetric nonpolar *I*–*V* hysteresis into origin-asymmetric bipolar *I*–*V* hysteresis and vice versa.

The origin-asymmetric bipolar resistive switching is considered to be caused by the valence change of cations in the oxide [1, 39, 41]. This is indeed the redox reaction. In several combinations of the oxide and the metal electrode, the averaged free enthalpy of oxygen segregation is close to the free reaction enthalpy of oxide precipitation from the metal [58]. Therefore, one may conjecture that a large current density could easily drive local chemical reactions such as electro-oxidation [59, 60] and electroreduction [61], and the reactions may be drastically accelerated by a local Joule heating owing to the strong current inhomogeneity. By considering the defects chemistry [62], the direction of the polarity (clockwise or counterclockwise) of the bipolar switchings is well explained. Another mechanism of the valence change is the doping-induced Mott-Hubbard transition (called simply *Mott transition*), that is, the electron-correlation-driven MI transition [37, 63, 64].

Interestingly, when the size of the system becomes smaller, the implication of the Mott transition resembles more to that of the redox reaction.

16.2.3
Phase–Delay Response – Memristor

Origin symmetric bipolar $I-V$ hystereses have been reported recently [57, 65]. The unique hysteresis curve is explained as due to the history-dependent features. $I-V$ relation is expressed as $V(t) = R[q(t)]I(t)$, where $\dot{q}(t) = f[I(t)]$, t is time, and f is an arbitrary functional of the current [66]. This $R[q(t)]$ is now known as *memristor*, which means that the resistance depends on the past states (memory) through which the system has evolved.

It is argued that the memristive behavior is caused by a rearrangement of charged ions and defects under the effect of external electric field. The mechanism is similar to that of the origin asymmetric bipolar $I-V$ hystereses, but the history dependence is more dominant than the redox reaction. We denote here memristor as it is used in the original definition [65]. This original memristor is often confused with the (c)-type resistive switching in Figure 16.3, but the apparent difference is that there are two NDRs in the memristor-type hysteresis (this is sometimes called a *figure-of-eight loop*). One NDR, which is indicated by the thick black arrows in Figure 16.3d, is quite unique. When the absolute value of the voltage is increased, this NDR does not appear; however, when the absolute value of the voltage is decreased, this NDR is observed. This is hardly explained by a naive redox reaction, for which the reaction occurs whenever the voltage exceeds a threshold value. However, this original and pure memristor-type hysteresis is not very appropriate for the nonvolatile memory. For better retention, the other types of hystereses are incorporated, which makes the hysteresis of practical memristors rather similar to the origin asymmetric bipolar types. Another possible way to fit up the nonvolatility to the original memristors is to use the temperature inhomogeneity. Given that the diffusion of vacancies or ions are nonlinearly accelerated by the Joule heating, then the vacancies or ions can diffuse less when the Joule heating is smaller; this is seen to be nonvolatile for a scale of time, and the hysteresis is kept origin symmetric.

16.2.4
Electroforming

As described earlier, the resistive switching hardly occurs when the metal/oxide/metal sandwich is in the featureless initial state, as schematically drawn in the left panel of Figure 16.5. Initiation to enable the resistive switching is called *electroforming* or *simply forming* [4–8]. After the electroforming, many types of "inhomogeneity" are created in the oxides (middle panel of Figure 16.5), which are manifested by the various kinds of $I-V$ hystereses classified in Figure 16.3.

The variety of the inhomogeneity is caused by several factors and their combinations: some of them are the degrees of nonstoichiometry of oxides, difference of the electron affinities of oxides and metal electrodes, structural/electronic instability of

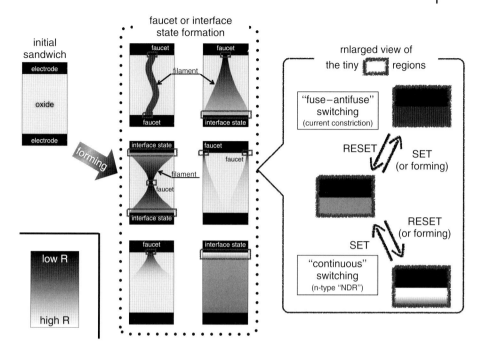

Figure 16.5 Schematic illustration of the electroforming. Blackness of the electrodes and oxides corresponds to the smallness of the resistance, as indicated by the grayscale at the bottom-left corner. By the electroforming, inhomogeneity is created in the oxide bulk, and some of the structures are called *filament*. Essential structures are marked by the open squares of thick gray lines in the middle panel, that is, faucets and interface states. The change of resistance at the faucet and interface state is responsible for the resistive switching. See text for details.

oxides, as well as heat capacity/conductivity of oxides and metal electrodes [41]. The gas atmosphere, especially oxygen and moisture, is also responsible [54, 67]. With regard to the gas atmosphere, it should be noted that some kinds of oxides and metal electrodes can behave as reservoirs of those gases [54]. Joule heat generated during the electroforming provides the system with the thermal energy, which is necessary for the oxides to give rise to the inhomogeneous structure. The heating is controlled by the so-called compliance [68, 69] and also by the parasitic current components [70, 71]. Here, the compliance current/voltage is the maximum current/voltage that can be intentionally set in order to limit the heat generation and dielectric breakdown during the electroforming.

The electroforming reduces the resistance of the sandwich if the electric bias generates a significant amount of ionic defects (e.g., oxygen vacancies and metal interstitials) in the oxides and the defects form the conducting paths [40]; meanwhile, the electroforming increases the resistance of the sandwich if the defects are assembled at the metal/oxide interface rather uniformly to form a Schottky barrier [45–47]. Thus, the most essential point of the electroforming for the subsequent resistive switching is not the whole structure of the inhomogeneity but is the *tiny* region, as is indicated by the open squares of thick gray lines in the middle panel of Figure 16.5. The tiny current-constricting structure is called *faucet* owing to the similarity to the tap of electric current [49], while the "interface state" (described in the following sections) denotes the voltage drop region at the interface [1]. As shown in the right panel of Figure 16.5, if the fuse–antifuse-type changes of resistance (i.e., local MI transition) occurred in the tiny region, the resistance switching becomes either (a)- or (b)-type of Figure 16.3. On the other hand, if the continuous movement of ionic defects is dominant in the tiny region, the resistance switching becomes either (c)- or (d)-type of Figure 16.3. Therefore, it should be noted that, even in the presence of clear filament structures in the oxide, the resistance switching can become the memristor-type or the bipolar continuous type.

From the application point of view, eliminating the electroforming process is an important and an inevitable problem. There are several reasons: large power consumption and instability of the electroforming process, as well as the accompanied gas release, which may induce physical damages on several semiconductor devices on the same chip. The schematics in Figure 16.5 tell us that an approach to engineer a ReRAM device without electroforming is to eliminate the unnecessary filament and bulk regions while leaving the essential faucets and interface-state regions. For that purpose, a natural approach is to thin the oxide or make the size of the device as small as that of a faucet [72]. There have been several reports along with this approach suggesting indeed the electroforming-free ReRAM; however, these devices tend to be degraded rapidly with the switching cycles probably due to the lack of a reservoir for the mobile ionic defects. A breakthrough idea on this issue is eagerly anticipated.

As discussed above, we have categorized the resistance switching into four types: (a) – (d) in Figure 16.3. Each of them has strengths and limitations, which are all interesting to be reviewed here. However, because of the limitation of our ability and the spaces, we refrain from discussing each example in detail. Instead, in the following section, we focus on a possible mechanism of (c)-type resistive switching, that is, "bipolar continuous switching," and in brief its possibility for applications.

16.3
Bipolar Continuous Switching

In this section, semiconducting perovskite oxides $Pr_{1-x}Ca_xMnO_3$ (PCMO) and Nb-doped $SrTiO_3$ are brought into focus as model materials of the bipolar continuous switching [(c)-type resistive switching in the classification of Figure 16.3].

16.3.1
Continuous Switching at Interface

As discussed earlier, the resistive switchings can be classified into two types in terms of the uniformity of the current path: one is the fuse–antifuse switching, and the other is the continuous switching. They are further classified into origin symmetric and origin asymmetric switching, depending on the balance of drift/diffusion of charge migration and that of temperature/potential inhomogeneity.

The (c)-type continuous resistive switching is caused by the drift of ions or charged vacancies under the uniform electric current and inhomogeneity of the electric field. The uniform electric current is evidenced by the area dependence of the junction resistance; the junction resistance is inversely proportional to the junction area, indicating that switching takes place over the whole area of the junction, that is, the entire interface [73]. Therefore, this resistive switching is often referred to as *homogeneous interfacial resistive switching*.

In the bipolar continuous switching, the switching characteristics generally depend on the electrode materials [1, 46, 74]. Figure 16.6 shows $I-V$ curves for $M/Pr_{1-x}Ca_xMnO_3/SrRuO_3$ (M/PCMO/SRO) with $x = 0.3$ and $M/SrTi_{0.99}Nb_{0.01}O_3/Ag$ (M/Nb:STO/Ag) devices, where PCMO and STO are p- and n-type semiconductors that are the most popular materials for bipolar continuous switching [1]. $M = Ti$, Au, or SRO is the top electrode with the work function of ~ 4.3, ~ 5.1, and ~ 5.3 eV, respectively. The SRO and Ag are bottom electrodes, which form ohmic contacts with PCMO and Nb:STO, respectively. For the p-type PCMO cells, as the work function of M decreases, the contact resistance between M and PCMO increases, and the Ti/PCMO interface shows rectification behavior in the $I-V$ curves. On the other hand, for the n-type Nb:STO cells, as the work function of M increases, the contact resistance between M and Nb:STO increases, and the Au/Nb:STO and SRO/Nb:STO interfaces show rectification behavior in the $I-V$ curves.

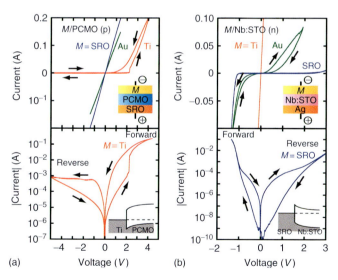

Figure 16.6 $I-V$ curves for (a) p-type $M/Pr_{1-x}Ca_xMnO_3/SrRuO_3$ (M/PCMO/SRO) with $x = 0.3$ and (b) n-type $M/SrTi_{0.99}Nb_{0.01}O_3/Ag$ (M/Nb:STO/Ag) cells ($M = Ti$, Au, and SRO). Current rectification with hysteresis is seen in the $I-V$ curves for the Ti/PCMO/SRO cell and that with opposite polarity is seen in Au/Nb:STO/Ag and SRO/Nb:STO/Ag cells. (Source: Reprint with permission from Ref. [1] © 2008 Elsevier.)

The rectification of current is due to a Schottky-like barrier at the interface, as shown in the insets of the lower panels of Figure 16.6. In addition to the rectification, $I-V$ curves for Ti/PCMO, Au/PCMO, and SRO/Nb:STO interfaces exhibit hystereses indicative of resistive switchings. However, the ohmic $I-V$ curves for Au/PCMO, SRO/PCMO, and Ti/Nb:STO interfaces show no resistive switching. These results suggest that the Schottky-like barrier plays a key role in the (c)-type resistive switching. Since this is a bipolar valence change switching, the polarity of the applied bias relative to the Schottky bias is important. The resistance changes from HRS to LRS only when a forward bias voltage is applied to the interface, and it changes from LRS to HRS only when a reverse bias voltage is applied, as shown in the lower panels of Figure 16.6.

16.3.2
Valence Change of Transition-Metal Cations

As mentioned earlier, the bipolar continuous switching is considered to be caused by the valence change of cations in oxides. In such a case, the valence of cations may be a crucial parameter for controlling the resistive switching characteristics. In fact, it has been reported that the resistive switching characteristics of the Ti/PCMO devices depend on the Mn valence [75]. Figure 16.7 shows the Ca-composition, x, dependence of resistive switching ratios, R_H/R_L, for the Ti/Pr$_{1-x}$Ca$_x$MnO$_3$ devices, where R_H and R_L are resistances of high- and low-resistance states, respectively. Assuming no oxygen vacancy in the Pr$_{1-x}$Ca$_x$MnO$_3$ layer, the Mn valence increases from 3+ to 4+, as the Ca-composition x increases from 0 to 1, respectively. The resistive switching ratio R_H/R_L increases with increasing x and shows a maximum value at around $x=0.4$. As x increases above 0.4, R_H/R_L decreases and becomes smaller than 2 for $x > 0.8$. This clear x dependence of R_H/R_L evidences the importance of the valence of cations for the bipolar continuous switching.

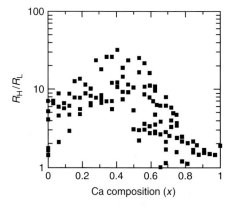

Figure 16.7 Ca-composition, x, dependence of resistive switching ratio R_H/R_L of Ti/Pr$_{1-x}$Ca$_x$MnO$_3$ junctions measured at room temperature. R_H/R_L clearly depends on x, indicating that the Mn valence is a crucial parameter for controlling the resistive switching characteristics.

In the resistive switching process, the oxygen-vacancy density at the interface between a metal electrode and an oxide is changed because of the electric-field-induced drift of oxygen vacancies or ions. As a result, the change in the oxygen-vacancy density causes the valence change of cations at the interface. This can be confirmed by a cross-sectional transmission electron microscope (TEM) measurement with an electron energy-loss spectroscopy (EELS) [50, 75]. Figure 16.8 shows cross-sectional TEM images of the Ti/PCMO junctions with $x = 0.5$ [75]. At the cross section of the as-prepared junction of a Ti electrode and a PCMO layer (before applying electric field), there was an amorphous TiO_y (a-TiO_y) layer with a thickness of at most 1 nm. After applying electric fields above 5 V, the thickness of the a-TiO_y layer was increased to $\sim$10 nm, as seen in Figure 16.8b. The formation of a-TiO_y layer indicates an electromigration that oxygen ions are drifted from the PCMO layer into the Ti electrode.

Moreover, EELS measurements of the Mn-L edge confirmed the valence change of the Mn site. Figure 16.8c shows the Mn-L edge EELS spectra obtained at different positions in the PCMO layer, as indicated in Figure 16.8b. The peak intensity ratio of Mn-L_3 and Mn-L_2, $I(L_3)/I(L_2)$, was decreased with increasing the distance from the a-TiO_y/PCMO boundary. This means that the Mn valence near the interface is smaller than that away from the interface.

16.3.3
Alteration of the Barrier Characteristics

Electric field drift of the oxygen ions and oxygen deficiencies causes the valence change of the transition metal as well as the lattice disorders, both of which lead to the change in the electronic states of the transition-metal oxides.

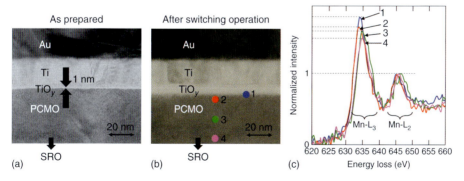

Figure 16.8 Cross-sectional TEM images of a Ti/PCMO junction observed (a) before and (b) after applying electric field at room temperature. Before the field application, a thin amorphous TiO_y layer (<1 nm) was confirmed between the Ti and PCMO layers. After the application of voltage stress, the thickness of the amorphous TiO_y layer was increased to $\sim$10 nm. (c) Mn-L edge EELS spectra at several positions around the Ti/PCMO junction. The positions are indicated in (b). Each spectrum was normalized by the intensity of the Mn-L_2 peak. (Source: Adapted with permission from Ref. [75] © 2009 APS.)

Figure 16.9 shows optical absorption spectra of oxygen-deficient and oxygenated PCMO films [75]. The oxygen-deficient PCMO film has a larger absorption gap compared with that of the oxygenated one. In a Mn-O-Mn chain in PCMO, the lack of the oxygen suppresses the overlap of the neighboring Mn e_g orbitals. Then, the effective e_g band width is reduced and the optical absorption gap is increased [76, 77].

The TEM and EELS experiments (Figure 16.8) indicate that the electromigration of oxygen vacancies generates the oxygen-deficient PCMO layer in the vicinity of the interface between the a-TiO$_y$ and PCMO. The I–V curves suggest the preferable direction of the oxygen electromigration. In the Ti/PCMO junction, when a positive voltage bias was applied to the PCMO layer, the resistance state was converted from the HRS to the LRS, as shown in Figure 16.6. In this case, positively charged oxygen vacancies are expected to drift from the PCMO layer to the a-TiO$_y$ layer. During the switching from the LRS to the HRS, oxygen vacancies are expected to drift back to the PCMO layer from the a-TiO$_y$ layer. Therefore, the number of oxygen vacancies in PCMO near the interface is larger in HRS than in LRS. In HRS, the highly oxygen-deficient PCMO layer has a narrower e_g band, leading to a larger bandgap. This could act as an effective barrier for the hole-carrier conduction as shown in Figure 16.10. On the other hand, when the oxygen vacancies in PCMO are less, the bandgap is smaller. Then, the hole carriers could flow through a thin depletion layer via a tunneling process. This is LRS as depicted in Figure 16.10. The change of the barrier at the interface may alternate the hole-carrier conduction from the tunneling to the thermionic emission. This barrier width change was confirmed by a capacitance (C) measurement [78]. The value of C is given by $C = \varepsilon_{TiO}\varepsilon_{PCMO}S/(\varepsilon_{PCMO}W_a + \varepsilon_{TiO}W_d)$, where ε_{TiO} and ε_{PCMO} are the dielectric constant of a-TiO$_x$ and PCMO layers, respectively, S the cell area, W_a the thickness of the a-TiO$_x$ layer, and W_d the depletion layer width in the PCMO layer. The effective barrier width W is $W = W_a + W_d$. Assuming that ε_{TiO},

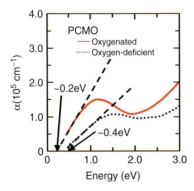

Figure 16.9 Optical absorption spectra of oxygen-deficient (dashed line) and oxygenated (solid line) PCMO films on STO substrates measured at room temperature. The broken straight lines are fits to the rising part of the absorption spectra, and the optical absorption gap was estimated from the intersection of the broken line at the optical absorption coefficient $\alpha = 0$. (Source: Adapted with permission from Ref. [75] © 2009 APS.)

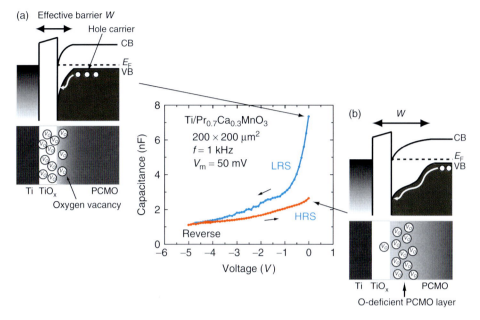

Figure 16.10 C–V curves under reverse bias for a Ti/PCMO$(x=0.3)$/SRO device. The hysteretic characteristic indicates that the barrier width (W_d) at the Ti/PCMO interface is altered by applying an electric field. The insets show possible band diagrams of Ti/PCMO interface (a) in a low-resistance state (LRS) and (b) in a high-resistance state (HRS). The work function of Ti is ~4.3 eV, that of p-type PCMO for $x<0.5$ is 4.4–4.8 eV, and the bandgap of 0.2–0.6 eV. Therefore, a Schottky-like barrier can be formed at the interface. In the HRS, as the oxygen-deficient PCMO layer at the interface has a larger bandgap, the hole-carrier conduction is interfered. (Source: Adapted with permission from Ref. [78] © 2005 SPIE.)

$\varepsilon_{\text{PCMO}}$, and W_a do not change, the change in C is attributable to the alteration of W_d. As shown in Figure 16.10, the C–V curve for the Ti/PCMO shows hysteretic behavior, that is, the capacitive switching, and the value of C in the LRS is larger than that in the HRS. This suggests that W_d in the LRS is narrower than that in the HRS, consistent with the possible band diagram of Ti/PCMO interface in HRS and LRS, as shown in Figure 16.10.

16.4
Toward Device Applications and Summary

Owing to the continued advances in information technology, there is always an ever-growing requirement for nonvolatile memory devices with faster (less than a few nanoseconds) bitwise access and with smaller cell size (less than 20 nm) for storing one bit. Practically, the emerging nonvolatile memories are to be compared with state-of-the-art memories such as the flash memory and DRAM. A typical charge-storage-type memory DRAM is currently in the 30 nm range with a cell size of $6F^2$, where F is the minimum feature size on a given process. Because it requires

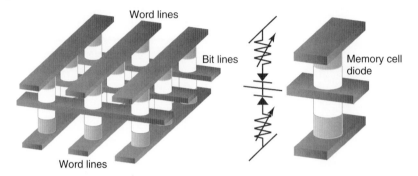

Figure 16.11 Schematic diagrams of stacked passive cross-point memory array.

a three-dimensional structure of capacitors, the thickness of its dielectric limits the scalability. This is actually a chance for new types of memories to edge into, and a number of alternative nonvolatile memories of non-charge-storage types have been explored. ReRAM has attracted considerable attention, because of its simple memory cell structure with high scalability as well as its fast switching speed [1, 39, 79].

The ReRAM memory cell has a capacitor-like structure composed of insulating or semiconducting metal oxides. Because of such a simple structure, highly scalable passive cross-point and multilevel stacking memory arrays with rows and columns of word and bit lines have been proposed [80]. For those array structures, one memory cell has an area of about $4F^2$, which is much smaller than that of DRAM. In every write/read operation, a bias voltage is applied on a memory cell through the selected word and bit lines. To ensure the reliability of the write/read operation, a selection element such as a diode or highly nonlinear resistor has to be connected to a memory cell in series to prevent signal from bypassing through neighboring "ON"-state memory cells, that is, the so-called parasitic-path problem or the sneak current problem [39, 81].

In response to the requirement, oxide p–n diodes and Schottky diodes compatible with the metal oxides of ReRAM memory cell have been proposed [27]. However, for the ReRAM of origin symmetric fuse–antifuse type memory cell [(b)-type of Figure 16.3] with sub-100 nm scaling, the forward current densities are very large to be handled by the proposed diode selectors. Thus, for the origin symmetric fuse–antifuse type ReRAM with the passive cross-point array structure to be realized, it is essential to develop a diode selector that has a high forward current density ($>10^6$ A/cm^2) and/or to suppress the reset current density ($\ll 1\,\mu$A) [82]. On the other hand, for the bipolar continuous resistive switching, especially, "interfacial Schottky resistive switching," the memory cell is accompanied by the self-formed Schottky barrier, which can be used as the selection element [81]. This means that the passive cross-point memory array can be realized without stacking any selection element. This is an advantage of the interfacial Schottky resistive switching memory cell over the origin-symmetric fuse–antifuse memory cell.

Another critical issue for the future development of ReRAM is the reliability, such as data retention and memory endurance (the number of erase and write cycles).

The variation of characteristics from cell to cell as well as from chip to chip is also an important issue to be controlled. A data retention time over 10 years can be extrapolated from the retention characteristics measured at high temperatures [73, 83, 84], and the memory endurance over 10^{11} cycles has been demonstrated for the Ta_2O_5/TaO_2 memory cell [85]. These reported characteristics seem to be enough for an alternative to the flash memory. However, little has been investigated on the characteristic variation. A method to realize the device characteristics uniform over a large area, and an elucidation of the fatigue mechanism are immediately worked out satisfactory for the future development of large-capacity ReRAM.

In summary, we provided a one-look review of the recent understandings of the resistive switchings by classifying their $I-V$ characteristics. They were roughly classified into fuse–antifuse type and continuous type, depending on the *locality* of the switching in space and time compared with the scale of device and bias sweeping time. Each of them was further classified into origin-symmetric and origin-asymmetric types, depending on the dominance of electric-field drift (potential inhomogeneity) or thermal diffusion (temperature inhomogeneity). For each category, a representative example was shown; especially, the continuous bipolar switching with alternating Schottky barrier was discussed in detail. A possible device structure when applied in a practical electronic device was introduced with regard to the ultimate scalability of the ReRAM device. We hope that this concise review will help readers to understand these seemingly complicated phenomena, and will contribute to light a direction toward a viable development of new memory devices.

Acknowledgments

This work was supported in part by Industrial Technology Research Grant Program from NEDO and by Grant-in-Aid for Scientific Research (No. 22360280) from JSPS. The authors are grateful to T. Fujii, S. Asanuma, H. Shima, M. J. Rozenberg, M. Kawasaki, and Y. Tokura for their helpful discussions.

References

1. Sawa, A. (2008) Resistive switching in transition metal oxides. *Mater. Today*, **11**, 28.

2. Oxley, D.P. (1977) Electroforming, switching and memory effects in oxide thin films. *Electrocomp. Sci. Technol.*, **3**, 217.

3. Chudnovskii, F.A., Odynets, L.L., Pergament, A.L., and Stefanovich, G.B. (1996) Electroforming and switching in oxides of transition metals: the role of metal-insulator transition in the switching mechanism. *J. Solid State Chem.*, **122**, 95.

4. Kreynina, G.S. (1962) Volt-ampere characteristics of composite metal-dielectric-metal cathodes. *Radio Eng. Electron. Phys.*, **7**, 1949.

5. Hickmott, T.W. (1962) Low-frequency negative resistance in thin anodic oxide films. *J. Appl. Phys.*, **33**, 2669.

6. Gibbons, J.F. and Beadle, W.E. (1964) Switching properties of thin NiO films. *Solid-State Electron.*, **7**, 785.

7. Chopra, K.L. (1965) Avalanche-induced negative resistance in thin oxide films. *J. Appl. Phys.*, **36**, 184.

8. Simmons, J.G. and Verderber, R.R. (1967) New conduction and reversible memory phenomena in thin insulating films. *Proc. R. Soc. London, Ser. A*, **301**, 77.

9. Dearnaley, G., Stoneham, A.M., and Morgan, D.V. (1970) Electrical phenomena in amorphous oxide films. *Rep. Prog. Phys.*, **33** 1129.

10. Biederman, H. (1976) Metal-insulator-metal sandwich structures with anomalous properties. *Vacuum*, **26**, 513.

11. Pagnia, H. and Sotnik, N. (1988) Bistable switching in electroformed metal–insulator–metal devices. *Phys. Status Solidi A*, **108**, 11.

12. Bednorz, J.G. and Muller, K.A. (1986) *Z. Phys. B*, **64**, 189.

13. Ueno, K. and Koshida, N. (1999) Light-emissive nonvolatile memory effects in porous silicon diodes. *Appl. Phys. Lett.*, **74**, 93.

14. Hickmott, T.W. (2000) Voltage-dependent dielectric breakdown and voltage-controlled negative resistance in anodized $Al–Al_2O_3–Au$ diodes. *J. Appl. Phys.*, **88**, 2805.

15. Stefanovich, G., Pergament, A., and Stefanovich, D. (2000) Electrical switching and Mott transition in VO_2. *J. Phys.: Condens. Matter*, **12**, 8837.

16. Hu, J., Snell, A.J., Hajto, J., Rose, M.J., and Edmiston, W. (2001) Field-induced anomalous changes in Cr/a-Si:H/V thin film structures. *Thin Solids Films*, **396**, 240.

17. Tulina, N.A., Ionov, A.M., and Chaika, A.N. (2001) Reversible electrical switching at the $Bi_2Sr_2CaCu_2O_{8+y}$surface in the normal metal–$Bi_2Sr_2CaCu_2O_{8+y}$ single crystal heterojunction. *Physica C*, **366**, 23.

18. Thurstans, R.E. and Oxley, D.P. (2002) The electroformed metal–insulator–metal structure: a comprehensive model. *J. Phys. D*, **35**, 802.

19. Baikalov, A., Wang, Y.Q., Shen, B., Lorenz, B., Tsui, S., Sun, Y.Y., Xue, Y.Y., and Chu, C.W. (2003) Field-driven hysteretic and reversible resistive switch at the $Ag–Pr_{0.7}Ca_{0.3}MnO_3$ interface. *Appl. Phys. Lett.*, **83**, 957.

20. Gravano, S., Amr, E., Gould, R.D., and Abu Samra, M. (2003) Monte Carlo simulation of current–voltage characteristics in metal–insulator–metal thin film structures. *Thin Solids Films*, **433**, 321.

21. Rozenberg, M.J., Inoue, I.H., and Sánchez, M.J. (2004) Nonvolatile memory with multilevel switching: a basic model. *Phys. Rev. Lett.*, **92**, 178302.

22. Beck, A., Bednorz, J.G., Gerber, C., Rossel, C., and Widmer, D. (2000) Reproducible switching effect in thin oxide films for memory applications. *Appl. Phys. Lett.*, **77**, 139.

23. Liu, S.Q., Wu, N.J., and Ignatiev, A. (2000) Electric-pulse-induced reversible resistance change effect in magnetoresistive films. *Appl. Phys. Lett.*, **76**, 2749.

24. Hosoi, Y., Tamai, Y., Ohnishi, T., Ishihara, K., Shibuya, T., Inoue, Y., Yamazaki, S., Nakano, T., Ohnishi, S., Awaya, N., Inoue, I.H., Shima, H., Akinaga, H., Takagi, H., Akoh, H., and Tokura, Y. (2006) High speed unipolar switching resistance RAM (RRAM) technology. Tech. Dig. – International Electron Devices Meeting, p. 793.

25. Hsu, S.T., Li, T.K., and Awaya, N. (2007) Resistance random access memory switching mechanism. *J. Appl. Phys.*, **101**, 024517.

26. Seo, S., Seo, S., Lee, M.J., Seo, D.H., Jeoung, E.J., Suh, D.-S., Joung, Y.S., Yoo, I.K., Hwang, I.R., Kim, S.H., Byun, I.S., Kim, J.-S., Choi, J.S., and Park, B.H. (2004) Reproducible resistive switching in polycrystalline NiO films. *Appl. Phys. Lett.*, **85**, 5655.

27. Lee, M.-J., Seo, S., Kim, D.-C., Ahn, S.-E., Seo, D.H., Yoo, I.-K., Baek, I.-G., Kim, D.-S., Byun, I.-S., Kim, S.-H., Hwang, I.-R., Kim, J.-S., Jeon, S.-H., and Park, B.H. (2007) A low-temperature-grown oxide diode as a new switch element for high-density, nonvolatile memories. *Adv. Mater.*, **19**, 73.

28. Rohde, C., Choi, B.J., Jeong, D.S., Choi, S., Zhao, J.-S., and Hwang, C.S. (2005) Identification of a determining parameter for resistive switching resistive switching of TiO_2 thin films. *Appl. Phys. Lett.*, **86**, 262907.

29. Fors, R., Khartsev, S.I., and Grishin, A.M. (2005) Giant resistive switching in metal–insulator–manganite junctions: evidence for Mott transition. *Phys. Rev. B*, **71**, 045305.

30. Szot, K., Speier, W., Bihlmayer, G., and Waser, R. (2006) Switching the electrical resistance of individual dislocations in single-crystalline $SrTiO_3$. *Nat. Mater.*, **5**, 312.

31. Kim, K.M., Choi, B.J., Koo, B.W., Choi, S., Jeong, D.S., and Hwang, C.S. (2006) Resistive switching in $Pt/Al_2O_3/TiO_2/Ru$ stacked structures electrochem. *Solid-State. Lett.*, **9**, G343.

32. You, Y.-H., So, B.-S., Hwang, J.-H., Cho, W., Lee, S.S., Chung, T.-M., Kim, C.G., and An, K.-S. (2006) Impedance spectroscopy characterization of resistive switching NiO thin films prepared through atomic layer deposition. *Appl. Phys. Lett.*, **89**, 222105.

33. Jeong, D.S., Schroeder, H., and Waser, R. (2006) Impedance spectroscopy of TiO_2 thin films showing resistive switching. *Appl. Phys. Lett.*, **89**, 082909.

34. Jeong, D.S., Schroeder, H., and Waser, R. (2007) Coexistence of bipolar and unipolar resistive switching behaviors in a Pt/TiO2/Pt stack electrochem. *Solid-State Lett.*, **10**, G51.

35. Sato, Y., Kinoshita, K., Aoki, M., and Sugiyama, Y. (2007) Consideration of switching mechanism of binary metal oxide resistive junctions using a thermal reaction model. *Appl. Phys. Lett.*, **90**, 033503.

36. Oka, T. and Nagaosa, N. (2005) Interfaces of correlated electron systems: proposed mechanism for colossal electroresistance. *Phys. Rev. Lett.*, **95**, 266403.

37. Rozenberg, M.J., Inoue, I.H., and Sánchez, M.J. (2006) Strong electron correlation effects in non-volatile electronic memory devices. *Appl. Phys. Lett.*, **88**, 033510.

38. Jeong, D.S., Choi, B.J., and Hwang, C.S. (2006) Study of the negative resistance phenomenon in transition metal oxide films from a statistical mechanics point of view. *J. Appl. Phys.*, **100**, 113724.

39. Waser, R., Dittmann, R., Staikov, G., and Szot, K. (2009) Redox-based resistive switching memories – nanoionic mechanisms, prospects, and challenges. *Adv. Mater.*, **21**, 2632.

40. Kim, K.M., Jeong, D.S., and Hwang, C.S. (2011) Nanofilamentary resistance switching in binary oxide system; a review on the present status and outlook. *Nanotechnology*, **22**, 254002.

41. Yang, J.J., Inoue, I.H., Mikolajick, T., and Hwang, C.S. (2012) Metal oxide memories based on thermochemical and valence. *MRS Bull.*, **37**, 131.

42. Ridley, B.K. (1963) Specific negative resistance in solids. *Proc. Phys. Soc.*, **82**, 954.

43. Jäger, D., Baumann, H., and Symanczyk, R. (1986) Experimental observation of spatial structures due to current filament formation in silicon PIN diodes. *Phys. Lett. A*, **117**, 141.

44. Schöll, E. (2001) *Nonlinear Spatio-Temporal Dynamics and Chaos in Semiconductors*, Cambridge Univ. Press, Cambridge.

45. Pan, Z. and Shum, K. (2000) Demonstration of III–V semiconductor-based nonvolatile memory devices. *Appl. Phys. Lett.*, **76**, 505.

46. Sawa, A., Fujii, T., Kawasaki, M., and Tokura, Y. (2004) Hysteretic current–voltage characteristics and resistive switching at a rectifying $Ti/Pr_{0.7}Ca_{0.3}MnO_3$ interface. *Appl. Phys. Lett.*, **85**, 4073.

47. Smits, J.H.A., Meskers, S.C.J., Janssen, R.A.J., Marsman, A.W., and de Leeuw, D.M. (2005) Electrically rewritable memory cells from poly(3-hexylthiophene) schottky diodes. *Adv. Mater.*, **17**, 1169.

48. Odagawa, A., Sato, H., Inoue, I.H., Akoh, H., Kawasaki, M., and Tokura, Y. (2004) Colossal electroresistance of a $Pr_{0.7}Ca_{0.3}MnO_3$ thin film at room temperature. *Phys. Rev. B*, **70**, 224403.

49. Inoue, I.H., Yasuda, S., Akinaga, H., and Takagi, H. (2008) Nonpolar resistive switching of metal/binary-transition-metal oxides/metal sandwiches: homogeneous/inhomogeneous transition of current distribution. *Phys. Rev. B*, **77**, 035105.

50. Shono, K., Kawano, H., Yokota, T., and Gomi, M. (2008) Origin of negative differential resistance observed on

bipolar resistive switching device with Ti/Pr$_{0.7}$Ca$_{0.3}$MnO$_3$/Pt structure. *Appl. Phys. Express*, **1**, 055002.

51. Adler, D., Henisch, H.K., and Mott, N.F. (1978) The mechanism of threshold switching in amorphous alloys. *Rev. Mod. Phys.*, **50**, 209.

52. Lee, H.-Y., Chen, P.-S., Wang, C.-C., Maikap, S., Tzeng, P.-J., Lin, C.-H., Lee, L.-S., and Tsai, M.-J. (2007) *Jpn. J. Appl. Phys.*, **46**, 2175.

53. Shima, H., Takano, F., Akinaga, H., Tamai, Y., Inoue, I.H., and Takagi, H. (2007) *Appl. Phys. Lett.*, **91**, 012901.

54. Goux, L., Czarnecki, P., Chen, Y.Y., Pantisano, L., Wang, X.P., Degraeve, R., Govoreanu, B., Jurczak, M., Wouters, D.J., and Altimime, L. (2010) Evidences of oxygen-mediated resistive-switching mechanism in TiN/HfO$_2$/Pt cells. *Appl. Phys. Lett.*, **97**, 243509.

55. Shen, W., Dittmann, R., and Waser, R. (2010) Reversible alternation between bipolar and unipolar nonpolar resistive switching resistive switching in polycrystalline barium strontium titanate thin films. *J. Appl. Phys.*, **107**, 094506.

56. Wang, S.-Y., Lee, D.-Y., Huang, T.-Y., Wu, J.-W., and Tseng, T.-Y. (2010) Controllable oxygen vacancies to enhance resistive switching resistive switching performance in a ZrO$_2$-based RRAM with embedded Mo layer. *Nanotechnology*, **21**, 495201.

57. Yang, J.J., Pickett, M.D., Li, X., Ohlberg, D.A.A., Stewart, D.R., and Williams, R.S. (2008) Memristive switching mechanism for metal/oxide/metal nanodevices. *Nat. Nanotechnol.*, **3**, 429.

58. Pippel, E., Woltersdorf, J., Gegner, J., and Kirchheim, R. (2000) *Acta Mater.*, **48**, 2571.

59. Schmidt, T., Martel, R., Sandstrom, R.L., and Avouris, Ph. (1998) *Appl. Phys. Lett.*, **73**, 2173.

60. Martel, R., Schmidt, T., Sandstrom, R.L., and Avouris, Ph. (1999) *J. Vac. Sci. Technol. A*, **17**, 1451.

61. Abraham, M.M., Boatner, L.A., Christie, W.H., and Modine, F.A. (1984) *J. Solid State Chem.*, **51**, 1.

62. Kofstad, P.K. (1972) *Nonstoichiometry, Diffusion and Electrical Conductivity in Binary Metal Oxides*, Science & Technology

of Materials. John Wiley & Sons, Inc., New York.

63. Inoue, I.H. and Rozenberg, M.J. (2008) Taming the mott transition for a novel mott transistor. *Adv. Funct. Mater.*, **18**, 2289.

64. Vaju, C., Cario, L., Corraze, B., Janod, E., Dubost, V., Cren, T., Roditchev, D., Braithwaite, D., and Chauvet, O. (2008) Electric-pulse-driven electronic phase separation, insulator–metal transition, and possible superconductivity in a mott insulator. *Adv. Mater.*, **20**, 2760.

65. Strukov, D.B., Snider, G.S., Stewart, D.R., and Williams, R.S. (2008) The missing memristor found. *Nature*, **453**, 80.

66. Pershin, Y.V. and Di Ventra, M. (2011) Memory effects in complex materials and nanoscale systems. *Adv. Phys.*, **60**, 145.

67. Jeong, D.S., Schroeder, H., Breuer, U., and Waser, R. (2008) Characteristic electroforming behavior in Pt/TiO$_2$/Pt resistive switching cells depending on atmosphere. *J. Appl. Phys.*, **104**, 123716.

68. Kim, K.M. and Hwang, C.S. (2009) *Appl. Phys. Lett.*, **94**, 122109.

69. Chang, S.H., Lee, J.S., Chae, S.C., Lee, S.B., Liu, C., Kahng, B., Kim, D.-W., and Noh, T.W. (2009) *Phys. Rev. Lett.*, **102**, 026801.

70. Ielmini, D., Cagli, C., and Nardi, F. (2009) *Appl. Phys. Lett.*, **94**, 063511.

71. Song, S.J., Kim, K.M., Kim, G.H., Lee, M.H., Seok, J.Y., Jung, R., and Hwang, C.S. (2010) *Appl. Phys. Lett.*, **96**, 112904.

72. Ogimoto, T., Tamai, Y., Kawasaki, M., and Tokura, Y. (2007) *Appl. Phys. Lett.*, **90**, 143515.

73. Sim, H., Choi, H., Lee, D., Chang, M., Choi, D., Son, Y., Lee, E.H., Kim, W., Park, Y., Yoo, I.K., and Hwang, H. (2005) Tech. Dig. - Int. Electron Devices Meet., p. 777.

74. Tsubouchi, K., Ohkubo, I., Kumigashira, H., Oshima, M., Matsumoto, Y., Itaka, K., Ohnishi, T., Lippmaa, M., and Koinuma, H. (2006) *Adv. Mater.*,

75. Asanuma, S., Akoh, H., Yamada, H., and Sawa, A. (2009) *Phys. Rev. B*, **80**, 235113.

76. Imada, M., Fujimori, A., and Tokura, Y. (1998) *Rev. Mod. Phys.*, **70**, 1039.

77. Ju, H.L., Gopalakrishnan, J., Peng, J.L., Li, Q., Xiong, G.C., Venkatesan, T., and

Greene, R.L. (1995) *Phys. Rev. B*, **51**, 6143.

78. Sawa, A., Fujii, T., Kawasaki, M., and Tokura, Y. (2005) *Proc. SPIE*, **5932**, 59322C.

79. Waser, R. and Aono, M. (2007) *Nat. Mater.*, **6**, 833.

80. Baek, I.G., Lee, M.S., Seo, S., Lee, M.J., Seo, D.H., Suh, D.-S., Park, J.C., Park, S.O., Kim, H.S., Yoo, I.K., Chung, U.-I., and Moon, J.T. (2005) Tech. Dig. - Int. Electron Devices Meet., p. 750.

81. Lee, J., Jo, M., Seong, D.-J., Shin, J., and Hwang, H. (2011) *Microelectron. Eng.*, **88**, 1113.

82. Ielmini, D., Bruchhaus, R., and Waser, R. (2011) *Phase Transit.*, **7**, 570.

83. Tsunoda, T., Kinoshita, K., Hoshiro, H., Yamazaki, Y., Iizuka, T., Ito, Y., Takahashi, A., Okano, A., Sato, Y., Fukano, T., Aoki, M., and Sugiyama, Y. (2005) Tech. Dig. - Int. Electron Devices Meet., p. 767.

84. Muraoka, S., Osano, K., Kanazawa, Y., Mitani, S., Fujii, S., Katayama, K., Katoh, Y., Wei, Z., Mikawa, T., Arita, K., Kawashima, Y., Azuma, R., Kawai, K., Shimakawa, K., Odagawa, A., and Takagi, T. (2005) Tech. Dig. - Int. Electron Devices Meet., p. 779.

85. Lee, M.-J., Lee, C.B., Lee, D., Lee, S.R., Chang, M., Hur, J.H., Kim, Y.-B., Kim, C.-J., Seo, D.H., Seo, S., Chung, U.-I., Yoo, I.-K., and Kim, K. (2011) *Nat. Mater.*, **10**, 625.

Index

Functional Metal Oxides: New Science and Novel Applications, First Edition.
Edited by Satishchandra B. Ogale, Thirumalai V. Venkatesan, and Mark G. Blamire.
© 2013 Wiley-VCH Verlag GmbH & Co. KGaA. Published 2013 by Wiley-VCH Verlag GmbH & Co. KGaA.